Casebooks
in Earth Sciences
Series Editor: R.N. Ginsburg

B.J. Katz (Ed.)

Petroleum Source Rocks

With 274 Figures, Some in Color and 40 Tables

Springer-Verlag
Berlin Heidelberg New York
London Paris Tokyo
Hong Kong Barcelona
Budapest

Editor

Dr. Barry J. Katz
Texaco Inc.
E & P Technology Department
3901 Briarpark
Houston, TX 77042
USA

Series Editor

Dr. Robert N. Ginsburg
Marine Geology & Geophysics
R.S.M.A.S.
University of Miami
4900 Rickenbacker Causeway
Miami, FL 33149-1098
USA

ISBN 3-540-57864-1 Springer-Verlag Berlin Heidelberg New York
ISBN 0-387-57864-1 Springer-Verlag New York Berlin Heidelberg

Library of Congress Cataloging-in-Publication Data. Petroleum source rocks/Barry Katz (editor). p. cm. – (Casebooks in earth sciecnes) Includes bibliographical references and index. ISBN 0-387-57864-1 1. Petroleum-Geology. I. Katz, Barry Jay, 1953– . II, Series. TN870.5.P4793 1994 553.2′8-dc20 94-28601

Typesetting: Macmillan India Ltd. Bangalore 25

SPIN: 10423551 32/3130/SPS - 5 4 3 2 1 0 - Printed on acid-free paper

Series Preface

The case history approach has an impressive record of success in a variety of disciplines. Collections of case histories, casebooks, are now widely used in all sorts of specialties other than in their familiar application to law and medicine. The case method had its formal beginning at Harvard in 1871 when Christopher Lagdell developed it as a means of teaching. It was so successful in teaching law that it was soon adopted in medical education, and the collection of cases provided the raw material for research on various diseases. Subsequently, the case history approach spread to such varied fields as business, psychology, management, and economics, and there are over 100 books in print that use this approach.

The idea for a series of Casebooks in Earth Sciences grew from my experience in organizing and editing a collection of examples of one variety of sedimentary deposits. The project began as an effort to bring some order to a large number of descriptions of these deposits that were so varied in presentation and terminology that even specialists found them difficult to compare and analyze. Thus, from the beginning, it was evident that something more than a simple collection of papers was needed. Accordingly, the nearly fifty contributors worked together with George de Vries Klein and me to establish a standard format for presenting the case histories. We clarified the terminology and some basic concepts, and when the drafts of the cases were completed we met to discuss and review them. When the collection was ready to submit to the publisher, and I was searching for an appropriate subtitle, a perceptive colleague. R. Michael Lloyd pointed out that it was a collection of case histories comparable in principle to the familiar casebooks of law and medicine. After this casebook [Tidal Deposits (1975)] was published and accorded a warm reception, I realized that the same approach could well be applied to many other subjects in earth science.

It is the aim of this new series, Casebooks in Earth Sciences, to apply the discipline of compiling and organizing truly representative case histories to accomplish various objectives: establish a collection of case histories for both reference and teaching; clarify terminology and basic concepts; stimulate and facilitate synthesis and classification; and encourage the identification of new questions and new approaches. There are no restrictions on the subject matter for the casebook series save that they concern earth science. However, it is clear that the most appropriate subjects are those that are largely descriptive. Just as there are no fixed boundaries on subject matter, so is the format and approach of individual volumes open to the discretion of the editors working with their contributors. Most casebooks will of necessity be communal efforts with one or more editors working with a group of contributors. However, it is also likely that a collection of case histories could be assembled by one person drawing on a combination of personal experience and the literature.

Clearly the case history approach has been successful in a wide range of disciplines. The systematic application of this proven method to earth science subjects holds the promise of producing valuable new resources for teaching and research.

Miami, Florida
August, 1994

Robert N. Ginsburg
Series Editor

Preface

This book is about petroleum source rocks, their geochemical characteristics, and depositional environments. The focus of this volume is largely restricted to those rocks which are dominated by either type I or type II kerogens and source primarily liquid hydrocarbons. It is aimed at a broad spectrum of individuals including those actively involved in petroleum exploration, organic geochemists, and sedimentologists studying depositional systems and processes. Each of these groups will view this volume differently. The petroleum explorationist will find information necessary for the proper selection of a geochemical analog, thus assisting with the development of a viable exploration play as well as assisting with the quantification of available hydrocarbon reserves. The organic geochemist will find this volume a handy reference where organic geochemical data have been compiled on a group of source rocks, eliminating the need for a time-consuming search through the literature for available data. Such a database can be used to further refine the significance attributed to isotopic and biomarker compositions of oils and bitumens. While the sedimentologist may learn more about the processes which control the incorporation of organic matter into the rock record and the dynamics of the processes based on the variability of the organic facies within these studied rocks.

Unlike reservoirs, source rocks are not generally field-specific and their study is commonly viewed at a more basinal scale. Consequently, this book is not based on field studies but on broader, regional investigations. Specific case studies may incorporate both outcrop and subsurface information. Note that also unlike reservoirs, source rocks are not the primary objective of a petroleum exploration program and the capture of any information associated with these unique rocks is commonly secondary.

Much of our current understanding of petroleum source rocks has been evolving over the past two decades. This has largely been driven by the general acceptance of an organic origin for petroleum and advancements in analytical techniques. These analytical advancements have resulted in both a better understanding of the threshold criteria for the presence of source rocks as well as the detailed characteristics of these rocks at both the molecular and isotopic level. Much of this information, however, has been trapped within large corporate databases or dispersed throughout much of the literature. The case studies presented in this volume represent both the release of new and previously unpublished information and the compilation of data from a broad spectrum of previously published works.

Although petroleum source rocks are unique in the rock record, the geologic data support the concept that they were deposited within a broad spectrum of depositional environments. While assembling this collection of case studies, an attempt was made to sample many of these environments from both the marine and lacustrine realms. In addition, both carbonates and claystone source rocks are included. Although potential and/or effective source rocks have been documented from the Precambrian through the Pleistocene, the stratigraphic units presented in this volume range in age from the Devonian through the Eocene.

It is hoped that through this compilation of case studies, which present both the geochemical characteristics and the current understanding of the depositional histories and environments of these units, not only will a series of source rock analogs be made available, but the concepts associated with their deposition and formation will be clarified, and new research will be initiated to further refine our understanding of these unique rocks. In addition, a better understanding of how geochemical attributes vary in time and space in known source rocks may also result in better sampling procedures in the future, leading ultimately to databases which more accurately describe these organically enriched rocks.

I would also like to acknowledge Dr. Robert Ginsburg, the series editor, who suggested that this volume on petroleum source rocks be compiled as a companion to the two previous volumes which describe petroleum reservoirs, *Carbonate Petroleum Reservoirs* by Perry Roehl and Phil Choquette and *Sandstone Petroleum Reservoirs* by John H. Barwis, John G. McPherson, and Joseph R.J. Studlick.

An undertaking of this nature requires the assistance and cooperation of many. This includes not only the authors who have contributed their time to prepare and revise each of the chapters, but their organizations which permitted the release of much data which have been previously unpublished, and the many reviewers who have added significantly to the technical merit of each contribution. To the following reviewers I offer my thanks and that of the authors:

R. Burwood
A.R. Caroll
F.G. Christiansen
J.A. Curiale
A.R. Daly
W.E. Dean
H. von der Dick
R.J. Hwang
S. Imbus
M. Kruge
G.B. Newton
J.G. Palacas
S. Palmer
M.A. Pasley
K.E. Peters
T.G. Powell
C.R. Robison
V.D. Robison
J. Rullkötter
L.R. Snowdon
J.R. Stonebraker
S.C. Teerman

In addition, I would like to thank Texaco Inc. for logistical support.

Spring 1994
Houston, Texas

Barry J. Katz
Texaco Inc.

Contents

List of Contributors

Barnard, P.C.
Simon Petroleum Technology Limited, Llandudno, Gwynedd LL30 1SA, UK

Bohacs, K.M.
Exxon Production Research Company, 3120 Buffalo Speedway, Houston, TX 77252, USA

Bordenave, M.L.
Total, 24 cours Michelet, Cedex 47, 92069 Paris la Défense, France

Burwood, R.
FINA Exploration and Production, FINA Research, Zone Industrielle C, 7181 Seneffe (Feluy), Belgium

Carrigan, W.J.
Saudi Arabian Oil Company, Lab R&D Center, P.O. Box 62, Dhahran, 31311, Saudi Arabia

Cole, G.A.
Saudi Arabian Oil Company, Lab R&D Center, P.O. Box 62, Dhahran, 31311, Saudi Arabia

Colling, E.L.
Saudi Arabian Oil Company, Lab R&D Center, P.O. Box 62, Dhahran, 31311, Saudi Arabia

Cooper, B.S.
B.S. Cooper and Associates, Northern Cottage, Appleton Le Moors, York YO6 6TF, UK

De Witte, S.M.
FINA Exploration and Production, FINA Research, Zone Industrielle C, 7181 Seneffe (Feluy), Belgium

Dias, J.L.
Petrobrás/Depex, Av. República do Chile 65, Rio de Janeiro, RJ 20035, Brazil

Erazo, W.Z.
Petroproduction, Filial de Petroecuador, Guayaquil, Ecuador

Fowler, M.G.
Institute of Sedimentary and Petroleum Geology, 3303-33rd Street N.W., Calgary, Alberta T2L2A7, Canada

Grabowski, Jr., G.J.
Exxon Exploration Company, P.O. Box 4778, Houston, TX 77210-4778, USA

Isaksen, G.H.
Exxon Production Research Company, 3120 Buffalo Speedway, Houston, TX 77252, USA

Jiang Renqi
Research Institute of Petroleum Exploration and Development, P.O. Box 910, Beijing 100083, People's Republic of China

Jones, P.J.
Saudi Arabian Oil Company, Lab R&D Center, P.O. Box 62, Dhahran, 31311, Saudi Arabia

Katz, B.J.
Texaco Inc., E&P Technology Department, 3901 Briarpark, Houston, TX 77042, USA

Kelley, P.A.
Texaco E&P Technology Department, 3901 Briarpark, Houston, TX 77042, USA

Koutsoukos, E.A.M.
Petrobrás/Cenpes/Divex, Ilha do Fundão, CEP 21949-900, Rio de Janeiro, RJ, Brazil

Li Desheng
Research Institute of Petroleum Exploration and Development, P.O. Box 910, Beijing 100083, People's Republic of China

McAlpine, K.D.
Atlantic Geoscience Centre, Bedford Institute of Oceanography, P.O. Box 1006, Dartmouth, Nova Scotia B2Y 4A2, Canada

Mello, M.R.
Petrobrás/Cenpes/Divex, Ilha do Fundão, CEP 21949-900, Rio de Janeiro, RJ, Brazil

Mertani, B.
P.T. Caltex Pacific Indonesia, Rumbai, Indonesia

Mycke, B.
FINA Exploration and Production, FINA Research, Zone Industrielle C, 7181 Seneffe (Feluy), Belgium

Paulet, J.
FINA Exploration and Production, FINA Research, Zone Industrielle C, 7181 Seneffe (Feluy), Belgium

Robison, V.D.
Texaco E&P Technology Department, 3901 Briarpark, Houston, TX 77042, USA

Telnaes, N.
Norsk Hydro Petroleum Research Centre, Bergen, Norway

Trindade, L.A.F.
Petrobrás/Cenpes/Divex, Ilha do Fundão, CEP 21949-900, Rio de Janeiro, RJ, Brazil

Williams, H.H.
National Research Authority, Amman, Jordan

Petroleum Source Rocks – an Introductory Overview

B.J. Katz[1]

Introduction

Historically, petroleum exploration has relied on the identification of structural targets displaying four-way closure. This approach to exploration has its roots in the century-old anticlinal theory (T.S. Hunt 1862). Exploration has evolved to incorporate information on reservoir properties, taking into consideration both primary and secondary porosity development. This information on the volume of rock under structural closure in association with pore volume and estimated hydrocarbon recovery factors has commonly been used to estimate the hydrocarbon potential of a prospect. Such an approach assumes that hydrocarbon charge is not a limiting factor (i.e., hydrocarbons will be present as long as closure and porosity exist).

Over the past two decades there has been increasing interest in the availability of hydrocarbon charge through a better understanding of petroleum geochemistry and the identification and characterization of many hydrocarbon source rocks. The significance of this information to a successful exploration program was pointed out by Sluijk and Parker (1986) who showed a 125% increase in exploration efficiency when geochemical input is included in an exploration program. The importance of geochemistry is further highlighted when the exploration postmortems for such areas as the Norton basin (offshore Alaska), the Texas continental shelf, and the Marajo basin (Brazil) are examined. These studies have revealed either a lack of adequate source rock or the wrong type of source rock.

Although most, if not all, major exploration programs now incorporate efforts to identify a source rock, sample availability tends to limit the success of these analytical programs. Samples are more often available from structural highs and shallow horizons rather than the deeper and more basinal positions where effective petroleum source rocks are typically present. Commonly, this results in the use of geologic and geochemical analogs as substitutes for analytical data or as a means of extrapolating source rock information from sampled localities across a basin. For such an approach to be valid, however, a meaningful database must exist upon which these analogies may be based. Unfortunately, little detailed organic geochemical information is publicly available on many of the more significant hydrocarbon source rocks and where available these data are typically dispersed throughout the literature and require significant compilation time in order to fully characterize them and understand their depositional setting. Published accounts of source rocks normally provide little more than an acknowledgment of the source rock potential of a stratigraphic unit. Occasionally, sufficient data are provided to geochemically establish the presence of a source but rarely are data presented to establish the unit's oil- or gas-prone tendencies, let alone their detailed molecular or isotopic character. Where data are presented, this information may only include a single average organic carbon, generation potential and hydrogen index value. Only recently has any consideration been given to organic facies variations (Curiale and Odermatt 1989; Belin and Brosse 1992; Curiale et al. 1992) within these source units and how this stratigraphic and areal variability impacts resource assessment and correlations between oils and sources (Katz et al. 1993).

The following brief discussion is presented as a general overview attempting to present the criteria commonly used to establish the presence of a petroleum source rock and the varied conditions under which these rocks were deposited.

Source Rock Attributes

Petroleum source rocks are unique. By definition, they either have or had the capability to generate

[1]Texaco Inc., 3901 Briarpark, Houston, Texas 77042, USA

sufficient quantities of hydrocarbons at the appropriate levels of thermal maturity to saturate the pore network of the rock unit and permit oil expulsion, allowing the development of a commercial hydrocarbon accumulation. With the work of Ronov (1958) it became evident that the ability of a rock to generate and expel hydrocarbons is dependent on the quantity of organic matter present. While studying the basins of the Siberian platform, Ronov noted that the rocks within petroliferous provinces displayed higher average levels of organic matter enrichment than those of nonpetroliferous provinces. He concluded that it was this difference in organic matter enrichment that determined whether or not petroleum would be present. He further noted, after examining nonreservoir rocks within these basins, that it appeared that the threshold of organic enrichment for a petroleum source rock was approximately 1.4 wt.% organic carbon, the average value for shales within the petroliferous provinces. This level of organic enrichment is nearly three times the 0.5 wt.% organic carbon average for shales within the nonpetroliferous provinces.

It is interesting to note that it is the lower value of Ronov which has commonly been presented as the threshold for effective shale petroleum source rocks (J.M. Hunt 1979; Tissot and Welte 1984). Further complicating the matter of source rock definition was the observation that nonreservoir carbonate rocks typically contained less organic matter than shales. This has led some authors to further reduce the critical source rock threshold in carbonates to 0.3 wt.% (J. M. Hunt 1967). These low threshold values would indicate that hydrocarbon source rocks are largely ubiquitous and should never be a major limiting factor within exploratory programs.

Subsequent studies, however, have shown that potential and/or effective petroleum source rocks contain a minimum of 1.0 wt.% organic carbon, independent of lithology (Bissada 1982), thus supporting the original conclusions of Ronov. It has been further observed that most source rock units display significant variability with respect to organic enrichment, with some stratigraphic intervals within these source rock sequences displaying levels of organic enrichment below this apparent threshold value (see, for example, Mancini et al. 1993).

It should also be noted that there is geochemical evidence which suggests that there is a point of diminishing return with respect to the effectiveness of a petroleum source rock based on the level of organic enrichment. At elevated levels of organic enrichment, associated with coals and some carbonaceous shales, there appears to be hydrocarbon retention through self-absorption by the organic matrix (Youtcheff et al. 1983), thus reducing the source rock effectiveness of these units. It is the heavy-end products (i.e., the more oil-like material) which display the greatest degree of retention. These oil-like hydrocarbons may be released at more advanced levels of thermal maturity as gas following their thermal cracking (Katz et al. 1990).

In addition, since different types of organic matter have the ability to yield different quantities and types of hydrocarbons (Tissot et al. 1974), it became apparent that the level of organic enrichment was not, by itself, an adequate means of establishing the presence of a petroleum source. A more effective means of establishing hydrocarbon source rock potential was by directly measuring the ability of a rock to generate hydrocarbons through a simulation of the maturation process. This was established through the use of various pyrolysis techniques (see Barker and Wang 1988) and was standardized through the use of the "Rock-Eval" instrument (Espitalié et al. 1977). This pyrolysis approach accounted not only for differences in organic enrichment, but for variations in the different abilities of the kerogen types to yield hydrocarbons as well as any mineral matrix effects including catalysis and retention (Katz 1983).

An analysis of fine-grained rocks containing a minimum of 1.0 wt.% organic carbon reveals that rocks capable of acting as a source for commercial quantities of hydrocarbons yield, upon pyrolysis, above-average quantities of hydrocarbons (i.e., $S_1 + S_2 > 2.5$ mg HC/g rock; Bissada 1982), with those rocks considered good or excellent sources yielding greater than 6 mg HC/g rock (Peters 1986). This apparent threshold is independent of source rock lithology. And, as above in the organic carbon data set, significant variability may exist within any given source rock unit (i.e., not all samples from a petroleum source exhibit the same source rock potential).

It has been suggested by some authors (see, for example, Demaison and Huizinga 1991) that these and other proposed geochemical source rock thresholds may be reduced and that the thickness or volume of a rock can compensate for low levels of organic enrichment and/or low generation potentials. Such an argument is commonly presented for many deltaic sequences which, although

petroleum-bearing, commonly appear to lack rocks in the penetrated section with either above-average levels of organic enrichment or generation potential. One such approach utilizes the SPI or source potential index (Demaison and Huizinga 1991) as a qualitative and quantitative measure of the ability of a rock to generate hydrocarbons. However, available geochemical data do not support such an argument. Geochemical studies in such diverse places as the North Sea (Leythaeuser et al. 1983) and the Wind River basin (Katz and Liro 1993) have shown that expulsion efficiency decreases with increasing gross source rock thickness. This is particularly the case if coarse-grained interbeds are lacking. Under such circumstances generated hydrocarbons may actually be retained within the thick shale sequence. These hydrocarbons are, therefore, not contributory. Consequently, source rock volume and thickness cannot be used to compensate for either low levels of organic enrichment or generation potential.

Exceptions to these analytical thresholds may occur, however, when the rocks display an advanced level of thermal maturity. At more elevated levels of maturity neither the measured organic carbon content (Raiswell and Berner 1987) nor the total generation potential (Daly and Edman 1987) reflects the sample's initial characteristics. Both measures of source rock potential have been reduced through hydrocarbon generation and expulsion. This reduction is greatest in rocks which originally contained oil-prone organic matter (Daly and Edman 1987). Thus, exhausted source rocks (i.e., those which have matured beyond the oil window) may display levels of organic enrichment and generation potential below the stated thresholds for potential and/or effective source rocks.

In addition, hydrocarbon source rocks are characterized by the type of organic matter they contain. Tissot et al. (1974) originally identified three primary types of organic matter. Type I organic matter is hydrogen enriched and strongly oil-prone. Hydrocarbon yields from the kerogen as high as 80% by weight may be obtained. The type I kerogen macromolecule includes a substantial number of long-chained functional groups. Type II organic matter is defined as being moderately hydrogen enriched with its primary products being both oil and gas. These kerogens may yield up to 60% hydrocarbons by weight. The type II kerogen macromolecule tends to include more polyaromatic nuclei than type I kerogens and also includes a higher proportion of heteroatomic compounds including ketones and carboxylic acids. Type III organic matter is defined as being hydrogen depleted and gas-prone. These kerogen macromolecules are built largely of aromatic compounds with short-chained functional groups attached. Many of these functional groups are oxygenated including phenols, quinones, and acids.

Subsequently, two additional kerogen types have been added. These are type IV or residual organic matter which is largely inert and incapable of hydrocarbon generation (Tissot et al. 1979) and type II-S, a sulfur-enriched form of type II kerogen (Orr 1986). As a consequence of the abundance of carbon-sulfur bonds in type II-S kerogen, it has been proposed that this type of organic matter may generate oil earlier than those containing type II organic matter. The crude oil generated by this kerogen would typically be sour (i.e., high sulfur).

Commonly, organic matter is characterized through an examination of the elemental composition of isolated kerogen. Immature type I kerogens typically display atomic H/C ratios greater than 1.45 and atomic O/C ratios less than 0.10. Immature type II kerogens display average atomic H/C ratios of 1.25 with O/C ratios of ~ 0.15. Immature type II-S kerogens display similar H/C and O/C ratios but contain at least 6% organic sulfur. Immature type III organic matter displays average atomic H/C ratios of 0.80 and O/C values of ~ 0.18. Type IV organic matter displays H/C values less than 0.65 and variable O/C ratios, depending on the degree of oxidation.

Although less definitive, the character of the organic matter can also be established through an examination of the hydrogen and oxygen indices, where these indices are substituted for the H/C and O/C ratios, respectively. The hydrogen index is the ratio between generable hydrocarbons (S_2 peak) and organic carbon content and is expressed as mg HC/g TOC. The oxygen index is the ratio between pyrolytically derived CO_2 (S_3 peak) and organic carbon content and is expressed as mg CO_2/g TOC. Unlike the atomic ratios obtained on isolated kerogens, these indices are strongly influenced by the level of organic enrichment and the mineral matrix (Katz 1983). Hydrogen index values associated with any given kerogen type tend to increase with increasing organic carbon content. Hydrogen index values for the same kerogen type also tend to be higher in carbonate rocks than shales at comparable levels of organic enrichment and thermal maturity (Katz 1983). The hydrogen index may also be impacted by heavy-end hydrocarbon staining

and/or contamination (Clementz 1979). This is a result of cross-contamination of the S_2 peak by heavy hydrocarbons and asphaltenes. Oxygen index values tend to decrease with increasing carbon content. These values tend to be more elevated in carbonate rocks than in clay-rich rocks.

Immature type I organic matter commonly displays hydrogen index values greater than 750 mg HC/g TOC with oxygen index values typically below 30 mg CO_2/g TOC. Immature type II and II-S organic matter display average hydrogen index values of about 600 mg HC/g TOC with oxygen index values of ~ 50 mg CO_2/g TOC. Immature type III organic matter displays average hydrogen index values of ~ 125 mg HC/g TOC, with oxygen index values ranging upward to ~ 175 mg CO_2/g TOC. Type IV organic matter displays hydrogen index values of less than 50 mg HC/g TOC and variable, but commonly elevated, oxygen index values.

The type of organic matter can also be established by a visual examination of the kerogen. Type I organic matter typically appears as finely disseminated amorphous material which fluoresces under ultraviolet light. This material typically has an algal and/or bacterial origin. Some type I kerogens contain material with an identifiable algal or bacterial structure, including the lacustrine alga *Botryococcus braunii* and the marine Ordovician microfossil *Gloeocapsomorpha prisca*. Type II organic matter appears visually in several different forms. This material may appear as allochthonous material, principally as exinites consisting of spores, pollen grains, and phytoplankton cysts, and as cuticles from plant leaves and stems. Alternatively, this material may be autochthonous and derived principally from bacterially reworked phyto- and zooplankton. This bacterially reworked material will also appear largely as finely disseminated amorphous organic matter which fluoresces under ultraviolet light. As with type II material, there is more than one form of type III organic matter. Type III material often appears as vitrinite, a structured woody plant derivative. Type III material may also appear as finely disseminated amorphous material which does not fluoresce under ultraviolet light. This material forms through the degradation and/or oxidation of the other maceral types. Type IV organic matter commonly appears as black, structured organic matter in transmitted light and displays high reflectivity.

These bulk methods provide information on the oil and gas proneness of a sample or source rock. For example, Saxby and Shibaoka (1986) provide a means of estimating the relative oil and gas yield based on the elemental composition of isolated kerogens. Unfortunately, however, these bulk methods do not provide specific information on the nature of the generated products at the appropriate levels of thermal maturity. A more specific characterization of hydrocarbon products may be accomplished through the use of pyrolysis-gas chromatography (Giraud 1970). This assessment may be either qualitative or quantitative. Qualitative examination of specific macerals has revealed distinct products for each maceral type (Larter and Douglas 1980). For example, alginites yield pyrograms dominated by alkane-alkene doublets with only a minor contribution from aromatic species. In contrast, vitrinites produce pyrograms dominated by aromatic compounds and may also contain significant quantities of phenolic compounds, with lesser amounts of straight-chain moieties.

A further analysis of pyrolysis-gas chromatograms reveals that the nature of generated products varies within individual kerogen "types" (Fig. 1). Type I and type II kerogens may display a waxy character as is the case for the lacustrine Green River (western United States) and Pematang (central Sumatra) formations or may display a nonwaxy fingerprint as is the case for the marine Hanifa (Arabian basin) and Kimmeridge Clay (North Sea basin) formations. And, as with the other kerogen types, type III material may result in different chromatographic signatures. One group of pyrograms, such as that obtained from the Talang Akar Formation (Ardjuna basin), displays a largely aromatic signature similar to that of pure vitrinite maceral. The other group of chromatograms, as obtained on the Banquereau Formation (Scotian Shelf) displays a less well-defined chromatographic signature. This later pattern appears to be associated with the presence of poorly preserved marine organic matter.

Depositional Settings

Organically enriched, oil-prone sediments are deposited under rather restricted and unique depositional conditions where large volumes of organic matter are incorporated into the sedimentary rock record. Considerable discussion in the literature has attempted to focus on the role of either anoxia (Demaison and Moore 1980) or organic productivity

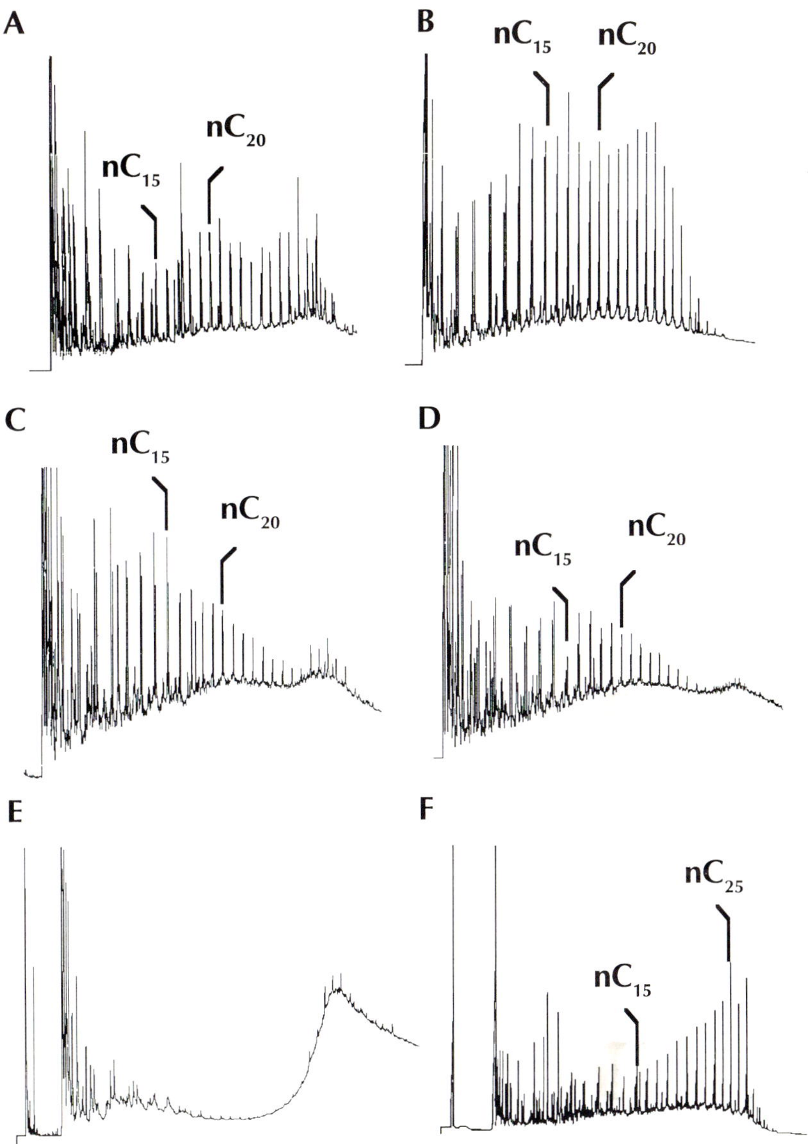

Fig. 1A-F. Representative pyrolysis-gas chromatograms for the Green River Shale (**A** type I kerogen, Utah), the Brown Shale (**B** type II kerogen, central Sumatra), the Hanifa Formation (**C** type I kerogen, Saudia Arabia), the Kimmeridge Clay (**D** type II kerogen, North Sea), Banquereau Shale (**E** type III kerogen, Scotian Shelf), and Talang Akar Formation (**F** type III kerogen, Ardjuna basin)

(Parrish 1982) as a primary control of these organically enriched, oil-prone sediments, with additional discussion focusing on the role that sedimentation rate plays in their formation (Müller and Suess 1979). Yet, there appears to be no general consensus as to the causal mechanism. In part, this is the result of the interrelationship between anoxia, productivity, and sedimentation rate, and the associated feedback mechanisms between them.

In general, there are four primary depositional settings under which oil-prone source rocks are deposited (Meissner et al. 1984). These are: (1) across flooded, broad continental shelves and epicontinental seas; (2) within the area where the oxygen-minimum zone impinges on the sediment-water interface; (3) within regions of active upwelling; and (4) within structurally isolated basins. This is not to imply that these settings are universally associated with source rock deposition. In fact, there appears to be a strong latitudinal dependence (Deutsch 1965; Tamrazyan and Ovnatanov 1983). Source rock development within these settings tends to be favored at lower latitudes as a consequence of conditions which favor both productivity (longer growing season) and preservation (less seasonality). The factors favoring this development in these depositional settings are summarized below.

Flooded continental shelves favor source rock development because of: (1) the potential for elevated levels of productivity through the introduction of nutrients from previously developed and exposed soil horizons; (2) the lack of clastic dilution through the trapping of sediment in more up-dip positions; and (3) the potential for water column stratification and enhanced preservation, if a strong pycnocline is established and if low energy conditions exist (i.e., low wind stresses). Within such settings a further examination of the distribution of organic matter reveals that bathymetric relief is also an important controlling factor. Higher levels of organic enrichment tend to be associated with bathymetric lows (Ettensohn 1984). This relationship is probably a consequence of both hydrodynamic processes and enhanced preservation potential under more oxygen-depleted conditions which may develop in these slightly more isolated lows which are often associated with poorer circulation.

Within the oxygen-minimum zone source rock development occurs principally through more efficient preservation of available organic matter. Although the mechanism for the formation of the oxygen-minimum zone has not been fully established, it appears to be associated with the attainment of neutral buoyancy by organic matter, permitting extended residence time and consumption of available free oxygen in the water column (Karl 1982). The depletion of free oxygen within this zone results in enhanced preservation of any remaining or newly introduced organic matter. The position of the oxygen-minimum zone is largely dependent on temperature and salinity which influence seawater density. The intensity of the oxygen-minimum zone is controlled by the availability of organic matter, initial oxygen content, and the nature of the current system which permits oxygen resupply. The importance of the oxygen-minimum zone as an environment associated with source rock deposition is controlled by its intensity, the bathymetric configuration of the basin, and the availability of organic matter. Not all oxygen-minimum zones result in the deposition of source rock quality material. For example, in the modern Gulf of Mexico organic matter deposition, although slightly elevated within a weak oxygen-minimum zone, does not obtain sufficient levels of organic enrichment to be considered of potential source quality (Jones 1983). In addition, the oxygen minimum zone is volumetrically most important where the bathymetric gradient is lowest in the area of impingement (Cornford 1979). This tends to favor continental shelves as opposed to continental slopes.

Source rock deposition within upwelling regions is largely a consequence of the associated elevated levels of productivity (Kruijs and Barron 1990), the resulting elevated sedimentation rates, and the associated reduction in free oxygen. The higher levels of productivity are a consequence of nutrient enrichment and resupply. Nutrient enrichment and resupply occur through the upwelling process itself, where longshore winds result in a net offshore transport of surface waters and their subsequent replacement by deeper nutrient enriched waters, from a depth of between 400 and 1000 m. These deeper waters are nutrient enriched through organic matter demineralization (decomposition). As a consequence of the zonal wind patterns, the Earth's rotation and resulting Coriolis force, upwelling is more commonly associated with western continental margins.

Source rock development within isolated basins is largely a consequence of enhanced preservation which has resulted from water column stratification and associated anoxia. Such conditions may develop in either marine or lacustrine settings and are particularly effective when the oxic portion of the water column is reduced and the anoxic-oxic interface approaches the photic zone. Stratification within these basins may result from either of two primary mechanisms. One relies on the maintenance of a warmer water cap on a cooler water layer. This temperature difference need not be more than a few degrees provided that the seasonal temperature differences are minimal and that wind stresses are also minimized. Such conditions are typically associated with tropical settings and, in the case of lakes, low altitudes. A second mechanism relies on the maintenance of salinity contrasts (Pratt 1984) where fresher water rests on more saline water. These conditions may develop in regions where a positive water balance exists (Brukner-Wein et al. 1990), where salt diapir dissolution occurs at or near the sediment-water interface (Trabant and Presley 1978), or where there are subsurface spring discharges which contain elevated total dissolved solids (Brunskill and Ludlam 1969).

Summary

Petroleum source rocks play a vital role in the formation of petroleum accumulations and are as important as reservoirs, seals, and traps. These rocks are unique and can be identified and characterized through a series of analytical approaches. These rocks have developed under a wide range of depositional conditions. All too often, however, source rocks are not identified prior to or during exploration, or are poorly characterized because of limited sampling. This results in the use of analogs based commonly on the nature of the gross depositional setting. The successful selection and utilization of an analog require a better understanding of the specifics of these depositional systems and the geochemical attributes of identified petroleum source rocks. Such detailed information will ultimately lead to a better understanding of the variability in time and space of these rocks and the significance of various geochemical attributes, including isotopic and biomarker compositions.

Acknowledgments. Permission to publish was obtained from Texaco Inc. Assistance with the preparation of this manuscript was provided by G. Mayfield and V.D. Robison.

References

Barker C, Wang L (1988) Applications of pyrolysis in petroleum geochemistry: a bibliography. J Anal Appl Pyrolysis 13: 9–61

Belin S, Brosse E (1992) Petrographical and geochemical study of a Kimmeridgian organic sequence (Yorkshire area, UK). Rev Inst Français du Pétrole 47: 711–725

Bissada KK (1982) Geochemical constraints on petroleum generation and migration – a review. Proc 2nd ASCOPE Conf, Manila, Oct, 1981, pp 69–87

Brukner-Wein A, Hetényi M, Vetö I (1990) Organic geochemistry of an anoxic cycle: a case history from the Oligocene section, Hungary. Org Geochem 15: 123–130

Brunskill GJ, Ludlam SD (1969) Fayetteville Green Lake, New York. I. Physical and chemical limnology. Limnol Oceanogr 14: 817–829

Clementz DM (1979) Effect of oil and bitumen saturation on source-rock pyrolysis. Am Assoc Pet Geol Bull 63: 2227–2232

Cornford C (1979) Organic deposition at a continental rise: organic geochemical interpretation and synthesis at DSDP Site 397, eastern North Atlantic. In: van Rad U, Ryan WBF et al. (eds) Initial reports of the Deep Sea Drilling Project. US Government Printing Office, Washington, DC, 47(1): 506–510

Curiale JA, Odermatt JR (1989) Short-term biomarker variability in the Monterey Formation, Santa Maria basin. Org Geochem 14: 1–13

Curiale JA, Cole RD, Witmer RJ (1992) Application of organic geochemistry to sequence stratigraphic analysis: Four Corners Platform Area, New Mexico, USA. Org Geochem 19: 53–75

Daly AR, Edman JD (1987) Loss of organic carbon from source rocks during thermal maturation. Am Assoc Pet Geol Bull 71: 546 (Abstrt)

Demaison G, Huizinga BJ (1991) Genetic classification of petroleum systems. Am Assoc Pet Geol Bull 75: 1626–1643

Demaison GJ, Moore GT (1980) Anoxic environments and source bed genesis. Am Assoc Pet Geol Bull 64: 1179–1209

Deutsch ER (1965) The paleolatitude of Tertiary oil fields. J Geophys Res 70: 5193–5203

Espitalié J, Laporte JL, Madec J, Marquis F, Leplat P, Paulet J, Boutefeau A (1977) Méthode rapide de caractérisation des roches mères de leur potential pétrolier et de leur degré d'évolution. Rev Inst Français du Pétrole 33: 23–42

Ettensohn FR (1984) The nature and effects of water depth during deposition of the Devonian-Mississippian black shale. Proc 1984 Eastern Oil Shale Symp, University of Kentucky, Lexington, pp 333–346

Giraud A (1970) Application of pyrolysis and gas chromatography to geochemical characterization of kerogen in sedimentary rocks. Am Assoc Pet Geol Bull 54: 439–451

Hunt JM (1967) The origin of petroleum in carbonate rocks. In: Chilingar GV, Bissell HJ, Fairbridge, RW (eds) Carbonate rocks. Elsevier, New York, pp 225–251

Hunt JM (1979) Petroleum geochemistry and geology. WH Freeman, San Francisco, 617 pp

Hunt TS (1862) Notes on the history of petroleum or rock oil. Smithsonian Institution Annu Rep for 1861, pp 319–329

Jones RW (1983) Organic matter characteristics near the shelf-slope boundary. In: Stanley DJ, Moore GT (eds) The shelf-break: critical interface on continental margins. Soc Econ Paleontol Mineral, Tulsa, Spec Publ 33: 391–405

Karl DM (1982) Microbial transformations of organic matter at oceanic interfaces: a review and prospectus. E∅S, Trans Am Geophys Union 63: 138–140

Katz BJ (1983) Limitations of 'Rock-Eval' pyrolysis for typing organic matter. Org Geochem 4: 195–199

Katz BJ, Liro LM (1993) The Waltman Shale Member, Fort Union Formation, Wind River basin: a Paleocene clastic lacustrine source system. In: Keefer WR, Metzger WJ, Godwin LH (eds) Oil and gas and other resources of the Wind River Basin Wyoming. Wyoming Geol Assoc, Casper, Wyoming, Spec Symp 1993, pp 163–174

Katz BJ, Royle RA, Mertani B (1990) Southeast Asian and southwest Pacific coals contribution to the petroleum resource base. Proc 19th Indon Pet Assoc Ann Conv, Jakarta, Oct, 1990, 1: 299–329

Katz BJ, Breaux TM, Colling EL, Darnell LM, Elrod LW, Jorjorian T, Royle RA, Robison VD, Szymczyk HM, Trostle JL, Wicks JP (1993) Implications of stratigraphic variability of source rocks. In: Katz BJ, Pratt LM (eds) Source rocks in a sequence stratigraphic framework. Am Assoc Pet Geol, Tulsa, Stud Geol 37: 5–16

Kruijs E, Barron E (1990) Climate model prediction of paleoproductivity and potential source-rock distribution. In: Huc AY (ed) Deposition of organic facies. Am Assoc Pet Geol, Tulsa, Stud Geol 30: 195–216

Larter SR, Douglas AG (1980) A pyrolysis-gas chromatographic method for kerogen typing. In: Douglas AG, Maxwell JR (eds) Advances in organic geochemistry, 1979. Pergamon Press, Oxford, pp 579–584

Leythaeuser D, Bjorøy M, Mackenzie AS, Schaefer AG, and Altebaumer FJ (1983) Recognition of migration and its effects within two coreholes in shale/sandstone sequences from Svalbard, Norway. In: Bjorøy M, Bjørlykke K, Eggen S, Elvsborg A, Finstaed KG, Grønneberg T, Haegh T, Skaar FE (eds) Advances in organic geochemistry, 1981. John Wiley, Chichester, pp 136–146

Mancini EA, Tew BH, Mink R M (1993). Petroleum source rock potential of Mesozoic condensed section deposits of southwest Alabama. In: Katz BJ, Pratt LM (eds) Source rocks in a sequence stratigraphic sequence. Am Assoc Pet Geol, Tulsa, Stud Geol 37: 147–162

Meissner FF, Woodward J, Clayton JL (1984) Stratigraphic relationships and distribution of source rocks in the greater Rocky Mountain region. In: Woodward J, Meissner FF, Clayton JL (eds) Hydrocarbon source rocks of the greater Rocky Mountain region. Rocky Mountain Association of Geologists, Assoc Geol, Denver, pp 1–34

Müller PJ, Suess E (1979) Productivity, sedimentation rate and sedimentary organic carbon in the oceans. I. Organic carbon preservation. Deep Sea Res 26A: 1347–1362

Orr WL (1986) Kerogen/asphaltene/sulfur relationships in sulfur-rich Monterey oils. Org Geochem 10: 499–515

Parrish JT (1982) Upwelling and petroleum source beds with reference to the Paleozoic. Am Assoc Pet Geol Bull 66: 750–774

Peters KE (1986) Guidelines for evaluation of petroleum source rock using programmed pyrolysis. Am Assoc Pet Geol Bull 70: 318–329

Pratt LM (1984) Influence of paleoenvironmental factors on preservation of organic matter in middle Cretaceous Greenhorn Formation, Pueblo, Colorado. Am Assoc Pet Geol Bull 68: 1146–1159

Raiswell R, Berner RA (1987) Organic carbon loses during burial and thermal maturation of normal marine shales. Geology 15: 853–856

Ronov AB (1958) Organic carbon in sedimentary rocks (in relation to presence of petroleum). Geochemistry 5: 510–536 (English translation)

Saxby JD, Shibaoka M (1986) Coal and coal macerals as source rocks for oil and gas. Appl Geochem 1: 25–36

Sluijk D, Parker JR (1986) Comparison of pre-drilling predictions with postdrilling outcomes, using Shell's prospect appraisal system. In: Rice DD (ed) Oil and gas assessment – methods and applications. Am Assoc Pet Geol, Tulsa, Stud Geol 21: 55–58

Tamrazyan GP, Ovnatanov ST (1983) Global latitudinal belts of hydrocarbon and coal accumulation. Dokl Akad Nauk SSSR 271: 171–173 (English translation)

Tissot BP, Welte DH (1984) Petroleum formation and occurrence. Springer, Berlin Heidelberg New York, 699 pp

Tissot B, Durand B, Espitalié J, Combaz A (1974) Influence of nature and diagenesis of organic matter in formation of petroleum. Am Assoc Pet Geol Bull 58: 499–506

Tissot B, Deroo G, Herbin JP (1979) Organic matter in Cretaceous sediments of the North Atlantic: contribution to sedimentology and paleogeography. In: Talwani M, Hay W, Ryan WBF (eds) Deep drilling results in the Atlantic Ocean: continental margins and paleoenvironment. Am Geophys Union, Washington, DC, Maurice Ewing Series 3: 362–374

Trabant PK, Presley BJ (1978) Orca basin, anoxic depression on the continental slope, northwest Gulf of Mexico. In: Bouma AH, Moore GT, Coleman JM (eds) Framework, facies, and oil trapping characteristics of the upper continental margin. Am Assoc Pet Geol, Tulsa, Stud Geol 7: 303–311

Youtcheff JS, Given PH, Baset Z, Sudaram MS (1983) The mode of association of alkanes with coals. Org Geochem 5: 157–164

The Exshaw Formation: a Devonian/Mississippian Hydrocarbon Source in the Western Canada Basin

V.D. Robison[1]

Abstract

The Devonian-Mississippian period marks a time of extensive development of black shale deposition within epeiric seaways. These source systems have been attributed with generating a large portion of the proven reserves around the world. One of these source systems, the Exshaw Formation of the Western Canada basin, is examined in this chapter.

The Exshaw Formation consists of two members, a lower shale-dominated member and an upper siltstone to silty limestone member. The lower member contains the organic-rich, oil-prone, black shale facies of the Exshaw Formation. The Exshaw black shales were deposited beneath an upwelling zone that extended along the western margin of the Devonian/Mississippian craton.

During deposition of the Exshaw Formation shales bottom waters apparently varied between dysaerobic to anoxic and the seaway may at times have been stratified. Variations in the preservation potential appears to have been the primary influence on the degree of organic enrichment and the oil-proneness of the preserved organic matter. Maximum burial depths in the Western Canada basin were reached in the Eocene and subsequent uplift and erosion indicates that the Exshaw black shales are at present not generating and expelling hydrocarbons.

The Exshaw Formation has been suggested as one of the primary sources for the Early Cretaceous heavy oil deposits in the Mannville Group. The Exshaw may also have contributed locally to several Devonian and Mississippian reservoired oils. The volumetric significance of the generated and trapped hydrocarbons from the Exshaw Formation is uncertain as a result of mixing with hydrocarbons from other source systems in each accumulation. It is clear, however, that the Exshaw Formation black shales are one of the primary sources in the Western Canada basin.

Introduction

The Western Canada basin, one of the most prolific hydrocarbon provinces in the world, occupies an area of approximately 1.4×10^6 km^2. Both conventional and nonconventional heavy oil occurs in significant reserves in the basin in two major sedimentary wedges. The first sedimentary package includes Cambrian to middle Jurassic, marine rocks deposited on a relatively stable platform margin that graded westward into a passive margin. The overlying late Jurassic to Tertiary sediments were deposited in a foreland basin, east of the highlands created by the collision and accretion of continents onto the western margin of North America (Podruski et al. 1988).

The presence of several potential source systems together with a relatively mild tectonic history that resulted in a westward gently dipping basin has led to a number of play concepts in this mature province. One of the source systems in the Western Canada basin is the Devonian/Mississippian Exshaw Formation (Creaney and Allan 1990; Allan and Creaney 1991). The Exshaw Fm. is one of a number of black shales that were deposited in the epeiric seaways of North America, South America and Russia during the late Devonian and early Mississippian. Source systems of this age have been attributed with generating 8% of the world's reserves (Klemme and Ulmishek 1991).

Stratigraphic Section

The earliest reference to the Exshaw was by McConnell (1887), who noted the presence of a black shale. Warren (1937) formally named the

[1]Texaco Inc., 3901 Briarpark, Houston, Texas 77042, USA

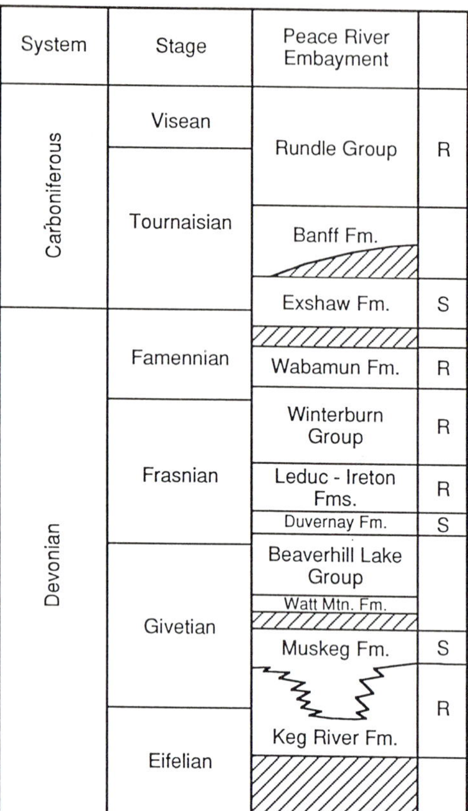

Fig. 1. Generalized stratigraphic column for the Devonian and Carboniferous systems of the Western Canada basin showing the identified source rocks and sampled reservoirs in the Peace River Arch area. *S*, source rock; *R*, sampled reservoirs

Formation and designated Jura Creek as the type locality. Several studies have since described in detail the type section and several of the other exposures of the Exshaw Formation in the foothills and mountains (Macqueen and Sandberg 1970; Richards and Higgins 1988 and references therein).

The Exshaw Formation overlies the Palliser Formation in the outcrop and the Wabamun Group in the subsurface (Fig. 1). Where it is exposed in outcrop, the Exshaw and Palliser Formations' contact is sharp and characterized by a black shale at the base of the Exshaw Formation with local occurrences of basal sandstone a few cm thick (Warren 1937; Harker and McLaren 1958; Macqueen and Sandberg 1970; Richards and Higgins 1988; Savoy 1990; and Workum 1991).

At the type section, the Exshaw Formation is approximately 47 m thick and consists of a lower, shale-dominated member and a siltstone to silty limestone upper member (Richards and Higgins 1988). The shaly, lower member, which is 9.3 m thick at the type section (Richards and Higgins 1988), is the focus of this chapter.

This lower member of the Exshaw Formation was deposited during a transgression that resulted in significant onlap onto the underlying Palliser Formation and Wabamun Group (Johnson et al. 1985; Johnson and Sandberg 1988; Sandberg et al. 1988) during the Famennian. A thin, basal sandstone is thought to represent a reworked, initial lag deposit of the transgression (Workum 1991). The shales of this lower member are typically planar laminated, brownish to brownish black, calcareous to slightly calcareous. At the type section, Richards and Higgins (1988) have suggested that these shales were deposited in moderately deep water (below storm wave base but < 300 m) in an anaerobic, density and chemically stratified pericratonic sea.

The Devonian/Carboniferous boundary has not been exactly established within this sequence but appears to occur within the lower shale member (Savoy 1990). It is not clear whether the section is continuous across the boundary. The shales of the lower member apparently represent two distinct facies (Richards and Higgins 1988) and the break between these two may mark the boundary. The lower portion of the black shale has been described as a noncalcareous, anomalously radioactive black shale with very high organic carbon contents, while the upper portion of the shale is a calcareous black shale with lower levels of organic enrichment (Richards and Higgins 1988).

The overlying siltstone to silty limestone member of the Exshaw Formation represents deposition which occurred in the Tournaisian in a shallowing and regressing sea (Savoy 1990). Anoxic waters had retreated and deposition apparently occurred within a normally oxygenated epicontinental sea.

The Exshaw Formation is contemporaneous with the lower and middle members of the Bakken Formation (Leenheer 1984). An equivalent member to the upper, organic-rich shale member of the Bakken Formation is not developed in either the outcrop or subsurface penetrations of the Exshaw Formation. This upper shale member of the Bakken Formation is apparently age-equivalent to the Banff Formation (Leenheer 1984; Richards and Higgins 1988).

Geologic and Paleogeographic Setting

The Exshaw Formation was deposited in a broad epeiric sea that covered much of western North

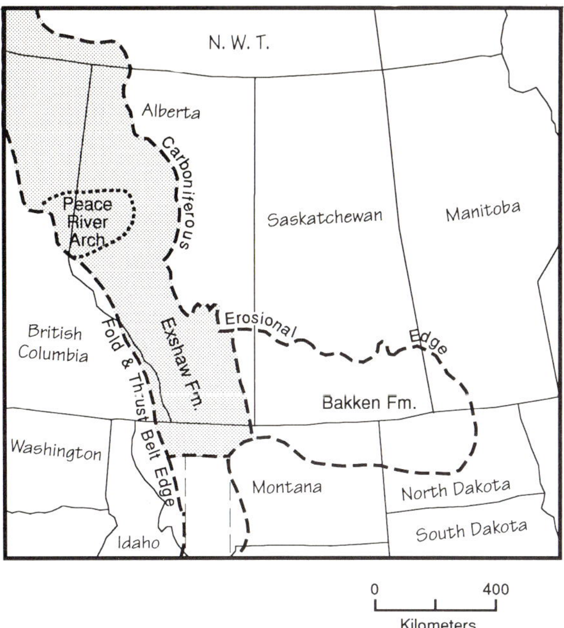

Fig. 2. Generalized map showing the distribution of Late Devonian and Early Mississippian black shale deposits. The siltstones of the Exshaw Formation and the Sappington Member of the Three Forks Formation are separated by an area of erosion leading Sandberg (1967) to extend usage of the Exshaw Formation into Montana to the first appearance of the siltstones of the Sappington. The Exshaw Formation shales are equivalent to the lower Bakken Shale in the Canadian Williston basin (Leenheer 1984)

America in the Late Devonian and Early Mississippian. Present-day extent of the Exshaw Formation extends from the outcrop in the thrust and fold belt of the Canadian Rockies eastward across Alberta and southern Saskatchewan (Fig. 2). The original western limit of the Devonian and Mississippian strata is not known due to Jurassic through Paleogene orogenic activity which displaced strata by up to 200 km eastward (Bally et al. 1966; Cook et al. 1988). The southern extent of the Exshaw Formation is typically placed near the Canada-United States border. Age-equivalent units of the Exshaw Formation include the Bakken Formation and the Sappington Member of the Three Forks Formation (Macqueen and Sandberg 1970; Leenheer 1984).

During the Late Devonian and Early Mississippian, the Exshaw seaway lay along the western margin of Euramerica (Witzke and Heckel 1988; Savoy 1990). Savoy (1990) suggests that the area could have been between 10°S to 12°N and therefore either arid tropical to humid equatorial climatic conditions existed. Scotese (1984, 1986) and Scotese et al. (1985) place western Canada 8° to 12°N of the equator in the Late Devonian, based primarily on paleomagnetic data. Based on the distribution of climate sensitive facies (carbonates, evaporites, coals, phosphorites, etc.) in western Euramerica, Witzke and Heckel (1988) suggest a southerly arid tropical belt setting for western Canada. Their reconstructions place the Exshaw seaway near 5° to 10°S of the equator. This position is similar to Early Carboniferous reconstructions presented by Tarling (1980) and Smith et al. (1981).

Parrish (1982) suggested that black shale deposition in the Late Devonian-Early Carboniferous epicontinental seaways may have been the result of upwelling. Upwelling is consistent with the occurrence of high biogenic silica, phosphate, and organic carbon contents (Macqueen and Sandberg 1970; Macdonald 1985; Richards and Higgins 1988) in the black shale portion of the Exshaw Formation. Deposition is generally considered to have occurred in dysaerobic to anaerobic sediments in a geographically extensive, stratified seaway (Macqueen and Sandberg 1970; Richards and Higgins 1988; Savoy 1990). The oxygen-depleted bottom waters invoked for deposition of the black shale facies of the Exshaw Formation may have resulted as a consequence of an expansion of the oxygen minimum layer into the epeiric seaway in association with the Famennian transgression. Oxygen-depleted bottom waters could also be maintained and may have been formed in response to high levels of organic productivity driven by upwelling.

Simulations of the atmospheric climate from a general circulation model suggest that surface winds all along the western margin of the continent were oriented appropriately to initiate upwelling (Fig. 3). Uncertainties as to the latitudinal position of the seaway in the Late Devonian should not affect the development of upwelling because the seaway was along a western-facing margin and was located at low latitudes. Predictions as to the circulation of the system, however, are greatly influenced by the latitudinal distribution. It has been suggested by several authors that the seaway may have been stratified (Macqueen and Sandberg 1970; Richards and Higgins 1988; Savoy 1990). The nature of this stratification is dependent on whether the seaway was in a tropical arid belt or a humid equatorial setting.

In an equatorial setting, the potential exists for stratification of the water column due to freshwater discharge as Pratt (1984) suggested for the

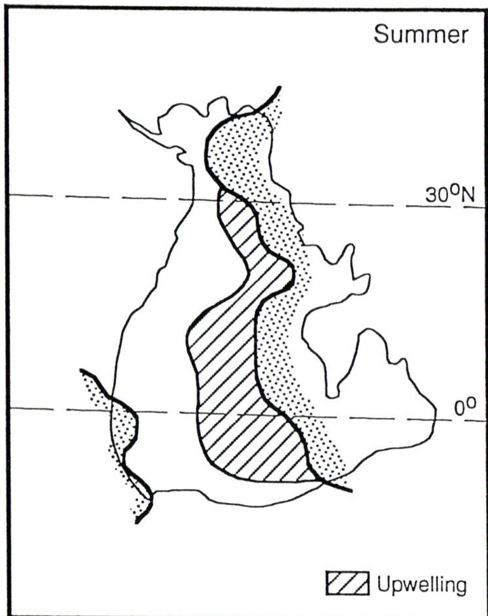

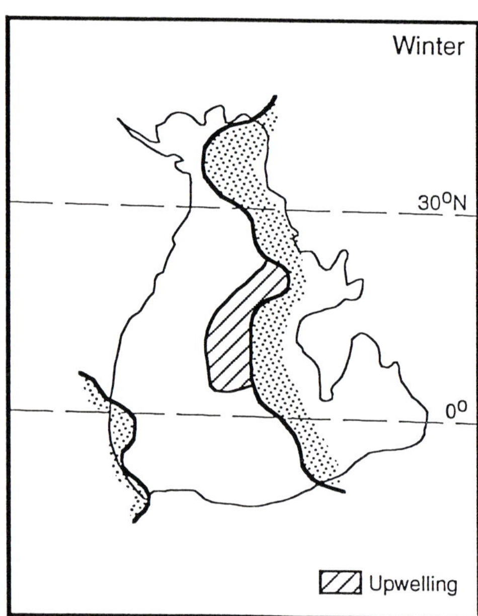

Fig. 3. Modeled summer and winter upwelling in the Late Devonian epeiric seaway. Upwelling areas are derived from a simulation using the National Center for Atmospheric Research's CCM0 atmospheric circulation model. Reconstruction for 360 Ma B.P. (after Scotese 1986)

Cretaceous Greenhorn Formation of the Western Interior Seaway. The lower salinity, "fresher" water cap could restrict overturn of the water column leading to a stagnation and development of anoxic bottom waters (Arthur et al. 1987). A salinity-stratified water column can also occur along an arid margin where a shallow shelf is present for bottom water formation. Witzke and Heckel (1988) proposed such a mechanism for the Exshaw where salinity stratification promoted the development of bottom anoxia. Alternatively, along a tropical arid continental margin a thermally stratified water column rather than a salinity stratified water column may have developed (Herbert and Fischer 1986; Tyson and Pearson 1991).

Source Rock Characterization

Organic Enrichment and Organic Matter Type

The development of a starved basin where upwelling developed and the bottom waters may have been anoxic to dysaerobic created an ideal setting for deposition of the organic-rich portions of the Exshaw Formation black shales. These black shales of the lower member of the Exshaw Formation contain the bulk of the organic matter found in the formation and are the source for a significant amount of the oil generated and trapped in the Western Canada basin. Samples from this study include black shale samples of the Exshaw Formation from the Peace River Arch Area (Table 1) which exhibit organic carbon values within this lower member that range from 0.9 up to 14% by weight (Fig. 4). Creaney and Allan (1990) report that organic content in the Exshaw Formation can be up to 20 wt. %. Total sulfur values for Exshaw Formation samples range from 0.7 to nearly 4 wt. %, suggesting that conditions varied from euxinic to normal marine conditions (Leventhal 1987; Berner and Raiswell 1983) at the time of deposition (Fig. 4).

The Exshaw Formation exhibits excellent generation potential, as determined by Rock-Eval pyrolysis, and range from 1.8 to 107 mg hydrocarbons/g rock (Fig. 5). Most of these values are well above the 2.5 and 6 mg HC/g rock limits identified as necessary to classify a rock as a good or excellent source, respectively (Bissada 1982; Tissot and Welte 1984) and indicate the significant potential of the Exshaw Formation for hydrocarbon generation.

Associated hydrogen index values of 286 to 730 indicate the significant liquid hydrocarbon potential of the Exshaw Formation (Fig. 6). Many of these higher hydrogen index values fall along the traditional type I curve on a modified van Krevelen diagram (Espitalie et al. 1977), suggesting

Table 1. Geochemical data for Exshaw Formation samples analyzed in this chapter

Locality (1)	2	3	4	5	6	7	8	9	10	11	12
16-30-77-25W5(3)	1.8–5.9	290–386	0.80	75–80	0.41–0.71	41–53	1.57–1.88	23	1.26	32/19/49	0.25
14-31-80-23W5(2)	2.2–4.2	303–364	0.75	75	0.68–0.83	54–65	1.56–1.59	21	1.43	37/22/41	0.26
13-13-78-26W5(4)	3.2–6.2	410–713	0.75	75–80	0.66–0.69	49–68	1.46–1.80	14	1.31	34/18/48	0.30
6-2-79-22W5(4)	2.1–13.0	336–731	0.75	80	0.42–0.73	59–63	1.65–1.91	15	1.25	38/20/42	0.24
16-17-80-23W5(1)	3.3	409	0.70	85	0.77	48	1.86	18	1.28	31/22/46	0.19
1-27-80-24W5(2)	1.8–6.1	321–367	0.85	75	0.86–1.02	46–64	1.63–1.68	12	1.41	32/20/48	0.26
14-30-80-23W5(3)	2.0–13.9	286–730	0.75	75–80	0.89–0.99	53–73	1.54–1.89	16	1.44	35/17/48	0.18
10-28-77-26W5(1)	4.4	562	0.80	75	0.56	56	1.68				
13-9MU-79-22W5(1)	4.7	489	0.65	75	0.69	58	1.48				
8-16-79-22W5(1)	3.7	536	0.65	90	0.54	62	1.53	25	1.43	34/19/47	0.26
6-19-80-23W5(3)	0.9–9.8	287–585	0.70	80–90	0.59–0.86	53–74	1.84–1.91	16	1.35	33/20/47	0.27
10-23-80-24W5(2)	1.0–2.5	336–435	0.65	90–95	0.86–1.06	65–74	1.59–1.72	18	1.38	38/21/41	0.28
16-36-80-24W5(2)	4.6–10.0	414–601	0.70	85–90	0.48–0.75	53–68	1.63–1.81	18	1.25	38/19/43	0.22
12-28-77-26W5(3)	1.9–2.8	304–408	0.80	75	0.56–0.78	61–69	1.72–1.76	21	1.21	32/18/50	0.26
14-2-80-23W5(1)	3.2	358	0.65	85	0.91	77	1.65				

1 Number of samples analyzed.
2 Range of total organic carbon contents.
3 Range of hydrogen indices.
4 Estimated vitrinite reflectance equivalent based on maturation modeling.
5 Percentage of fluorescent amorphous organic matter in isolated kerogen based on visual kerogen analysis.
6 Saturated/aromatic hydrocarbons.
7 % Nonhydrocarbons.
8 Pristane/phytane.
9 % Tricyclics –C_{19} through C_{29} tricyclics as a percentage of total identified components in the m/z 191 fragmentogram (Table 3).
10 [C_{33} 17α(H),21β(H)-homohopane (22S) + C_{33} 17α(H),21β(H)-homohopane (22R)]/[C_{34} 17α(H),21β(H)-homohopane (22S) + C_{34} 17α(H),21β(H)-homohopane (22R)].
11 14α,17α-Cholestane (20R)/24-methyl-14α,17α-cholestane (20R)/24-ethyl-14α,17α-cholestane (20R).
12 Sterane/hopane-[C_{27} + C_{28} + C_{29} ($\alpha\alpha$ & $\beta\beta$ R + S) steranes]/[norhopane + hopane + C_{31}–C_{35} homohopanes].

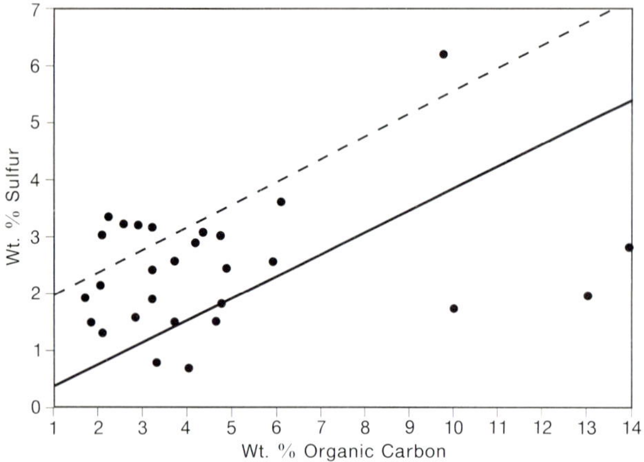

Fig. 4. Comparison of organic enrichment and sulfur content for samples from the lower black shale member of the Exshaw Formation

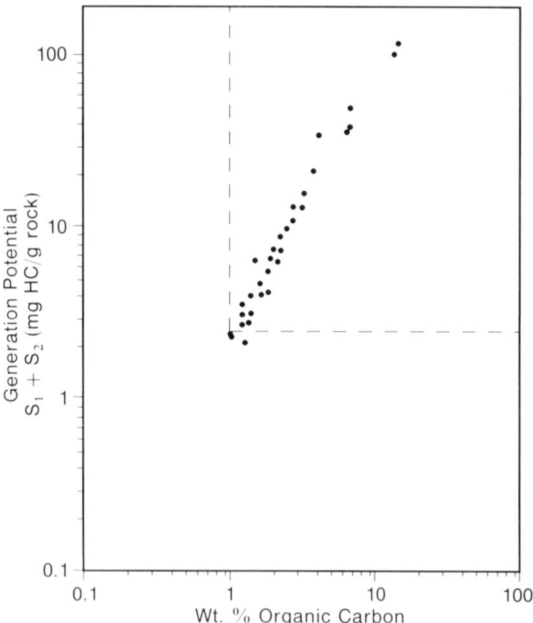

Fig. 5. Comparison of organic enrichment and generation potential. *Dashed lines* mark limits of potential and/or effective source rocks (Bissada 1982)

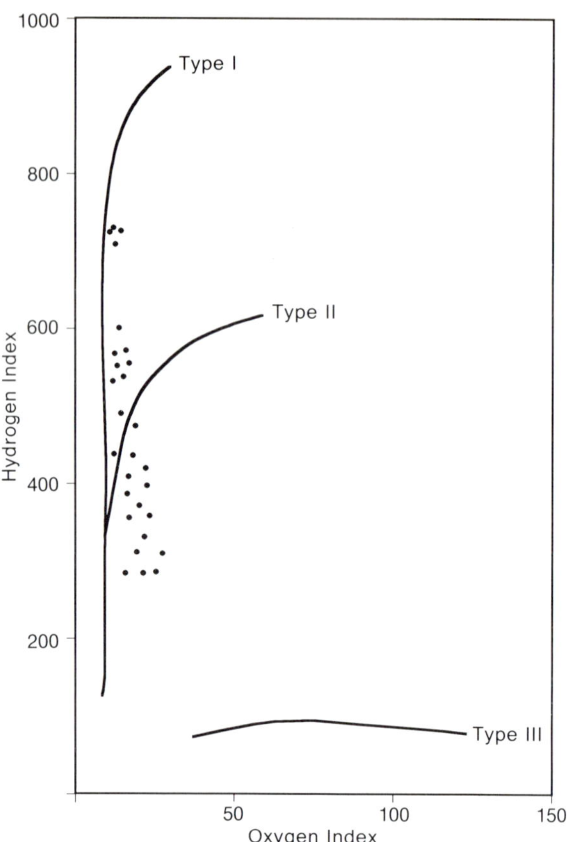

Fig. 6. Modified van Krevelen diagram for Exshaw Formation samples analyzed in this study. Samples with higher hydrogen indices are, in general, from highly laminated black shales. Samples with lower hydrogen indices are either from rocks that exhibit some degree of bioturbation or have experienced higher levels of thermal maturity

that excellent conditions existed at the time of deposition for preservation of marine sapropels and that there was only limited terrestrial input.

Visual kerogen analyses support the conclusion that little detrital organic matter was incorporated into the Exshaw Formation sediments. The Exshaw Formation contains predominantly fluorescent, amorphous organic matter (75–95%; Table 1). Creaney and Allan (1990) have also reported *Tasmanites* from several Exshaw Formation kerogens. A minor nonfluorescent maceral population (5–25%) in many of the examined samples from this study may be related to degradation of a portion of the organic matter at the time of deposition or alternatively, in other samples may be related to present-day thermal maturities equivalent to the middle of the oil window.

Savoy (1990) reports that Exshaw Formation kerogens recovered from outcrop samples contain up to several percent vitrinite. This is in contrast to the isolated kerogens examined as part of this study where no definitively identified vitrinite was observed. Most of organic matter in the samples examined by Savoy (1990) is described as being spent (i.e., high levels of thermal maturation), with

Table 2. Organic and total sulfur enrichment and hydrocarbon generation potential for the Exshaw Formation from selected wells in the Peace River Arch area. Note the increasing organic carbon and hydrogen enrichment with depth. Several wells in the area exhibit this trend which may be related to the facies described by Richards and Higgins (1988). See text

Well locality	Depth (m)	TOC	% Sulfur	S_2	HI
13-13-78-26W5	2052.7	6.03	2.65	33.46	555
	2059.6	5.98	2.53	33.85	566
	2063.4	3.15	3.18	12.91	410
	2066.0	6.18	3.93	44.04	713
14-30-80-23W5	1736.1	2.16	3.36	6.25	289
	1742.2	2.04	3.05	5.83	286
	1745.2	13.94	2.77	107.19	730
6-2-79-22W5	1708.25	2.09	2.22	7.72	369
	1714.1	2.55	3.30	9.28	336
	1718.0	4.00	0.67	34.80	731
	1721.3	13.02	1.91	94.63	727

high organic carbon and low organic hydrogen contents. This suggests that vitrinite may have been misidentified at these high levels of maturity and the described organic matter may represent unidentifiable amorphous organic matter or high reflectivity bitumenites.

A trend in both organic enrichment and organic matter type can be observed in several of the wells examined. In several wells, samples from the base of the section are organically enriched and have higher organic hydrogen contents than samples up section (Table 2). Organic carbon values up to 14 wt. % are found in the lower portion of the section with associated hydrogen index values above 700. Stratigraphically younger samples contain less organic matter and have significantly lower hydrogen index values. These overlying shales still represent an oil-prone, potential source system but the potential for generation is significantly lower than the basal part of the section.

This stratigraphic variability may correspond to the distinctions made by Richards and Higgins (1988) in the outcrop where they described two distinct lithologic facies in the lower member of the Exshaw Formation. The lower black shale member is divided into an upper and lower unit, where the diagnostic features of the lower unit are: thin planar lamination, few macroscopic benthos, anomalous radioactivity, and a high organic carbon content. The upper unit contains inarticulate brachiopods and is bioturbated, suggesting deposition in a more oxygenated environment. The data to make the distinction between these two shale facies in the subsurface was not available for this study. However, it is suggested that the trend observed in

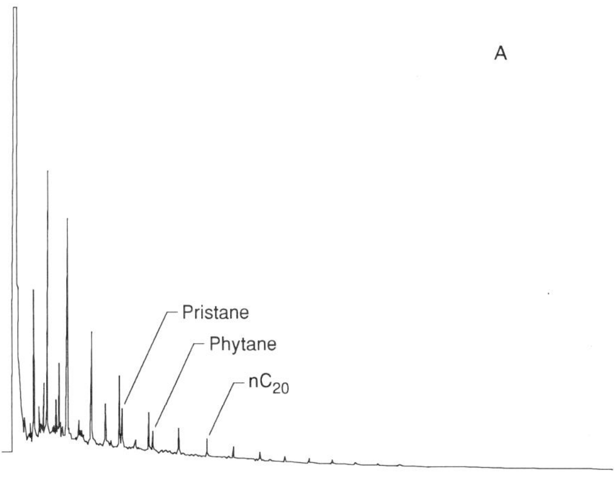

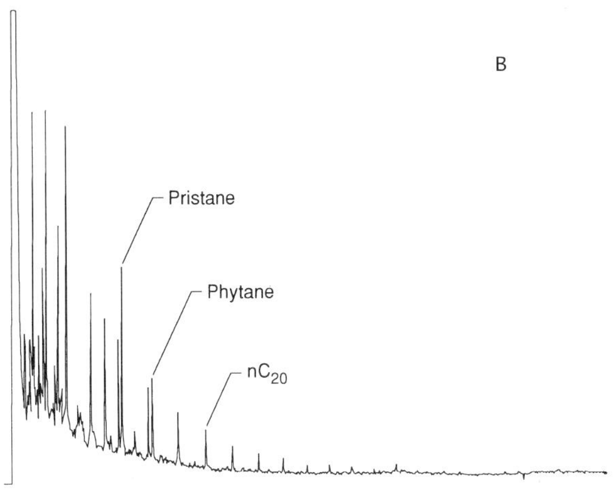

Fig. 7A, B. Typical saturated hydrocarbon chromatograms. **A** is from 1745.2 m in the 14-31-80-23W5 well. **B** is from 1714.1 m in the 6-2-GW79-22W5 well. See Table 2 for additional data. Both samples exhibit present day thermal maturities near the top of the oil window

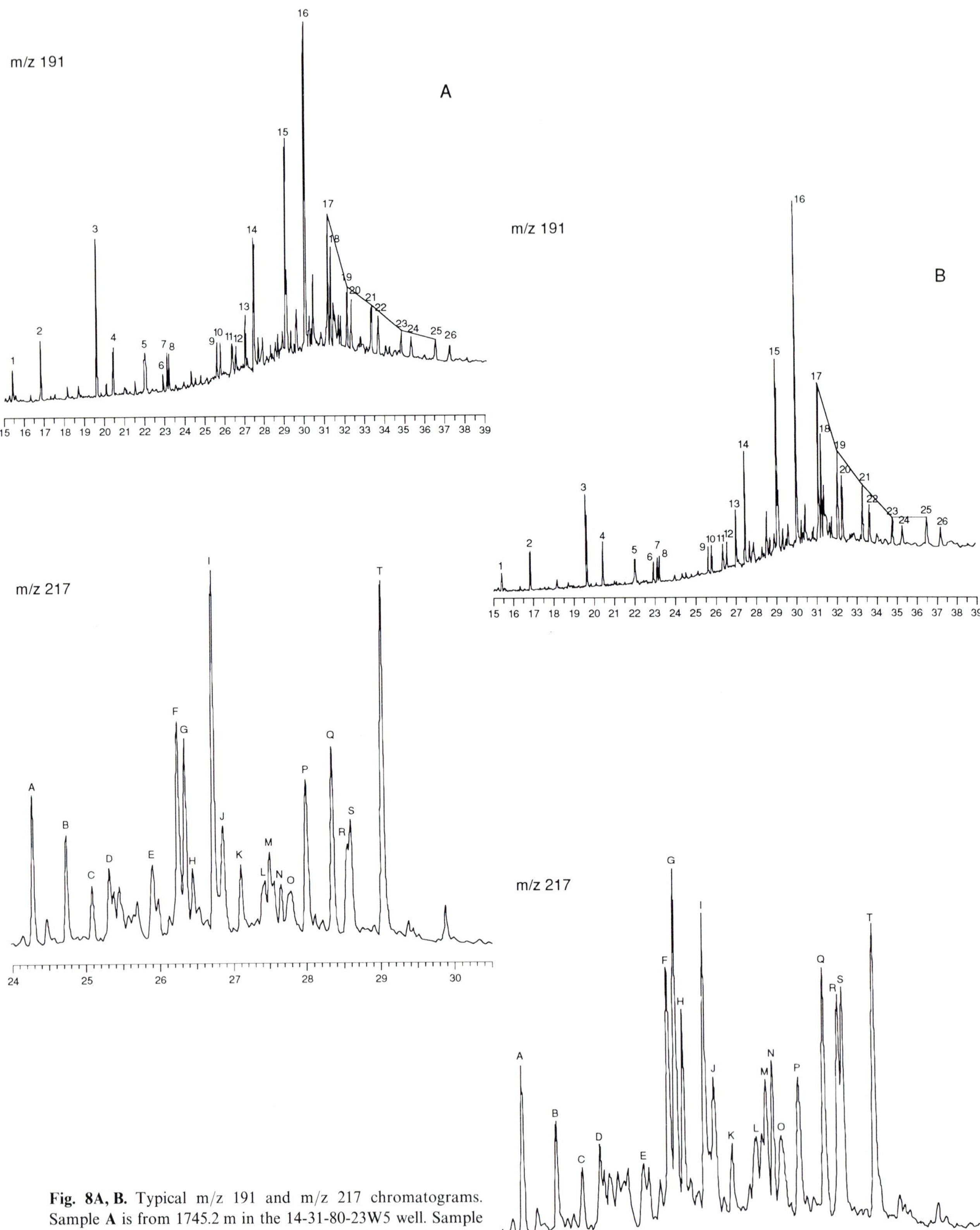

Fig. 8A, B. Typical m/z 191 and m/z 217 chromatograms. Sample **A** is from 1745.2 m in the 14-31-80-23W5 well. Sample **B** is from 1714.1 m in the 6-2-GW79-22W5 well. See Table 2 for additional data. See Tables 3 and 4 for peak identifications the organic matter is geologically controlled and may be related to facies described by Richards and Higgins (1988). Samples yielding higher generation potentials and associated hydrogen indices do exhibit less evidence of bioturbation than samples with lower hydrogen indices and are in most instances highly laminated. These samples are also the most organically enriched and are found in the lower part of the Exshaw black shales, suggesting that anoxia may have been less common through time.

Bitumen Characterization

Extracts of the Exshaw Formation are dominated by non-hydrocarbons and typically exhibit low saturated/aromatic hydrocarbon ratios even in samples that are thermally mature (Table 1). $C_{15}+$-saturated hydrocarbon chromatograms exhibit very few peaks beyond nC_{25}, suggesting little higher plant input into the primarily marine algal/bacterial organic matter (Fig. 7). Pristane/phytane ratios are all greater than 1, suggesting that the bottom waters were not anoxic at the time of deposition (Didyk et al. 1978). While maturity does affect this ratio (ten Haven et al. 1987), it does not appear to be a significant factor in this sample suite (Table 1). CPI values were not calculated for most Exshaw Fm. extracts because of the limited abundance of the $> nC_{25}$ components.

Biomarker distributions are similar for all of the examined Exshaw Fm. extracts. The tricyclics are dominated by the C_{23} terpane and are subordinate to the hopanes. The Tm/Ts ratio varies between 2 and 4 in immature to marginally mature samples from the Peace River Arch area. Hopane is consistently the largest peak in the m/z 191 chromatogram and the extended hopanes exhibit a smooth pseudo-harmonic decrease with increasing length of the side chains (Fig. 8).

Some variability is observed in the sterane distributions due primarily to differences in maturity, as exhibited by the C_{29} 5α,14α,17α 20S/20R sterane and the C_{29} 5α,14β,17β 20R + 20S/5α,14α,17α 20R sterane ratios. The C_{29} steranes are more abundant

Table 3. Identification of terpanes in the m/z 191 chromatograms

Peak	compound
1	C_{20} Tricyclic terpane
2	C_{21} Tricyclic terpane
3	C_{23} Tricyclic terpane
4	C_{24} Tricyclic terpane
5	C_{25} Tricyclic terpane
6	C_{24} Tetracyclic terpane
7	C_{26} Tricyclic terpane A
8	C_{26} Tricyclic terpane B
9	C_{28} Tricyclic terpane A
10	C_{28} Tricyclic terpane B
11	C_{29} Tricyclic terpane A
12	C_{29} Tricyclic terpane B
13	C_{27} 18α(H)-22,29,30-Trisnorhopane (Ts)
14	C_{27} 17α(H)-23,29,30-Trisnorhopane (Tm)
15	C_{29} 17α(H),21β(H)-30-Norhopane
16	C_{30} 17α(H),21β(H)-Hopane
17	C_{31} 17α(H),21β(H)-Homohopane (22S)
18	C_{31} 17α(H),21β(H)-Homohopane (22R)
19	C_{32} 17α(H),21β(H)-Homohopane (22S)
20	C_{32} 17α(H),21β(H)-Homohopane (22R)
21	C_{33} 17α(H),21β(H)-Homohopane (22S)
22	C_{33} 17α(H),21β(H)-Homohopane (22R)
23	C_{34} 17α(H),21β(H)-Homohopane (22S)
24	C_{34} 17α(H),21β(H)-Homohopane (22R)
25	C_{35} 17α(H),21β(H)-Homohopane (22S)
26	C_{35} 17α(H),21β(H)-Homohopane (22R)

Table 4. Identification of steranes in the m/z 217 chromatograms

Peak	compound assignment
A	13β,17α-Diacholestane (20S)
B	13β,17α-Diacholestane (20R)
C	13α,17β-Diacholestane (20S)
D	13α,17β-Diacholestane (20R) + 24-Methyl-13β,17α-Diacholestane (20S)
E	24-Methyl-13β,17α-Diachloestane (20R)
F	14α,17α-Cholestane (20S)
G	14β,17β-Cholestane (20R) + 24-Ethyl-13β,17α-Diacholestane (20S)
H	14β,17β-Cholestane (20S) + 24-Methyl-13α,17β-Diacholestane (20R)
I	14α,17α-Cholestane (20R)
J	24-Ethyl-13β,17α-Diacholestane (20R)
K	24-Ethyl-13α,17β-Diacholestane (20S)
L	24-Methyl-14α,17α-Cholestane (20S)
M	24-Methyl-14β,17β-Cholestane (20R) + 24-Ethyl-13α,17β-Diacholestane (20R)
N	24-Methyl-14β,17β-Cholestane (20S)
O	24-Propyl-13α,17β-Diacholestane (20S)
P	24-Methyl-14α,17α-Cholestane (20R)
Q	24-Ethyl-14α,17α-Cholestane (20S)
R	24-Ethyl-14β,17β-Cholestane (20R)
S	24-Ethyl-14β,17β-Cholestane (20S)
T	24-Ethyl-14α,17α-Cholestane (20R)

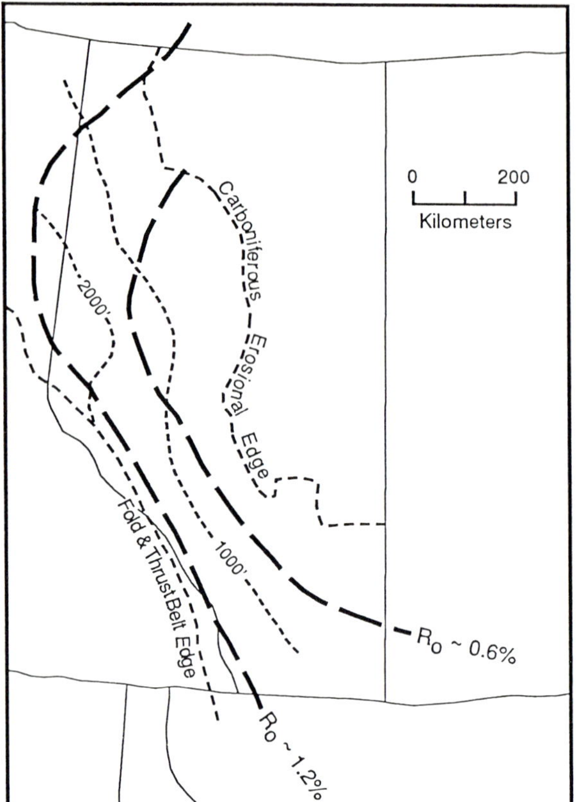

Fig. 9. Generalized present-day levels of thermal maturity for the Exshaw Formation from the Western Canada basin. Light *dashed lines* show estimates of the total Carboniferous section and *heavy dashed lines* mark equivalent present-day vitrinite reflectance values. Map is compilation of data from the present chapter as well as Jones et al. (1986), Creaney and Allan (1990), Majorowicz et al. (1990). Present-day levels of thermal maturity do not necessarily relate to current burial depths due to uplift and erosion since the Eocene and redistribution of heat flow due to hydrodynamics of the Western Canada basin

than both the C_{27} and C_{28} steranes (Fig. 8). Such a distribution is typical of Paleozoic strata (Grantham and Wakefield 1988). Carbon isotope values for the $C_{15}+$ bitumen extracts exhibit a relatively small range in variation from -29.67 to -27.26‰ (PDB).

Levels of Thermal Maturity

In the Western Canada basin, thermal maturities increase westward toward the deep basin and thrust front (Fig. 9). Samples from this study yield estimates of present-day levels of thermal maturity ranging from near the top to the middle of the oil window (approximate vitrinite reflectance (Ro) equivalent of 0.75–1.00%) based on thermal alteration indices (TAI) and pyrolysis T_{max} values (Table 5). Estimates of maturity based on the character of the saturated hydrocarbon chromatograms and selected biomarker ratios are consistent with the Exshaw Formation, having experienced maturities equivalent to the top of the oil-window in the Peace River Arch area.

Maximum burial for the Western Canada basin in Alberta occurred in the Eocene and the basin has been uplifted and eroded since that time (Jones et al. 1986; Majorowicz et al. 1990). This indicates that the maturities recorded in the Exshaw Formation samples represent older thermal events and do not necessarily indicate present-day generation and expulsion. Heat transfer in the Western Canada basin of Alberta has been affected at a regional scale by hydrodynamics, and several authors have suggested that present-day heat flows in the basin are not related to the paleotemperature field during the time of maximum burial in the Eocene (Beaumont et al. 1985; Majorowicz et al. 1985; Majorowicz et al. 1990; Jessop 1992).

Oil to Source Correlations

The most comprehensive oil-to-oil and oil-to-source correlations in the Western Canada basin have been reported by Allan and Creaney (1991) and Creaney and Allan (1990). They suggest that little oil generated by the Exshaw Formation has been retained in reservoirs of near-equivalent age. The Exshaw Formation is thought to have contributed primarily to the extensive reserves of the Early Cretaceous reservoirs known in the Western Canada basin. In their estimates, some oil in the underlying Wabamun and overlying Banff Formations can be attributed to the Exshaw Formation but these represent only minor accumulations compared to the reserves found in Early Cretaceous reservoirs that can be attributed to the Exshaw Formation.

The Exshaw Formation has been correlated to the Early Cretaceous Mannville Formation deposits. Allan and Creaney (1991) suggest that these heavy oil deposits represent mixing from both the Devonian/Mississippian Exshaw Formation and the Jurassic Nordegg Member of the Fernie Group. Jones et al. (1986) reached similar conclusions about the source of the heavy, biodegraded oils in

Table 5. Thermal maturity data

Well locality	Depth (m)	TAI	T_{max}	$\alpha\alpha S/\alpha\alpha R$ Steranes	$\beta\beta R + S/\alpha\alpha R$ Steranes
16-30-77-25W5	2022.5	2.7	453	0.82	1.36
14-31-8-23W5	1736.1	2.6	443	0.85	1.39
	1742.2		452		
	1745.2	2.65	451	0.92	1.46
13-13-78-26W5	2059.6	2.7	461		
6-2-79-22W5	1708.25		458	0.73	1.47
	1718.0	2.75	450		
16-17-80-23W5	1736.25	2.65	452	0.88	1.21
	1741.4		445		
1-27-80-24W5	1772.4	2.8	450	0.82	1.52
14-30-80-23W5	1729.3	2.65	447	1.03	1.82

TAI scale – 2.4–2.6 Initial generation – Ro equivalent ~ 0.5–0.7
– 2.6–3.2 Main stage of generation and expulsion – Ro equivalent ~ 0.7–1.3.
T_{max} from Rock-Eval pyrolysis.
$\alpha\alpha S/\alpha\alpha R$ Steranes = 24-ethyl-14α,17α-cholestane (20S)/24-ethyl-14α,17α-cholestane (20R).

$$\beta\beta R + S/\alpha\alpha R \text{ Steranes} = \frac{\text{24-ethyl-14}\beta\text{,17}\beta\text{-cholestane (20R)} + \text{24-ethyl-14}\beta\text{,17}\beta\text{-cholestane (20S)}}{\text{24-ethyl-14}\alpha\text{,17}\alpha\text{-cholestane (20R).}}$$

Alberta, suggesting that the bulk of the Mannville Group oils migrated from Exshaw Formation sources to the south and west with some minor contributions from Jurassic and Lower Cretaceous shales.

In contrast to these studies, Brooks et al. (1989) indicate that the Exshaw and Nordegg do not appear to have contributed to the Mannville Group oils. Heavy oils in the Western Canada basin were found to contain 28,30-bisnorhopanes (Brooks et al. 1989). These compounds were not detected in either the Exshaw or the Nordegg (Brooks et al. 1987; Osadetz et al. 1992) leading Brooks et al. (1989) to conclude that these sources had not contributed to the massive tar deposits. However, the absence of these compounds in Exshaw Formation bitumens does not exclude the possibility that the Exshaw contributed to these heavy oil deposits.

While much of the generated and expelled products from the Exshaw Formation may have migrated into significantly younger strata (Allan and Creaney 1991; Creaney and Allan 1990), several Devonian reservoired oils can be attributed to the Exshaw Fm. Several oils reservoired in stratigraphically older Devonian formations in the Peace River Arch area can be attributed to the Exshaw Fm.

At least two other source units are present in the Peace River Arch area. They are the Devonian Muskeg and Duvernay Formations (Fig. 1). Both formations exhibit geochemical characteristics that are unique and provide criteria which can be used to correlate the oils to their primary sources. Some of the most diagnostic criteria in both formations are the homohopane distributions and the stable carbon isotopic signatures. Extracts from the Duvernay Fm. in the Peace River Arch area yield chromatograms that exhibit C_{33} homohopanes in greater abundances than either the C_{32} or C_{34} homohopanes (Fig. 10). The Duvernay Fm. is also isotopically lighter than the Devonian reservoired oils, opposite of what would be expected if the Duvernay Fm. was the primary source for the oils (Fig. 11).

In contrast, the Muskeg Fm. exhibits an unusual homohopane distribution with C_{34} homohopanes dominating over both the C_{33} and C_{35} homohopanes in the m/z 191 chromatograms (Fig. 10). In addition, the stable carbon isotopic signature of the Muskeg Fm. extracts are considerably heavier than many of the Devonian reservoired oils (Fig. 11).

In general, two families of oils are apparent from the suite of studied samples (Robison 1992); (Table 6). One family appears to have been generated from the Muskeg Fm. while the second family appears to be most closely related to the Exshaw Fm. Some mixing of oils is probable and it is likely that the Duvernay Fm. contributed, in part, to the oils attributed to the Exshaw Fm. Osadetz et al.

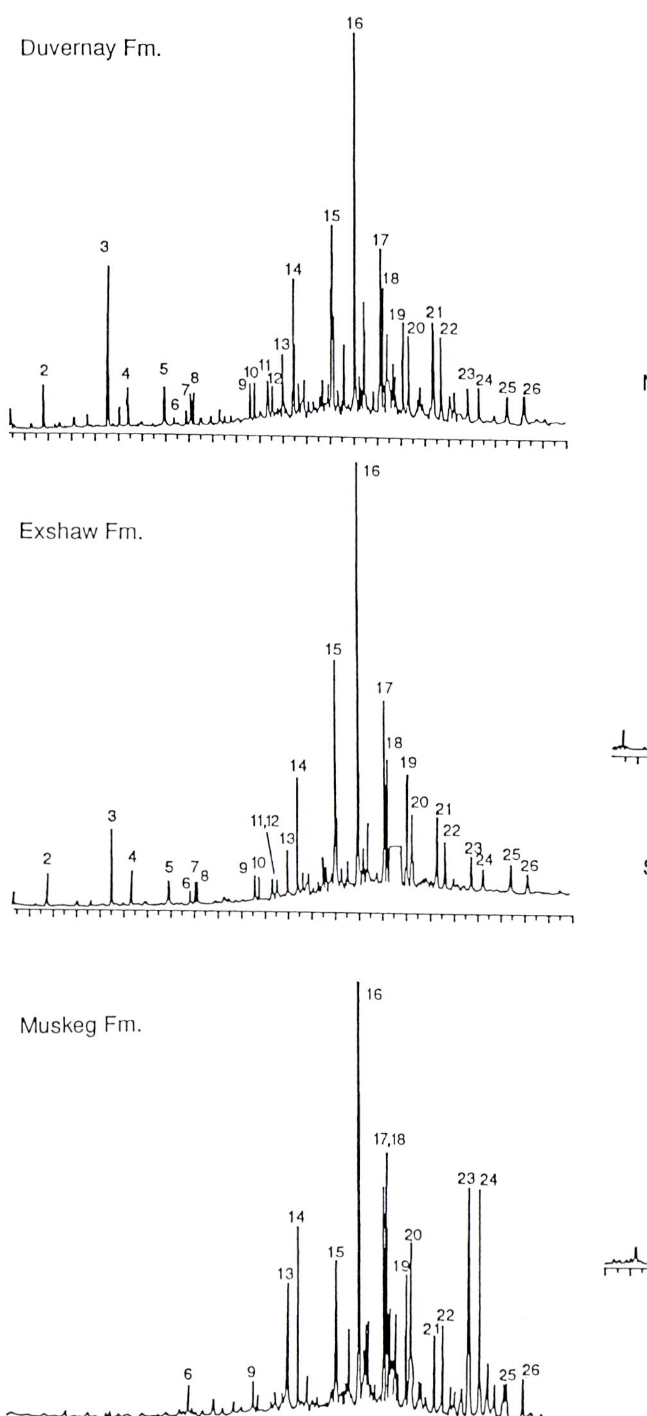

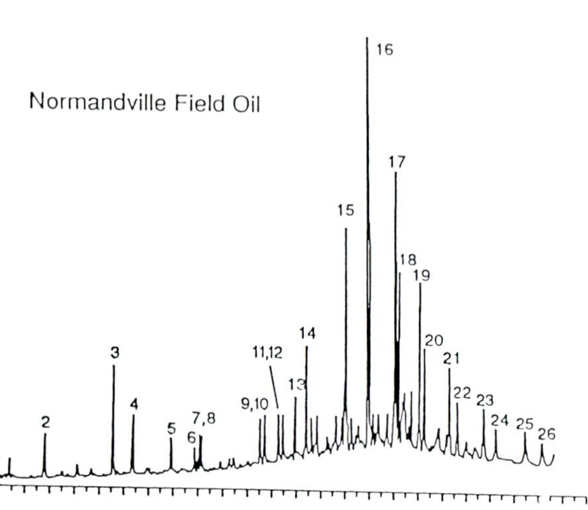

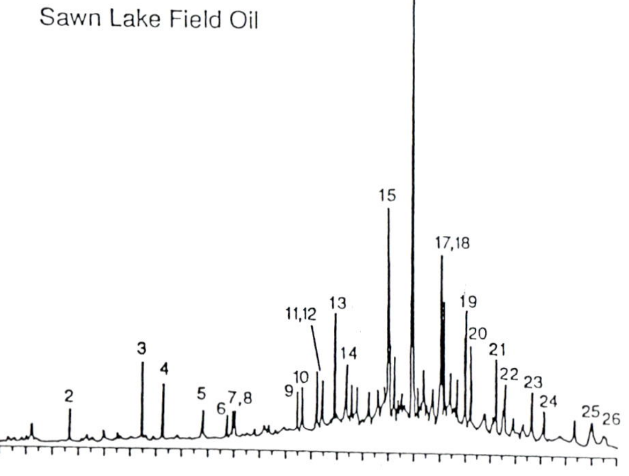

Fig. 10. Typical m/z 191 chromatograms for Muskeg, Duvernay, and Exshaw Formation bitumens and two oils from the Peace River Arch area. The two oils have been correlated by Robison (1992) to the Exshaw Formation. Note the predominance of the C_{34} homohopanes and near absence of tricyclic terpanes in the Muskeg Formation and the abundance of the C_{33} homohopanes relative to the C_{32} and C_{34} homohopanes in the Duvernay Formation

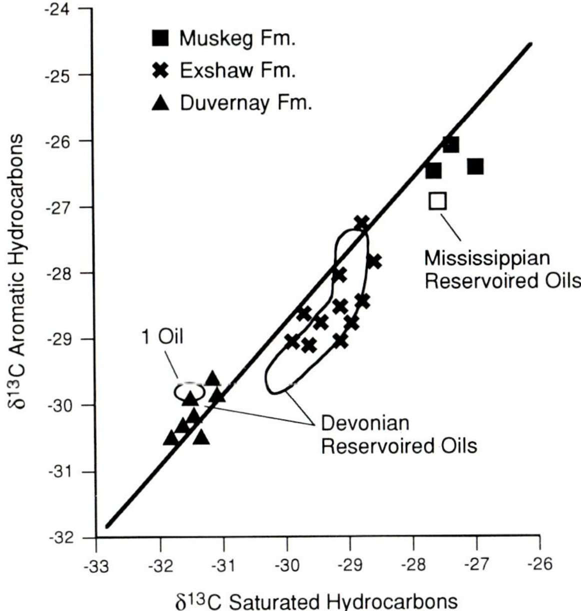

Fig. 11. Plot of $\delta^{13}C$ saturated vs. aromatic hydrocarbons for Muskeg, Duvernay, and Exshaw Formation bitumens and oils from the Peace River Arch area. *Circled areas* represent range of carbon isotopic values recorded for each family of oils

(1992) presented similar data for Devonian and Carboniferous source systems and reached similar conclusions in correlating oils and their potential source rocks in the Canadian portion of the Williston basin.

In a basin as mature as the Western Canada basin, it could be typically assumed that the oil-source relationships are well established. However, this is not always true. Several authors (Jones et al. 1986; Creaney and Allan 1990; Allan and Creaney 1991) have attributed the massive heavy oil deposits of the basin in part to the Exshaw Fm. In contrast, very little oil reservoired in near age-equivalent strata has been attributed to the Exshaw Fm. (Creaney and Allan 1990; Allan and Creaney 1991). Both of these assumptions have been questioned (Brooks et al. 1987; Brooks et al. 1989; Osadetz et al. 1992), and it is clear that the true significance and volumetrics of the generative portion of the Exshaw Fm. are still not known.

Migration pathways in the Western Canada basin are assumed to be relatively simple, and it is often cited as a classic example of a basin where long range migration can be demonstrated (Demaison and Huizinga 1991). This is evident for much of the Exshaw sourced oil if it is assumed that the Exshaw has contributed to the Mannville heavy oil deposits. However, in areas such as the Peace River Arch, it becomes apparent that migration can be complicated over short distances with Exshaw-generated oils migrating into stratigraphically older, but structurally higher, Devonian reservoirs in the Peace River Arch area.

Discussion

Based on visual kerogen assessment, biological marker studies, and stable carbon isotopic values, the type of organic matter preserved in the analyzed Exshaw Formation samples does not appear to vary significantly. Distributions of tricyclic terpanes, hopanes, and steranes are similar throughout the section (Table 1). Organic productivity in the Exshaw seaway may have been a primary control on organic enrichment and generation potential (Parrish 1982). However, from the available dataset a clear distinction among the controls on productivity and preservation cannot be made.

Preservation of organic matter can be influenced by the rate of production of organic detritus, degradation of organic matter due to oxidation and exposure, and the rate of burial and removal of the organic matter from these environments. Several authors have suggested the development of a stratified water column and the subsequent development of anoxia as being the primary controls on preservation in the Exshaw Formation (Macqueen and Sandberg 1970; Richards and Higgins 1988; Savoy 1990). Samples containing higher concentrations of organic carbon with higher relative organic hydrogen enrichments are, in general, highly laminated and exhibit less evidence of bioturbation than samples with lower organic carbon contents and associated hydrogen indices. This does not, however, imply that the development of anoxic bottom waters is directly responsible for the elevated organic carbon contents.

Decomposition rates of organic matter under oxic and anoxic conditions have been shown to be similar (Westrich and Berner 1984; Henrichs and Reeburgh 1987; Lee 1992), indicating that the development of anoxia itself does not result in sequestering quantities of organic matter greater than in oxygenated sediments. Recent studies have suggested that preservation of organic matter may be primarily influenced by the exclusion of bacterial grazers in anoxic environments rather than preservation being directly related to the degree of anoxia

Table 6. Peace River Arch area oils analyzed in this study. Devonian reservoired oils have been correlated to the Exshaw Formation and Mississippian reservoired oils are attributed to the Muskeg Formation

Locality	Field	Reservoir age	1	2	3	4	5	6	7	8	9
3-11-96-6W5	Panny	Devonian	3.45	19.0	1.71	17.7	0.97	1.27	60	31/16/53	0.15
10-25-95-5W5	Panny East	Devonian	3.75	15.8	1.74	15.4	0.92	1.37	60	36/19/45	0.14
3-20-95-6W5	Panny South	Devonian	3.76	18.1	1.72	17.7	0.56	1.36	68	39/18/43	0.14
10-1-88-9W5	Red Earth	Devonian	2.60	24.0	1.69	20.7	0.59	1.47	65	39/21/40	0.13
?	Red Earth	Devonian	3.21	20.0	2.12	19.8	0.87	1.26	64	34/23/43	0.11
5-22-89-3W5	Trout Mtn	Devonian	3.76	20.5	1.77	16.6	0.58	1.43	68	37/18/45	0.13
14-8-89-3W5	Trout Mtn	Devonian	2.63	12.2	1.65	23.4	0.69	1.26	65	29/24/47	0.13
6-30-91-12W5	Sawn Lake	Devonian	3.26	17.2	1.79	20.5	0.60	1.46	65	37/21/42	0.14
2-19-91-12W5	Sawn Lake	Devonian	3.57	18.9	1.79	19.8	0.59	1.33	65	37/21/42	0.10
4-31-91-12W5	Sawn Lake	Devonian	4.76	19.5	1.73	19.4	0.82	1.24	65	37/21/42	0.14
9-4-91-5W5	Kidney	Devonian	3.81	15.9	1.79	20.7	0.70	1.31	65	36/21/43	0.13
15-23-91-5W5	Kidney	Devonian	3.58	17.6	1.73	19.2	0.97	1.20	66	33/19/48	0.10
14-23-74-1W6	Puskwaskau	Devonian	1.84	10.6	1.61	24.9	1.71	1.31	55	28/35/37	0.16
1-26-74-1W6	Puskwaskau	Devonian	1.75	10.9	1.64	24.4	1.45	1.46	70	34/23/43	0.18
9-9-77-25W5	Eaglesham	Devonian	2.24	12.2	1.81	20.2	1.29	1.41	63	22/20/58	0.12
10-14-77-25W5	Eaglesham	Devonian	2.75	10.5	1.67	19.0	1.47	1.38	60	27/24/49	0.13
16-30-77-25W5	Eaglesham	Devonian	1.96	17.2	1.55	22.6	1.64	1.41	54	32/15/53	0.16
10-28-77-26W5	Eaglesham	Mississippian	1.36	15.3	1.74	6.5	1.36	0.72	47	35/20/45	0.11
12-28-77-26W5	Eaglesham	Mississippian	1.39	19.7	1.86	6.4	1.44	0.64	48	28/25/47	0.08
13-9-79-22W5	Normandville	Devonian	2.18	14.6	1.72	25.1	1.15	1.25	66	17/23/60	0.16
14-15-79-22W5	Normandville	Devonian	2.26	14.2	1.77	23.9	1.36	1.36	70	31/18/51	0.15
8-16-79-22W5	Normandville	Devonian	2.34	25.8	1.82	20.4	1.51	1.38	53	37/21/42	0.15
6-19-80-23W5	Tangent	Devonian	1.61	18.7	1.65	19.4	2.16	1.34	53	33/18/49	0.16
10-23-80-24W5	Tangent	Devonian	1.41	19.4	1.59	20.7	2.44	1.42	51	34/18/48	0.20
16-36-80-24W5	Tangent	Devonian	1.90	16.2	1.67	21.1	1.86	1.29	56	25/21/54	0.17

1 Saturated/aromatic hydrocarbons.
2 % Nonhydrocarbons.
3 Pristane/phytane.
4 % Tricyclics-C_{19} through C_{29} tricyclics as a percentage of total identified components in the m/z 191 fragmentogram (Table 3).
5 C_{27} 17α(H)-23,29,30-Trisnorhopane (Tm)/C_{27} 18α(H)-22,29,30-trisnorhopane (Ts).
6 [C_{33} 17α(H),21β(H)-Homohopane (22S) + C_{33} 17α(H),21β(H)-homohopane (22R)]/[C_{34} 17α(H),21β(H)-homohopane (22S) + C_{34} 17α(H),21β(H)-homohopane (22R)].
7 % Diasteranes-diasteranes as a percentage of total identified components in the m/z 217 fragmentogram (Table 4).
8 14α,17α-Cholestane (20R)/24-methyl-14α,17α-cholestane (20R)/24-ethyl-14α,17α-cholestane (20R).
9 Sterane/hopane – [C_{27} + C_{28} + C_{29} (αα & ββR + S) steranes]/[norhopane + hopane + C_{31}-C_{35} homohopanes].

that is developed in the water column and sediments (Lee 1992). This suggests that in the Exshaw Formation seaway burrowing organisms may have periodically been excluded from the environment due to reduced oxygen content in the bottom waters.

Whatever the mechanism, the lithology and geochemical attributes of the lower black shale member indicate an environment conducive to the production and preservation of organic matter. Sulfur contents and pristane/phytane ratios do not vary significantly throughout the organic-rich portion of the Exshaw (Table 1), and suggest that the sediments were depleted in oxygen but may have alternated between anoxic and dysaerobic conditions.

Periodic stratification of the Exshaw seaway is suggested by the presence of organic-rich, laminated black shales, suggesting the development of anoxia and a stable water column. It appears, however, that stratification and the development of anoxia may have varied through time in the Exshaw seaway. Stratification and anoxia appear to have been more prevalent during deposition of the lower portion of the black shale member where higher organic carbon contents, higher hydrogen indices, and a higher frequency of laminated sediments are seen.

Summary

The Exshaw Formation is a major source of hydrocarbons in the Western Canada basin. A reliable estimate of the volumes of hydrocarbons generated is not possible due to the uncertainties in the degree of mixing of Exshaw-sourced oils with hydrocarbons from one of the several other source systems in the basin. The Exshaw may have been a primary source for the heavy oil deposits of Alberta reservoired in the Mannville Group. The Exshaw appears to have also contributed locally to Devonian and Mississippian age reservoirs.

The Exshaw Formation consists of a lower black shale member and an upper siltstone to silty limestone member. The lower shaly member was deposited during a transgression that onlapped the underlying Palliser Formation and Wabamun Group. The upper member represents deposition in a shallowing and regressing seaway. All of the source potential in the Exshaw Formation occurs in the lower, transgressive, shaly portion of the unit.

The black shale lower member of the Exshaw Formation was deposited in a broad epeiric seaway beneath a zone of active upwelling. Several geologic and geochemical parameters suggest that the bottom waters varied from dysaerobic to anoxic and that the seaway may have occasionally been stratified. Highest organic carbon contents and the most oil-prone organic matter are found in the basal laminated portions of the Exshaw Formation black shales. Shales overlying the lower member, which vary from laminated to bioturbated, are not as oil-prone or organically enriched, suggesting that the development of a stratified seaway and anoxia in the bottom waters was not as frequent as in the lower portion of the member.

Acknowledgments. The author thanks Texaco for permission to publish. Jon Dudley collected samples used in this investigation during his time at Texaco Canada. The author extends his appreciation to Lloyd Snowdon and Barry Katz for their review and helpful comments on an earlier version of this chapter. Analytical support was provided by the geochemical research laboratories at the Texaco E&P Technology Department.

References

Allan J, Creaney S (1991) Oil families of the Western Canada basin. Bull Can Petrol Geol 39: 107–122

Arthur MA, Schlanger SO, Jenkyns HC (1987) The Cenomanian-Turonian oceanic anoxic event, II: Palaeoceanographic controls on organic-matter production and preservation. In: Brooks J, Fleet AJ (eds) Marine petroleum source rocks, Geol Soc, London, Spec Publ 26: 401–420

Bally AW, Gordy PL, Stewart GA, (1966) Structure, seismic data, and orogenic evolution of the southern Canadian Rocky Mountains. Bull Can Petrol Geol 14: 337–381

Beaumont CR, Boutilier AS, MacKenzie AS, Rullkötter J (1985) Isomerization and aromatization of hydrocarbons and the paleothermometry and burial history of the Alberta basin. Am Assoc Petrol Geol Bull 69: 546–566

Berner RA, Raiswell R (1983) Burial of organic carbon and pyrite sulfur in sediments over Phanerozoic time: a new theory. Geochim Cosmochim Acta 47: 855–862

Bissada, KK, (1982) Geochemical constraints on petroleum generation and migration – a review. Proc 2nd ASCOPE Conf, Manila, Oct, 1981, pp 69–87

Brooks PW, Snowdon LR, Osadetz KG (1987) Carlson CG, Christopher JE (eds) Families of oils in southeastern Saskatchewan. 5th North Dakota Geol Soc and Saskatchewan Geol Soc Williston Basin Int Symp Proc, pp 253–264

Brooks PW, Fowler MG, MacQueen RW (1989) Biomarker geochemistry of Cretaceous oil sands, heavy oils and Paleozoic carbonate trend bitumens, Western Canada basin. In: Meyer RF, Wiggins EJ (eds) 4th UNITAR/UNDP Int Conf on heavy crudes and tar sands 2: 593–606

Cook FA, Green AG, Simony PS, Price RA, Parrish RR, Milkereit B, Gordy PL, Brown RL, Coflin KC, Patenaude C (1988) Lithoprobe seismic reflection structure of the southeastern Canadian Cordillera: initial results. Tectonics 7: 157–180

Creaney S, Allan J (1990) Hydrocarbon generation and migration in the Western Canada sedimentary basin. In: Brooks J (ed) Classic petroleum provinces. Geol Soc, London, Spec Publ 50: 189–202

Demaison G, Huizinga BJ (1991) Genetic classification of petroleum systems. Am Assoc Petrol Geol Bull 75: 1626–1643

Didyk BM, Simoneit BRT, Brassell SC, Eglinton G (1978) Organic geochemical indicators of paleoenvironmental conditions of sedimentation. Nature 272: 216–222

Espitalie J, Laporte JL, Madec M, Marquis F, Leplat P, Poulet J, Boutefeu A (1977) Méthode rapide de caractérisation des roches mères de leur potentiel pétrolier et de leur degré d'évolution. Rev Inst Français du Pétrole 32: 23–42

Grantham PJ, Wakefield LL (1988) Variations in sterane carbon number distributions of marine source rock derived crude oils through geological time. Org Geochem 12: 61–73

Harker P, McLaren DJ (1958) The Devonian-Mississippian boundary in the Alberta Rocky Mountains. In: Goodman AJ (ed) Jurassic and Carboniferous of western Canada. Am Assoc Petrol Geol John Andrew Allan Mem Vol, pp 244–259

Henrichs SM, Reeburgh WS (1987) Anaerobic mineralization of marine sediment organic matter: rates and the role of anaerobic processes in the oceanic carbon economy. Geomicrobiol J 5: 191–237

Herbert TD, Fischer AG (1986) Milankovitch climatic origin of mid-Cretaceous black shale rhythms in central Italy. Nature 321: 739–743

Jessop AM (1992) Thermal input from the basement of the Western Canada sedimentary basin. Bull Can Petrol Geol 40: 198–206

Johnson JG, Sandberg CA (1988) Devonian eustatic events in the western United States and their biostratigraphic responses. In: McMillan NJ, Embry AF, Glass DJ (eds) Devonian of the world. Can Soc Petrol, Calgary, Geol Mem 14: 9–22

Johnson JG, Klapper G, Sandberg CA (1985) Devonian eustatic fluctuations in Euramerica. Geol Soc Am Bull 96: 567–587

Jones FW, Majorowicz JA, Linville A, Osadetz KG (1986) The relationship of hydrocarbon occurrences to geothermal gradients and time-temperature indices in Mesozoic formations of southern Alberta. Bull Can Petrol Geol 34: 226–239

Klemme HD, Ulmishek GF (1991) Effective petroleum source rocks of the world: stratigraphic distribution and controlling depositional factors. Am Assoc Petrol Geol Bull 75: 1809–1851

Lee C (1992) Controls on organic carbon preservation: the use of stratified water bodies to compare intrinsic rates of decomposition in oxic and anoxic systems. Geochim Cosmochim Acta 56: 3323–3335

Leenheer MJ (1984) Mississippian Bakken and equivalent formations as source rocks in the Western Canada basin. Org Geochem 6: 521–533

Leventhal JS (1987) C and S relationships in Devonian shales from the Appalachian basin as an indicator of environment of deposition. Am J Sci 287: 33–49

Macqueen RW, Sandberg CA (1970) Stratigraphy, age, and interregional correlation of the Exshaw Formation, Alberta Rocky Mountains. Bull Can Petrol Geol 18: 32–66

Majorowicz JA, Rahman M, Jones FW, McMillan NJ (1985) The paleogeothermal and present thermal regimes of the Alberta basin and their significance for petroleum occurrences. Bull Can Petrol Geol 33: 12–21

Majorowicz JA, Jones FW, Ertman ME, Osadetz KG, Stasiuk LD (1990) Relationship between thermal maturation gradients, geothermal gradients and estimates of thickness of the eroded foreland section, southern Alberta Plains, Canada. Mar Petrol Geol 7: 138–152

McConnell RG (1887) Report on the geological structure of a portion of the Rocky Mountains. Geol Surv Can Annu Rep 2: 1–41

Macdonald DE (1985) Sedimentary phosphate rock in Alberta and southeastern British Columbia: resource potential, the industry, technology and research needs. CIM Bull 81: 46–52

Osadetz KG, Brooks PW, Snowdon LR (1992) Oil families and their sources in Canadian Williston basin, (southeastern Saskatchewan and southwestern Manitoba). Bull Can Petrol Geol 40: 254–273

Parrish JT (1982) Upwelling and petroleum source beds, with reference to the Paleozoic. Am Assoc Petrol Geol Bull 66: 750–774

Podruski JA, Barclay JE, Hamblin AP, Lee PJ, Osadetz KG, Procter RM, Taylor GC (1988) Conventional oil resources of western Canada, Part 1: Resource endowment. Geol Surv Can Pap 87–26

Pratt LM (1984) Influence of paleoenvironmental factors on preservation of organic matter in Middle Cretaceous Greenhorn Formation, Pueblo, Colorado. Am Assoc Geol Bull 68: 1146–1159

Richards BC, Higgins AC (1988) Devonian-Carboniferous boundary beds of the Palliser and Exshaw formations at Jura Creek, Rocky Mountains, southwestern Alberta. In: McMillan NJ, Embry AF, Glass DJ (eds) Devonian of the world. Can Soc Petrol, Calgary, Geol Mem 14: 399–412

Robison VD (1992) Oil/Source correlations of Devonian and Mississippian reservoired oils in the Peace River Arch area of the Western Canada Basin. Am Assoc Petrol Geol Annu Conv Prog pp 110–111

Sandberg CA (1967) Exshaw Formation of Devonian and Mississippian age in northwestern Montana. In: Sandberg CA (ed) Changes in stratigraphic nomenclature by the U.S. Geological Survey, 1966. US Geol Surv Bull 1253: A39–A41

Sandberg CA, Poole FG, Johnson JG (1988) Upper Devonian of western United States. In: McMillan NJ, Embry AF, Glass DJ (eds) Devonian of the world, Can Soc Petrol Geol, Calgary, Mem 14: 183–220

Savoy LE (1990) Sedimentary record of Devonian-Mississippian carbonate and black shale systems, southernmost Canadian Rockies and adjacent Montana. PhD Diss, Syracuse Univ, 226 pp

Scotese CR (1984) Paleozoic paleomagnetism and the assembly of Pangea. In: Van der Voo R, Scotese CR, Bonhommet N (eds) Plate reconstruction from Paleozoic paleomagnetism. Am Geophys Union, Washington, DC, Geodyn Ser 12: 1–10

Scotese CR (1986) Basal Devonian, middle Emsian, and early Famennian maps. In: Roy S (ed) The Devonian: a portfolio of maps, 1978–1986. The Devonian Inst, Anchorage, pp 3–15

Scotese CR, Van der Voo R, Barrett SF (1985) Silurian and Devonian base maps. Philos Trans R Soc Lond, Ser B, pp 57–77

Smith AG, Hurley AM, Briden JC (1981) Phanerozoic palaeocontinental world maps. Univ Press, Cambridge 102 pp

Tarling DH (1980) Upper Paleozoic continental distributions based on palaeomagnetic studies. In: Panchen AL (ed) The terrestrial environment and the origin of land vertebrates. Academic Press, London, pp 11–37

ten Haven HL, de Leeuw JW, Rullkötter J, Sinninghe-Damsté JS (1987) Restricted utility of pristane/phytane ratio as a paleoenvironmental indicator. Nature 330: 641–643

Tissot BP, Welte DH (1984) Petroleum formation and occurrence. Springer, Berlin Heidelberg New York, 699 pp

Tyson RV, Pearson TH (1991) Modern and ancient continental shelf anoxia: an overview. In: Tyson RV, Pearson TH (eds) Modern and ancient continental shelf anoxia. Geol Soc, London, Spec Pub 58: 1–24

Warren PS (1937) Age of the Exshaw Shale in the Canadian Rockies. Am J Sci 33: 454–457

Westrich JT, Berner RA (1984) The role of sedimentary organic matter in bacterial sulfate reduction: the G model tested. Limnol Oceanogr 29: 236–249

Witzke BJ, Heckel PH (1988) Paleoclimatic indicators and inferred Devonian paleolatitudes of Euramerica. In: McMillan NJ, Embry AF, Glass DJ (eds) Devonian of the world. Can Soc Petrol Geol, London, Mem 14: 49–63

Workum RH (1991) Peace River Arch Wabamun dolomite, tectonic or subaerial karst? Bull Can Petrol Geol 39: 54–56

Geological Controls of Source Rock Geochemistry Through Relative Sea Level; Triassic, Barents Sea

G.H. Isaksen and K.M. Bohacs[1]

Abstract

Lower to Middle Triassic mudrock cores from the Svalis Dome (Dia-Structure); (Barents Sea, Norway) demonstrate the control of depositional conditions on the physical and chemical properties of mudrocks deposited in offshore and shelfal environments. Five cores (112 m) of Smithian to Late Anisian age were analyzed in detail to establish their environment of deposition. The details of the depositional environment were revealed by changes in lithology, bedding geometry, sedimentary structures, and trace-fossil assemblages, as well as bulk and molecular geochemical parameters. The redox condition at the sediment/water interface in these offshore and shelfal environments is likely to have been dysoxic. Inorganic and bulk-organic geochemical indicators (major and minor elements, trace metals, natural radioelements; organic carbon, HI, OI) provided additional parameters to define the facies within each depositional environment. Ni/Ni + V and Al_2O_3/TOC ratios are both good indirect measures of the quality of the organic matter; each facies has a characteristic range of these ratios.

Saturate and aromatic hydrocarbons show an early mature to immature molecular distribution. Estimates of maturity from biomarkers suggest a vitrinite reflectance value of 0.4 %Ro, whereas measured vitrinite reflectance values are in the range of 0.3 to 0.5 %Ro. Pristane/phytane ratios support the interpreted dysoxic to anoxic conditions near the sediment-water interface and within the sedimentary column. Triterpanes are present in near-equal or greater quantities than steranes, with hopane/sterane ratios of 1.1 to 3. Sterane distributions show a predominance of C_{29} regular steranes and appreciable quantities of C_{30} desmethylsteranes, indicating mostly marine-algal organic matter in the kerogen.

This study demonstrates the integration of molecular geochemistry with sequence stratigraphy. Steranes and triterpanes show predictable variations within each systems tract. C_{30} desmethylsteranes show a quantitative increase in absolute concentration that correlate with the second-order rise in sea level during the Lower to Middle Triassic.

Introduction

The objective of this chapter is to investigate the physical and chemical characteristics of Lower to Middle Triassic potential source rocks in the Barents Sea as a function of depositional environment and sequence stratigraphy (Fig. 1). The study involved physical description, lithofacies prediction, sequence-stratigraphic analysis, and chemical characterization of core samples collected from the Svalis Dome (Dia-Structure) by IKU (Norwegian Continental Shelf Institute) as part of their 1986 Shallow Drilling Project. The sequence stratigraphy serves as the framework for predicting the areal extent of the physical parameters of the environment that are then tied to their chemical character. The energy and oxygen conditions of the sedimentary environments were interpreted from lithology, bedding, sedimentary structures, stacking styles of the strata and body, and ichnofossils. Samples of each facies of the fine-grained rocks were collected on a close spacing (cm scale) to capture the inorganic and organic geochemical variation among facies as well as the character of the rock fabric as observed by thin-section analyses. The rocks are interpreted as normal-marine siliciclastics within a ramp-shelf depositional setting.

Previous Studies

The major structural elements in the study area include the Loppa High (trending SW-NE), the

[1]Exxon Production Research Company, 3120 Buffalo Speedway, Houston, Texas, 77252, USA

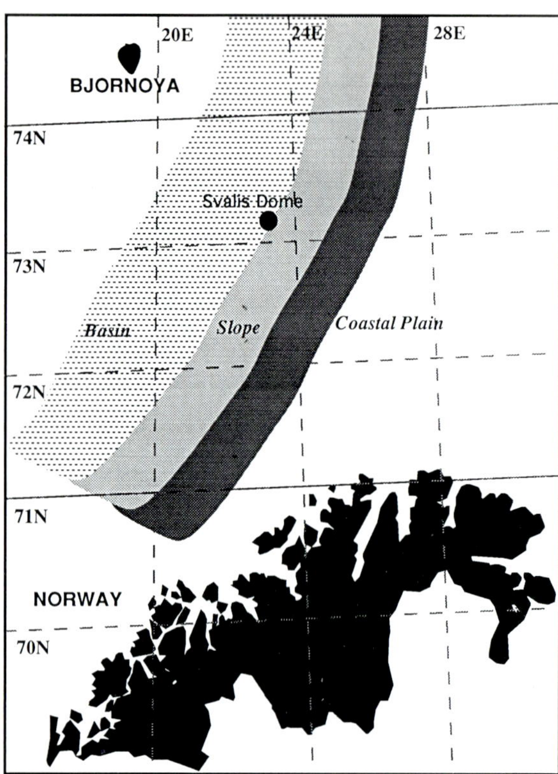

Fig. 1. Schematic representation of the depositional environments in the southern Barents Sea area during early to middle Triassic

Bjørnøya Basin to the west, and the Bjarmeland Platform and Maud Basin to the east and northeast. These have been described by Kristoffersen and Elverøi (1978) and Gabrielsen et al. (1990). Mangerud and Rømuld (1991) reported on the palynological assemblages of the Svalis Dome cores 7323/07-U-04 and 7323/07-U-01, classifying them as Late Spathian-Early Anisian and Middle Anisian, respectively. A sequence stratigraphic study of Anisian strata with the objective of defining semistratigraphic plays, was reported by Rasmussen et al. (1993). Rønnevik and Jacobsen (1984) interpret regional seismic data of the Triassic, whereas Van Veen et al. (1993) present sequence stratigraphic analyses of the Triassic on a regional scale in the Norwegian Barents Sea as well as paleogeographic reconstructions. There have been no publications on detailed geochemistry in this part of the Barents Sea. The nearest Triassic outcrops are on Børnøya (Stappen High), 200 km to the northwest. The organic geochemistry of Bjørnøya has been described by Bjorøy et al. (1983) and Isaksen (1985).

Analytical Conditions

Rock samples were extracted with a 9:1 mixture of methylene chloride and methanol. After deasphaltening (15 times excess pentane) the compound classes were separated by a Waters HPLC system. Gas-chromatographic (GC) analyses were carried out by on-column injection into a Carlo Erba Series 4160 GC coupled to an Extrel Mass Spectrometer Quadrapole Detector Model 400. The column used was a 60-m DB-5 (dimethyl polysilixane stationary phase) with an inner diameter of 0.32 mm and a film thickness of 0.25 μm. The temperature was started at 75 °C and ramped at a rate of 2.5 °C per minute up to 310 °C. Quadrapole mass spectrometry was carried out in electron-impact ionization mode with an ionization energy of 70 eV. Ion-source and interface temperatures were held at 200 and 300 °C, respectively. A-ring substituted methyl steranes were monitored by GC/MS/MS-CAD (Collision Activated Dissociation) using a 60-m DB-1 column with 0.32 mm inner diameter and 0.25 μm film thickness. Major oxides were analyzed from whole rock samples by X-ray flourescence spectrometry (XRF) using a Philips PW 1600 simultaneous XRF. Total uranium was measured by delayed neutron counting (DNC). Samples were weighed into 10 cc vials, sealed, subjected to irradiation, and counted in an 8 detector neutron counter. Rare earth elements (including Th) were analyzed by first treating the rock samples with mixtures of 10 ml HF and 10 ml $HClO_4$ followed by analysis on a Sciex Elan Model 250 inductively coupled plasma mass spectrometer (ICP-MS). XRF, DNC and ICP-MS analyses were performed by X-Ray Assay Laboratories, Ontario, Canada.

Geological Setting

During the early to middle Triassic, the present-day Barents Sea area was covered by a relatively shallow epicontinental seaway that formed a west-opening, two-armed bight off the paleo-Pacific Ocean (Fig. 2); (Ziegler 1988; van Veen et al. 1993). One major source of detritus was the evolving Ural Mountains to the northeast (Green et al. 1986). Provenance studies by Bergan and Knarud (1993) suggest sediment sources hundreds of kilometers to the east-southeast, within the Hercynian orogenic belt of Novaya Zemlya and the Urals.

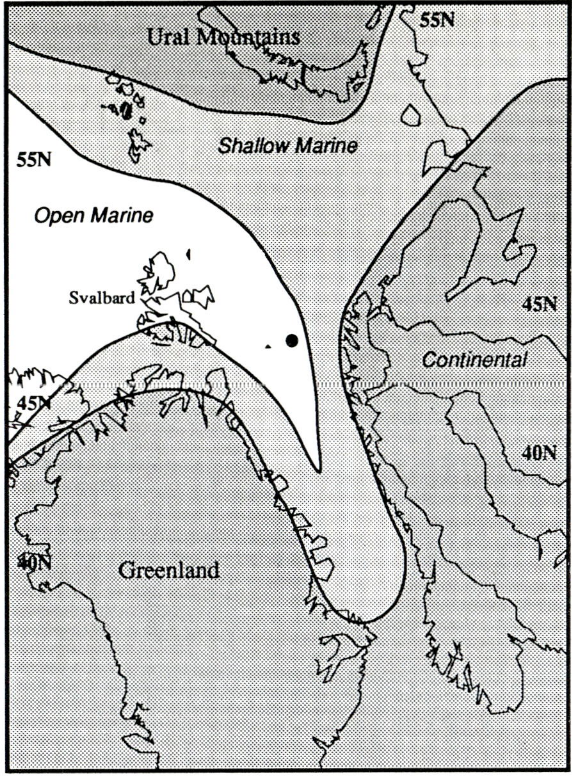

Fig. 2. Paleogeographic map of the early to middle Triassic (231–250 Ma) showing depositional environments on a plate-reconstructed base

Consequently, an extensive alluvial and coastal plain existed in the area from the Timan Pechora basin in the east to the Svalis Dome area in the west. The sedimentary influence of this westerly prograding coastal plain probably reached the Svalis Dome area in the Late Triassic. Concurrently, the progressive northward drift of the Arctic-area plates throughout the Paleozoic continued into the Mesozoic. During the Triassic, the area of Svalbard, Bjørnøya, and the Svalis Dome moved into the temperate climate zone, and was located approximately between 45° and 50° N (Ziegler 1988).

The rocks studied herein were collected from shallow sea-bottom cores taken by the Norwegian Continental Shelf Institute (IKU) from the Smithian to Late Anisian section at the Svalis Dome locality (73° 15′ N, 23° 20′ E; within block 7323/7) in the western Barents Sea. At this locality the Triassic section intersects the seafloor at a high angle (Fig. 3) (Gabrielsen et al. 1990; Mangerud and Rømuld 1991) due to halokinesis-related tilting along the flanks of a Late Carboniferous – Permian salt dome. The salt dome has a diameter of approximately 35 km. With its location on the flank of the Loppa High, the main doming probably occurred during Early Tertiary tectonic events and uplifts of this structural high. Due to Late Tertiary erosion, rocks of Permian through Late Cretaceous age subcrop along the margins of the dome (Gabrielsen et al. 1990).

Rock Attributes

Five cores, totalling 112 m of Smithian to Late Anisian age, that contained the most organic-rich rocks, have been described and sampled in detail to establish the relation between inorganic and organic attributes and relative sea level. Details of the depositional environment were revealed by changes in lithology, bedding geometry, sedimentary structures, body- and trace-fossil assemblages, as well as organic geochemical parameters. Depositional settings ranged from the oxic lower-shoreface-offshore transition environment to dysoxic-anoxic distal open-marine-shelf environment.

The oxic facies of the rock strata was interpreted from the trace fossil assemblages (see below) and the size, depth of penetration, and tiering relation of burrows. This lithofacies was also characterized by an abundance of ammonites and by the occurrence of numerous bivalves and a diverse trace-fossil assemblage. We attribute these observations to an open-marine depositional environment. Episodic anoxia, stirred up by high-energy events, resulted in thinly interbedded dysoxic/anoxic strata as the most stressed environments. Further evidence for this includes numerous low-relief scours, graded beds, turbidite sequences, thin-wave and wave-current ripple beds, and high-energy planar-parallel beds. Thus, the bottom oxygen level tracks fairly well with the energy level, with the most distal, quiet environments being the most prone to having low-oxygen conditions.

There is an overall organization of the bedding and lithology into thickening, coarsening-upward packages which are interpreted as the distal expression of shoreline parasequences. The most proximal deposits record a wave-dominated shoreline system, with wave ripples, high-energy planar beds, and burrowed to churned beds in the offshore-lower-shoreface transition environment. These parasequences range in thickness from 1 to 9 m

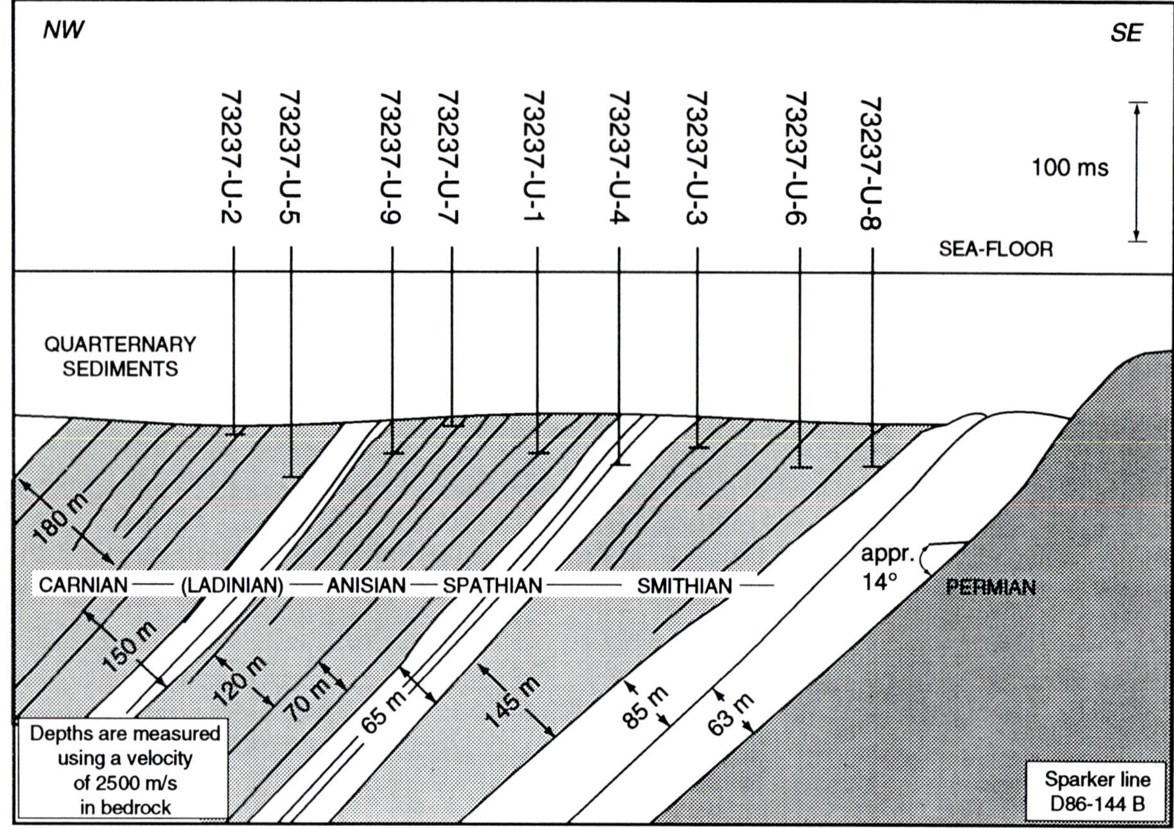

Fig. 3. Schematic representation of analogue sparker line across the Svalis Dome locality. The major reflectors within the Triassic section are shown. The Triassic strata are overlain by Quarternary sediment fill and bounded to the southeast by a salt-dome of Permian age. *Vertical lines* represent positions of shallow wells drilled into the Triassic. The data was collected by IKU (Norwegian Continental Shelf Institute) during their 1986 shallow drilling program

(mean = 3.88 m, standard deviation = 2.25 m, N = 26) and are well expressed by the stacking of the various distal marine depositional environments.

The sequence stratigraphy of the cores was worked out by examining the stacking patterns of the parasequences along with the bulk palynology and geochemical properties of the rocks. Lowstands are marked by a fairly sharp base overlain by aggradationally stacked parasequences with low gamma-ray values and generally low organic-carbon contents (TOC values around 2%). The tops of the lowstands are generally sharp transgressive surfaces, often marked by phosphatic lags. The transgressive systems tracts contain retrogradationally stacked parasequences with increasing gamma-ray values and organic carbon content (up to 10% TOC) up to the mid-sequence downlap surface. Highstands contain progradationally stacked parasequences with overall decreasing organic carbon content. These stacking patterns are illustrated in Fig. 4. The changes in energy and oxygen levels in the environment controls not only the stacking of lithologies and of the beds, but also the organic and inorganic geochemistry at the parasequence scale (discussed later).

Trace-Fossil Assemblages

Trace fossils (burrows, tracks, trails, etc.) are especially important data for the reconstruction of the depositional environment as they record the activities of the native organisms. The organisms perceive not only physical conditions such as bottom energy level and substrate strength, but also are sensitive to chemical conditions such as oxygen level, pH, and nutrient supply. The traces of animal activity generally provide a more detailed record of water chemistry than many geochemical indices

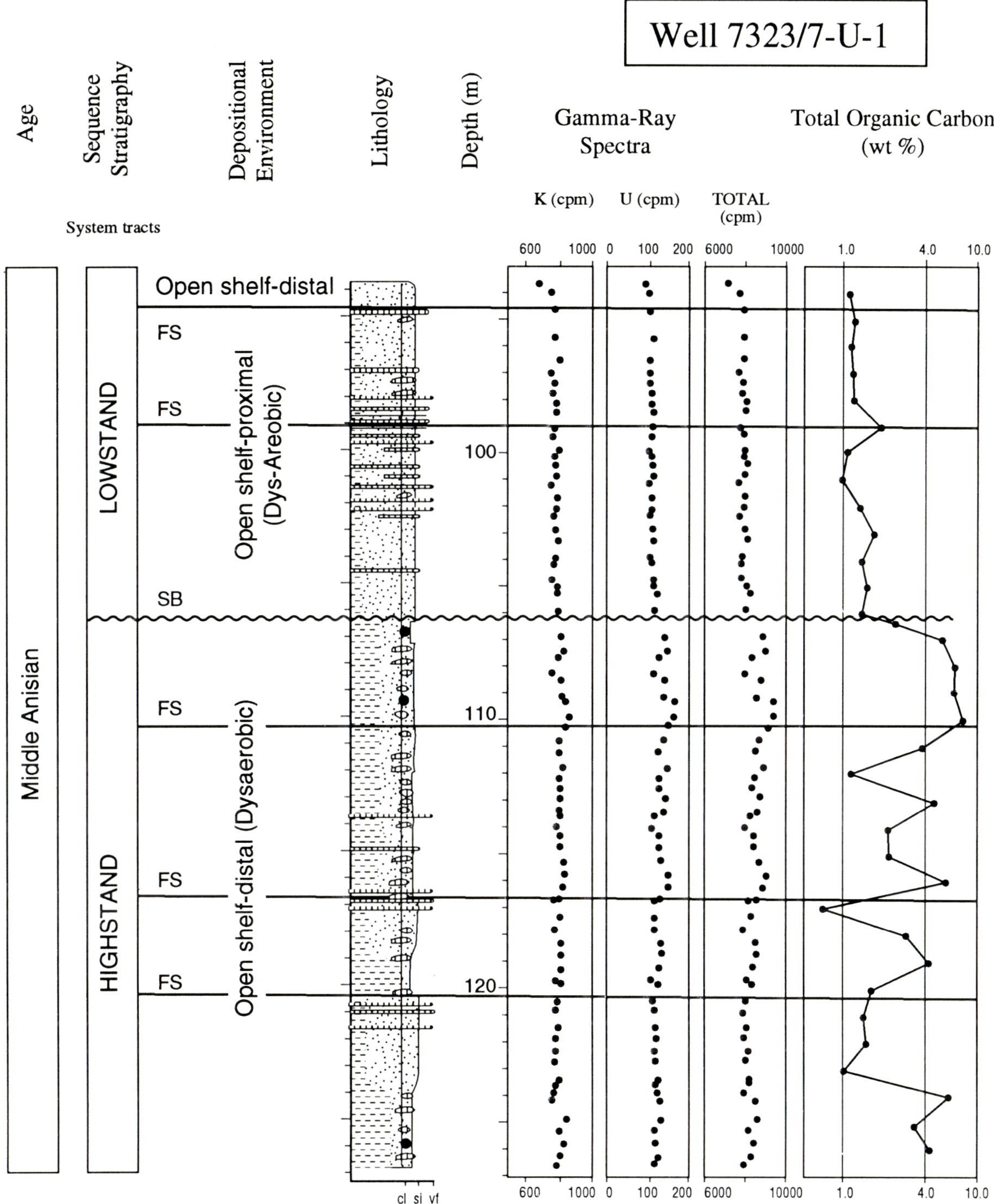

Fig. 4a. Lithofacies, depositional environments, sequence stratigraphic boundaries, and chemical attributes of the 7323/7-U-1 core. *FS* Flooding surface (interpreted); *SB* sequence boundary (interpreted). Lithology and gamma-ray spectral data courtesy IKU

(Sarvda and Bottjer 1986). The mudrocks show mostly *Helminthoides* burrows as small mono-specific, horizontal mining and grazing traces, with a very shallow depth of penetration. In this setting, there is a rapid decrease in the number of ichnogenera in the environments interpreted as dysoxic and dysoxic/anoxic. This restriction in the depth of biological reworking ("biologic irrigation")

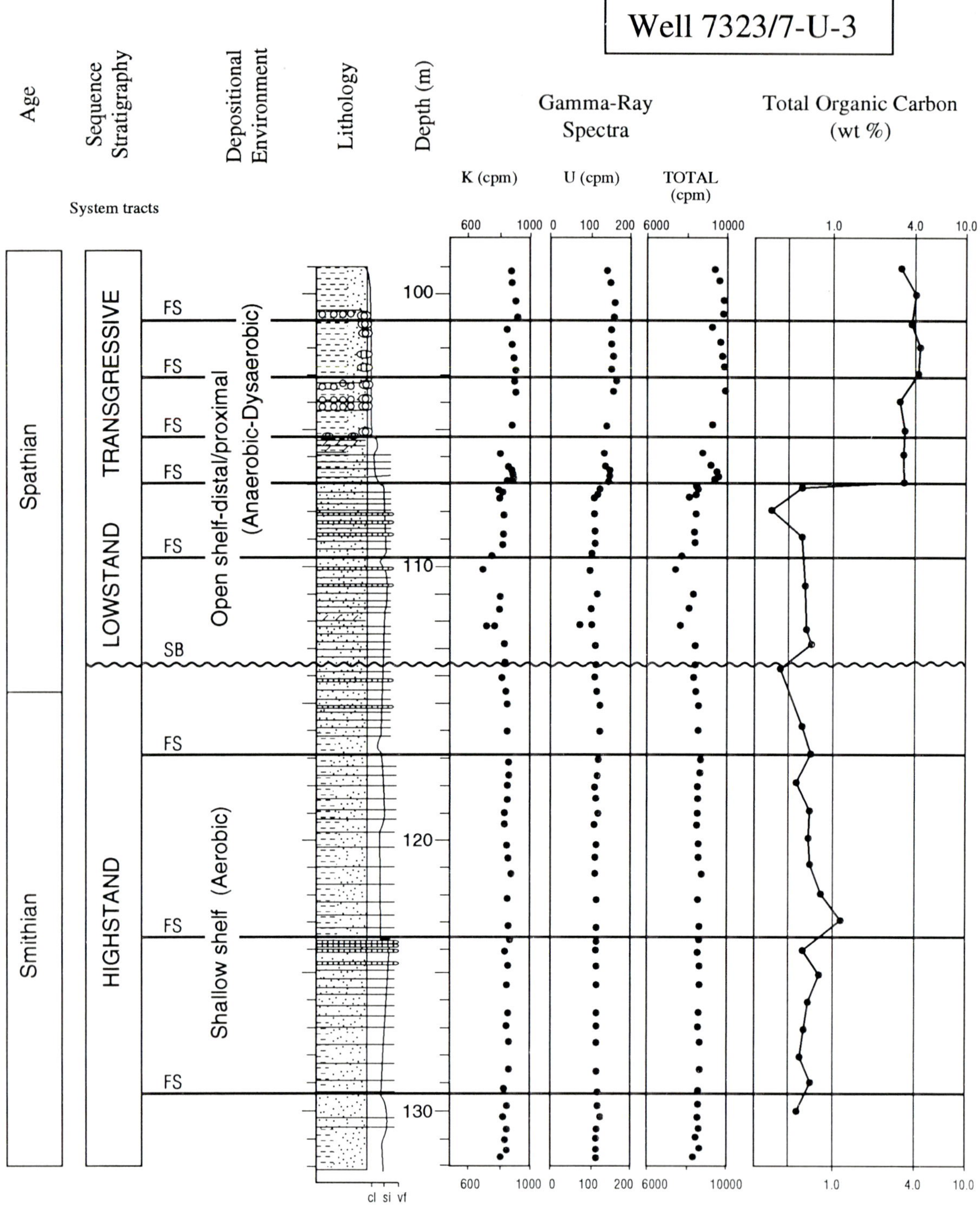

Fig. 4b. Lithofacies, depositional environments, sequence stratigraphic boundaries, and chemical attributes of the 7323/7-U-3 core. *FS* Flooding surface (interpreted); *SB* sequence boundary (interpreted). Lithology and gamma-ray spectral data courtesy IKU

contributes greatly to the preservation of organic matter delivered to the bottom. Higher total organic carbon contents (Fig. 13) and algal/amorphous organic matter are present in these mudrocks.

Thin Section Observations

Polished thin sections of 34 mudrocks were examined under transmitted, reflected, and ultraviolet

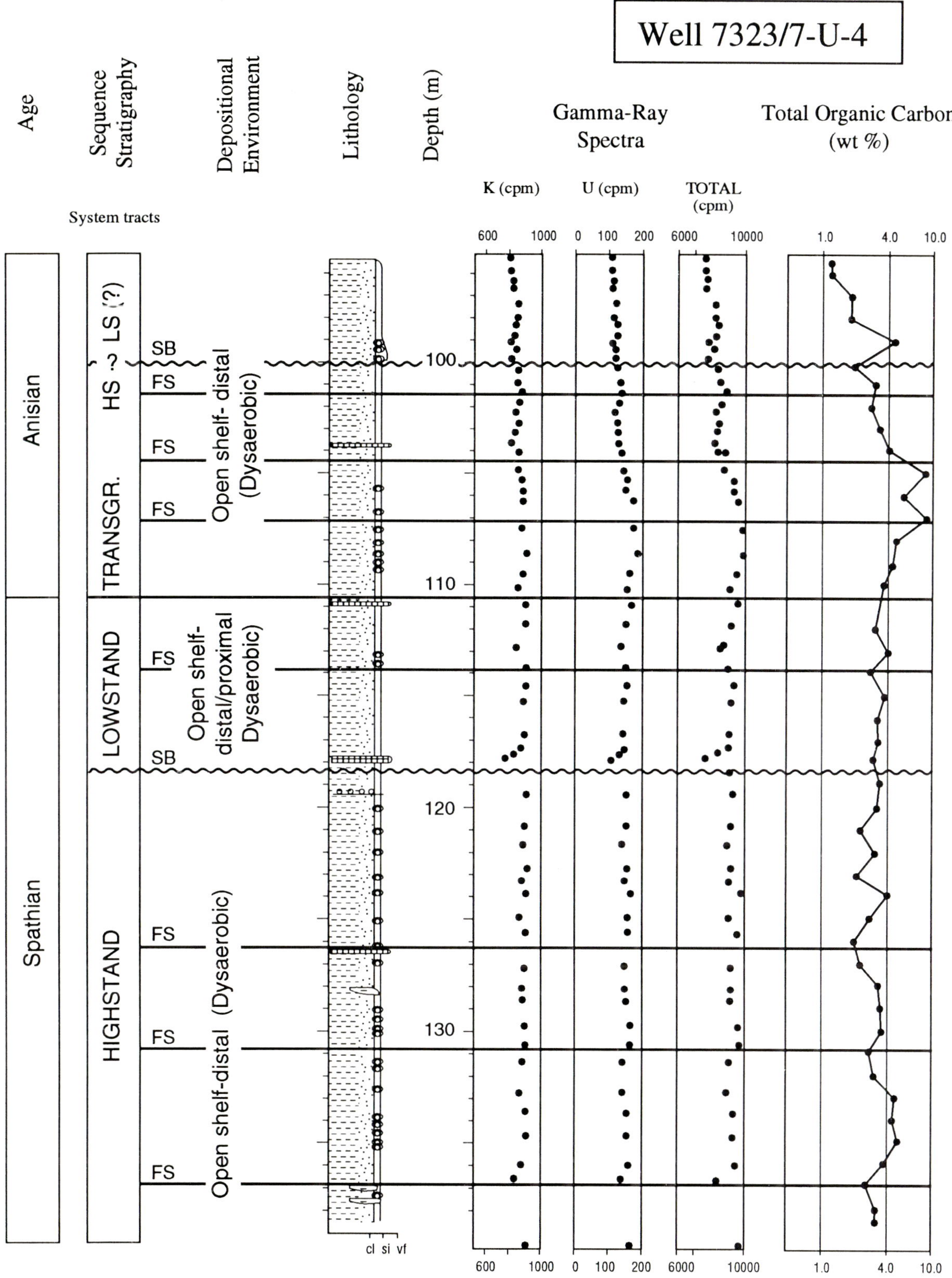

Fig. 4c. Lithofacies, depositional environments, sequence stratigraphic boundaries and chemical attributes of the 7323/7-U-4 core. *FS* Flooding surface (interpreted); *SB* sequence boundary (interpreted). Lithology and gamma-ray spectral data courtesy IKU

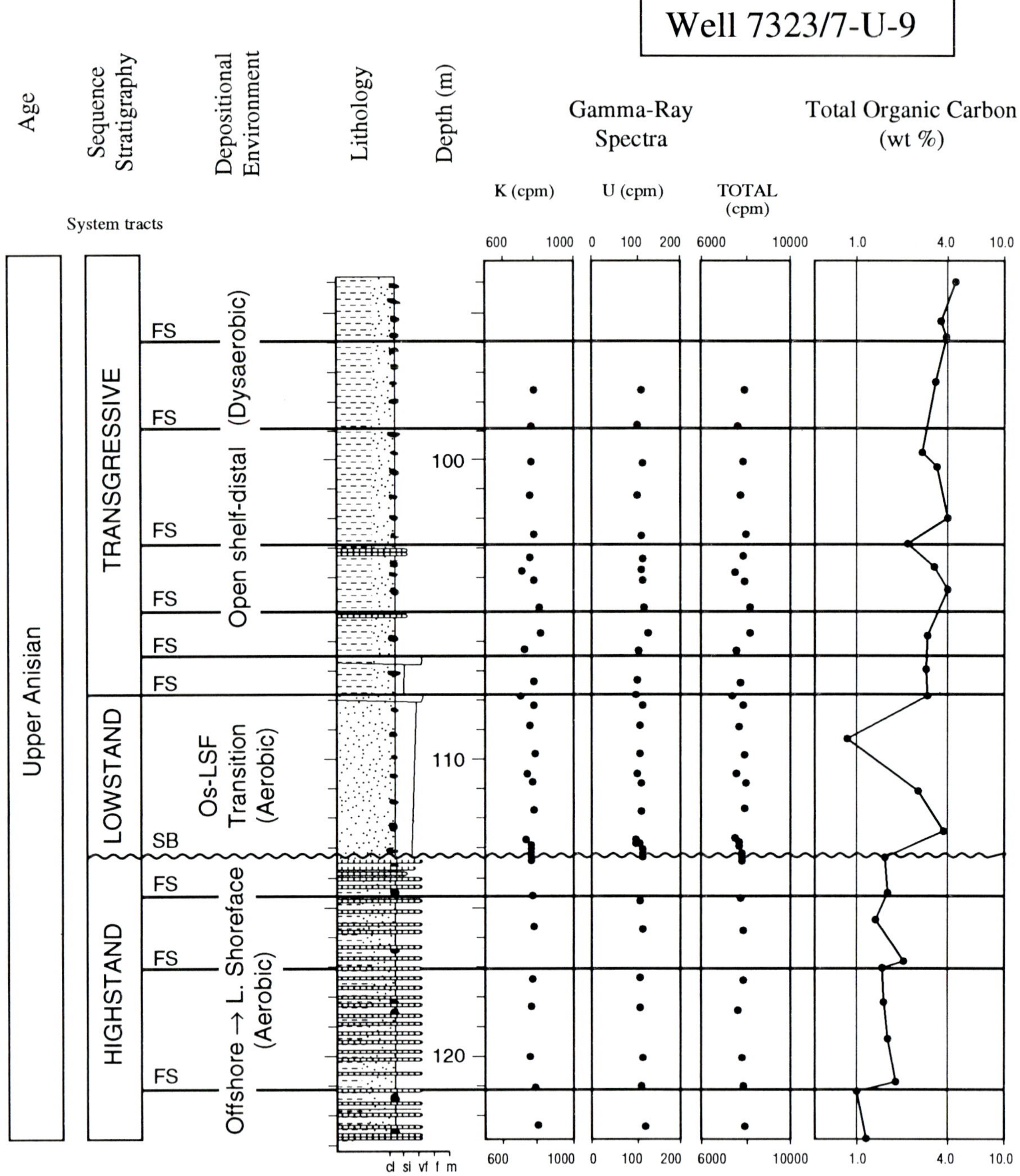

Fig. 4d. Lithofacies, depositional environments, sequence stratigraphic boundaries, and chemical attributes of the 7323/7-U-9 core. *FS* Flooding surface (interpreted); *SB* sequence boundary (interpreted); *Os* offshore; *LSF* lower shoreface. Lithology and gamma-ray spectral data courtesy IKU

light to ascertain the details of the depositional conditions. These samples represent fairly typical marine mudrocks with good covariance between the macroscopically observed facies and their microscopic aspects. Thin section observations were interpreted together with the observations of lithofacies and trace-fossil assemblage variations. Thus, with decreasing levels of energy and oxygen in the depositional environment:

1. The silt content decreases.
2. The silt-sized particles become progressively less concentrated into laminae and more randomly scattered.
3. The portion of the rock composed of pellets and peloids increases.
4. The content of obviously structured woody organic matter decreases.
5. The organic matter becomes increasingly smaller in size, but better connected (going from scattered wisps to continuous connected networks).

a

b

Fig. 5a,b. Thin sections of rock samples typical of **a** open-marine distal, dysoxic mudrock and **b** proximal, dysoxic mudrock

6. The clay content increases and the overall organization of the clay fabric increases (this happens on a lamina-by-lamina basis, not on a bulk scale).

Overall, the organic matter occurs as thin discontinuous to continuous networks of very fine-grained material aligned with bedding and a thin film around clay-mineral-rich pellets in the facies we have classified as a distal, open-marine dysoxic environment (Fig. 5a). A few microfossils were observed in thin section; these included agglutinated foraminifera (crushed) and (?) pyritized sponge spicules. Pyrite is common to abundant in almost all thin sections, occurring as clumped and dispersed framboids.

There is a general parallel fabric, mostly defined by alignment of clay minerals and organic macerals during compaction. Silt grains occur throughout the sections randomly scattered and floating in the groundmass. The common occurrence of silt-sized grains (of both quartz and feldspar) in all environments points to the significant contribution of wind-blown detritus to this system. This is in accord with the observed semi-arid to arid environments on the surrounding landmasses and with the low-rainfall levels predicted by climate modeling (Frakes 1979). The more proximal environments contain silt and very-fine sand concentrated in distinct beds with sharp bases (probably scoured) and distinct size grading even at the thin-section scale (Fig. 5b).

Inorganic Geochemistry

Physical and chemical properties demonstrated a high degree of covariance. Several indices, notably Th/U and authigenic U, can be derived from well logs (spectral gamma ray), and are of particular significance to petroleum exploration as they represent techniques of remotely sensing source-rock quality that might profitably be used in conjunction with remotely sensed source-rock richness (ΔLogR) (Passey et al. 1990). Th and U serve as indirect measures of source-rock quality (Schmoker 1981). Their particular value lies in their robustness; they are easy to measure, and they are indices that may not be seriously affected by outcrop weathering or sample aging. Inorganic geochemical data are listed in Table 1.

Thorium/Uranium

The ratio of thorium to uranium (Th/U) can be an indicator of depositional environment and geochemical-lithologic facies of epicontinental marine shales (Adams and Weaver 1958). It also has been used as an indicator of the original Eh of sedimentary systems (Zelt 1985). Samples from more proximal, oxic environments cluster at low values of Th/U and TOC; more dysoxic samples lie at lower values of Th/U. The best discriminator of depositional environment in the present case is a simple crossplot of thorium versus uranium (Fig. 6). The three fields in this figure may be delimited that separate the offshore, open-marine shelf-proximal, and open-marine shelf-distal environments (see Fig. 3). This type of information may be used to interpret details of the depositional environment of the mudrocks from spectral gamma-ray logs. These boundaries separating the environments are not necessarily universally applicable and should be calibrated locally when possible.

Authigenic Uranium

Another method used to ascertain the geochemical conditions of deposition is the calculation of authigenic uranium content (Myers and Wignall 1987). Sediments deposited under anoxic conditions are found to be consistently enriched in authigenic uranium. The authigenic uranium content is estimated by assuming that the detrital uranium content is directly proportional to the thorium content (assumed to be immobile and entirely detrital). The equation used is: $U_{auth} = U_{measured} - [Th_{measured}/3]$. Figure 7 shows the variation of TOC with U_{auth} content: an upper limit on source-rock richness follows the relation: $TOC > U_{auth}/2.25$. {TOC in %; U_{auth} in ppm}. The more proximal, oxic samples cluster in the lower left; samples with $U_{auth} > 3$ ppm are generally dysoxic. Source-rock quality, using the hydrogen index, increases with increasing authigenic uranium content (Fig. 7). Hence, one can use the diagonal line in this plot as representative of the minimum hydrogen index predicted from authigenic uranium.

Detritus/TOC

Source-rock quality (oil-proneness) may be viewed as proportional to the iron-to-sulfur ratio within

Table 1. Inorganic geochemical data

Sample	Well	Depth (m)	TOC (%)	HI	Al_2O_3 %	SiO_2 %	K_2O %	CaO %	Fe_2O_3 %	V ppm	Ni ppm	Th ppm	U ppm	Th/U	Al_2O_3/TOC	Ni/Ni + V	Authi U ppm	Envir.	Redox	SEQ Stratigr.
1	W7323/7-U1	99.06	1.27	78	18	57.5	2.04	0.59	7.39	180	98	6.3	2.3	2.74	14.17	0.35	0.20	OMP	DYS	LS
2	W7323/7-U1	104.85	2.26	192	20.8	52.5	2.28	0.82	6.58	400	66	4.5	2.7	1.67	9.20	0.14	1.20	OMD	DYS	LS
3	W7323/7-U1	115.82	6.25	393	17.6	49.2	2.7	1.78	7.03	1100	74	7.1	4.4	1.61	2.82	0.06	2.03	OMD	DYS	HS
4	W7323/7-U1	117.04	2.99	287	19.1	52.2	2.41	1.94	6.57	540	77	6.7	5.5	1.22	6.39	0.12	3.27	OMD	DYS-OXIC	HS
5	W7323/7-U1	118.87	4.14	331	18.3	51.1	2.59	2.01	6.6	400	74	5.8	3.4	1.71	4.42	0.16	1.47	OMD	DYS	HS
6	W7323/7-U1	121.92	1.22	113	20.6	52	2.2	0.66	7.49	190	84	3.4	3.7	0.92	16.89	0.31	2.57	OMD	DYS	HS
7	W7323/7-U1	126.19	4.49	352	17.7	50.6	2.74	2.52	6.61	430	60	7	11.2	0.63	3.94	0.12	8.87	OMD	DYS	HS
8	W7323/7-U3	101.80	3.81	307	14.3	53.5	2.96	3.09	6.29	1100	150	11.1	10.4	1.07	3.75	0.12	6.70	OMD	DYS	TST
9	W7323/7-U3	104.85	3.71	343	13.3	51.1	2.78	5.08	7.59	150	39	9.4	7.8	1.21	3.58	0.21	4.67	OMD	DYS	TST
10	W7323/7-U3	106.98	2.4	329	14.4	56.8	2.72	4.07	6.52	790	76	9.4	7	1.34	6.00	0.09	3.87	OMD	DYS	TST
11	W7323/7-U4	99.36	2.33	330	13.7	51.7	2.78	7.33	5.08	190	41	9.3	2.8	3.32	5.88	0.18	0.00	OMD	DYS-ANOX	LS
12	W7323/7-U4	112.78	2.15	394	8.24	31.8	1.59	17.1	6.61	110	32	4.8	6.2	0.77	3.83	0.23	4.60	OMD	DYS	LS
13	W7323/7-U4	117.35	2.34	216	n.a.	n.a.	n.a.	n.a.	n.a.	200	54	7.5	6.7	1.12	n.a.	0.21	4.20	OMD	DYS	LS
14	W7323/7-U4	119.79	2.87	179	16.4	52.1	2.99	2.3	7.51	200	47	9.5	7.8	1.22	5.71	0.19	4.63	OMD	DYS	HS
15	W7323/7-U4	130.76	2.85	258	15.1	54.5	2.82	2.76	6.81	370	43	7.7	8	0.96	5.30	0.10	5.43	OMD	DYS	LS
16	W7323/7-U4	135.03	5.47	365	16.3	52.3	3.01	1.76	7.3	220	43	8.1	8.7	0.93	2.98	0.16	6.00	OMD	DYS	HS
17	W7323/7-U7	99.36	0.88	175	14.4	62.9	1.74	3.44	5.15	170	67	4.2	2.2	1.91	13.36	0.28	0.80	OS	OXIC	UNKN
18	W7323/7-U7	101.19	1.57	89	20.7	52.7	2.49	0.46	6.6	390	92	5.3	4.1	1.29	13.18	0.19	2.33	OMP	DYS-OXIC	UNKN
19	W7323/7-U9	104.55	2.75	382	16.4	57.4	2.27	2.13	5.24	250	59	3.6	3.4	1.06	5.96	0.19	2.20	OS	OXIC	TST
20	W7323/7-U9	112.47	1.21	238	14.3	66.8	1.87	0.68	4.77	170	61	5	2.7	1.85	11.82	0.26	1.03	OS	OXIC	LS
21	W7323/7-U9	114.30	1.21	143	16.9	60.5	2.15	0.45	6.34	200	79	4.9	2.5	1.96	13.97	0.28	0.87	OS	DYS-OXIC	HS
22	W7323/7-U9	117.96	0.91	156	16.1	56.5	1.95	3.18	7.43	190	83	4	2.5	1.60	17.69	0.30	1.17	OS	OXIC	HS
23	W7323/7-U9	120.09	1.06	139	18.5	55.7	2.29	0.67	8.44	220	87	4.8	2.7	1.78	17.45	0.28	1.10	OS	OXIC	HS
24	W7323/7-U9	95.71	2.34	209	18.2	53	2.37	3.2	6.34	340	57	3.7	3.8	0.97	7.78	0.14	2.57	OMD	OXIC	TST
25	W7323/7-U9	102.11	2.6	287	18.8	53	2.1	2.61	5.87	260	55	4.9	5.9	0.83	7.23	0.17	4.27	OMD	DYS-ANOX	TST
26	W7323/7-U9	109.42	2.11	191	12.3	67.3	1.63	2.03	4.85	160	49	3.3	3.2	1.03	5.83	0.23	2.10	OS	OXIC	LS
27	W7323/7-U9	116.74	0.98	135	15.3	60.8	2.93	0.85	7.17	180	76	4.7	2.6	1.81	15.61	0.30	1.03	OS	OXIC	HS

OMP: Open marine proximal.
OMD: Open marine distal.
OS: Offshore marine.

DYS: Dysoxic
ANOX: Anoxic.
Auth: Authigenic.

LS: Lowstand system tract.
HS: Highstand systems tract.
TST: Transgressive systems tract.
UNKN: Unknown.

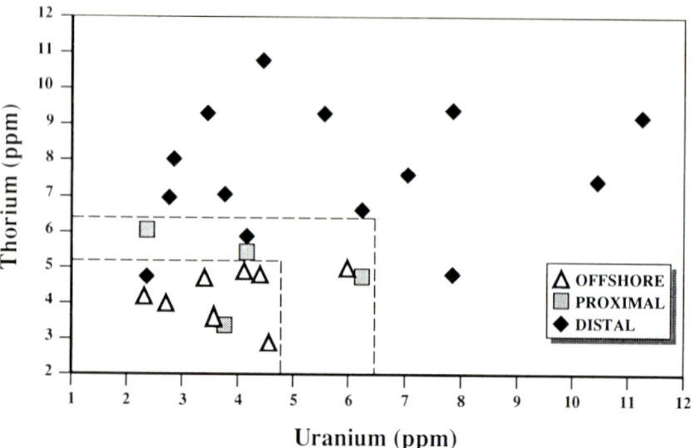

Fig. 6. Thorium versus uranium as an aid in predicting depositional environment from well-logs

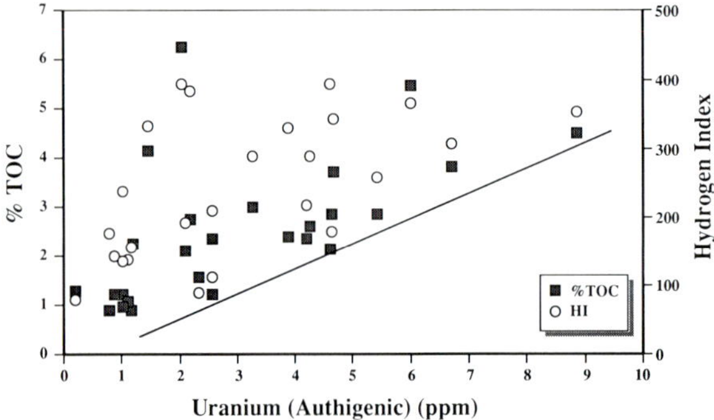

Fig. 7. Authigenic uranium versus hydrogen index and %TOC as an aid in predicting source-rock quality from well-logs. The *diagonal line* represents the minimum HI and TOC predicted from authigenic uranium

a particular mudrock. This ratio portrays the proportion of sulfur that is combined into pyrite (and taken out of the organic-chemical system) versus the excess sulfur available to be bound in the kerogen (and produce sour oil). In normal marine sediments (those deposited in oxygen-containing bottom waters), pyrite formation is limited mainly by the amount and reactivity of detrital iron minerals and organic matter buried in the sediment. As a result, pyrite sulfur and organic carbon correlate positively with one another (Berner 1984).

The Fe/S ratio may be transformed to a more easily measured and intuitively interpreted parameter: detritus/TOC. This transformation is based on two key observations: (1) the proportion of sulfur to carbon is relatively constant (3:1) in marine mudrocks (Sweeney 1972; Berner and Raiswell 1983; Leventhal 1983; Fisher and Hudson 1987) and (2) the iron in fine-grained sediments is strongly associated with the detrital clay (Al_2O_3) content (Curtis 1987).

Figure 8 shows a strong clustering of the poorer potential source rocks (more proximal and oxic environments) at high values of Al_2O_3/TOC. The cluster of distal, dysoxic environment samples point to higher source-rock quality at lower values of Al_2O_3/TOC. This parameter has the advantage of being mappable with fairly robust geochemical analyses, and may be calculated in some cases based solely on well-log response. Total organic carbon content may be estimated by using ΔLogR (Passey et al. 1990), and the detrital content may be derived from the K/U ratio or alumina-activation clay logging tools.

Nickel and Vanadium

Nickel and vanadium are preferentially concentrated in tetrapyrrole (metallo-organic) complexes in organic matter under anoxic conditions (Lewan and Maynard 1982). Tetrapyrrole complexes are most likely derived from chlorophyll and heme pigment precursors in living matter (Corwin 1959). The quantity of metallo-organic complexes preserved

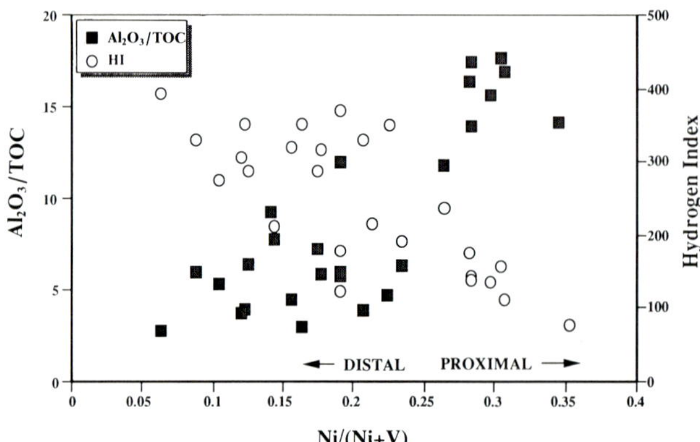

Fig. 8. Assessment of the organic facies and depositional environments using nickel, vanadium, Al_2O_3, %TOC, and hydrogen index. The distal, dysoxic environments have lower values of Ni/(Ni + V). The Al_2O_3 content is a representation of the detrital-clay content in samples (see text for detailed explanation)

in an organic accumulation of sediment is a function of their exposure time to oxic conditions and the oxygen levels of the oxic conditions encountered. Thus, nickel and vanadium contents will be high in organic matter that has had a minimal exposure to oxygen-rich conditions: that which either has settled through a dysoxic water column or has been rapidly buried to anoxic conditions in the sediment. The relative proportion of nickel and vanadium (calculated as Ni/Ni + V) is determined by the chemical conditions of the depositional environment of the potential source rock (Lewan and Maynard 1982). With decreasing oxidation potential and concomitant increase in the importance of sulfate-reducing bacteria in the environment, the ratio Ni/Ni + V decreases. More specifically, Ni^{2+} is available for metallation of tetrapyrroles and vanadium exists primarily as anionic V, leading to a dominance of Ni(II) porphyrins (Lewan and Maynard 1982; van Berkel et al. 1989). The more distal, dysoxic environments should have the lower values of Ni/Ni + V. Figure 8 corroborates this hypothesis: TOC content is inversely proportional to the Ni/Ni + V ratio, with a strong clustering of the more proximal, oxic-environment samples at higher values of the ratio. A plot of hydrogen index (HI) against Ni/Ni + V also follows this trend (Fig. 8). Another indirect measure of source quality that is influenced by the Eh conditions in the sediment, Al_2O_3/TOC, also shows a good separation of proximal, oxic environments at higher values of Ni/Ni + V and Al_2O_3 (Fig. 8 and Table 1). A similar trend was observed with K_2O/U. In all these cases, a Ni/Ni + V value of greater than 0.26 corresponds to proximal, oxic depositional environments.

Organic Geochemistry

The tilted orientation of the Svalis Dome strata (Fig. 3) and positioning of the core samples permitted

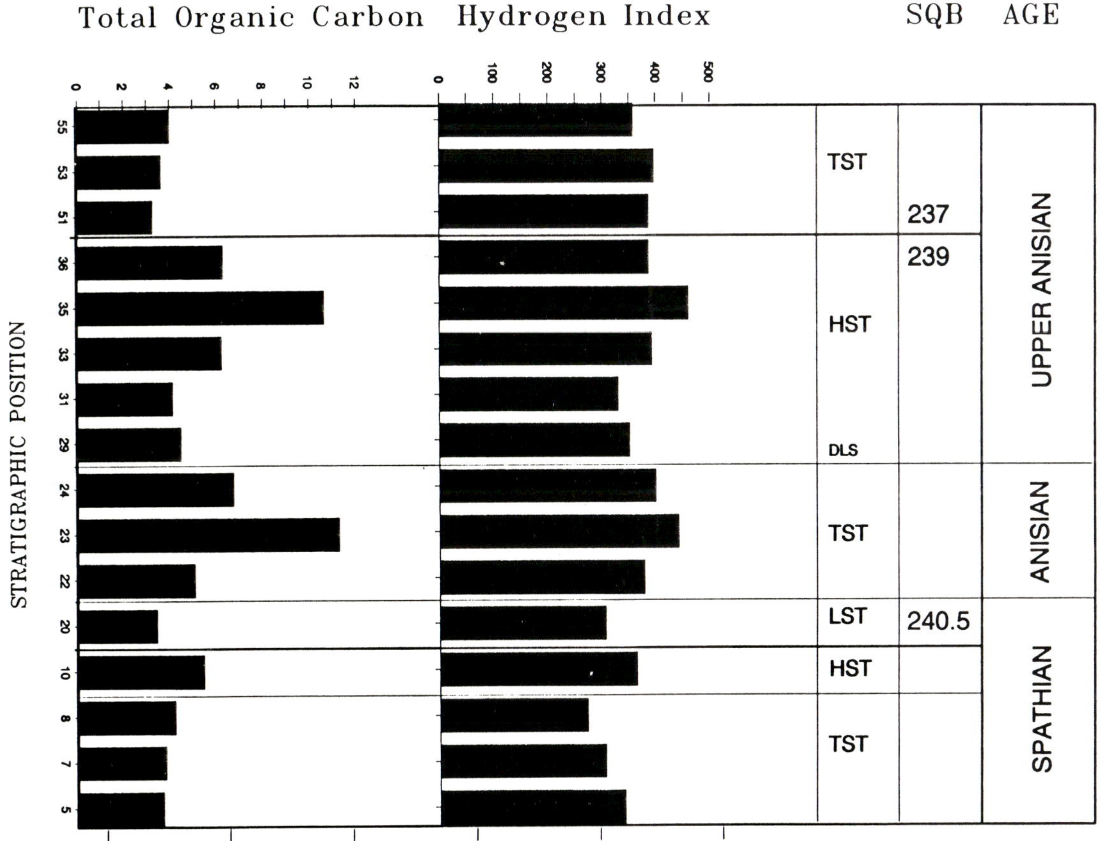

Fig. 9. Organic richness and quality of potential source rocks of Spathian through Late Anisian age positioned within the interpreted sequence stratigraphic framework. *TST* Transgressive systems tract; *HST* highstand systems tract; *LST* lowstand systems tract; *DLS* downlap surface

Table 2. Organic geochemical parameters for samples selected for solvent extraction and biomarker analyses

Sample no.	55	53	51	36	35	33	31	29	24	23	22	20	10	8	7	5
Well	7323/ 7U9	7323/ 7U9	7323/ 7U9	7323/ 7U1	7323/ 7U1	7323/ 7U1	7323/ 7U1	7323/ 7U1	7323/ 7U4	7323/ 7U4	7323/ 7U4	7323/ 7U4	7323/ 7U4	7323/ 7U3	7323/ 7U3	7323/ 7U3
Depth (m)	99.5	103.3	106.2	107.5	109.9	116.1	119.1	126.4	106.96	108.1	110.1	115.45	135.05	100.3	102	105
% TOC	4.02	3.65	3.28	6.3	10.65	6.25	4.14	4.49	6.76	11.29	5.07	3.45	5.47	4.23	3.81	3.71
S1	0.16	0.26	0.29	0.5	0.99	0.61	0.45	0.32	0.81	0.92	0.74	0.38	0.84	0.53	0.48	0.63
S2	14.72	14.51	12.72	24.4	49.07	24.61	13.72	15.84	27.04	49.93	19.23	10.62	19.97	11.55	11.71	12.74
S3	1.05	1.19	1.04	1.11	1.68	1.41	1.04	1.26	2.25	2.05	1.14	1.05	0.85	0.77	0.94	1.19
HI	358	397	387	387	460	393	331	352	400	442	379	307	365	273	307	343
OI	26	32	31	17	15	22	25	28	33	18	22	30	15	18	24	32
T_{max}	429	426	428	424	422	425	428	428	421	421	421	424	421	417	422	420
% Ro	n.a.	n.a.	n.a.	n.a.	n.a	0.49	0.45	n.a.	n.a.	n.a.	n.a.	n.a.	0.39	n.a.	0.37	0.35
Rock extract (g)	50.0	50.0	50.0	52.0	50.0	50.0	50.0	50.0	50.0	46.8	44.0	50.0	50.0	54.0	45.0	50.0
TOC (g)	2.0	1.8	1.6	3.3	5.3	3.1	2.1	2.2	3.4	5.3	2.2	1.7	2.7	2.3	1.7	1.9
mg EOM	66.5	28.4	33.9	40.4	51.6	37.8	27.6	66.4	42	90.4	94.4	78.3	70.6	52.3	84.3	86.5
Sat % EOM	9.5	7.5	11.6	9.5	4.9	2.7	15.2	6.6	10.8	4.3	4.8	5.7	6.1	5.6	4.4	5.3
Arom % EOM	7.0	5.9	7.5	8.0	1.9	7.1	11.5	6.5	11.2	4.7	5.3	6.3	7.1	7.9	4.9	5.8
NSO % EOM	11.1	10.3	14.8	12.9	13.7	11.0	15.2	10.8	20.8	8.6	8.1	10.9	10.3	11.1	7.5	8.7
Asph (mg)	23.0	14.3	14.9	18.4	24.8	16.3	12.9	21.8	20.8	32.8	37.0	27.8	22.9	23.2	31.2	31.0
dC13 SAT	– 30.29	– 30.36	– 30.3	– 30.45	– 30.88	– 30.74	– 30.47	– 30.66	– 31.43	– 31.6	– 31.47	– 31.22	– 32.32	– 30.77	– 30.54	– 30.15
dC13 AROM	– 30.19	– 28.84	– 29.17	– 29.69	– 30.32	– 30.25	– 29.3	– 30.44	– 30.48	– 30.1	– 30.4	– 31	– 32.25	– 30.6	– 30.23	– 30
Steranes (1)[a]	5137	1652	4569	2323	1779	1419	2730	1546	4009	1658	1704	2257	1654	1027	1099	2075
Steranes (2)[a]	2627	970	2649	1374	966	895	1123	829	2227	799	789	1175	844	496	601	1025
Diasteranes (3)[a]	592	131	390	453	309	174	476	198	722	310	218	446	408	222	236	586
Triterpanes (4)[a]	7034	3742	6527	5815	5202	4175	5240	4405	9431	3171	3090	4827	3731	1951	1697	2268
% 20S	11.5	7.1	9.6	13.0	13.3	11.7	18.5	11.0	18.4	21.2	12.6	17.8	14.3	17.0	21.2	23.3
% C27abb	21.0	28.1	21.0	27.5	29.2	30.8	27.7	27.5	27.0	32.1	31.2	35.8	41.8	37.1	37.0	36.4
% C28abb	30.2	31.6	30.7	22.3	23.8	19.1	24.9	23.8	31.0	25.9	23.5	24.6	29.5	18.0	23.0	22.8
% C29abb	48.8	40.3	48.3	50.2	47.1	50.1	47.3	48.8	42.0	42.0	45.4	39.6	28.8	44.9	39.9	40.8
% C32ab22S	32.1	27.1	29.7	38.0	31.6	29.7	38.6	33.2	38.0	37.6	29.1	36.1	40.1	26.5	37.4	34.9
Ts/Tm	0.75	0.61	0.76	0.49	0.38	0.82	1.02	0.52	0.55	0.82	0.81	0.54	0.5	0.4	0.69	0.74
% C30Moretane	26.2	21.8	15.4	24.9	25.5	23.4	24.1	25.7	20.3	19.8	19.0	21.4	21.2	19.5	20.3	18.4
Hopane/sterane	1.4	2.3	1.4	2.5	2.9	2.9	1.9	2.9	2.4	1.9	1.8	2.1	2.3	1.9	1.5	1.1

[a]ppm of extractable organic matter (EOM): 1 Steranes from m/z 217. 2 Steranes from m/z 218. 3 Diasteranes form m/z 259. 4 Triterpanes from m/z 191. n.a. = no analyses available.

investigation of organic facies variations within time-equivalent rocks deposited in proximal through to distal environments. Organic geochemical data are listed in Table 2.

TOC contents are in the range of 0.5 to 11.5 wt. %, whereas hydrogen indices range upwards to a maximum of 460. The best hydrocarbon-source potential is present between the 242 Ma and 238.5 Ma sequence boundaries (Spathian to Middle Anisian) (Fig. 9). The sedimentary package located between the 237 Ma and 231 Ma sequence boundaries also has very good hydrocarbon potential. The mudrocks with the best hydrocarbon-source potential were deposited under dysoxic conditions during the late transgressive and early highstand system tracts within a distal open-marine shelf environment.

Maturity

These rocks are interpreted as immature to early mature, based on T_{max} temperatures from Rock-Eval range of 418 to 435 °C (Table 2), and vitrinite reflectance values range from 0.3 to 0.5 % R_o. C_{15} + -saturated hydrocarbons show a low-maturity molecular distribution (Fig. 10), characterized by a high isoprenoid/*n*-alkane ratio, with pristane/*n*-C_{17} values ranging from 1.9 to 2.8, and phytane/*n*-C_{18} values ranging from 3.1 to 4.3. Steroid and triterpenoid hydrocarbons are more abundant relative to the *n*-alkanes of similar boiling point. An example of the steroid and triterpenoid molecular distribution within the offshore marine distal (dysoxic) depositional environment is also shown in Fig. 10. Biomarkers still have a significant

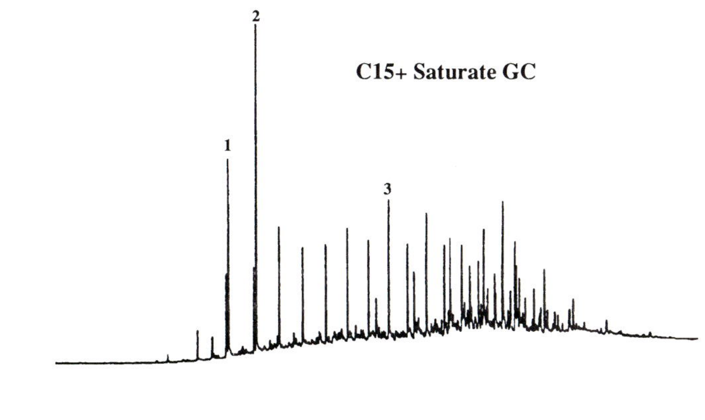

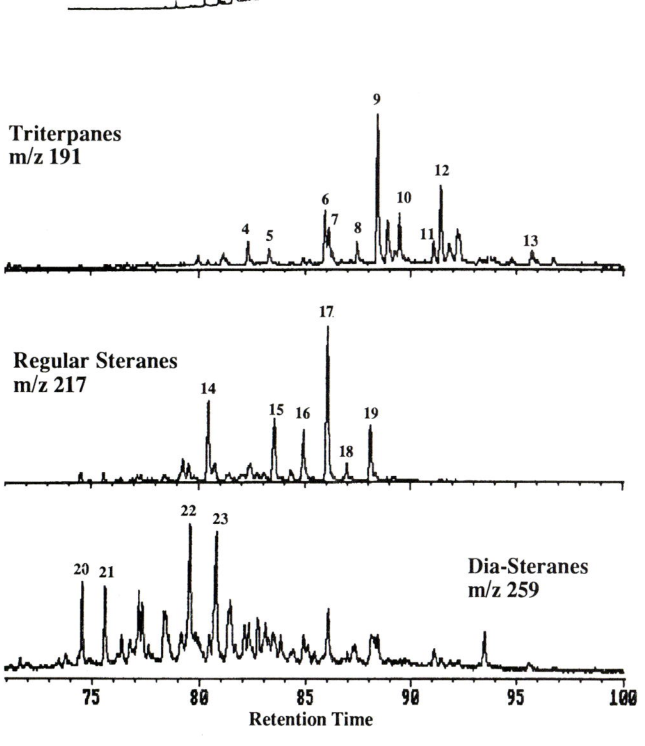

Fig. 10. Typical molecular distributions of mudrocks deposited within open-marine distal, dysoxic environment. The data shown is form well 7323/7-U-9 at 99 m. Key to peak identification: *1* pristane; *2* phytane; *3* nC_{24}; 4 Ts [18α(H)-22,29,30 trisnorneohopane]; *5* Tm [17α(H)-22,29,30 trisnorhopane]; *6* 17α(H),21β(H)-30 norhopane; *7* 18α(H) 30-norneohopane; *8* 17β(H), 21α(H)-30 normoretane; *9* 17α(H), 21β(H) hopane; *10* 17β(H), 21α(H) moretane; *11* 17α(H), 21β(H) homohopane 22S; *12* 17α(H), 21β(H) homohopane 22R; *13* 17α(H), 21β(H) bishomohopane 22R; *14* 5α(H),14α(H),17α(H) cholestane 20R; *15* 5α(H),14α(H),17α(H) 24-methylcholestane 20R; *16* 5β(H),14α(H),17α(H) 24-ethylcholestane; *17* 5α(H), 14α(H),17α(H) 24-ethylcholestane 20R; *18* 5α(H),14α(H), 17α(H) 24-propylcholestane; *19* 5α(H),14α(H),17α(H) 24-propylcholestane; *20* 13β(H),17α(H) diacholestane 20S; *21* 13β(H),17α(H) diacholestane 20R; *22* 13β(H),17α(H) 24-ethyl-diacholestane 20S; *23* 13β(H),17α(H) 24-ethyl-diacholestane 20R

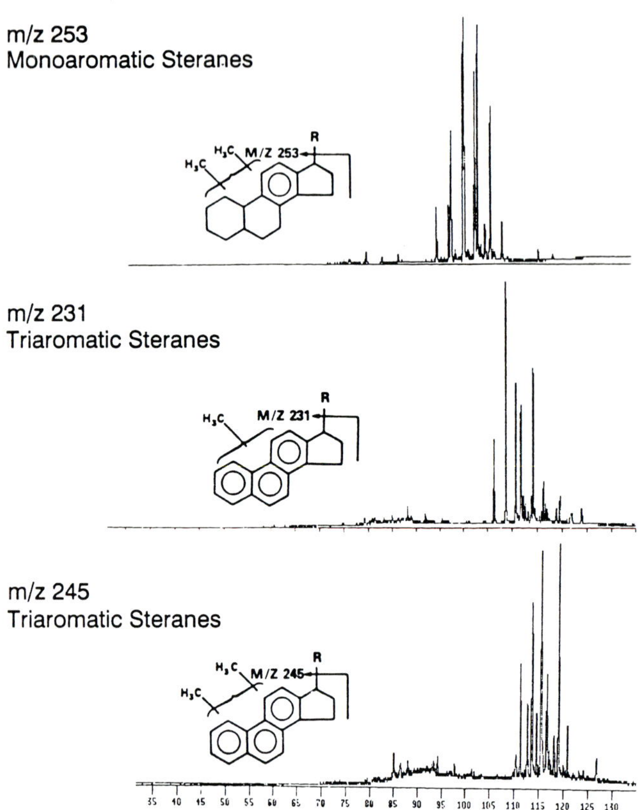

Fig. 11. Typical distribution of mono- and triaromatic steranes in samples from open shelf-distal depositional environments

portion of the biologically inherited isomers present, e.g., 5β(H) steranes, and 5α(H), 14β(H), 17β(H) 20 R steranes predominating over the 5α(H), 14β(H), 17β(H) 20S steranes.

The aromatization of steranes is usually monitored by the increase in triaromatic steranes along with a decrease in monoaromatic steranes (Mackenzie et al. 1981, 1982; ShiJiyang et al. 1982). Typical distributions of mono- and triaromatic steranes from the Svalis Dome locality (Fig. 11), show the low degree of side-chain cracking of the steranes. Aromatic steranes with carbon numbers around C_{26} to C_{29} predominate whereas the high maturity counterparts with C_{19} to C_{23} carbon atoms are nearly absent.

Organic Facies

The Svalis Dome samples have kerogens ranging from type II to type II-III, as shown in Figs. 12 and 13. Typing of organic matter from Rock Eval pyrolysis has some limitations (Katz 1983), but the hydrogen and oxygen indicies for these samples do not show significant mineral matrix effects (Table 1). Kerogen composition varies systematically with depositional environment (Fig. 13). Algal amorphous kerogen is most common in the transgressive systems tracts (most distal from the paleoshoreline, i.e., upper parts of cores 7323/7-U-3, 7323/7-U-4, and 7323/7-U-9) and more structured woody material is found within lowstand and late highstand system tracts (most proximal to the paleo-shoreline, i.e., the upper part

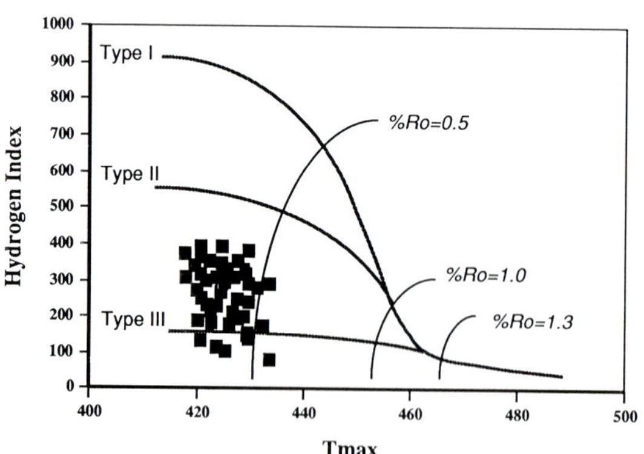

Fig. 12. Organic matter characterization as suggested by the hydrogen index and T_{max}

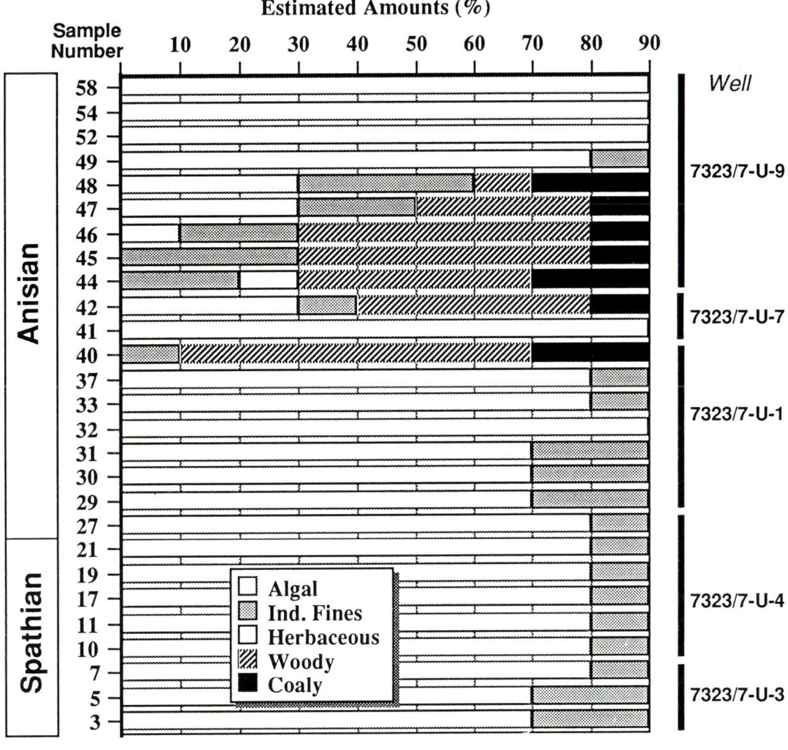

Fig. 13. Visual kerogen compositions for the Spathian to Anisian section. Note increase in terrigenous higher plant organic matter (woody and coaly) in the upper part of core 7323/7-U-1 and the lower part of core 7323/7-U-9, corresponding to lowstand and late highstand in relative sea level

of core 7323/7-U-1 and the lower part of core 7323/7-U-9). Amorphous material with lower hydrogen contents, together with hydrogen-rich algal material, would account for the observed type II to II-III kerogen from Rock-Eval pyrolysis. In the following we discuss how the geochemical attributes of the rocks co-vary within the interpreted sequence stratigraphic framework.

Smithian – Mid-Anisian

A high percentage of the organic matter is algal within the 242-238.5 Ma (Spathian to Middle Anisian) (Fig. 13) time period (from the 7323/7-U-3, U-4 and U-1 cores), in good agreement with the location of the most organic-rich rocks. The strata immediately below the 237 Ma sequence boundary were found to contain predominantly woody organic matter with subordinate amounts of coaly material, indeterminate fines, and algal material (samples 42 through 48). At the Svalis Dome locality, we have interpreted the paleoenvironment prior to the 237 Ma sequence boundary as an oxic offshore to lower shoreface setting, with alternating thin shales and very fine sandstone beds deposited during a highstand in sea level (Fig. 4).

The ensuing transgression in the Anisian reestablished a dysoxic open-shelf to distal-marine setting with predominantly algal organic matter (Fig. 14a) and with the terrigenous organic matter being trapped in the more proximal settings (Fig. 14b). The correlation between oxygen level and the marine transgression is probably a result of lower energy conditions and less bioturbation and scavaging of organic matter at the sea floor. The sterane and triterpane biomarkers vary from 4000 to 16 000 ppm of the extractable organic matter (EOM) (Fig. 15). There is an overall increase in the absolute amounts of C_{30} steranes with younger-age sediments within this time period. C_{30} steranes are thought to be indicators of marine algal organic matter (Moldowan et al. 1985; Mello et al. 1988a,b). This return to high levels of algal material preserved in the sedimentary column is also quite distinctly shown by the sterane biomarker distribution. C_{30} steranes reach their highest concentrations within the transgressive shale package in the upper parts of core 7323/7-U-9. This correlates well both with the interpreted systems tracts seen in cores and with the overall second-order rise in eustatic sea level (Haq et al. 1987) (Fig. 16).

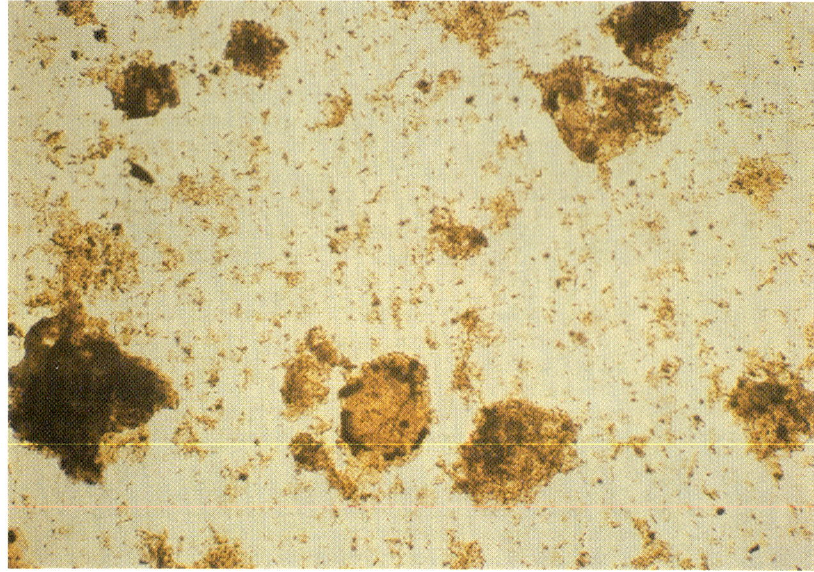
a

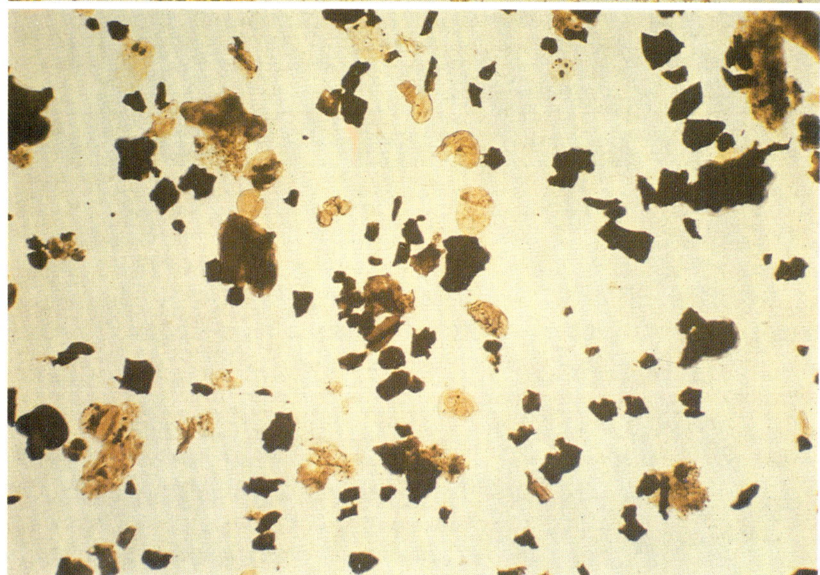
b

Fig. 14a, b. Visual kerogen photomicrographs showing predominance of marine algal organic matter in distal marine settings and predominance of woody organic matter in proximal marine settings. *Al* Algal; *F* indeterminate fines; *H* herbaceous; *W* woody; *C* coaly. The classification of organic matter types lists the most abundant macerals first and the least abundant macerals last

a) Spathian (242-240.5 Ma)

The Spathian section is characterized by overall low total amounts (400–7000 ppm of EOM) of steranes and triterpanes (Fig. 15) within the transgressive and highstand systems tract. There is a slight increase in sterane and triterpane concentrations going upward into younger rocks above the flooding surface approximately 8 m above the 242 Ma sequence boundary. We relate these trends to organic-matter production and preservation. A very distinct flooding surface occurs 8 m above the 242 Ma sequence boundary. It is most likely that this flooding event moved the shoreline eastward, trapped terrigenous organic matter further east, and allowed the development of dysoxic to anoxic conditions at the more distal Svalis Dome location.

The hydrogen indices correlate inversely with the triterpane/sterane ratio for the Spathian samples located within the transgressive systems tract (Table 2). This suggests that a portion of the triterpanes are of terrigenous higher-plant origin, because the terrigenous material would be expected to lower the hydrogen index (Tissot et al. 1974). This interpretation of the sterane and triterpane biomarkers is supported by the following observations: (1) the hydrogen indices decrease from 350 to

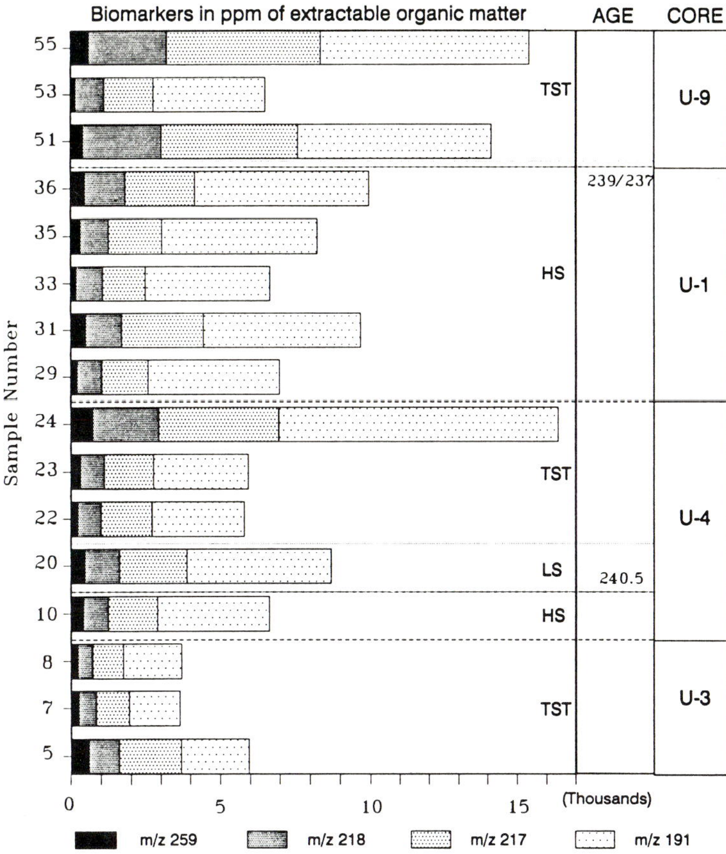

Fig. 15. Sterane and triterpane biomarker concentrations (in ppm of extractable organic matter) for the Lower Spathian to Upper Anisian sedimentary sequence. Samples are shown in their sequence stratigraphic framework with parasequences and sequence boundaries noted

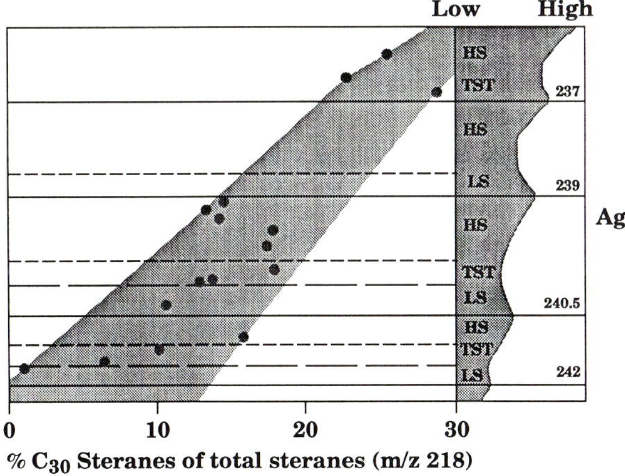

Fig. 16. Relative amounts of C_{30} desmethyl steranes, reflecting marine algal input, shown in the sequence stratigraphic framework from Lower Spathian to Upper Anisian. The C_{30} desmethyl steranes relative increase correlates with the second-order rise in sea level during the Lower to Middle Triassic

270 with slightly increasing total organic carbon values (Fig. 9) indicating deteriorating potential for liquid hydrocarbon generation from the kerogen; (2) the relative proportion of C_{30} moretane to C_{30} hopane increases, a trend correlating with increasing terrigenous organic matter [Rullkotter and Marzi (1988) noted higher moretane/hopane ratios in bitumens from hypersaline rocks; we suggest a correlation between moretane concentration and terrigenous higher plant organic matter]. Highest moretane/hopane ratios are seen in the highstand systems tract below the 240.5 Ma sequence boundary in the 7323/7-U-4 core (Fig. 17); (3) C_{29} steranes predominate over the corresponding C_{30}, C_{28} and C_{27} steranes. C_{29} steranes covary with terrigenous organic matter (e.g., Czochanska et al. 1988); (4) Although within a transgressive system tract the shoreline during the Spathian seems to have been closer to the Svalis Dome locality than during periods of transgression within the Anisian. Pristane/phytane ratios taken as indicators of oxygenation state are about 1.3, suggesting that anoxic conditions did not develop. The proximity of the shoreline interpreted from the sedimentary structures in the core supports this observation.

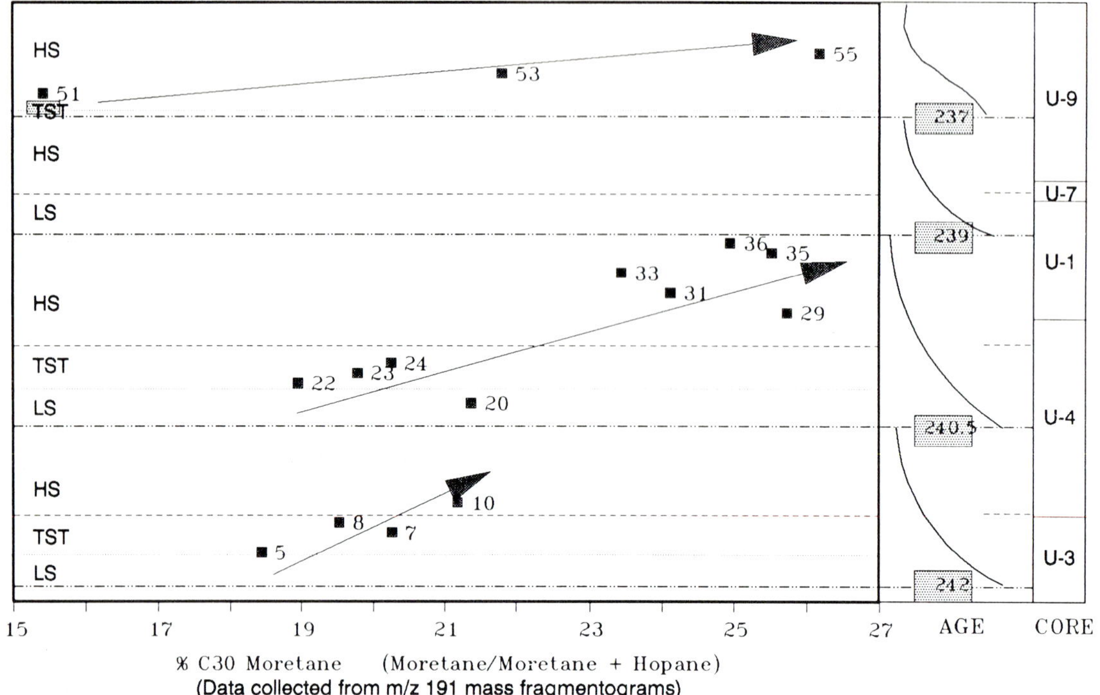

Fig. 17. C_{30} moretane variations within the Lower Spathian to Upper Anisian sequence stratigraphic framework. Increasing $\%C_{30}$ moretane from transgressive (TST) to highstand (HS) systems tracts may reflect increasing terrigenous higher plant organic matter input

b) Anisian (240.5-238.5 Ma)

Following the lowstand event above the 240 Ma sequence boundary, another transgressive systems tract was deposited. This time the NW/SE-trending shoreline (shown schematically in Fig. 1) was farther to the east than during the Spathian. The Anisian flooding event caused (1) trapping of terrigenous organic matter further to the east, (2) deeper marine waters and a more distal setting at the Svalis Dome locality, and (3) increased marine-algal organic-matter production, accumulation, and preservation in the Svalis Dome area, with resulting higher quality oil-prone kerogen (Fig. 13).

The Anisian transgressive system tract above the 240.5 Ma sequence boundary displays a covariance between total organic-carbon contents and hydrogen indices (Table 2). In addition, there is a positive correlation between the hydrogen indices and the triterpane/sterane ratio, suggesting that the triterpanes are mainly of bacterial rather than terrigenous origin. Terrigenous material would be expected to reduce the hydrogen index, whereas a bacterial contribution would provide a higher hydrogen supply.

Alternatively, one can speculate that the steranes, mainly contributed from planktonic organisms, would react more readily with hydrogen sulfide formed during microbial sulfate-reduction during diagenesis. Triterpanes of the hopanoid type could bypass the scavenging effect of highly reactive sulfide by remaining organically bound with bacterial biomass or still retained in living bacteria. There is also an increase in C_{30} steranes, specific markers for marine algal organic matter (Moldowan et al. 1985) (Fig. 16). Pristane/phytane ratios are low (0.86 to 0.68), suggesting a development of more restricted oxygen conditions (Didyk et al. 1978). [The bulk of pristane and phytane at this location is most likely form terrigenous organic matter, and not of bacterial origin, although Volkman and Maxwell (1986) have shown that Archaebacteria can also be sources of acyclic isoprenoids].

These data support that a northwest-trending shoreline had developed in front of a broad, low-lying delta plain area to the east. The transgression which occurred shortly after 240.5 Ma most likely flooded large areas of this eastern lowland, trapped terrigenous organic matter in the eastern provinces, and developed distal open-marine conditions in the

Svalis Dome area. We interpret that the dysoxic conditions resulted in the low pristane/phytane ratios observed as well as an increased potential for preservation of accumulated organic matter. Marine algal material together with bacterial organic matter now became the major input, with secondary amounts of terrigenous organic matter. This production, accumulation, and preservation scenario resulted in formation of the more oil-prone source rocks. These high-quality potential source rocks were formed during the late transgressive systems tract when the rate of eustatic sea-level rise was maximal. Thus, the eustatic rise results in movement of organic-rich facies (condensed section) further up onto the shelf (Loutit et al. 1988).

The ensuing highstand event is recorded in core 7323/7-U-1. From a sequence stratigraphic point of view one would expect depositional conditions to have remained favorable for formation of high-quality, oil-prone potential source rocks (e.g., Haq et al. 1987; Van Wagoner et al. 1990) as the paleoshoreline likely was situated a significant distance east of the Svalis Dome locality and trapping the bulk of terrigenous higher plant material brought into the marine environment. This interpretation supports the observed covariance between total organic carbon and hydrogen index values (Fig. 9). Triterpane and sterane biomarker concentrations are about average for all Svalis Dome samples, i.e., 7000 to 10 000 ppm of the EOM. In addition, the triterpane/sterane ratio remains high, in the range of 1.8 to 3.0. The early parts of this highstand systems tract show a slight decrease in the absolute amounts of the C_{30} steranes derived from marine algal organic matter input into the sedimentary column (Fig. 16), which is a function of marine algal production, preservation, and accumulation. Lithologically, there are some distinct periods within this highstand event with coarser clastic material (up to a very fine sand-grain size at 124 to 120 m and 119 to 117 m in the 7323/7-U-1 core); (Fig. 4a). However, a westward progradation of the Anisian shoreline and significantly higher contents of terrigenous input material is not supported by molecular evidence.

These very good source-forming conditions existed prior to a major drop in relative sea level and ceased with the development of the 239 Ma sequence boundary. No samples were selected for GC/MS biomarker characterization from the sedimentary package between the 239 and 237 Ma sequence boundaries because of the low contents of organic matter. This interval is represented by the upper 13 m of core 7323/7-U-1, the entire 7323/7-U-7 core, and the lower 10 m of core 7323/7-U-9. During this time, we believe the westward-prograding shoreline came very close to the Svalis Dome locality. Offshore to lower shoreface depositional environments are recorded immediately above the 237 Ma sequence boundary. This lowstand sedimentary package is characterized by high contents of structured woody organic matter and type III kerogen (Fig. 13).

c) Anisian (238.5-227 Ma)

The ensuing transgressive event above the 237 Ma sequence boundary trapped terrigenous organic matter updip and re-established distal open shelf and dysoxic conditions in the bottom waters. The environment of deposition, oxygenation state, and type of organic-matter input was similar to the conditions which prevailed during the transgressive systems tract of the 240.5 Ma sequence. However, we propose that the sedimentary package within the transgressive systems tract in core 7323/7-U-9 differed in that the water depth had now reached a maximum at the Svalis Dome locality. This is based on the distal marine shales described from the core and its geochemical attributes such as an overall increase in the absolute concentration of C_{30} steranes due to marine algal organic matter input (Fig. 16). Deeper marine conditions allowed for enhanced preservation of algal material in the distal shelf areas and consequently a higher content of steranes per gram of sediment or per mg of EOM (as observed for the 7323/7-U-9 core). The concentrations of C_{27} through C_{30} steranes, and C_{30} steranes in particular, are highest (up to 500 ppm in the EOM) in the upper parts of the 7323/7-U-9 core. These observations are in good agreement with the overall second-order rise in sea level throughout this time period (Haq et al. 1987).

GC/MS/MS analyses were performed in order to identify the type of C_{30} steranes present. Figure 18 shows the mass spectra of the C_{30} sterane peaks labeled 1 and 2, identified by their m/z 414 molecular parent ion and m/z 217 common fragment ion. The analytical difficulties in distinguishing between C_{30} des-methyl and C_{30} methyl steranes due to co-elution and use of quadrupole GC/MS necessitates the use of GC/MS/MS (Moldowan et al. 1985; Wolf et al. 1986b). GC/MS/MS analyses reveal that these compounds are predominantly C_{30} des-methyl steranes by monitoring the m/z 414 to m/z 217 transition and comparison with the m/z

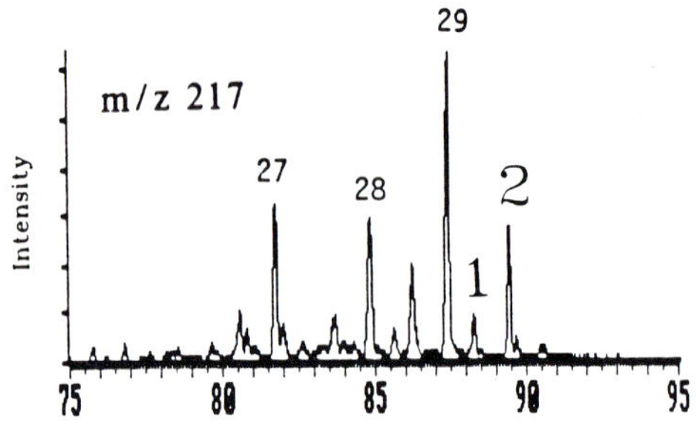

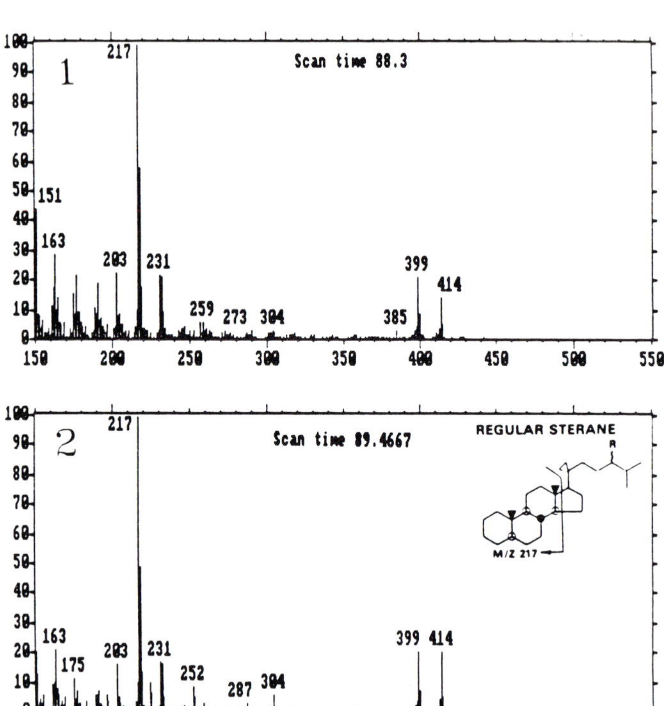

Fig 18. Mass spectra of C_{30} steranes. *Top* m/z 217 mass fragmentogram (see Fig. 10 for peak identification); *middle and bottom* mass spectra of peaks labeled *1* and *2*, confirming their nature as C_{30} steranes

414 to m/z 231 transition monitoring A-ring methyl substituted steranes (Fig. 19). Some A-ring methyl-steranes are present, but in quantities of around 12% of the des-methyl C_{30} steranes. The relative amounts of A-ring methylated steranes are greatest within the highstand systems tracts and lower within the transgressive systems tracts. These A-ring methylated steranes with the methyl group in the "4" position could have a dinosterane or 4-methyl-24-ethyl-cholestane configuration. Biological precursors of the 4-methyl-24-ethyl-cholestanes are unknown. Dinoflagellates are thought to be the precursors of the dinosteranes (Summons et al. 1987); thus the greater abundance of dinoflagellates in proximal versus distal marine settings could explain the methyl sterane distributions.

Exploration Significance

The demonstrated covariance of the lithofacies (including trace fossil assemblages) with the chemical properties of the mudrocks deposited in these offshore and shelfal marine environments enables the construction of depositional facies models to predict the organic matter content and quality from normal exploration data. This tie allows the explorer to constrain and predict more closely the nature of potential source rocks from seismic and well-log data. For the Triassic section, the interval between the 242 and 238.5 Ma sequence boundaries (Spathian to Middle Anisian) contains the highest quality potential source rocks, deposited in distal and transitional distal-to-proximal

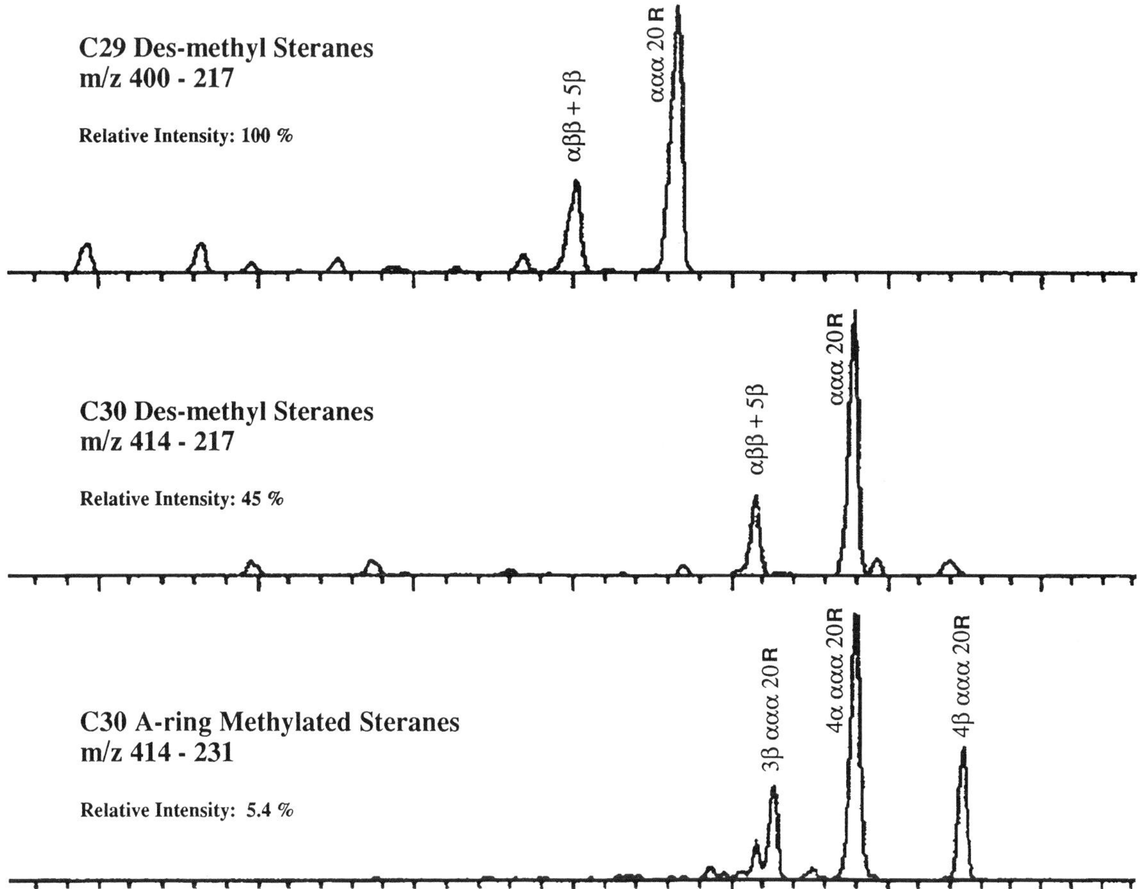

Fig. 19. Identification of C_{30} steranes as C_{30} des-methyl steranes using GC/MS/MS. The m/z 414 to 217 transition monitors C_{30} desmethyl steranes, whereas the m/z 414 to 231 transition monitors A-ring methyl substituted steranes. A-ring methyl substituted C_{30} steranes are present in quantities of around 12% of the C_{30} des-methyl steranes

open-marine-shelf environments under anoxic/dysoxic to dysoxic conditions. This strata is comprised of a minimum of 72 m of mudrocks with an average organic carbon content of 3.12% (range: 0.46 to 11.3%) and an average hydrogen index of 268 (range: 78 to 460). Oil-prone mudrocks are also present within the 14 m thick package 5 m above the 237 Ma sequence boundary. The interval with the best source potential in the Svalis Dome area correlates with the upper 15 meters of the Urd Formation and the lower 15 m of the Skuld Formation on Bjørnøya (Bjorøy et al. 1981; Isaksen 1985) and the Botneheia Formation (Anisian-Ladinian) on Svalbard (Mørk and Bjorøy 1984).

Shallow seismic lines from the area show the richest source rocks occurring in an interval with relatively continuous and parallel reflections with higher amplitude and slightly higher frequency content. The highstand systems tracts seen on sparker data (not included) are generally well layered, with more continuous reflections of higher amplitude and slightly higher frequency; the bedding in the lowstands tends to be more obscure.

The oxygen level in the water column and sediment interstices is a key control on the preservation of organic matter in the system. In this area, it appears that the oxygen level within the sediments covaried closely with the bottom energy levels and sedimentation rates.

The best source rocks are most likely to occur in distal open-marine-shelf environments in the 242 and 240.5 Ma sequences. Even if only the Triassic interval is mappable, this still high-grades a part of the section prone to having organic enrichment. With the limited number of sediment samples available together with seismic-facies mapping, it seems that the best Early to Middle Triassic oil-prone source rocks are developed within the area occupied by the deeper marine seaway extending from the open Arctic Ocean, across Svalbard and the western Barents Sea and south toward the northern Norwegian mainland (Fig. 2). The interval with the best source potential in the Svalis Dome area correlates with the upper 15 m of the Urd Formation and the lower 15 m of the Skuld Formation on Bjørnøya and the Botneheia Formation (Anisian-Ladinian) on Svalbard.

Summary

This chapter demonstrates that a better undertanding of the depositional environments can be achieved through geological and geochemical integration within a sequence stratigraphic framework. Variations in source rock potential are controlled by the depositional setting and constrained by the sequence stratigraphic framework. The depositional setting is discernable from outcrop, core, and well-log data; the sequence stratigraphy can be recognized from biostratigraphy, log, and seismic data. Details of the depositional environment are revealed by changes in lithology, bedding geometry, sedimentary structures, body and trace-fossil assemblages, as well as organic facies.

Lower to Middle Triassic strata cored on the Svalis Dome comprise rocks deposited in shallow to deep open-marine-shelf environments (Fig. 1) with oxic to dysoxic water conditions. The type and richness of the organic matter preserved in these strata are directly controlled by the depositional setting and vary systematically within the depositional sequences containing them. Even in this relatively distal environment, variations at the parasequence level are seen in the bedding, lithology, sedimentary structures, and organic matter character.

The sediment/water system appears to have never become completely anoxic: the most restricted condition is interpreted to have been a dysoxic water column over dysoxic to anoxic sediment interstitial water. $Ni/Ni+V$ and Al_2O_3/TOC ratios are both good indirect measures of the quality of the organic matter (hydrogen index); each facies has a characteristic range of these ratios. A lower bound on the organic richness of the rocks may be estimated from the uranium content: $TOC(\%) > [U(ppm) - 1]/2.5$. Organic-carbon content may be estimated from gamma-ray-spectral well logs.

Biomarker compounds display a molecular distribution typical of low-early thermal maturity, corresponding to a vitrinite reflectance value of 0.4 %Ro maximum. Measured vitrinite reflectance values are in the range of 0.3–0.5 %Ro. Source-facies indicators from biomarkers show appreciable contents of C_{30} steranes derived from marine algal material. The highest concentrations of sterane and triterpane biomarkers are found between the 237 and 231 Ma sequence boundaries (corresponding to the maximum transgression in this interval and equivalent to the lower parts of the Skuld Formation on Bjørnøya). C_{30} steranes show a quantitative increase in absolute concentration with the second-order rise in sea level during the Lower to Middle Triassic, and with the amount of marine algal organic matter deposited in distal marine settings. Biomarker compounds show a very good covariance with sea-level variations, proximity to the shoreline, and input of terrigenous or marine algal organic matter. These variations in molecular geochemistry are predictable within the seqence stratigraphic framework.

Acknowledgments. We would like to thank Esso Norge for permission to publish this work and help with the paleogeographic reconstruction, the Norwegian Continental Shelf Institute (IKU) for assistance during sample collection and lithological descriptions, Hans Rendall at IKU for spectral gamma-ray data, and the technical staff at Exxon Production Research Company for assistance with sample analyses. We are also grateful for the constructive comments from the reviewers Flemming G. Christiansen, Mark A. Pasley, and Barry Katz.

References

Adams JAS, Weaver CE (1958) Thorium-to-uranium ratios as indicators of sedimentary processes: example of concept of geochemical facies. Am Assoc Pet Geol Bull. 42 (2): 387–430

Bergan M, Knarud R (1993) Apparent changes in clastic mineralogy of the Triassic-Jurassic secession, Norwegian Barents Sea: possible implications for paleodrainage and subsidence. In: Vorren TO et al. (eds) Arctic geology and petroleum potential. Norw Petrol Soc (NPF) Spec Publ 2: 481–493, Elsevier, Amsterdam

Berner RA (1984) Sedimentary pyrite formation: an update. Geochem et Cosmochim Acta 48: 605–615

Berner RA, Raiswell R (1983) Burial of organic carbon and pyrite sulfur in sediments over Phanerozoic time: a new theory. Geochem et Cosmochim Acta 47: 855–862

Bjorøy M, Mørk A, Vigran JO (1981) Organic geochemical studies of the Devonian to Triassic succession on Bjørnøya and the implications for the Barents Shelf. In Bjorøy M, Bjørlykke K, Eggen S, Elvsborg A, Finstaed KG, Grønneberg T, Haegh T, Skaar FE (eds) Advances in organic geochemistry 1981, John Wiley, Chichester, pp 49–59

Corwin, AH (1959) Petroporphyrins. Proc 5th World Petroleum Congr Sec V, pp 131–142

Curtis CD (1987) Inorganic geochemistry and petroleum exploration. In Brooks J, Welte D (eds) Advances in petroleum geochemistry, Vol. 2. Academic Press, London, pp 91–140

Czochanska Z, Gilbert TD, Philp RP, Sheppard CM, Weston RJ, Wood T, Woolhouse AD (1988) Geochemical application of sterane and triterpane biomarkers to a description of oils from the Taranaki basin, New Zealand. Org Geochem 12: 123–135

Didyk BM, Simoneit BRT, Brassell SC, Eglinton G (1978) Organic geochemical indicators of paleoenvironmental conditions of sedimentation. Nature 272: 216–222

Fisher IStJ, Hudson JD (1987) Pyrite formation in Jurassic shales of contrasting biofacies. In: Brooks J, Fleet AJ (eds) Marine petroleum source rocks. Geol Soc, London, Spec Publ 26, pp 69–78

Frakes LA (1979) Climates through geological time. Elsevier, Amsterdam, 310 pp

Gabrielsen RH, Faerseth RH, Jensen LN, Kalheim JE, Riis F (1990) Structural elements of the Norwegian Continental Shelf. Part I: the Barents Sea region. Norw Petrol Directorate (NPD) Bull 6: 20, May 1990

Green AR, Kaplan AA, Vierbuchen RC (1986) Circum-Arctic Petroleum Potential. In: Halbouty MT (ed) Future petroleum provinces of the world. Am Assoc Pet Geol, Tulsa, Mem 40 pp 101–130

Haq BU, Hardenbol J, Vail PR (1987) The chronology of fluctuating sea level since the Triassic, Science 235: 1156–1167

Isaksen GH (1985) Petroleum geochemistry of some pre-Jurassic sediments in the western Barents Sea, vols. 1 and 2. Cand Sci Theses, Univ of Bergen, Bergen, Norway

Kristoffersen Y, Elverhøi A (1978) A diapir structure in Bjørnøyarenna. Nor. Polarinst. Aarbok 1977, pp. 189–198

Leventhal JS (1983) An interpretation of the carbon and sulphur relationships in the Black Sea sediments as indicators of the environment of deposition. Geochim Cosmochim Acta 47: 133–137

Lewan MD, Maynard JB (1982) Factors controlling the proportionality of vanadium and nickel in the bitumen of organic sedimentary rocks. Geochim Cosmochim Acta 46: 2547–2560

Loutit TS, Hardenbol J, Vail PR, Baum GR (1988) Condensed sections: the key to age determination and correlation of continental margin sequences, In: Wilgus CK et al. (eds) Sea-level changes, an integrated approach. Soc Econ Paleontol Mineral, Tulsa, Spec Publ 42: 183–213

Mackenzie AS, Hoffman CF, Maxwell JR (1981) Molecular parameters of maturation in Toarcian shales, Paris basin, France, III. Changes in aromatic steroid hydrocarbons. Geochim Cosmochim Acta. 45: 1345–1355

Mackenzie AS, Patience, RL, Yon DA, Maxwell JR (1982) The effect of maturation on the configurations of acyclic isoprenoid acids in sediments. Geochim Cosmochim Acta 46: 783–792

Mangerud G, Rømuld A (1991) Spathian-Anisian (Triassic) palynology at the Svalis Dome, SW Barents Sea. Rev Palaeobot Palynol 70: 199–216

Mello MR, Gaglianone PC, Brassell SC, Maxwell JR (1988a) Geochemical and biological marker assessment of depositional environments using Brazilian offshore oils. Mar Petrol Geol 5: 205–223

Mello MR, Telnoes N, Gaglianone PC, Chicarelli MI, Brassell SC, Maxwell JR (1988b) Organic geochemical characterization of depositional paleoenvironments of source rocks and oils in Brazilian marginal basin. In: Mattavelli L, Novelli L (eds) Advances in organic geochemistry 1987, Pergamon Press, Oxford pp 31–45

Moldowan JM, Seifert WK, Gallegos EJ, (1985) Relationship between petroleum composition and depositional environment of petroleum source rocks. Am Assoc Pet Geol Bull 69: 1225–1268

Mørk A, Bjorøy M (1984) Mesozoic source rocks on Svalbard. In: Spencer AM et al. (eds) Petroleum geology of the North European margin. Norw Petrol Soc, Graham and Trotman, London, pp 371–382

Myers KJ, Wignall PB (1987) Understanding Jurassic organic-rich mud rock - new concepts using gamma-ray spectrometry and paleoecology: examples from the Kimmeridge Clay of Dorset and the Jet Rock of Yorkshire, In: Legget JK (ed) Marine clastic environments: concepts and case studies, Graham and Trotman, London, pp 175–192

Passey QR, Creaney S, Kulla JB, Moretti FJ, Stroud JD (1990) A practical model for organic richness from porosity and resistivity logs. Am Assoc Pet Geol Bull 74: 1777–1794

Rasmussen A, Kristensen SE, Van Veen PM, Stølan T, Vail PT (1993) Use of sequence stratigraphy to define a semi-stratigraphic play in Anisian sequences, SW Barents Sea. In: Vorren TO et al. (eds) Arctic geology and petroleum potential. Norw Petrol Soc (NPF) Spec Publ 2: 439–455. Elsevier, Amsterdam

Rønnevik H, Jacobsen HP (1984) Structural highs and basins in the Western Barents Sea. In: Spencer AM et al. (eds) Petroleum geology of the North European margin. Norw Petrol Soc, Graham and Trotman, London, pp. 19–32

Sarvda CE, Bottjer DJ (1986) Trace-fossil model for reconstruction of paleo-oxygenation in bottom waters. Geology 14: 3–6

Schmoker JW (1981) Determination of organic matter content of Appalachian Devonian shales from gamma-ray logs. Am Assoc Pet Geol Bull 65: 1285–1298

ShiJiyang, Mackenzie AS, Alexander R, Eglinton G, Wolff GA, Maxwell JR (1982) Chemical Geology 35: 1–31

Summons RE, Volkman JK, Boreham CJ (1987) Dinosterane and other steroidal hydrocarbons of dinoflagellate origin in sediments and petroleum. Geochim Cosmochim Acta 51: 3075–3082

Sweeney RE (1972) Pyritization during diagenesis of marine sediments. PhD Thesis, Univ California. 184 pp

Tissot B, Durand J, Espitalie J, Combaz A (1974) Influence of nature and diagenesis of organic matter in formation of petroleum. Am Assoc Pet Geol Bull 58: 499–506

van Berkel GJ, Quirke JME, Filby RH (1989) The Henryville Bed of the New Albany shale – I: preliminary characterization of the nickel and vanadyl porphyrins in the bitumen. Org Geochem 14 (2): 119–128

van Veen PM, Skjold LJ, Kristensen SE, Rasmussen A, Gjelberg J, Stolan T (1993) Triassic sequence stratigraphy in the Barents Sea. In: Vorren TO et al. (eds) Arctic geology and petroleum potential. Norw Petrol Soc (NPF) Spec Publ 2: 439–455, Elsevier, Amsterdam van Wagoner JC, Mitchum RM, Campion KM, Rahmanian VD (1990) Siliciclastic sequence stratigraphy in well logs, cores and outcrops: concepts for high-resolution correlation of time and facies. AAPG Methods in Exploration Ser 7, Am Assoc Pet Geol, Tulsa, Oklahoma 55pp

Volkman JK, Maxwell JR (1986) Acyclic isoprenoids as biological markers. In: Johns RB (ed) Biological markers in the sedimentary record. Methods in geochem geophys 24: 1–42

Wolf GA, Lamb NA, Maxwell JR (1986b) The origin and fate of 4-methyl steroid hydrocarbons. II. Dehydration of stanols and occurrence of C_{30} 4-methyl steranes. Org Geochem 10: 965–974

Zelt FB (1985) Natural gamma-ray spectrometry, lithofacies, and depositional environments of selected upper Cretaceous marine mud rocks, western United States, including Tropic Shale and Tununk Member of Mancos Shale. PhD Diss, Princeton Univ, NJ p 334

Ziegler PA (1988) Evolution of the Arctic-Atlantic and the Western Tethys. Am Assoc Pet Geol Mem 43: 196

The Schistes Carton – the Lower Toarcian of the Paris Basin

B.J. Katz[1]

Abstract

The Schistes Carton was deposited within an epicontinental sea during the *falciferum* zone (early Toarcian), during a period of approximately 500 000 years. Available data suggest that deposition occurred under oxygen-depleted conditions. These shales typically obtain thicknesses between 10 and 20 m, with maximum values in excess of 50 m. The shales are dominated by illite and kaolinite, with only minor amounts of calcite and quartz. Total organic carbon content ranges from ~1 to ~12 wt.% organic carbon, with total generation potentials ranging from 0.5 to 85 mg HC/g rock. Thermally immature samples display hydrogen index values of ~600 mg HC/g TOC and atomic H/C ratios ~1.25. Visual kerogen analysis reveals that the samples are dominated by finely disseminated amorphous organic matter. The kerogen displays only a minor terrestrial background signal, i.e., only trace quantities of vitrinite are present. Studies of the associated bitumen fractions also reveal a predominantly marine character, with only a minor terrestrial overprint. Differences in the character of extracted bitumens from outcrop samples commonly ascribed to provincial differences appear more likely due to differences in the level of thermal maturity. The lack of significant amounts of vitrinite precludes its use in the determination of thermal maturity. Other thermal maturity indices, however, suggest that the onset of thermal hydrocarbon generation occurs within the Toarcian shales at a depth between 1500 and 2100 m, with the main stage of hydrocarbon generation and expulsion occurring at a depth of ~2400 m. Petroleum expulsion occurred primarily between 35 and 20 Ma B.P. Exploration for Schistes Carton-derived oils is largely limited by the unit's low level of thermal maturity.

[1]Texaco Inc., 3901 Briarpark, Houston, Texas 77042, USA

Introduction

The Schistes Carton was deposited during the early Toarcian, specifically during the *falciferum* or *serpentinus* zone a period of ~500 000 years (Jenkyns 1988) (Fig. 1); Deposition of this unit within the Paris basin (Fig. 2) was time-equivalent with other organic-rich, bituminous shales deposited throughout much of Europe (Fleet et al. 1987; Farrimond et al. 1989) with limited occurrences also being known in North America (Imlay 1952; Frebold 1957), South America (Riccardi 1983), Japan (Tanabe 1983), Australia (Crostella and Barter 1980), and Madagascar (Besairie and Collignon 1972).

The early Toarcian shales in the Paris basin are typically between 10 and 20 m thick, with maximum thicknesses in excess of 50 m (Tissot et al. 1971). The Lower Toarcian displays two primary depocenters (Espitalié et al. 1987); (Fig. 3). The unit is exposed in outcrop along the southern and eastern margin of the basin. It obtains a maximum depth of burial of slightly greater than 2 km in the center of the basin east of Paris (Fig. 4) (Durand et al. 1972). The current depths of burial do not represent maximum burial depths. Along the eastern margin of the basin more than 500 m of overburden may have been removed, while less than 100 m of Tertiary erosion appears to have occurred near the basin center (Tissot et al. 1971). There is also evidence that the amount of erosion was generally greater in the south than in the north (Mackenzie et al. 1980).

The unit is economically important both for its oil shale potential along the eastern margin of the basin and as a source rock for conventional hydrocarbon resources. Espitalié and Madec (1981) observed an average oil shale yield of greater than 30 kg oil/t rock, with individual horizons yielding as much as 80 kg oil/t rock. Greater than 100 MMBBL of initial recoverable conventional oil reserves have been attributed to this unit in reservoirs ranging in age from Keuper to Lower

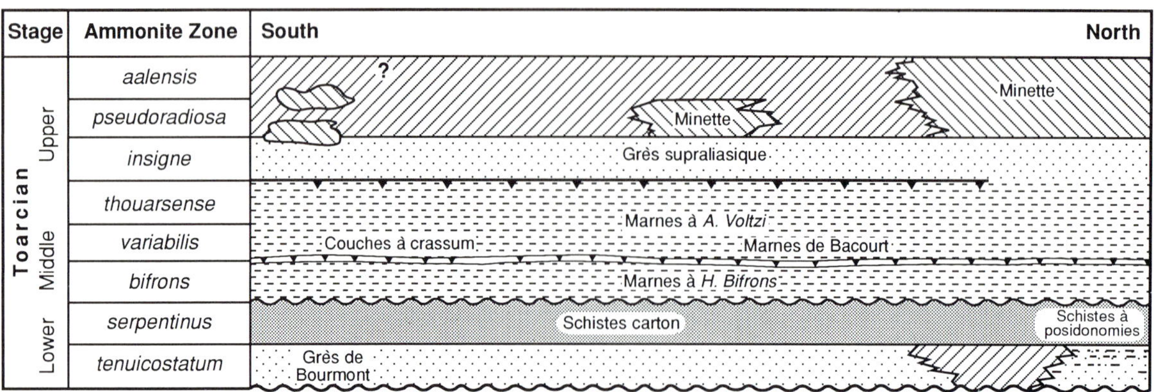

Fig. 1. Toarcian stratigraphic column of the Paris basin. (After Mouterde et al. 1980)

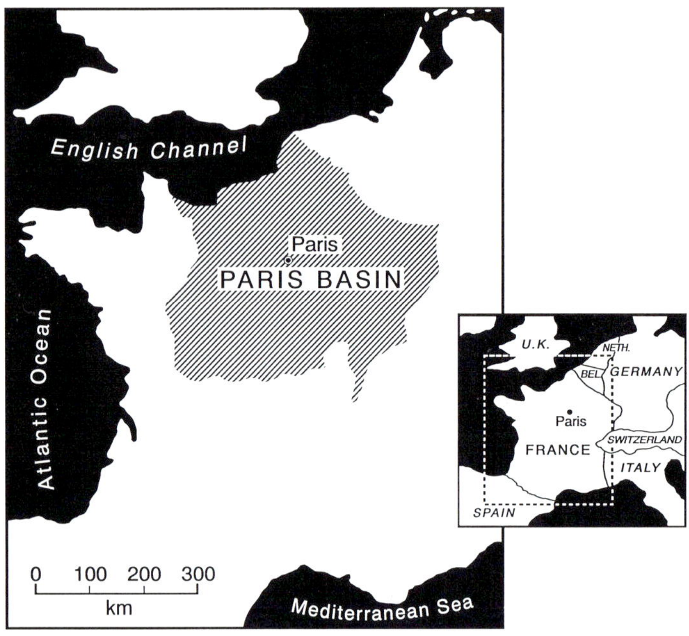

Fig. 2. Geographic index map for the Paris basin

Cretaceous (Espitalié et al. 1987). The lack of more substantial reserves appears to be largely a result of the unit's low levels of thermal maturity rather than due to limited generation potential (see discussion below).

Regional Setting

The Schistes Carton was deposited within an epicontinental sea adjacent to the northern Tethyan margin (Hollander et al. 1991). The basin was located at approximately 40 °N latitude. The Paris basin was one of a series of interconnected Liassic basins including the North Sea, NW German, SW German, and Chalhac basins. These basins displayed intermittent communication with both the Arctic and Tethys oceans as sea level fluctuated and gateways changed. However, differences in organic and inorganic geochemical character of the sediments in these various basins suggest that different depositional conditions existed within each of these European basins (Farrimond et al. 1989; Hollander et al. 1991). Deposition of these organic-rich shales occurred during a third-order rise in sea level (Hallam 1981; Bessereau and Guillocheau 1993).

A stratified basin was suggested by Prauss and Riegel (1989) and Hollander et al. (1991). Farrimond et al. (1989) suggested that this stratification

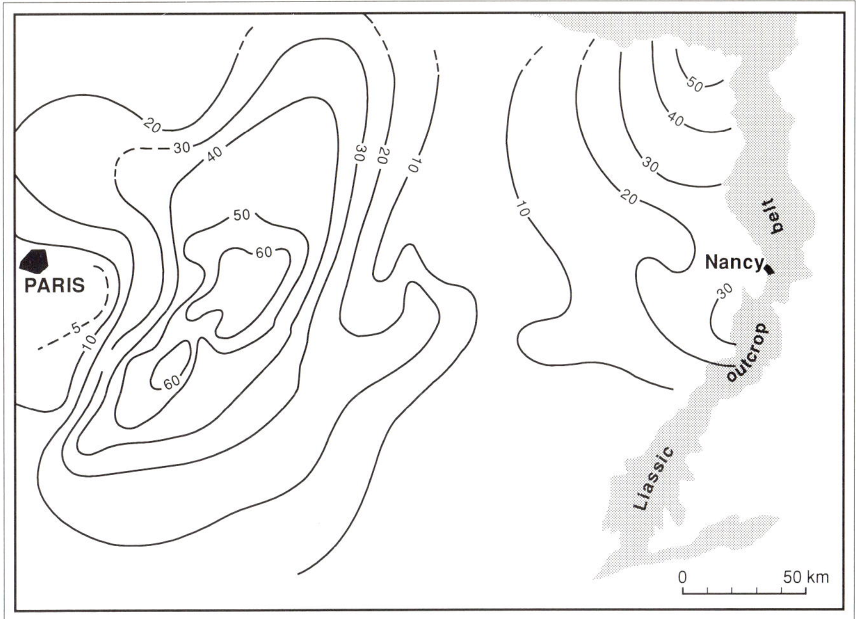

Fig. 3. Isopach map (m) of the Lower Toarcian – Schistes Carton – of the Paris basin. (After Espitalié et al. 1987)

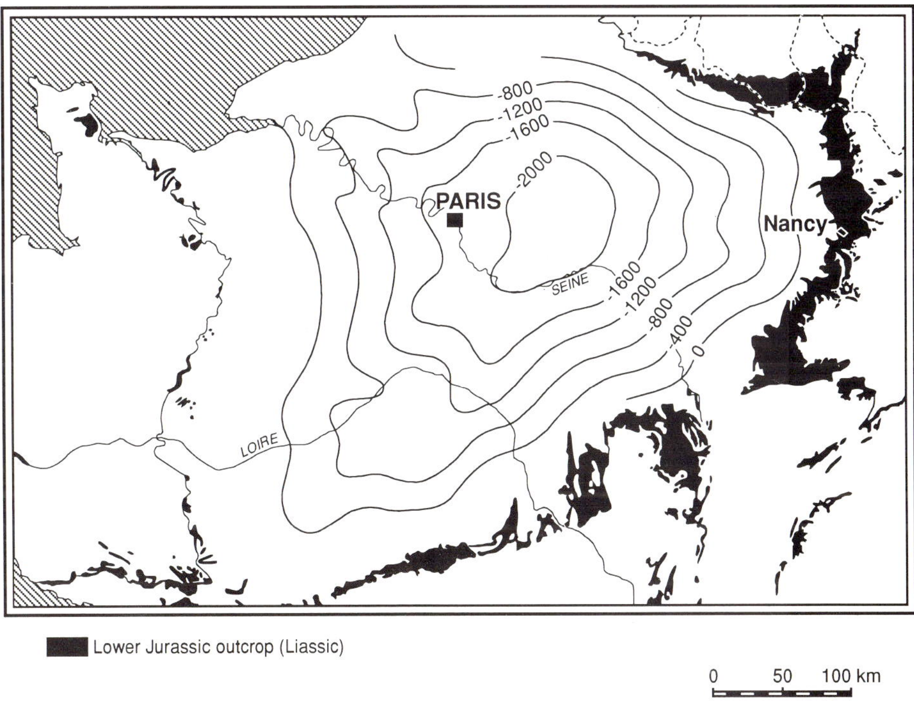

Fig. 4. Current depth (m) to the base of the Toarcian shales. *Shaded area* represents Liassic outcrop belt. (Tissot et al. 1971)

was induced by differences in salinity within the water column. Anoxic conditions at or near the sediment-water interface are also suggested by a lack of benthonic fauna, although a significant planktonic fossil assemblage is present within the unit.

Hollander et al. (1991) concluded that based on changes in matrix mineralogy reducing conditions were more intense during the initial deposition of the Toarcian shales, with conditions ranging from methanogenic to sulfate reducing. Hollander et al.

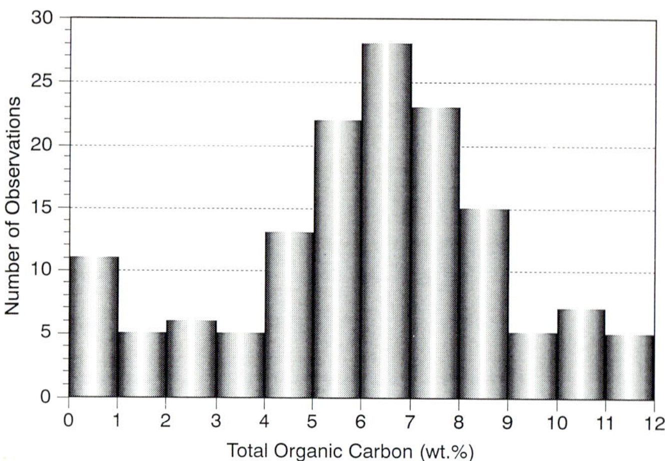

Fig. 5. Histogram of organic enrichment (total organic carbon) – Schistes Carton

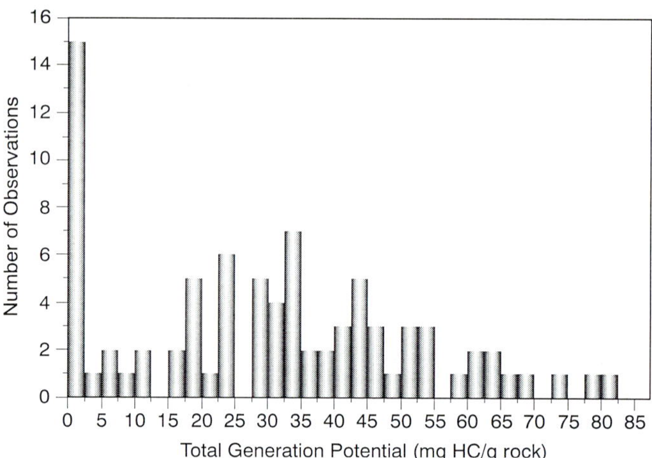

Fig. 6. Histogram of total generation potential ($S_1 + S_2$) – Schistes Carton

(1991) further noted that the geochemical data support an euxinic water column with the redox boundary approaching the photic zone during the initial phases of deposition making organic preservation efficiencies quite high.

Goy et al. (1978) had suggested, based on the abundance of calcareous nannofossil remains, that productivity was occasionally elevated. In contrast, Hollander et al. (1991) concluded, based on stable carbon isotope data, that productivity was generally low to moderate throughout the deposition of the Schistes Carton.

Source Rock Characteristics

Organic Richness and Generation Potential

The organic carbon content of the lower Toarcian shales range from slightly less than 1.0 to ~12.0 wt.% (Fig. 5), with a mean value of 6.17 wt.%. Consequently, the majority of the samples contain above-average levels of organic enrichment (TOC > 1.0 wt.%, Bissada 1982). Such levels of organic enrichment are considered one of the necessary prerequisites for a rock to be classified as a possible hydrocarbon source rock.

Rock-Eval pyrolysis reveals that the total generation potential ($S_1 + S_2$) of the section ranges ~0.5 to ~85 mg HC/g rock (Fig. 6), with a mean value of 30.4 mg HC/g rock. As with the organic carbon, the majority of these samples displayed above-average hydrocarbon yields ($S_1 + S_2 >$ 2.5 mg HC/g rock; Bissada 1982) and may, therefore, be considered representative of a hydrocarbon source rock. In fact, most of the samples exhibit yields consistent with good or excellent petroleum source rocks ($S_1 + S_2 > 6$ mg HC/g rock; Tissot and Welte 1984).

The shallow portion of the unit has also been examined for its oil shale potential. Espitalié and

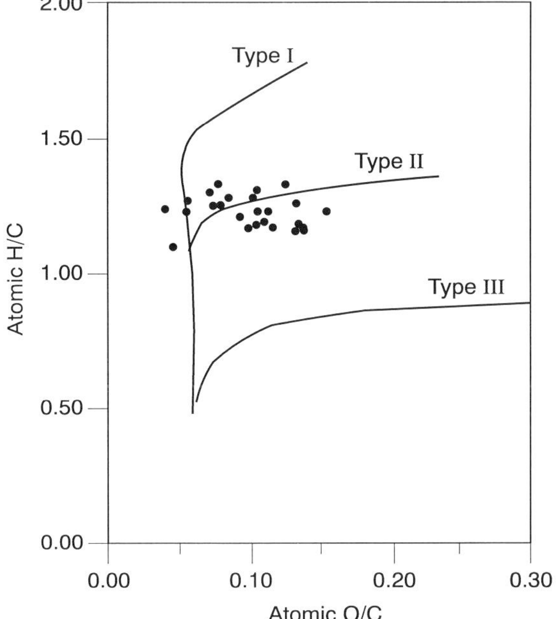

Fig. 7. Van Krevelen-type diagram displaying scatter about the type II reference curve

Madec (1981) estimated oil potential as high as 4.35 Mt/km^2 south of Nancy when both yield and thickness are considered, with an average yield of 1.7 Mt/km^2 for the entire region.

Kerogen Character and Composition

The Lower Toarcian of the Paris basin was originally used by Tissot et al. (1974) to define the type II reference curve on the van Krevelen diagram. Thermally immature kerogens typically display atomic H/C ratios of >1.20, ranging from ~1.15 to 1.30 (Fig. 7). These values show a progressive decrease with increasing burial, with an atomic H/C ratio of ~1.0 being obtained near the basin center at Bouchy (Durand et al. 1972). The atomic O/C ratios also undergo similar changes decreasing from 0.13 to 0.02.

Similarly, the type II reference curve for the modified van Krevelen diagram was established by Espitalié et al. (1977) using the Lower Toarcian of the Paris basin. Thermally immature whole-rock samples typically display hydrogen index values of ~600 mg HC/g TOC (Fig. 8). Within the Lower Toarcian sequence slightly higher hydrogen index values are observed in the more basal portion of the sequence (Fig. 9). Hollander et al. (1991) concluded that this was the result of more effective organic matter preservation during the initial phases of deposition when the redox boundary was at its shallowest position within the water column.

Visual kerogen analysis reveals minor variability in palynofacies (Huc, 1977). Structured algal material, largely *Tasmanites*, accounts for ~20% of the organic matter. Another 10% of the kerogen appears as a terrestrial background signal. This background signal is composed principally of woody debris and other recognizable higher plant components. The remainder of the organic matter appears as finely disseminated amorphous organic material (Hollander et al. 1991). It was further observed that with increasing organic carbon content there is an increase in the abundance of elongate organic elements attributed to algal bodies (Hollander et al. 1991). The kerogens also include a minor reworked component. This reworked component decreases from north to south (Alpern and Cheymol 1978).

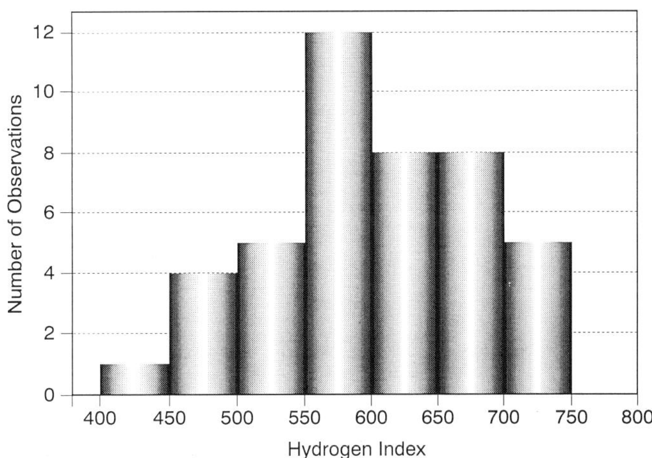

Fig. 8. Histogram of hydrogen index values – Schistes Carton

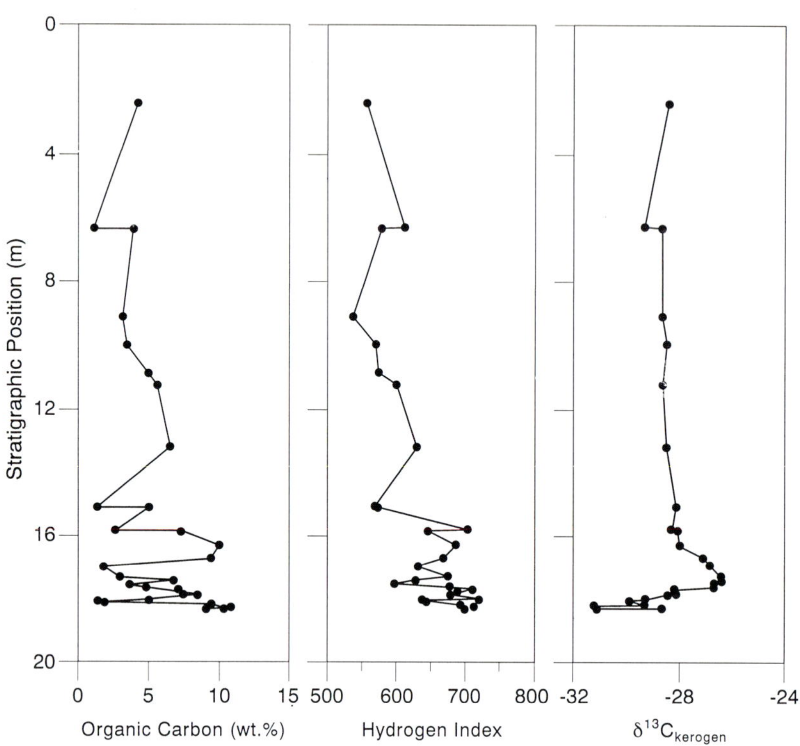

Fig. 9. Geochemical log for the Schistes Carton. (After Hollander et al. 1991)

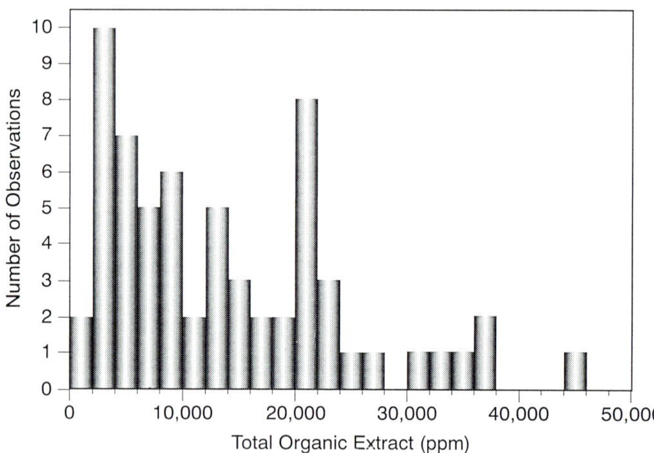

Fig. 10. Histogram of total organic extract (or bitumen) yields

The kerogen carbon isotopic composition ranges from a $\delta^{13}C$ value of -31.18 to -26.36‰ (relative to PDB). These values are depleted relative to Neogene marine kerogens but are comparable to most Mesozoic marine kerogens. Much of the internal variability in kerogen isotopic composition occurs within the lower part of the sequence (Fig. 9).

Bitumen Composition

The total organic extract yields of thermally immature samples of Schistes Carton samples ranged from ~1000 to ~45000 ppm, with most values being less than 15000 ppm (Fig. 10).

As would be expected for thermally immature material, the extract is dominated by nonhydrocarbon components (resins and asphaltenes). These components typically account for greater than 70% of the total extract (Fig. 11). There appears to be a suggestion of two relatively distinct populations. One of these populations is dominated by samples from the northern portion of the outcrop belt. This northern group of samples displays slightly elevated polar contents relative to the second group of samples. The second population which contains

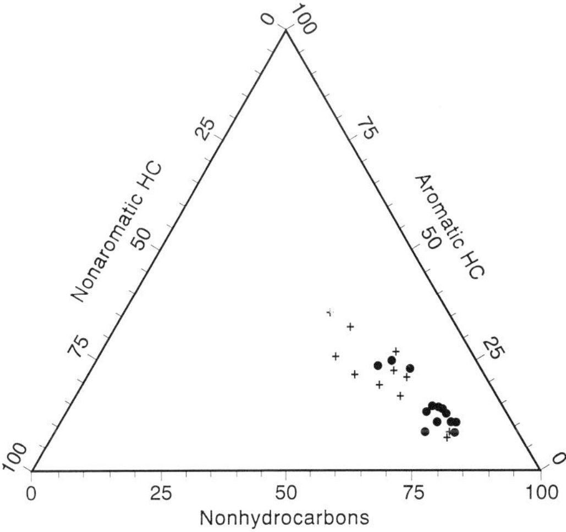

Fig. 11. Ternary plot displaying the bulk bitumen composition of the southern (+) and northern (●) provinces. (Data from Huc 1976)

a slightly greater percentage of hydrocarbons is composed of samples from the more southerly portion of the eastern outcrop belt (Fig. 11). Such a compositional trend is consistent with a slightly more advanced level of thermal maturity in the south (Mackenzie et al. 1980).

C_{15} +-saturated fraction chromatography of thermally immature samples also reveals differences between material obtained from the northern and southern portions of the basin. Material from the north typically displays a more waxy fingerprint, with an abundance of alkanes in the nC_{25} + range (Fig. 12A). Often this suite of material displays a bimodal *n*-alkane distribution, one modal value at nC_{17} and a second at either nC_{27} or nC_{29} (Fig. 13). These northern samples also display CPI (carbon preference index) values ranging from 1.4 to greater than 2.0 (Fig. 14A). In contrast, the chromatograms obtained on extracts from southern outcrop samples are unimodal, with a modal

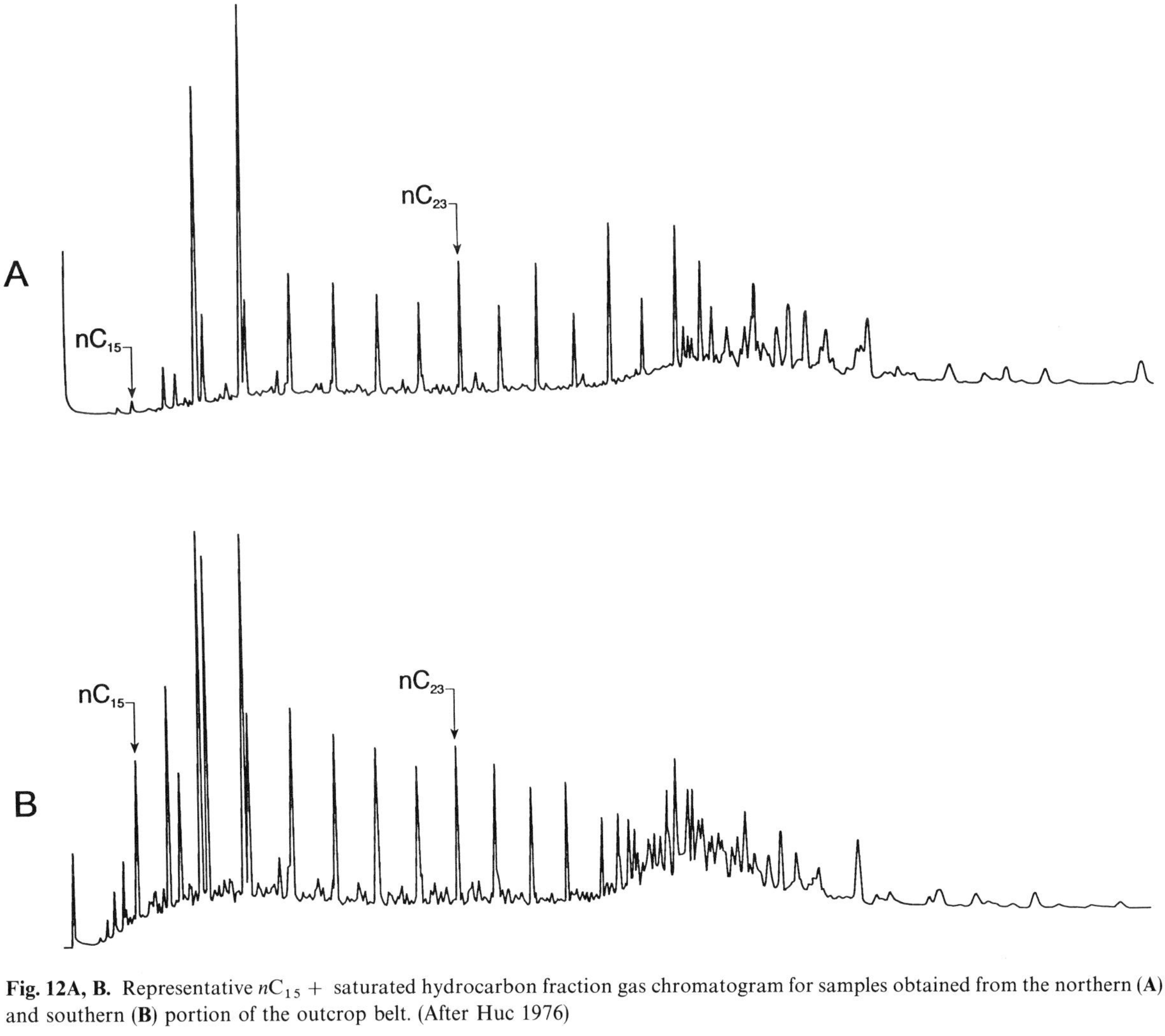

A

nC_{15}

nC_{23}

B

nC_{15}

nC_{23}

Fig. 12A, B. Representative nC_{15} + saturated hydrocarbon fraction gas chromatogram for samples obtained from the northern (**A**) and southern (**B**) portion of the outcrop belt. (After Huc 1976)

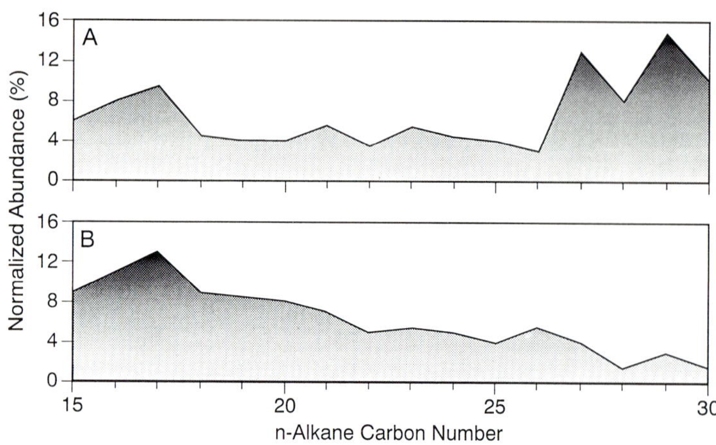

Fig. 13A, B. Normalized *n*-alkane distribution of thermally immature material from the northern (**A**) and southern (**B**) portion of the outcrop belt. (Data from Huc 1976)

value typically located at either nC_{15} or nC_{17}. These samples also display a harmonic decrease in *n*-alkane abundance with increasing carbon number (Fig. 13). In addition, the southern samples display CPI values less than 1.4 (Fig. 14A). These differences also appear to be the result of differences in thermal maturity and not significant differences in the nature of the precursor material. With depth the waxy character and bimodal nature of the *n*-alkane distribution also disappears and the *n*-alkane distribution displays a harmonic decrease in *n*-alkane abundance with increasing carbon number (Fig. 15).

A further examination of the $C_{15}+$-saturated fraction chromatograms of the immature material also reveals two populations based on pristane/phytane (Pr/Ph) ratios (Fig. 14B). In the northern province, the Pr/Ph values are typically less than 1.0. In contrast, the values in the south are commonly greater than 1.0. This change is also consistent with an increase in the level of thermal maturity toward the south rather than a change in organic provenance. If the previously noted changes in bitumen composition were the result of a greater terrestrial influence in the north it would be expected that those samples would display the more elevated Pr/Ph values. There is also an increase in the Pr/Ph ratio with increasing burial depth as well as a decrease in both the pristane/nC_{17} and the phytane/nC_{18} ratios with increasing level of thermal maturity (Fig. 16).

An examination of the biomarker components reveals a generally consistent pattern independent of the level of organic enrichment and hydrogen index (Hollander et al. 1991). The m/z 191 (Fig. 17A) ion chromatogram is dominated by the $17\alpha,21\beta$(H)-C_{30} hopane. The Tm/Ts ratio, which is dependent upon both thermal maturity and organic precursors, is typically about 4.0, at the low levels of maturity of the samples analyzed. Also at low levels of thermal maturity gammacerane is observed. Although Farrimond et al. (1989) concluded that the presence of gammacerane suggests elevated salinity levels within the lower portion of the water column within the Paris basin during deposition, it is unclear as to whether this compound is present in sufficient quantities to be considered a meaningful indicator of elevated salinity.

A further examination of the biomarker compounds reveals that the examined samples contain a significant proportion of diasteranes (Fig. 17B). Among the normal steranes the C_{27}/C_{29} sterane ratios are less than 1.0. The lack of significant terrestrial input into the system suggests that this ratio is being influenced by the nature of the algal input. The samples also contain significant amounts of 4-methylsteranes. The presence of these compounds suggests that at least part of the original biomass included dinoflagellates (Wolff et al. 1986). A comparison of the relative abundance of hopanes and steranes reveals a slightly greater abundance of steranes (i.e., hopane/sterane <1.0). The presence of the isoprenoid lycopane suggests that methanogenic bacteria were active contributors to the biomass (Brassell et al. 1981).

Mineralogy

The stratigraphic interval is dominated by dark-colored, moderately to highly laminated shales, which are dominated by illite and kaolinite. Calcite

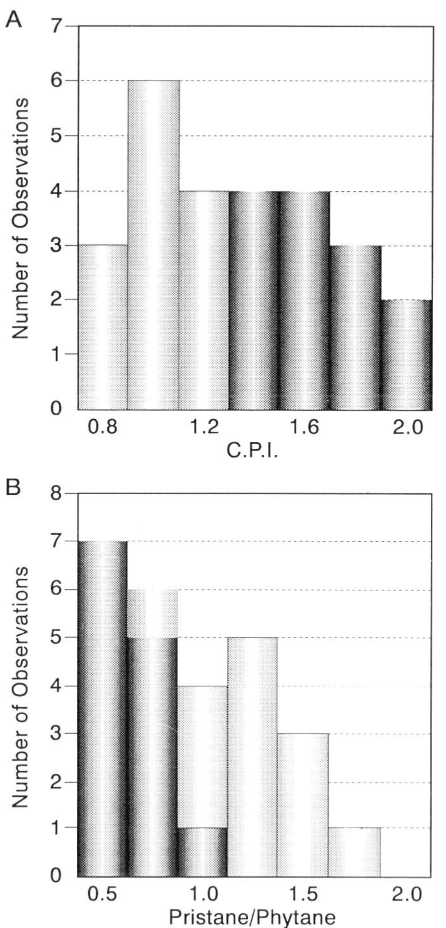

Fig. 14A, B. Histogram of carbon preference index (*C.P.I.*; **A**) and pristane/phytane (**B**) ratios obtained on Lower Toarcian outcrop samples. (Data from Huc 1976)

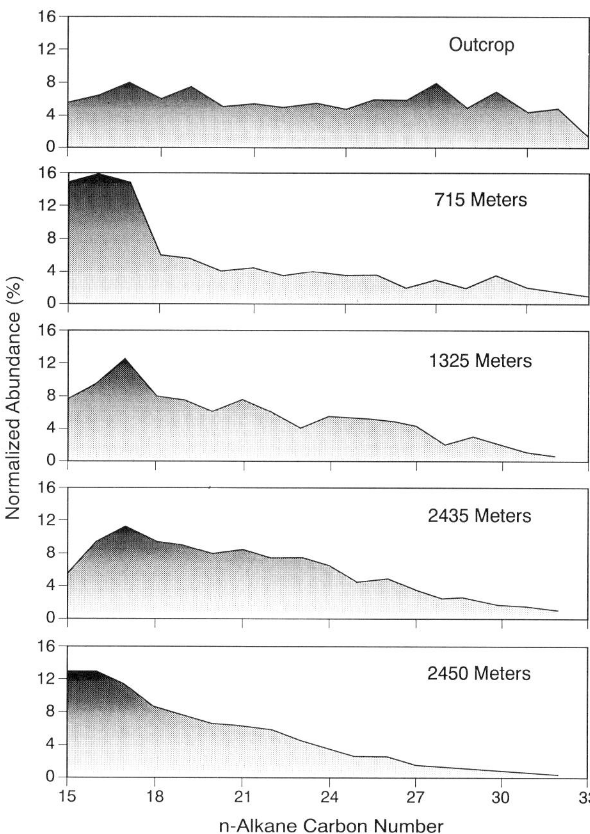

Fig. 15. Normalized *n*-alkane distribution as a function of depth of burial. (Data from Tissot et al. 1971)

accounts for between 10 and 30% of the rock with quartz accounting for between 5 and 20% of the rock matrix.

These shales are occasionally interrupted by Mg-rich calcitic and dolomitic concretion layers and carbonate beds. These carbonate-rich horizons tend to be more concentrated near the base of the sedimentary sequence, and are believed to be early diagenetic in origin (Hollander et al. 1991). This conclusion is supported by the observed bedding relationships between concretions and primary laminations. Hollander et al. (1991) have used the distribution of these carbonate layers and concretions to suggest that reducing conditions were slightly more intense within the water column during the initial phases of Toarcian shale deposition compared with the later phases of deposition when stratification was less well developed.

These shales also commonly contain between 4 and 9% pyrite. Pyrite is typically present as framboids, suggesting sulfate reduction within the shallow sediments.

Thermal Maturity

The thermal maturity of the Schistes Carton in the Paris basin has been addressed through the construction of a series composite profiles. These profiles incorporate many of the commonly accepted thermal maturation indices (e.g., bitumen content of the organic matter, transformation ratio, T_{max}, etc.).

Vitrinite reflectance is not one of the directly measured thermal maturity indices. Insufficient

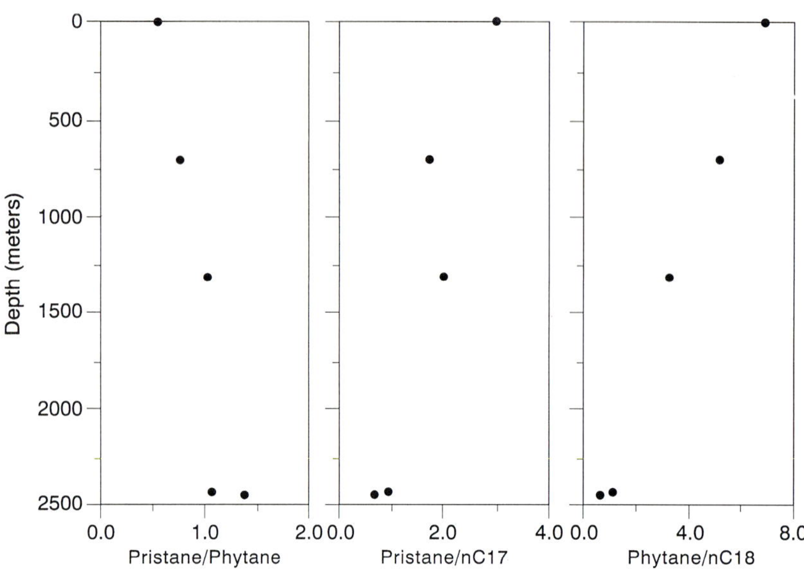

Fig. 16. Changes in the pristane/phytane and isoprenoid/*n*-alkane ratios as a function of burial depth. (Data from Tissot et al. 1971)

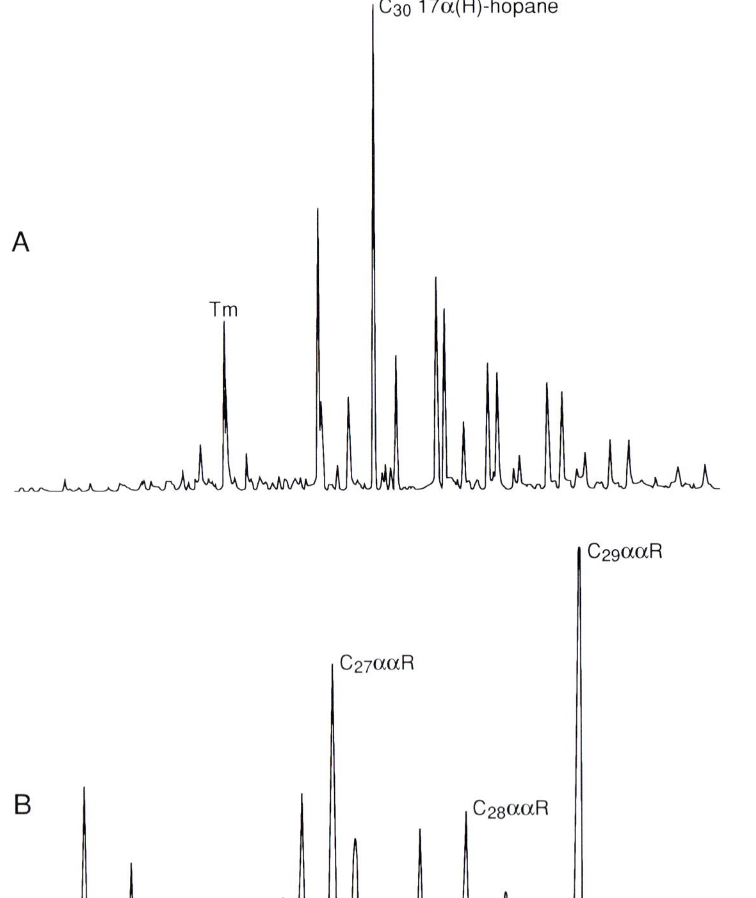

Fig. 17A, B. Representative m/z 191 (**A**) and m/z 217 (**B**) ion chromatograms. (After Hollander et al. 1991)

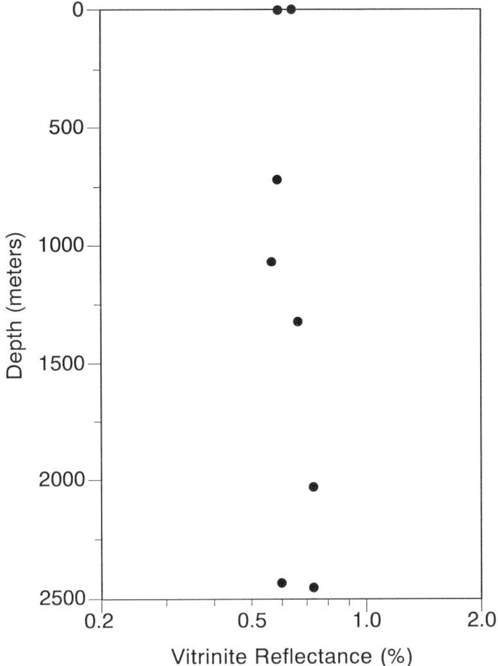

Fig. 18. Composite vitrinite reflectance profile for the Schistes Carton of the Paris Basin. (Data from Alpern and Cheymol 1978)

primary indigenous vitrinite appears to be present in the Schistes Carton in order to obtain valid mean reflectance values. Attempts by such individuals as Alpern and Cheymol (1978) to establish thermal maturity trends resulted in unacceptable profiles, where little change in reflectivity was observed with depth. Reported mean reflectance values for the Toarcian shales range from 0.58% at the surface to 0.73 at 2450 m (Fig. 18). It can be further noted that the elevated near-surface values suggest much more erosion within the basin than can be supported geologically. These elevated near-surface values suggest that these interpreted primary populations include a significant amount of recycled material of moderate maturity.

In contrast, there are significant changes in kerogen fluorescence parameters (Fig. 19) (Alpern and Cheymol 1978). For example, the red/green quotient, Q, appears to be highly sensitive to changes in the level of thermal maturity. These data place the top of the oil window at a depth of ~2400 m.

An examination of the relationship between T_{max} and depth reveals that the main stage of hydrocarbon generation and release ($T_{max} = 440\,°C$) is obtained at a depth of ~2300 m with the onset of thermal generation ($T_{max} = 435\,°C$) beginning at a depth of ~2100 m (Fig. 19). Similarly, the transformation ratio [$TR = S_1/(S_1 + S_2)$] places the top of the main phase of generation and expulsion (TR = 0.2) at a depth of ~2350 m (Fig. 19).

Artificial maturation studies have examined changes in the elemental composition of isolated kerogens. Landais et al. (1989) have suggested that

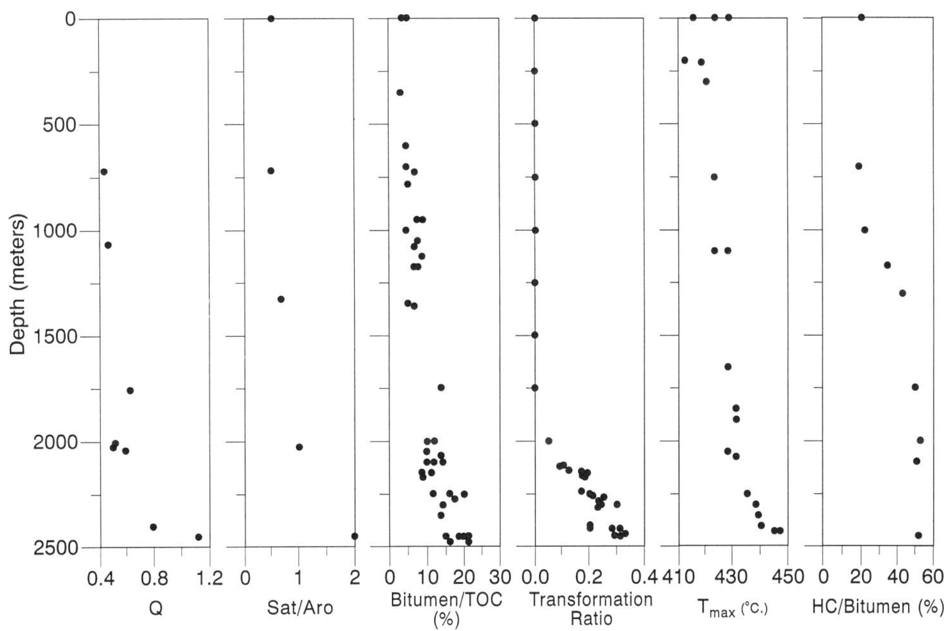

Fig. 19. Composite maturation profiles using a series of both kerogen- and bitumen-related maturation indices

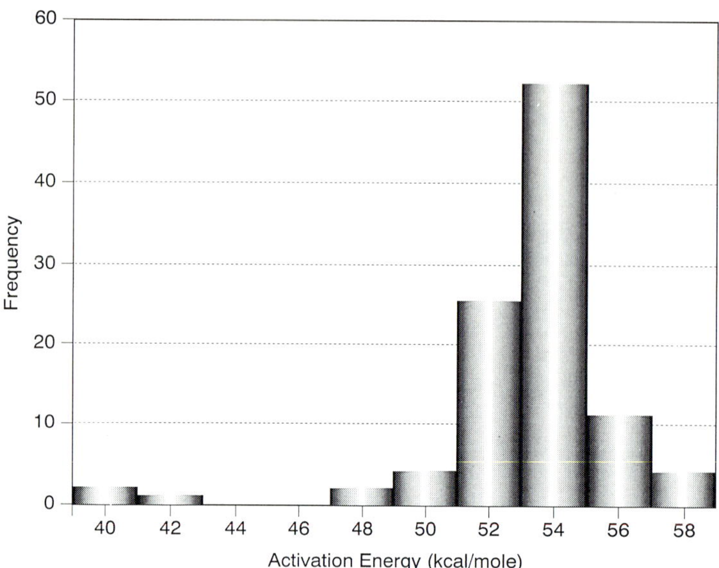

Fig. 20. Distribution of activation energies. (After Huc 1990)

the main stage of hydrocarbon generation and release occurs when the corrected organic carbon value [%C × 100/(%C + %H + %O)] is 85. Within the Paris basin such values are typically achieved in the Schistes Carton at a depth of ~2500 m.

In addition to the kerogen-related maturation indices, the thermal evolution of the bitumen fraction has also been examined. These changes are manifested at both the gross compositional scale and at the molecular level. The relative concentration of extractable material displays little variation to a depth of ~1500 m (Fig. 19). Below this depth there is an increase in total bitumen content largely as the result of an increase in hydrocarbon abundance (Fig. 19). Within this hydrocarbon fraction there is an increase in the relative abundance of saturated hydrocarbons relative to aromatic hydrocarbons with increasing depth (Fig. 19). Most of the molecular or biomarker maturity indices also increase over the depth range of 2000–2500 m, reflecting the onset of generation (Lichtfouse et al. 1994).

Kinetic studies have revealed a broad distribution of activation energies with a modal activation energy of ~54 kcal/mol (Fig. 20); (Huc 1990). Such broad distributions of activation energies are typical of marine kerogens and are in sharp contrast to the narrow distributions of lacustrine kerogens.

As a consequence of the limited depth of burial of the Lower Toarcian Shale (maximum burial depth of ~2500 m) the unit never extends beyond the oil window within the Paris basin.

Modeling by Poulet and Espitalié (1987) suggests that thermal hydrocarbon generation began 80 Ma B.P. with the onset of petroleum expulsion occurring ~35 Ma B.P. Progressive uplift of the basin since the early Miocene (20 Ma B.P.) has further restricted generation from these organically enriched shales, i.e., active hydrocarbon generation was terminated as a result of regional uplift and erosion.

Stratigraphic and Areal Variability

An examination of the distribution of organic matter within the Schistes Carton reveals a general decrease in organic carbon content from the base of the unit to its top across the basin (Fig. 9) (Bessereau et al. 1992). The lower section also appears to contain two organically enriched sections.

Reconstructions of average initial carbon content within the Lower Toarcian by Espitalié et al. (1987) reveals that there is areal variability as well as stratigraphic variability in richness (Fig. 21). The highest levels of organic enrichment appear to be generally associated with the isopach thicks of the unit (Bessereau et al. 1992). This relationship suggests that hydrodynamic properties of the organic matter may play a key role in its redistribution.

Pyrolysis-gas chromatography results indicate that there are minor facies variations within the Lower Toarcian (van Graas et al. 1981). Several of the samples display a more aromatic pyrolysis product. This difference in aromaticity appears to

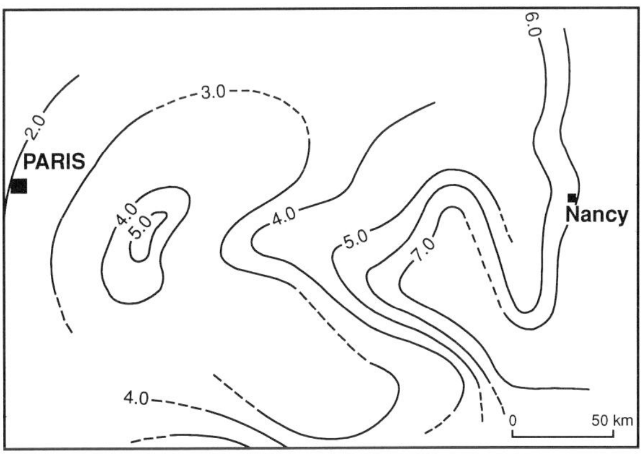

Fig. 21. Reconstructed average initial organic carbon content (wt. %) for the Lower Toarcian. (After Espitalié et al. 1987)

be the result of an organic facies variation. It is possible that this reflects a slight increase in the terrestrial (vitrinitic) signal.

Hydrocarbon Characteristics

The oils derived from the lower Toarcian shales and that derived from the underlying Hettangian/Sinemurian are quite similar (Poulet and Espitalié 1987). The major difference appears to be in the apparent level of maturity of the oils, as suggested by the various maturity sensitive indicators including bulk chemical composition, and the composition of both the saturated (i.e., n-alkane/isoprenoid ratios and biomarker indices) and aromatic hydrocarbon fractions.

The oils display API gravities ranging from 20° to 36°. Based on their bulk composition (Fig. 22A), these oils would be classified as paraffinic-naphthenic oils. Such oils are typical of a marine origin. The hydrocarbons generated by the Schistes Carton are of approximately equal proportions of naphthenes, aromatics, and normal and branched alkanes (Fig. 22B).

Gas chromatography of the saturated hydrocarbon fraction (Fig. 23) reveals a harmonic decrease in n-alkane abundance with increasing carbon number. An examination of the isoprenoids reveals a pristane/phytane ratio of ~ 1.4. The pristane/nC_{17} and phytane/nC_{18} ratios range from approximately 0.4 to 0.9 and appear to be a reflection of thermal maturity.

Stable isotopic composition reveals generally little variability. Whole-oil carbon isotopic com-

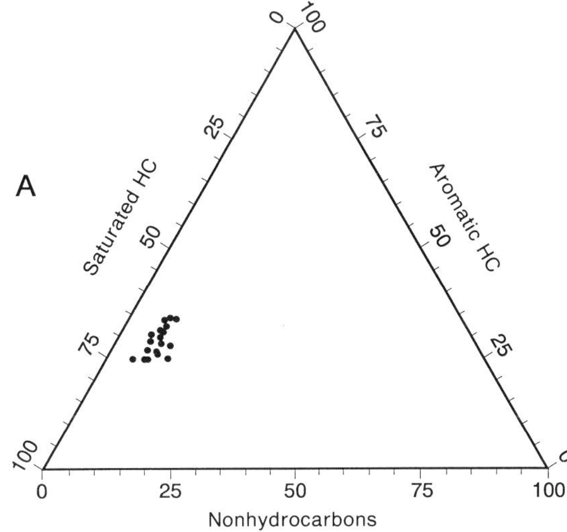

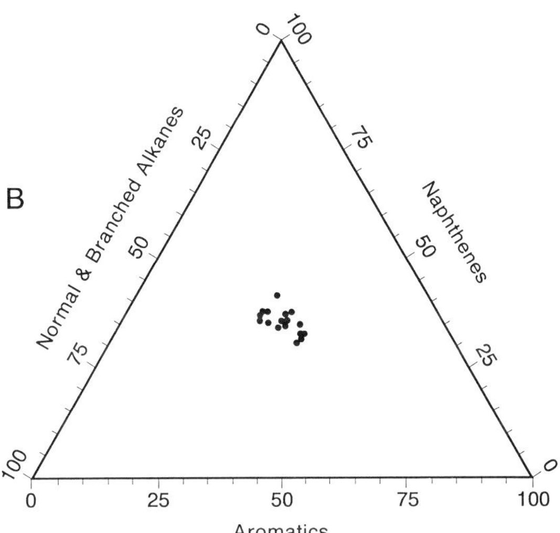

Fig. 22A, B. Ternary plot of bulk (**A**) and hydrocarbon (**B**) composition of oils believed to be derived from the Lower Toarcian section

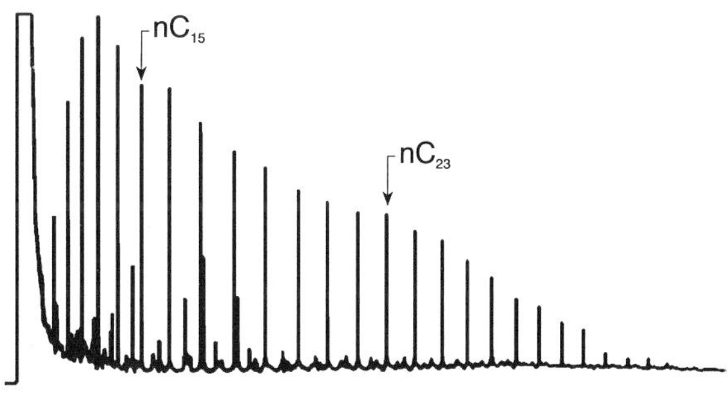

Fig. 23. Representative $nC_{15}+$ saturated fraction gas chromatogram of a Schistes Carton-derived oil. (After Espitalié et al. 1987)

position ranges from −31 to −30‰ (relative to PDB), with deuterium compositions ranging from −132 to −116‰ (relative to SMOW; Espitalié et al. 1987).

Exploration Strategy

Although the lower Toarcian shales of the Paris basin display excellent oil source rock potential, their effectiveness is limited by their generally low level of thermal maturity. Mature, generative portions of the shales are limited to the basin deep. However, Poulet and Espitalié (1987) note that there is clear evidence for lateral and vertical migration within the basin. Lateral migration is limited by both the petrophysical character of the section (i.e., lateral continuity of porous/permeable sandstones) and the regional hydrodynamic regime. Espitalié et al. (1988) estimate vertical movement within the Bray fault zone of about 2.5 km and lateral movement from the generative prism of between 10 and 15 km.

Summary

The Schistes Carton is an example of a marine source rock. It was deposited within an epicontinental sea associated with the Tethyan margin. Deposition occurred during a period of ~500 000 years (*falciferum* zone; early Toarcian). The levels of measured organic carbon and hydrogen enrichment, as well as generation potential, are typical of many other marine source rocks. The organic matter was previously used to define type II kerogen. The products generated are generally paraffinic-naphthenic crude oils.

The most significant variation in organic facies appears to be associated with organic enrichment. Stratigraphically, the most organically enriched material appears to be associated with the more basal part of the sequence. Areally, the most organically enriched material appears to be generally associated with the isopach thicks. This association suggests that hydrodynamic factors control the redistribution of organic matter and hence its final concentration. There is also evidence suggesting that there are minor differences in organic input. A limited number of samples display a slightly more aromatic product upon pyrolysis than the remainder of the available samples.

Exploration potential for Schistes Carton-derived oils is limited largely by the low level of thermal maturity of the Lower Toarcian shale. Most of the sequence is thermally immature. Generation and expulsion has occurred only within the basin center. Hydrocarbon generation was terminated during the Miocene as a consequence of uplift and erosion.

Acknowledgments. The author thanks Texaco Inc. for permission to publish this work. An earlier draft of this manuscript was reviewed by Dr. V.D. Robison. Graphics were prepared by Pauletta Kay.

References

Alpern B, Cheymol D (1978) Réflectance et fluorescence des organoclastes du Toarcien du Bassin de Paris en fonction de la profondeur et de la température. Rev Inst Français Pétrole 33: 515–535

Besairie H, Collignon M (1972) Géologie de Madagascar. I. Les terrains sédimentaires. Madagascar Ann Géol 35, 463 pp

Bessereau G, Guillocheau F (1993) Stratigraphie séquentielle et distribution de la matière organique dans Lias du Bassin de Paris. C R Acad Sci Sér II, 316: 1271–1278

Bessereau G, Huc AY, Carpentier B (1992) Distribution of organic matter in the Liassic series of the Paris basin: an example of organic heterogeneity in a source rock interval. In: Spencer AM (ed) Generation, accumulation and production of Europe's hydrocarbons II. Springer, Berlin Heidelberg New York, pp 117–125

Bissada KK (1982) Geochemical constraints on petroleum generation and migration – a review. Proc 2nd ASCOPE Conf, Manila, Oct, 1981, pp 69–87

Brassell SC, Wardroper AMK, Thomson ID, Maxwell JR, Eglinton G (1981) Specific acyclic isoprenoids as biological markers of methanogenic bacteria in marine sediments. Nature 290: 693–696

Crostella A, Barter T (1980) Triassic-Jurassic depositional history of the Dampier and Beagle sub-basins, northwest shelf of Australia. Aust Petrol Explor Assoc J 20: 25–33

Durand B, Espitalié J, Nicaise G, Combaz A (1972) Etude de la matière organique insoluble (kérogène) des argiles du Toarcien du Bassin de Paris. Rev Inst Français du Pétrole 27: 865–884

Espitalié J, Madec M (1981) Les schistes bitumineux du Toarcien de la bordure orientale du Bassin de Paris. Bull Cent Rech Explor Prod Elf-Aquitaine 5: 461–472

Espitalié J, Laporte JL, Madec J, Marquis F, Leplat P, Paulet J, Boutefeau, A (1977) Methode rapide de caracterisation des roches mères de leur potential pétrolier et de leur degré d'évolution. Rev Inst Français du Pétrole 33: 23–42

Espitalié J, Marquis F, Sage L, Barsony I (1987) Géochimie organique du Bassin de Paris. Rev Inst Français du Pétrole 42: 271–302

Espitalié J, Maxwell JR, Chenat Y, Marquis F (1988) Aspects of hydrocarbon migration in the Mesozoic in the Paris basin as deduced from an organic geochemical survey. Org Geochem 13: 467–481

Farrimond P, Eglinton G, Brassell SC, Jenkyns HC (1989) Toarcian anoxic event in Europe: an organic geochemical study. Mar Petrol Geol 6: 136–147

Fleet AJ, Clayton CJ, Jenkyns HC, Parkinson DN (1987) Liassic source rock deposition in western Europe. In: Brooks J, Glennie K (eds) Petroleum geology of North West Europe. Graham & Trotman, London, pp 59–70

Frebold H (1957) The Jurassic Fernie Group in the Canadian Rocky Mountains and foothills. Can Geol Surv Mem 287, 197 pp

Goy G, Noël D, Busson G (1978) Calcareous nannofossils in Lower Toarcian paper-shales from Paris basin (France): a contribution to their environmental knowledge. 10th Int Congr Sedimentol 1: 267–268 (Abst)

Hallam A (1981) A revised sea-level curve for the Early Jurassic. Geol Soc Lond J 138: 735–743

Hollander DJ, Bessereau G, Belin S, Huc AY, Houzay JP (1991) Organic matter in the early Toarcian Paris basin, France: a response to environmental changes. Rev Inst Français du Pétrole 46: 543–562

Huc AY (1976) Mise en evidence de provinces géochimiques dans les schistes bitumineux du Toarcien de l'est du Bassin de Paris. Rev Inst Français du Pétrole 31: 933–953

Huc AY (1977) Contribution de la géochimie organique a une esquisse paléoécologique des schistes bitumineux du Toarcien de l'est du bassin de Paris – étude de la matière organique insoluble (kérogènes). Rev Inst Français du Pétrole 32: 703–718

Huc AY (1990) Understanding organic facies: a key to improved quantitative petroleum evaluation of sedimentary basins. In: Huc AY (ed) Deposition of organic facies. Am Assoc Petrol Geol, Tulsa, Stud Geol 30: 1–11

Imlay RW (1952) Correlation of the Jurassic formations of North America, exclusive of Canada. Geol Soc Am Bull 63: 953–992

Jenkyns HC (1988) The Early Toarcian (Jurassic) anoxic event: stratigraphic, sedimentary, and geochemical evidence. Am J Sci 288: 101–151

Landais P, Monin J-C, Monthioux M, Poty B, Zaugg P (1989) Détermination expérimentale de l'évolution d'indicateurs de maturité de la matière organique. Application aux kérogènes du Toarcien du bassin de Paris. C R Acad Sci Sér II, 308: 1161–1166

Mackenzie AS, Patience RL, Maxwell JR, Vandenbroucke M, Durand B (1980) Molecular parameters of maturation in the Toarcian shales, Paris basin, France – 1. Changes in the configurations of acyclic isoprenoid alkanes, steranes and triterpanes. Geochim Cosmochim Acta 44: 1709–1721

Mouterde R, Tintant H, Allouc J, Gabilly J, Hanzo M, Lefavrais A, Rioult M, Delance J, Donze P, Ruget C (1980) Lias. In: Mégnien C, Mégnien F (eds) Synthèse géologique du Bassin de Paris – stratigraphie et paléogéographie. BRGM, Orléans, Mem 101: 75–123

Poulet M, Espitalié J (1987) Hydrocarbon migration in the Paris basin. In: Doligez B (ed) Migration of hydrocarbons in sedimentary basins. Éditions Technip, Paris, pp 131–171

Prauss M, Riegel W (1989) Evidence from phytoplankton associations for causes of black shale formation in epicontinental seas. Neues Jahrb Geol Paläeontol Monatsh 11: 671–682

Riccardi AC (1983) The Jurassic of Argentina and Chile. In: Moullade M, Nairn AEM (eds) The Phanerozoic geology of the world. II. The Mesozoic. Elsevier, Amsterdam, pp 201–263

Tanabe K (1983) Mode of life of an inoceramid bivalve from the Lower Jurassic of Japan. Neues Jahrb Geol Paläeontol Monatsh 7: 419–428

Tissot B, Califet-Debyser Y, Deroo G, Oudin JL (1971) Origin and evolution of hydrocarbons in early Toarcian shales, Paris basin. Am Assoc Petrol Geol Bull 55: 2177–2193

Tissot B, Durand B, Espitalié J, Combaz A (1974) Influence of nature and diagenesis of organic matter in formation of petroleum. Am Assoc Petrol Geol Bull 58: 499–506

Tissot BP, Welte DH (1984) Petroleum formation and occurrence, 2nd edn. Springer, Berlin Heidelberg New York, 699 pp

van Graas G, De Leeuw JW, Schenck PA (1981) Kerogen of Toarcian shales of the Paris basin. A study of its maturation by flash pyrolysis techniques. Geochim Cosmochim Acta 45: 2465–2474

Wolff GA, Lamb NA, Maxwell JR (1986) The origin and fate of 4-methyl steroid hydrocarbons I. Diagenesis of 4-methyl steranes. Geochim Cosmochim Acta 50: 335–342

Geochemistry of the Upper Jurassic Tuwaiq Mountain and Hanifa Formation Petroleum Source Rocks of Eastern Saudi Arabia

W.J. Carrigan, G.A. Cole, E.L. Colling, and P.J. Jones

Abstract

Thick, regionally extensive, laminated organic-rich lime mudstone units are present within the late Jurassic Hanifa and Tuwaiq Mountain Formations. These potential source rocks were deposited during the late Callovian to early Kimmeridgian within a relatively short-lived intra-shelf basin, the Arabian basin. This basin formed on the northeastern continental shelf of the Afro-Arabian plate, in what is now eastern Saudi Arabia, as a result of relative sea level rise in combination with differential subsidence and/or local structuring within the shelf. The Arabian basin was at least partially separated from the open neo-Tethys ocean by flanking paleo-highs composed of grainstone shoal/barrier island facies. The source rocks, defined as units having TOC > 1%, have an average TOC content of about 3%, with contents as high as 13%. Total pyrolytic yield ($S_1 + S_2$ from Rock-Eval) is as high as 88 mg HC/g rock, with an average yield of 25 mg HC/g rock, indicating excellent source rock potential. Hydrogen indices of thermally immature rocks are between 600 and 800 mg HC/g TOC, which indicates an oil-prone kerogen. The organic material is dominated by lamalginite, with subordinate amounts of vitrinite and inertinite, most of which is fluorescent. Kerogen from immature rocks isolated for elemental analysis plot as type II on a van Krevelen diagram. These results show that organic-rich units within the Hanifa and Tuwaiq Mountain Formations contain type II kerogen having excellent, oil-prone source rock potential.

Major oil accumulations occur in several late Jurassic carbonate reservoirs in Saudi Arabia. Representative oils from these reservoirs show very similar chromatographic and biomarker fingerprints to bitumen extracted from the Hanifa and Tuwaiq Mountain Formation source rocks, suggesting that the oils were derived from these source rocks. The similar stable carbon isotope ratios between the oils (avg. $\delta^{13}C = -26.6‰$) and the kerogen (avg. $\delta^{13}C = -26.4‰$) and bitumen (avg. $\delta^{13}C = -27.1‰$) is also consistent with an origin from the Hanifa and Tuwaiq Mountain Formations.

The thermal maturation history of the Hanifa and Tuwaiq Mountain source rocks was calculated using kinetic models. Results indicate that oil generation and expulsion began about 75 Ma B.P. in the eastern part of the basin. By 50 Ma B.P. the oil kitchens had expanded westward and oil had started filling the broad, gentle structures that had formed during the late Cretaceous. Today, the Hanifa and Tuwaiq Mountain source rocks east of the Ghawar structure have passed through the oil generation window. The source rocks in the basin center are still within the oil generation window. In the western part of the basin, the source rocks are either immature or just starting to enter the oil window.

Saudi Arabian Oil Company, Lab R&D Center, P.O. Box 62, Dhahran, 31311, Saudi Arabia

Introduction

The Arabian basin is without question one of the most prolific oil-producing regions of the world (see Fig. 1 for location of major oil fields in Saudi Arabia). Approximately 26% of the world's known oil reserves are found in Saudi Arabia (Oil & Gas Journal 1992), of which about two-thirds are found in Jurassic carbonate reservoirs. Optimal conditions existed for source rock, reservoir rock and cap rock development during the Jurassic. Several previous publications have described the regional tectonic and depositional settings of these important rock facies (Steineke et al. 1958; Powers et al. 1966; Murris 1980; Ayres et al. 1982; Beydoun 1991) and have attempted to correlate the various oil reservoirs with their source rocks (Ayres et al. 1982).

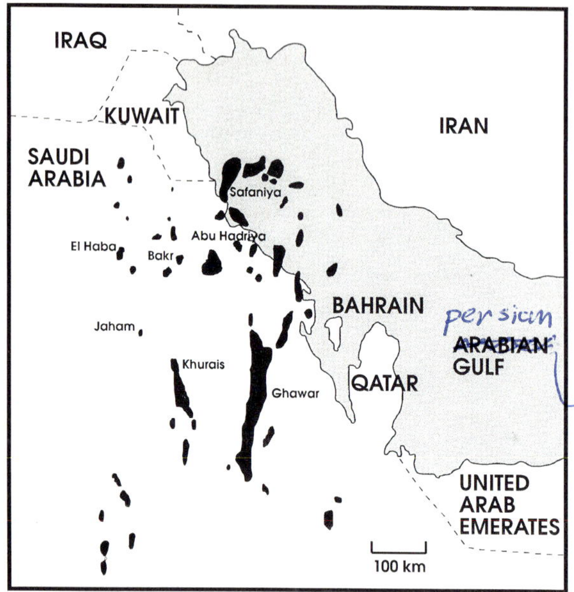

Fig. 1. Location map showing major oil fields of Eastern Saudi Arabia (After Beydoun 1991)

Potential source rocks are distributed throughout much of the Phanerozoic, and possibly the latest Precambrian, in the Arabian Plate region (Ala et al. 1980; Ayres et al. 1982; Alsharhan and Kendall 1986; Stoneley 1990; McGillivray and Husseini 1992; Mahmoud et al. 1992). Ayres et al. (1982) considered the Callovian to Oxfordian interval of the Tuwaiq Mountain and Hanifa Formations to be the principal source for the oil contained within the southern part of the main producing area of eastern Saudi Arabia. They also identified potential source rocks in the Jurassic Marrat and Dhruma Formations, and the early Cretaceous Sulaiy Formation. In general, these formations have much lower source potential than the Tuwaiq Mountain and Hanifa source rocks (Ayres et al. 1982). Other potential source rocks adjacent to Saudi Arabia, which may have contributed migrated oil, include the middle Jurassic Sargelu Formation and lower to middle Cretaceous Garau Formation of Kuwait and Iraq and the middle Cretaceous Kazhdumi Formation of Iran (Ala et al. 1980).

The primary objective of this chapter is to describe the geochemical characteristics of the Tuwaiq Mountain and Hanifa petroleum source rocks of eastern Saudi Arabia and to determine the timing of oil generation and expulsion from these rocks. Selected core and cuttings samples were analyzed for total organic carbon (TOC), Rock-Eval pyrolysis, bitumen content and bulk bitumen composition, carbon isotope ratios of kerogen and bitumen, H/C and O/C ratios of kerogen, and C_{15+} gas chromatography (GC) and gas chromatography/mass spectrometry (GCMS) of the bitumen. The carbon isotope, GC, and GCMS results are compared to results for typical Jurassic-reservoired oils from the Ghawar-Khurais area of eastern Saudi Arabia. Timing of oil generation and expulsion from the source rocks was determined by constructing burial history and thermal models, using the BasinMod (version 2.9) software package from Platte River Associates.

Geologic Setting

Throughout the Mesozoic, the northeast passive margin of the Afro-Arabian Plate evolved as an extensive continental shelf, several thousands of kilometers in length and about 2000 kilometers wide (Murris 1980; Beydoun 1991). It was bounded on the northeast by the open neo-Tethys ocean and on the west and south by basement outcrops of the Arabian Shield. During the Jurassic and much of the Cretaceous, deposition in the Arabian Gulf region was dominated by platform carbonates. Differential subsidence and/or local structuring within the shelf, led to the formation of relatively short-lived intra-shelf basins. These intra-shelf basins were characterized by flanking paleo-highs composed of regressive grainstone shoal/barrier-island facies grading into carbonaceous, organic-rich lime mudstone within the interiors of these basins (Langdon and Malecek 1987; McGuire et al. 1993).

Three such basins formed in the Arabian Gulf region. A deep water basin, the Gotnia basin (Fig. 2), existed in the northern gulf region from the middle Jurassic to the early Cretaceous (Murris 1980; Ayres et al. 1982). In eastern Qatar and the western part of the United Arab Emirates, a smaller, shallower basin, the Southern Gulf basin, existed from late Oxfordian to early Kimmeridgian time (Murris 1980; Alsharhan and Kendall 1986; Droste 1990, 1993). In eastern Saudi Arabia, an intra-shelf basin, the Arabian basin (Fig. 2), existed from late Callovian to early Kimmeridgian time (Ayres et al. 1982). Water depths within the Arabian basin were probably on the order of a few tens of meters (Ayres et al. 1982). At maximum water depth, however, storm wave base was almost certainly exceeded since the full spectrum of carbonate ramp deposition is observed from shallow water

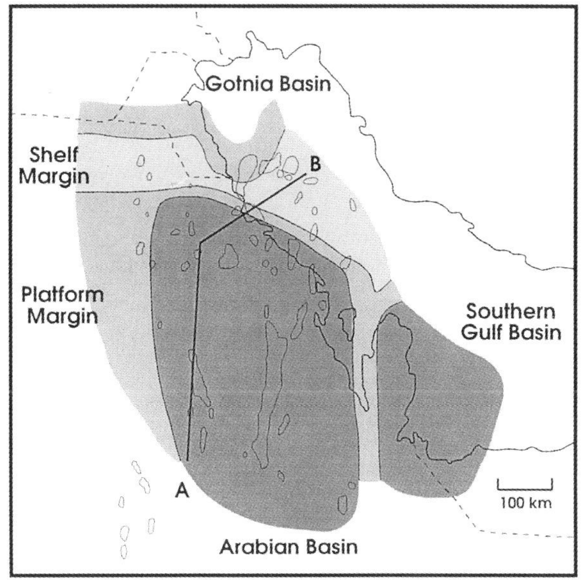

Fig. 2. Paleogeographic map of Eastern Saudi Arabia for late Callovian to late Oxfordian time. (After Murris 1980; Ayres et al. 1982; Alsharhan and Kendall 1986). Line A-B is cross-section shown in Fig. 3

shoaling upward grainstone belts, to deeper water storm deposition, grading ultimately into organic-rich limestone mudstones of very low-energy environments (McGuire et al. 1993).

Throughout the Jurassic, the Arabian Plate was situated in the tropical or subtropical zone (Beydoun 1991). The shallow water setting and the low latitude position of the Arabian basin implies that organic productivity on the shelf would have been high (Pelet 1983; Suess 1980). The silled margin separating the Arabian basin from the open Tethys ocean (Ayres et al. 1980) may have resulted in restricted vertical circulation of seawater. Restricted circulation and/or high organic productivity can create oxygen-deficient bottom waters, which will enhance the preservation of organic material. This is consistent with the preservation of a thick sequence of laminated, organically enriched (predominantly algal) sediments.

The location and paleogeographic setting of the Arabian basin during late Callovian to early Oxfordian is shown in Fig. 2 and a schematic cross-section is shown in Fig. 3. The stratigraphy of the

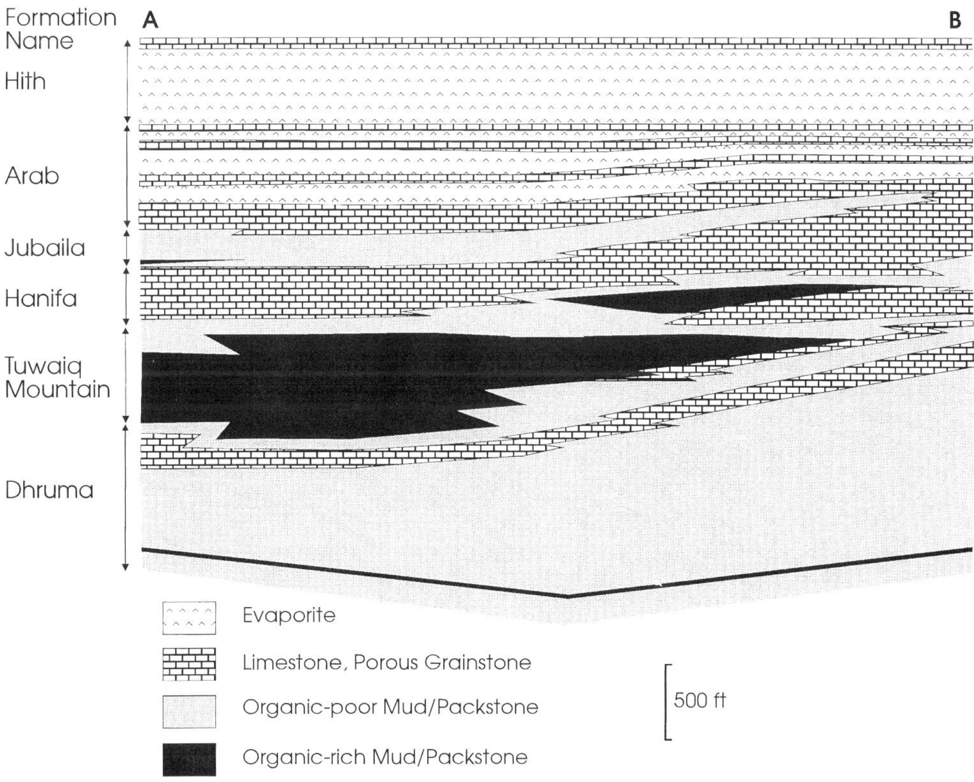

Fig. 3. Schematic cross-section of the Jurassic stratigraphic sedimentary package across the Arabian basin (*line A-B* in Fig. 2), showing the relationship between source, reservoir, and seal facies

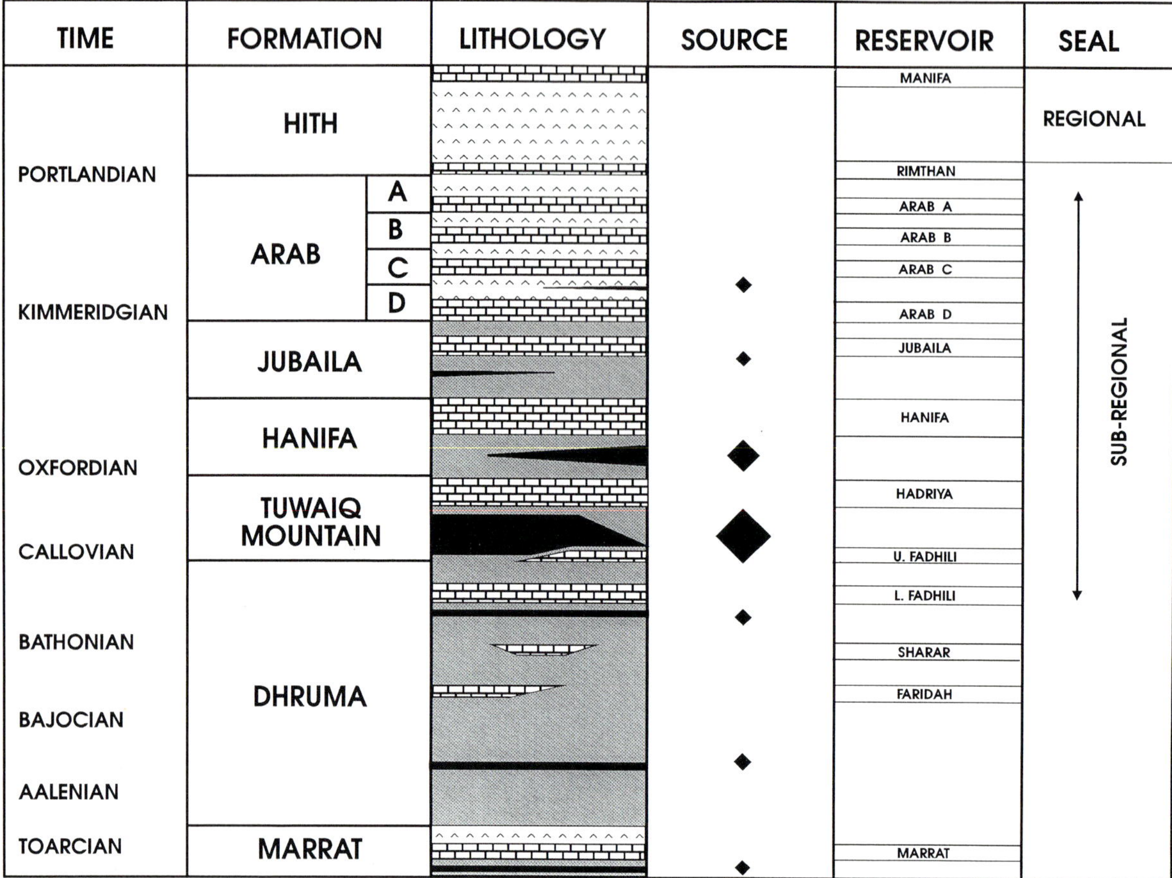

Fig. 4. Generalized stratigraphic column for the Jurassic sediments in Saudi Arabia showing ages, formation, lithology, source rocks, reservoirs, and seals

Jurassic sediments in Saudi Arabia has been described by Steineke et al. (1958), Powers et al. (1966), and Murris (1980). The approximate stratigraphic locations of the major source rocks and hydrocarbon reservoirs are shown on the Jurassic stratigraphic section (Fig. 4). It is important to note that many of the formations and reservoirs share the same name, but do not coincide exactly in the stratigraphic column. To avoid any possible confusion, the full name of the unit will be used where a potential conflict exists (e.g., Hanifa Formation versus Hanifa reservoir).

The lower to middle Jurassic Marrat and Dhruma Formations are composed essentially of broad sheets of pelletoidal-ooidal pack-grainstone alternating with argillaceous pelletoidal-bioclastic mud-wackestone. This "layer-cake" stratigraphy is typical of deposition on a carbonate ramp (Murris 1980). A major rise in relative sea level starting in the late Callovian, combined with differential subsidence, led to drowning of the platform interior and the formation of the Arabian basin (Murris 1980; Ayres et al. 1982). It was separated from the deeper, more extensive Gotnia basin to the north (Fig. 2), and from the shorter-lived Southern Gulf basin to the east, by shallow grainstone shoals and dolomitized facies. The resulting decrease in circulation of bottom waters in the Arabian basin created anoxic conditions, which allowed the preservation of organic-rich material over large parts of the basin.

The uniform sedimentary layering typical of the Dhruma Formation changed by the late Callovian to a clear separation of shallow water shelf facies to the north and deeper, basinal facies in the interior of the Arabian basin (Figs. 2, 3). Cyclical changes in sea level led to the development of a stacked sequence of shallowing upward carbonate platform deposits on the Northern Shelf (Upper Fadhili, Hadriya, and Hanifa reservoirs) (Figs. 3, 4). The high energy, shelf facies grainstones and packstones, which compose the reservoirs, thin to the south, and are replaced in the basin interior by alternating laminated, organic-rich lime mud-wackestones. These organic-rich basinal mudstones are considered to be contemporaneous with both the transgressive units deposited during deepening of the basin and the shoaling-upward reservoir units and, thus, are part of the transgressive and highstand systems tracts (McGuire et al. 1993). Moreover, the high organic productivity, as evidenced by the building of carbonate shoals, not only led to increased organics in the system, but also was the ultimate cause of restricted water flow, thereby aiding organic preservation.

In the southwestern part of the basin, the main source rock facies is found in the Tuwaiq Mountain Formation and is laterally equivalent to the Upper Fadhili and Hadriya reservoirs. The thickness of the source rock facies exceeds 500 feet in the center of the basin and thins laterally (Fig. 3). In the northeastern area, the source rock facies overlies the Hadriya reservoir.

Deposition in the uppermost portion of the Hanifa Formation consists of a thick shelf margin wedge of the lowstand systems tract, capped by a subaqueous anhydrite of limited extent, which was deposited during maximum restriction of the basin during the Hanifa low stand. The shelf margin wedge sediments are generally organic-poor, bioturbated lime wacke-packstones in the basinal facies and grade laterally into the basinward-most grainstone beds and boundstones (McGuire et al. 1993). The shelf margin wedge sediments accomplish the major portion of filling of the Arabian basin. Lime mudstones of the Jubaila Formation, which overlie the Hanifa reservoir grainstones, are generally organic-poor but are locally organic-rich.

Extensive shallow shelf carbonate facies returned during the Kimmeridgian with the deposition of the Arab Formation. The Arab Formation contains four shallowing upward cycles of grainstone-evaporite and, although the Arabian basin had been essentially filled, subtle local remnants of the basinal facies play an important role in the development of reservoir facies. These grainstones make up the most important reservoirs in Saudi Arabia — the Arab A, B, C, and D reservoirs (Fig. 4). The capping evaporite cycles provide the seal for these reservoirs. A thick sequence of evaporites, the Hith Formation (Fig. 4), concludes the Jurassic and forms the principal regional seal to the Jurassic-sourced oil.

Source Rock Characteristics

TOC and Rock-Eval Pyrolysis

Total organic carbon (TOC) and Rock-Eval pyrolysis were performed on more than 600 samples of core and drill cuttings. With increasing maturity, hydrocarbons are generated and expelled from the source rock, which leads to a reduction in the hydrocarbon (HC) potential of the remaining organic material in the source rock. Therefore, to accurately determine the original source potential, it is best to analyze samples of relatively immature rocks. For this reason, samples selected for analysis were mostly from thermally immature to early mature rocks. Although the immature sediments could not produce the reservoired oils, comparison with mature samples indicate that both probably had the same original source potential.

Organically enriched (TOC $> 1\%$) intervals occur throughout the Jurassic section, but the main source facies is located between the top of the Upper Fadhili reservoir and the base of the Hanifa reservoir (Figs. 3, 4) in the Tuwaiq Mountain Formation and lower Hanifa Formation. A second major source interval occurs in the upper part of the Hanifa Formation. The most organic-rich interval is located in the central part of the basin, where it may exceed 500 feet in thickness (Fig. 5). TOC contents tend to be cyclical, probably as a result of fluctuations in relative sea level (Droste 1990, 1993). This cyclicity is readily apparent on geophysical logs (e.g., Ayres et al. 1982; Droste 1990). In general, the base of each source rock cycle is the most organic-rich.

The source rock facies averages about 3% TOC, but values as high as 13% have been measured. Figure 6 shows a histogram of TOC data from the source rock facies ($>1\%$ TOC); values from nonsource rocks were excluded. In general, average TOC values from the Tuwaiq Mountain samples are slightly higher than samples from the Hanifa Formation. However, this is partly due to the fact that more of the thermally mature samples were from the Hanifa Formation and so represent residual values. The major source rock interval in the southwestern part of the basin is contained within the Tuwaiq Mountain Formation, whereas in the northeast it is contained mostly in the Hanifa Formation above the Hadriya reservoir (Figs. 3, 4). Thermal maturity increases towards the northeast and southeast (see Burial

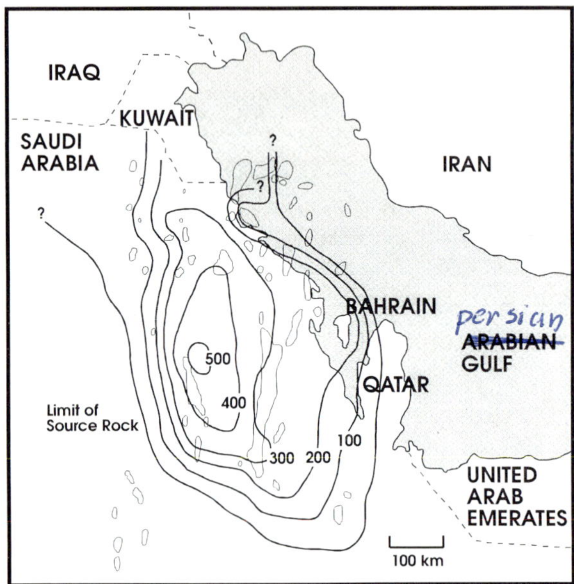

Fig. 5. Source rock isopach map showing thickness of source rock having greater than 1% TOC (contours in feet) (After Ayres et al. 1982)

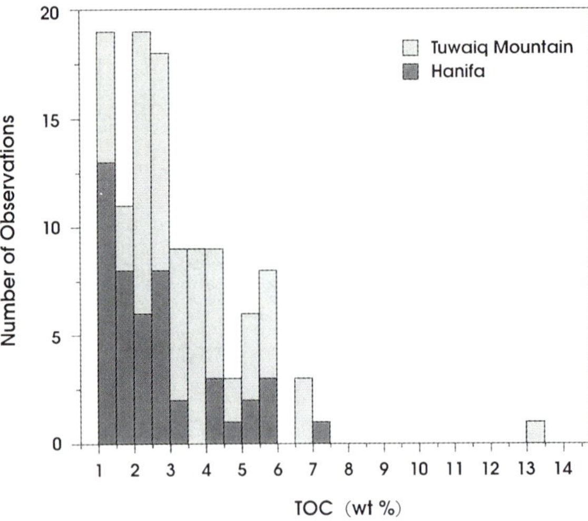

Fig. 6. Histogram of total organic carbon (% TOC) for source rock intervals within the Tuwaiq Mountain and Hanifa Formations. Only results > 1% TOC are shown; nonsource rocks are excluded

History/Thermal Modeling section), thus TOC would be expected to show a general decrease in this direction as hydrocarbons are generated and expelled.

The total Rock-Eval pyrolytic yield ($S_1 + S_2$) of all Hanifa and Tuwaiq Mountain samples ranges from below detection limits to 88.3 mg HC/g of rock. For organically enriched intervals (TOC > 1%), the average total pyrolytic yield is about 25 mg HC/g of rock (Fig. 7), indicating good to excellent source potential (Peters 1986). Pyrolysis results reveal a difference between the Hanifa and Tuwaiq Mountain source intervals. Most of the Tuwaiq Mountain samples, and some of the Hanifa samples, have hydrogen indices (HI) between 600 and 800 mg HC/g TOC, with S_2 values up to 79 mg HC/g of rock, indicating oil-prone source rock potential (Peters 1986). On an S_2 vs. TOC diagram (Langford and Blanc-Valleron 1990) these values plot along a line that has a slope derived HI of 640 mg HC/g TOC, which indicates an oil-prone type II kerogen (Fig. 8). Most of the Hanifa samples have HI values between 170 to 300 mg HC/g TOC with S_2 values up to 10 mg HC/g of rock, indicating a more gas-prone source potential (Peters 1986). The Hanifa samples generally plot along a line that gives a slope-derived HI of less than 300 mg HC/g TOC, which implies a type III kerogen (Fig. 8) (Langford and Blanc-Valleron 1990). These differences, like those of the TOC, probably reflect differences in thermal maturity. Samples which have HI values greater than 600 mg HC/g TOC have T_{max} values less than about 425 °C (< 0.5% vitrinite reflectance equivalence, VRe) (Fig. 9), whereas samples with HI values less than 300 have T_{max} values up to and exceeding 440 °C (0.75 to 0.8% VRe). This indicates that the source rock is essentially spent by about 0.85% VRe (Cole et al. 1994). Oxygen index data were not used to classify the kerogen type because of the carbonate matrix effect (Katz 1983).

Kerogen Composition

Petrographic examination of the organic-rich sediments of the Tuwaiq Mountain and Hanifa Formations shows that the organic matter is dominated by lamalginite (Cook and Sherwood 1991) with subordinate amounts of vitrinite and inertinite. Of the samples examined, lamalginite comprises 90–95% of the kerogen assemblages in the Tuwaiq Mountain Formation and 75–95% of the kerogen assemblages in the Hanifa Formation. The lamalginite consists of compressed leiospheres which form distinct, but fine laminae in peloidal carbonate packstone parallel to bedding. Most of the organic material (> 90% of Tuwaiq Mountain and 50–95% of Hanifa) is fluorescent, indicating oil-prone character. The Tuwaiq Mountain kerogen

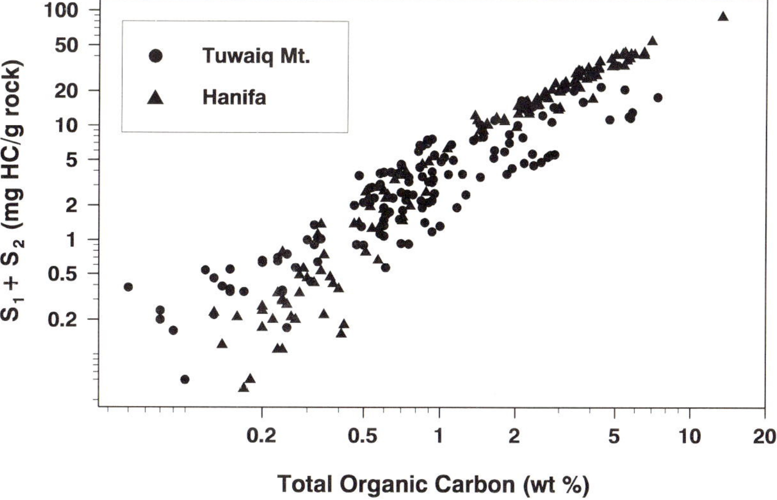

Fig. 7. Total hydrocarbon generation potential ($S_1 + S_2$) versus % TOC for all of the Tuwaiq Mountain and Hanifa Formation samples (includes both source and non-source rocks)

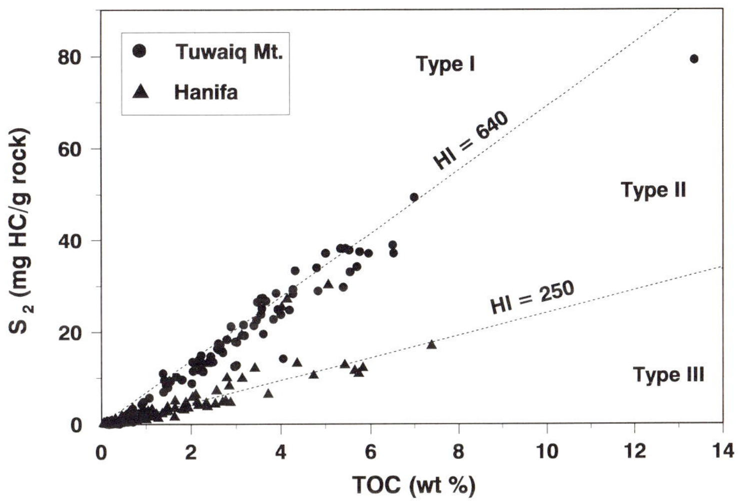

Fig. 8. S_2 pyrolytic yield versus % TOC for all of the Tuwaiq Mountain and Hanifa Formation samples. The group HI for the Tuwaiq Mountain and Hanifa data sets was determined from the slope of the regression line through the data (Langford and Blanc-Valleron 1990). The Tuwaiq Mountain samples are classed as type II, whereas the Hanifa samples are classified as type III

assemblages are very similar throughout the source rock. The Hanifa samples show more variation in kerogen constituents. Some Hanifa samples are dominated by lamalginite with subordinate bituminite, vitrinite, and inertinite, which is consistent with a marine origin. However, a few samples from the El Haba area (Fig. 1) consist of a more terrigenous type of input. These samples contain abundant cutinite and vitrinites that are mixed with lamalginites. This area is located near the margin of the basin and so this organic assemblage was more likely to have been deposited in a nearshore environment.

Kerogens isolated for elemental analysis have H/C ratios between 0.65 and 1.3 and O/C ratios between 0.03 and 0.12. Most of these values plot along the type II pathway on the van Krevelen diagram (Fig. 10). Most immature Tuwaiq Mountain and Hanifa samples have H/C ratios greater than 1.1. However, two exceptions are observed in samples from the El Haba area. These immature samples, which have H/C and O/C ratios of 0.9 and 0.7, respectively, plot closer to the type III pathway. Mature Hanifa samples from the Ghawar and Abu Hadriya area have H/C ratios between 0.65 and 0.8 and O/C ratios between 0.03 and 0.04. Kerogen classification is equivocal in this region of the van Krevelen diagram because H/C and O/C ratios decrease with increasing maturity. However, these mature samples appear to plot along the type II pathway. The carbon isotope composition of these kerogens have $\delta^{13}C$ values from -22.1 to -27.6‰ (Fig. 11). The average $\delta^{13}C$ value is -26.4‰ with a standard deviation of 1.1‰. The

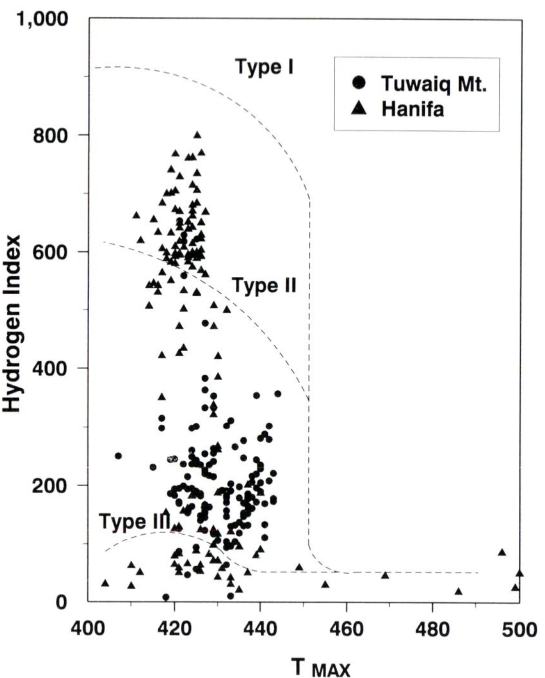

Fig. 9. Hydrocarbon generation potential (HI) versus maturity, estimated from T_{max}, for the Tuwaiq Mountain and Hanifa Formation source rock intervals. Tuwaiq Mountain results generally have higher HI values (up to 800) and lower T_{max} maturity ($< 425\,^{\circ}C$) than results for Hanifa Formation (HI < 350 with T_{max} up to and exceeding $440\,^{\circ}C$). Tuwaiq Mountain kerogen is classified as type II and Hanifa kerogen as type III. (Peters 1986)

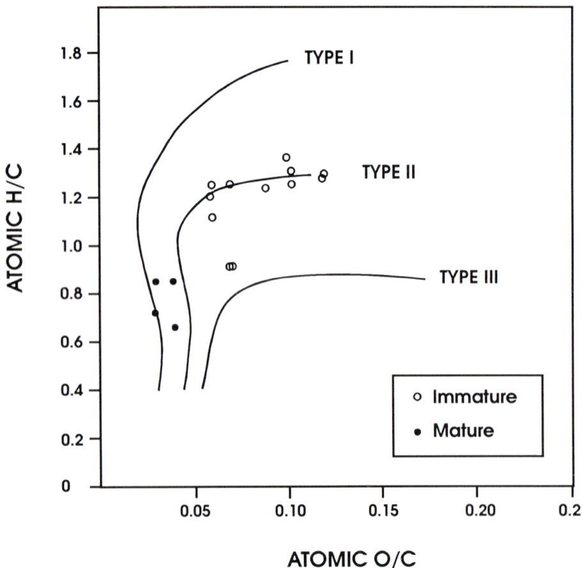

Fig. 10. Van Krevelen diagram showing H/C and O/C ratios of kerogens separated from Tuwaiq Mountain and Hanifa Formation source rocks. Most kerogens plot along the type II pathway, except for two immature Hanifa source rock samples from the El Haba field, which plot as type III kerogen

one anomalously heavy value of $-22.1‰$ comes from the El Haba area.

Bitumen Composition

Bitumen concentrations from thermally immature source rocks are mostly in the range of 4000 to 14 000 ppm. These bitumens consist of 4 to 17% saturates (av. 11%), 9 to 18% aromatics (av. 13%), and 69 to 84% resins plus asphaltenes (av. 76%). Carbon isotope ratios of the bitumen fall between -26.4 to $-27.8‰$ (Fig. 11). The average $\delta^{13}C$ value is $-27.1‰$ with a standard deviation of 0.4‰. The bitumens are, on average, 0.7‰ lighter than the kerogen.

C_{15+} gas chromatograms of the saturated hydrocarbon fractions from immature source rocks in the El Haba and Khurais areas (Fig. 12) show bimodal and sometimes trimodal distributions, which is typical for immature sediments. The bimodal distributions are generally due to two *n*-alkane maxima, whereas the trimodal distributions show an additional high abundance of steranes and triterpanes. Another common feature is the low pristane/phytane ratio (0.7 to 1.0). Pr/nC_{17} and Ph/nC_{18} ratios vary between 0.25 and 0.81 and between 0.31 and 1.05, respectively (Fig. 13). In general, these ratios decrease with increasing maturity. However, one immature sample from the El Haba area has low Pr/nC_{17} and Ph/nC_{18} ratios (0.25 and 0.25, respectively), which may indicate local variations in organic facies.

Results of gas chromatography/mass spectrometry for typical Tuwaiq Mountain and Hanifa source rock extracts are shown in Figs. 14 and 15. The m/z 191 (tricyclics and hopanes) mass fragmentograms (Fig. 14) show low to moderate tricyclic abundance and a prominent C_{24} tetracyclic peak. Ts/Tm ratios, which are both source- and maturity-dependent (Seifert and Moldowan 1978), are between 0.25 and 0.5 for immature sediments but are greater than 1 for mature sediments ($T_{max} > 440\,^{\circ}C$) (Cole et al. 1994). The full range of C_{31} to C_{35} homohopanes for some extracts (i.e., Tuwaiq Mtn. extract, Fig. 14) indicates reducing conditions and the occasional elevated C_{35} homohopane is commonly associated with marine carbonates and evaporites (Connan et al. 1986) and is considered to be an indicator of highly reducing marine conditions (Peters and Moldowan, 1991). The presence of hexahydrobenzohopanes (hexacyclic hopanoids), 2α-methyl-17α(H), 12β(H)-hopanes,

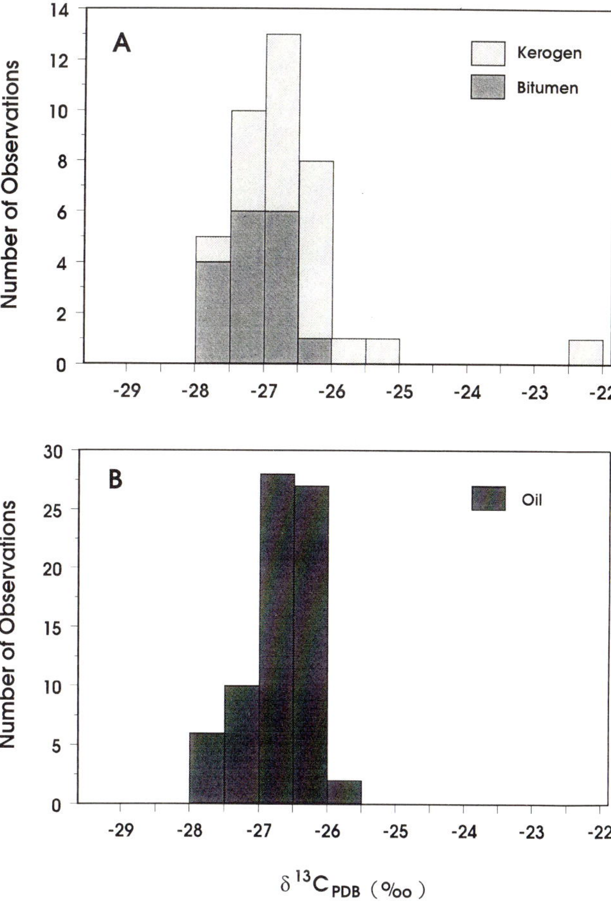

Fig. 11A,B. $\delta^{13}C$ histogram of **A** kerogen and bitumen from Tuwaiq Mountain and Hanifa source rocks, and **B** Jurassic-reservoired oils from eastern Saudi Arabia

and an extended norhopane series is also consistent with an anoxic carbonate or hypersaline depositional environment (Connan and Dessort 1987; Peters and Moldowan 1991; Summons and Walter 1990). In some cases the homohopane series is truncated at about C_{33} or C_{34} (i.e., Hanifa extract, Fig. 14), which may indicate fluctuating anoxic/dysoxic conditions during deposition.

The m/z 217 (steranes) mass fragmentograms (Fig. 15) display a wide range of diasterane abundances, which may reflect the varying clay content (usually less than about 2% but occasionally higher). Diasteranes have often been used to distinguish clay-poor carbonate environments (low diasteranes) from clay-rich clastic ones (high diasteranes) (i.e., Mello et al. 1988). However, Clark and Philp (1989) have observed abundant diasteranes in several clay-free carbonates and suggest that there may be other, as yet unknown, mechanisms for forming diasteranes. The $C_{29} > C_{27} \gg C_{28}$ sterane distribution (Fig. 16) and the presence of C_{30} steranes are consistent with a marine algal source for the organic material (Moldowan et al. 1985; Volkman 1986; Waples and Machihara 1991; Peters and Moldowan 1993).

Lithology/Mineralogy

The source rocks are thinly laminated (0.5 to 3 mm) dark gray to black peloidal carbonates composed of alternating laminae of peloidal packstone and peloidal grainstone. Peloids are composed of silt-sized, ovoid micritic lumps, which may be derived from reworking of lime muds or from micritization of shell material. Peloidal packstone laminae contain an organic-rich matrix, whereas peloidal grainstone laminae are grain-supported sediments cemented by finely crystalline equant sparry calcite. The major difference between organic-rich laminae and peloidal grainstone laminae is that the former have an organic-rich matrix instead of calcite cement. Most (> 70%) of the crystalline portion of the sediment is sparry calcite with minor amounts of dolomite. Accessory minerals such as anhydrite and quartz are also present. The clay content is

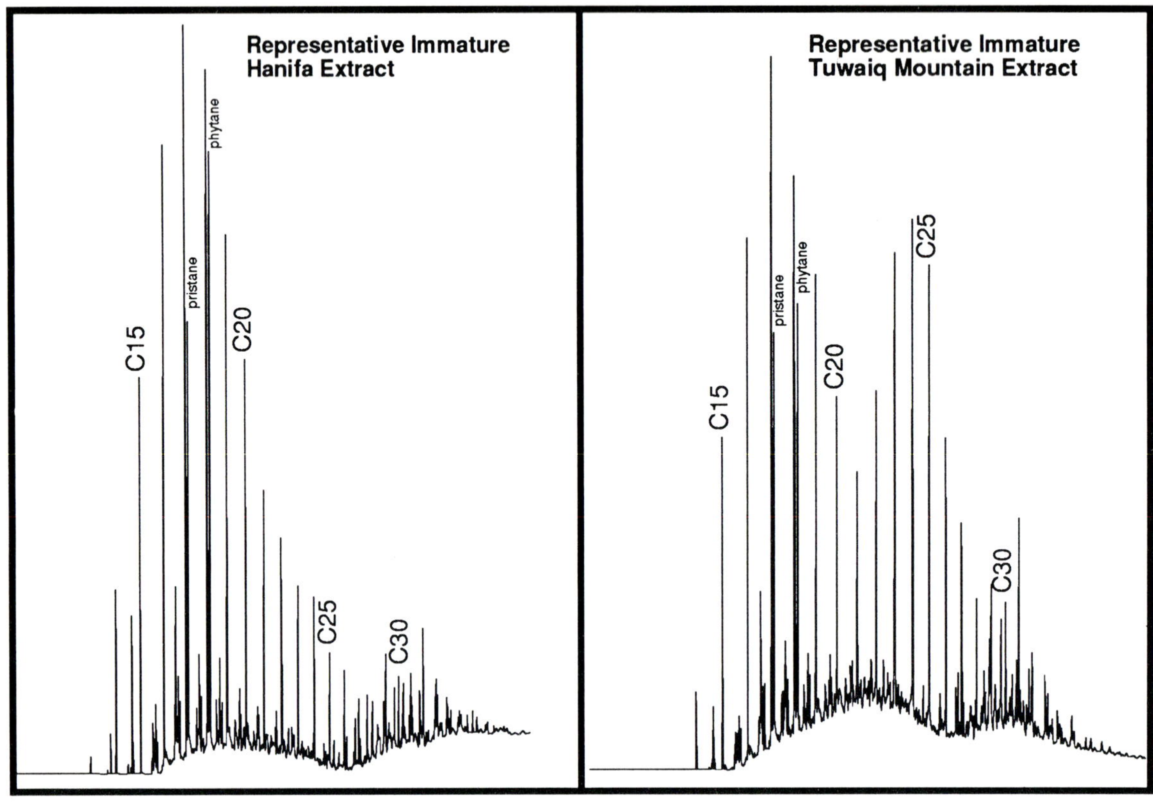

Fig. 12. C_{15+} gas chromatograms of representative immature Hanifa and Tuwaiq Mountain source rock extracts

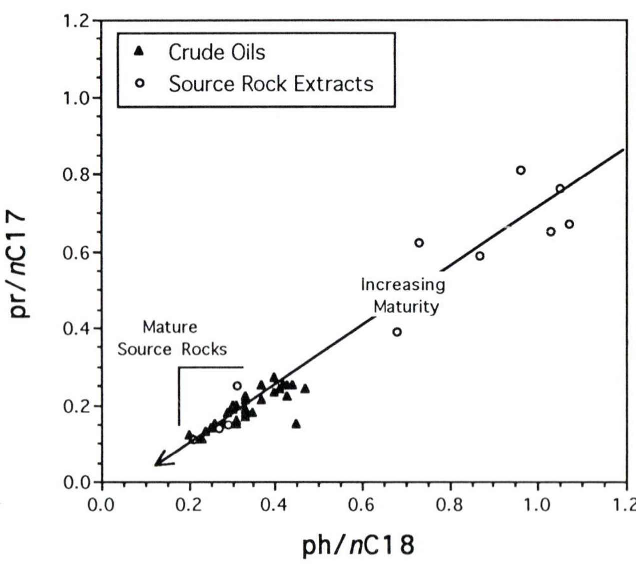

Fig. 13. Pristane/nC_{17} versus phytane/nC_{18} for Jurassic-reservoired oils and Hanifa and Tuwaiq Mountain source rocks. The oils and immature to mature source rock extracts show a common trend to lower ratios with increasing maturity, indicating a single group of oils derived from a single source rock type. An exception to this trend was observed for one immature source rock extract from the El Haba area, which had Pr/nC_{17} and Ph/nC_{18} ratios of 0.25 and 0.25, respectively

usually less than 2% and pyrite contents are typically 1 to 2%.

Fossils are present but are rarely abundant. Typical fossils include miliolid forams, sponge spicules, fish bones and scales, coccoliths, echinoid spines, ammonites, crustacea fragments, and various pelecypods. The well-preserved lamination and lack of burrowing suggest that bottom waters were anoxic or suboxic, although the presence of fossils indicates that at least occasionally the sea floor was

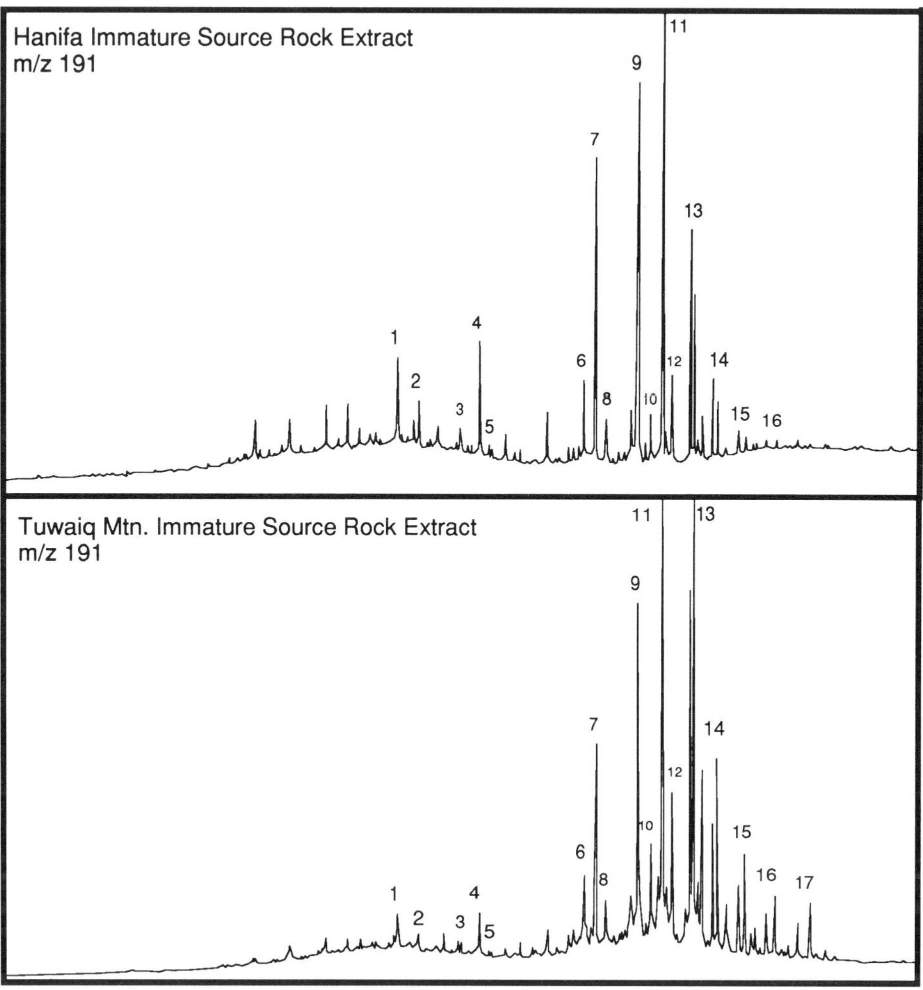

Fig. 14. Mass 191 biomarker fragmentograms for representative Hanifa and Tuwaiq Mountain source rock extracts (see Table 2 for identities of peaks)

not hostile to life. Carbonate mudstones of non-source rock facies are often well burrowed and contain a persistent benthonic fauna.

Oil-Oil and Oil-Source Correlation

As suggested by Ayres et al. (1982), the most likely source rock units for the Jurassic carbonate reservoired oil in the southern part of the area (i.e., Khurais-Ghawar area) are the organic-rich carbonates of the Jurassic Hanifa and Tuwaiq Mountain Formations. To determine the oil-source and oil-oil relationships, oils from Jurassic reservoirs were characterized using carbon isotopes ($\delta^{13}C$), gas chromatography (GC), and gas chromatography-mass spectrometry (GCMS). These results were then used to compare the oils with each other and to the prospective source units.

Significant oil accumulations are also present in Paleozoic and Cretaceous reservoirs, but are not included in this study. Oils in Paleozoic clastic reservoirs are distinctly different from Jurassic-reservoired oils and were probably generated from Silurian source rocks (Abu-Ali et al. 1991; McGillivary and Husseini 1992; Mahmoud et al. 1992). Carbonate and clastic Cretaceous reservoirs in the northeast offshore area (i.e., Safaniyah) may contain migrated oil generated from deeper water source rocks to the north or east (Ayres et al. 1982).

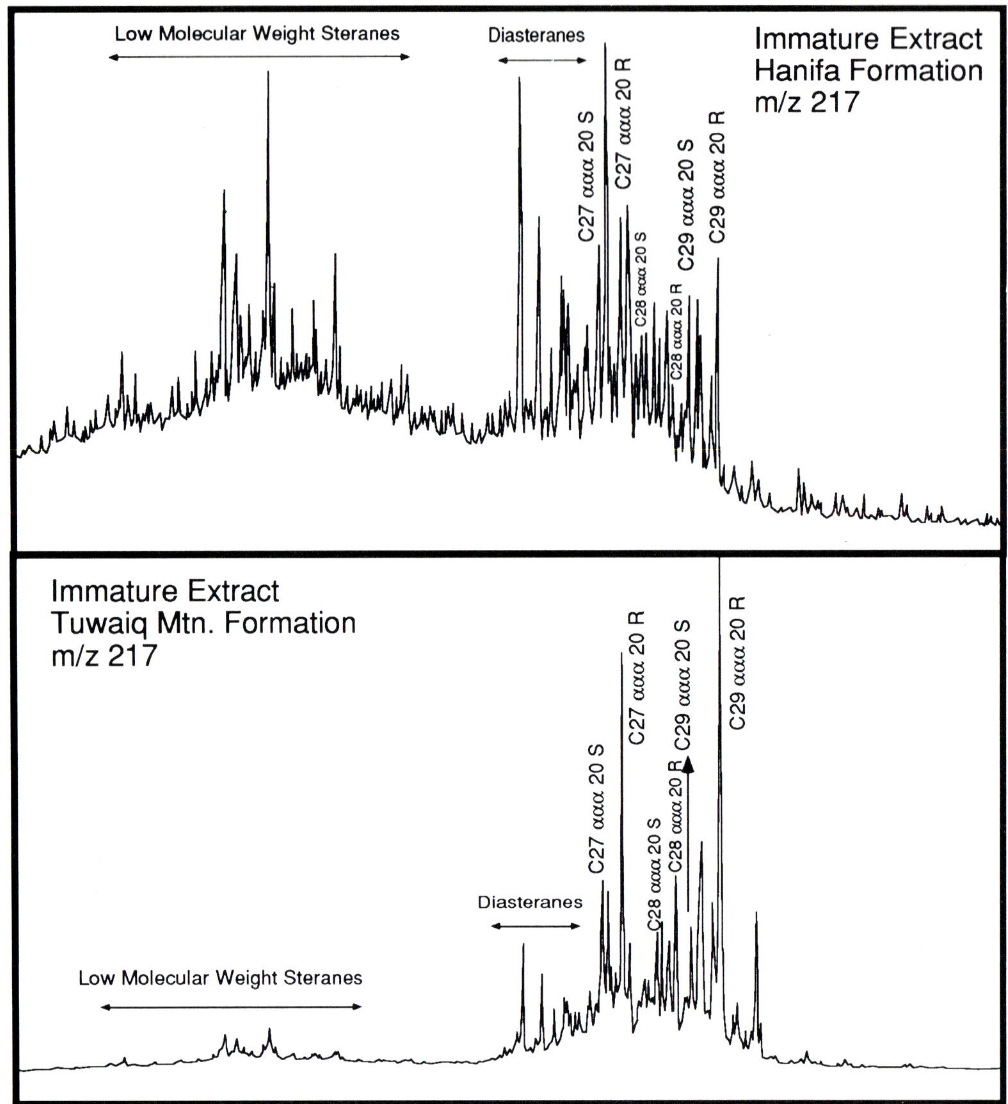

Fig. 15. Mass 217 biomarker fragmentograms for representative Hanifa and Tuwaiq Mountain source rock extracts

Stable carbon isotope ratios ($\delta^{13}C$) were determined for the whole oils and were compared to the kerogen and total soluble extracts from immature to mature Hanifa and Tuwaiq Mountain source units (Fig. 11). Oils from Jurassic reservoirs have an average carbon isotopic composition of −26.60‰ (standard deviation of 0.5‰), which range from −26.11‰ to −27.81‰. The Hanifa and Tuwaiq Mountain source rock kerogens have an average $\delta^{13}C$ of -26.4 ± 1.1‰. The conversion of kerogen to oil generally produces an oil with an isotopic value similar to or as much as 1‰ lighter (more negative value) than the kerogen (Clayton 1991a, b). Thus the isotopic values of the oils are consistent with an origin from the Hanifa and Tuwaiq Mountain source rocks.

On a gross compositional scale, the Arab- and Hanifa-reservoired oils are classified as aromatic-intermediate oils (Tissot and Welte 1984) as shown in Fig. 17. In most cases, these oils consist of 30–45% saturates, 30–45% aromatics, and 20–30% resins plus asphaltenes. Sulfur content is high (1 to 4 + %). To better define the relationships between the oils, saturate fraction C_{15+} gas chromatography was used to characterize oils from the Arab and Hanifa reservoirs. Regionally, the oils have similar chromatographic characteristics

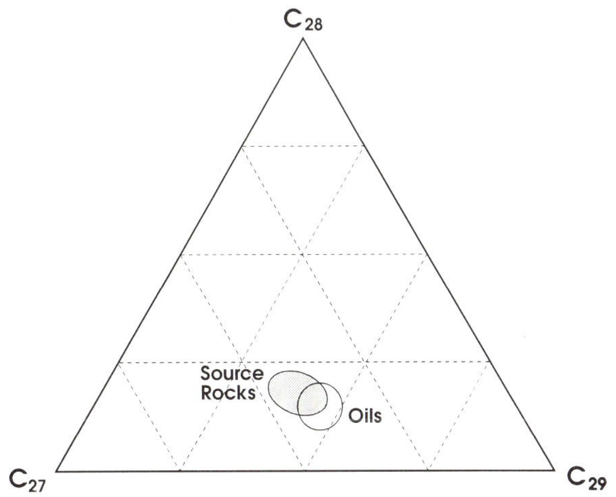

Fig. 16. $C_{27-28-29}$ steranes ternary diagram showing the close grouping between the oils and source rock extracts

(Fig. 18): (1) pristane/phytane ratios from 0.6 to 0.8; (2) pristane/n-C_{17} and phytane/n-C_{18} ratios plot along the same trend (Fig. 13); (3) isoprenoid distributions are similar; and (4) slight even preference in the C_{28} to C_{30} range (CPI < 1). Based on these chromatographic observations, the Arab and Hanifa oils are most likely derived from source rocks deposited in an anoxic, marine environment in a carbonate dominated system (Moldowan et al. 1985; Peters and Moldowan 1993) such as the Hanifa and Tuwaiq Mountain source rocks. Comparison of the chromatographic data of the oils with that of the Hanifa and Tuwaiq Mountain extracts show a high degree of similarity: low pristane/phytane ratios, similar isoprenoid distributions, and CPI < 1.

GCMS m/z 191 (tricyclics and hopanes) and m/z 217 (steranes) data were used to conduct a more detailed characterization of the oils and source rock extracts. Representative m/z 191 fragmentograms are shown in Fig. 19 for two Arab-reservoired oils. M/z 217 fragmentograms are shown in Fig. 20 for the same samples. Most oils are similar to each other in both fragmentogram fingerprints and in selected biomarker ratios. The tricyclic and hopane fingerprints (Fig. 19) are characteristic of oils derived from source rocks which were deposited in an anoxic, carbonate environment. Characteristic biomarkers that support this interpretation are the norhopane/hopane ratio of ≥ 1, the presence of the hexahydrobenzohopane and extended norhopane series (Connan and Dessort 1987; Peters and Moldowan 1991), high relative abundances of C_{24} tetracyclics (Connan et al. 1986), the presence of 2α-methyl-17α(H),21β(H)-hopane (Summons and Walter 1990), and the full range of C_{31} to C_{35} homohopanes (Peters and Moldowan 1991). The C_{29} dominated sterane distribution ($C_{29} > C_{27} \gg C_{28}$) (Fig. 20) is indicative of a marine algal, carbonate source rock. The C_{29} abundance is believed to be caused by a large contribution of brown or certain green algae in the source kerogen (Moldowan et al. 1985). There are variable amounts of diasteranes and variable amounts of low molecular weight steranes. C_{30} steranes, a marine indicator (Moldowan et al. 1985), are present, but are in low concentrations. Figure 16 shows the distribution of the C_{27} to C_{29} $\alpha\alpha\alpha$20R + 20S steranes after Huang and Meinschein (1979). All oils plot in the same region of the ternary plot and overlap the sterane distribution of extracts from the Hanifa and Tuwaiq Mountain source units.

GCMS biomarker patterns (m/z 191 and m/z 217) of extracts from source rocks from the immature to incipiently mature edge of the basin have the same biomarker compounds as observed in the oils. Similarities between the less mature biomarker fingerprints (i.e., Tm$\ll$Ts, unequilibrated hopanes, abundant moretanes, $\alpha\alpha\alpha$20R dominated steranes) of the extracts (Figs. 14 and 15) and the oils (Figs. 19 and 20) include: (1) the hexahydrobenzohopane and extended norhopane series of compounds; (2) high amounts of C_{24} tetracylic, relative to the C_{25} or C_{26} tricyclic terpanes; (3) the presence of 2α-methyl-17α(H),21β(H)-hopane; (4) a full range of C_{31} to C_{35} homohopanes; and (5) a similar C_{27}–C_{28}–C_{29} sterane distribution. The most significant differences between some of the source rock extracts and the oils are different tricyclic abundances and associated ratios, lower norhopane/hopane ratios, and the occasional truncation of the homohopanes after C_{33}. These differences are attributed to both the immaturity of these source rock samples and to minor facies variations within the basin. These facies variations may be related to fluctuating anoxic/dysoxic conditions during deposition or to proximity with the edge of the basin.

In summary, the Arab- and Hanifa-reservoired oils have similar bulk compositional, isotopic, chromatographic, and biomarker signatures. Because the overall source rock extract characteristics match the oils, there is a good correlation between the Jurassic-reservoired oils and the Hanifa and Tuwaiq Mountain source units.

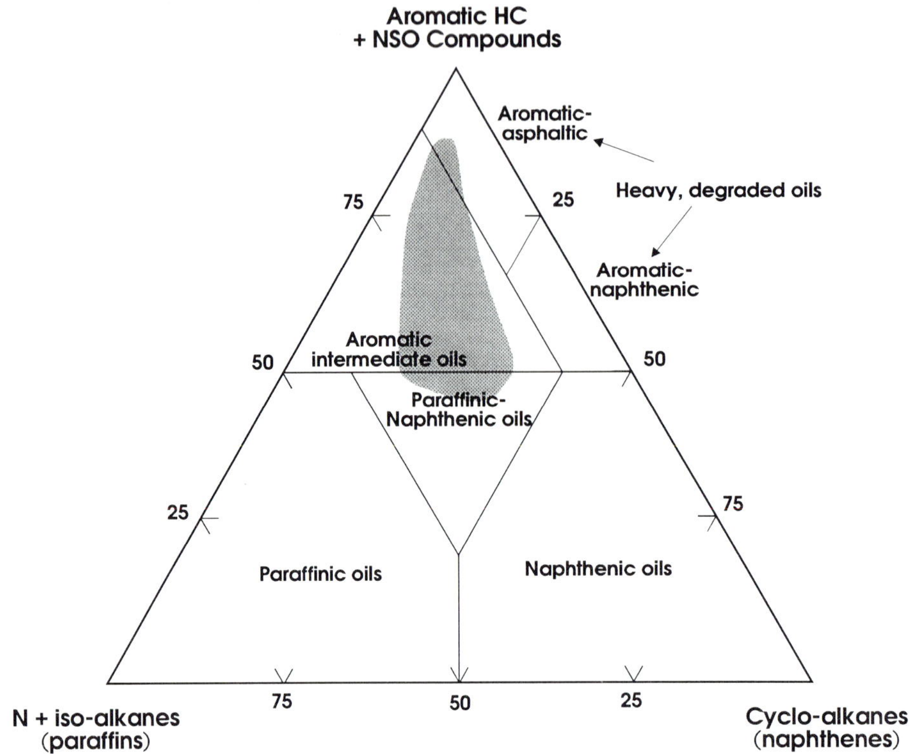

Fig. 17. Classification of oils based on bulk composition. Stippled area is typical composition of Arab and Hanifa-reservoired oils. (After Tissot and Welte 1984)

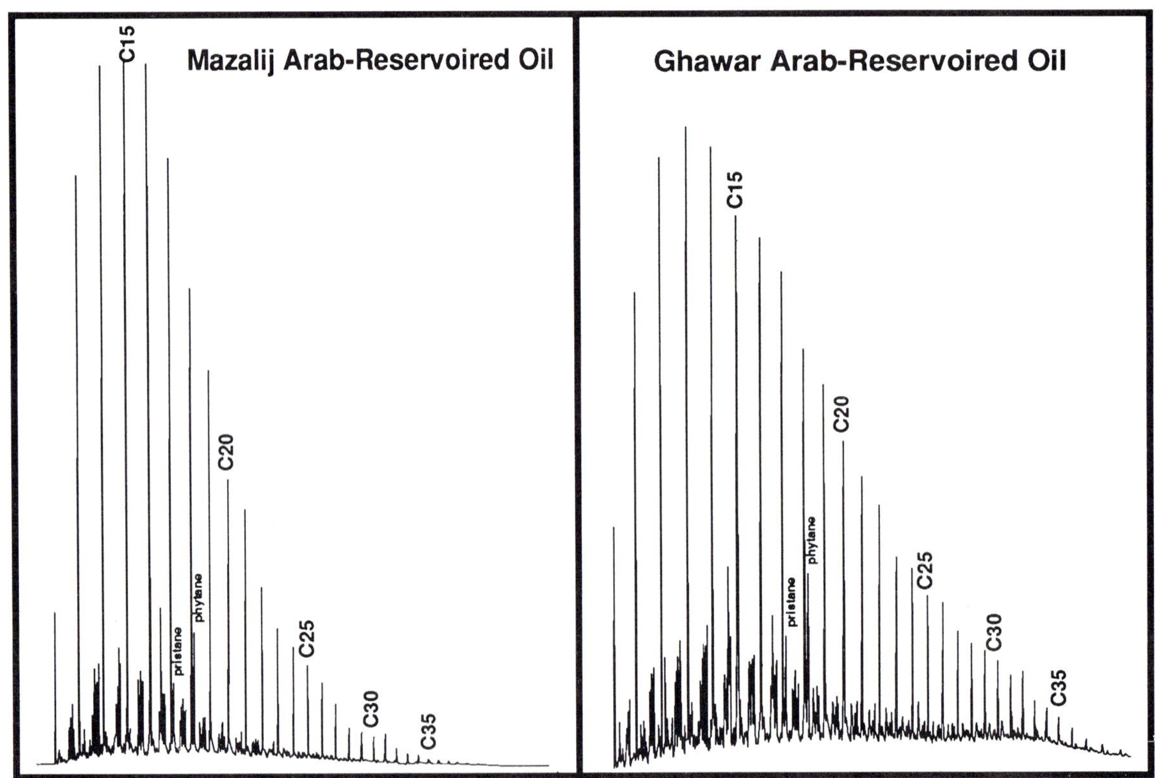

Fig. 18. C_{15+} gas chromatograms for representative Arab-reservoired oils

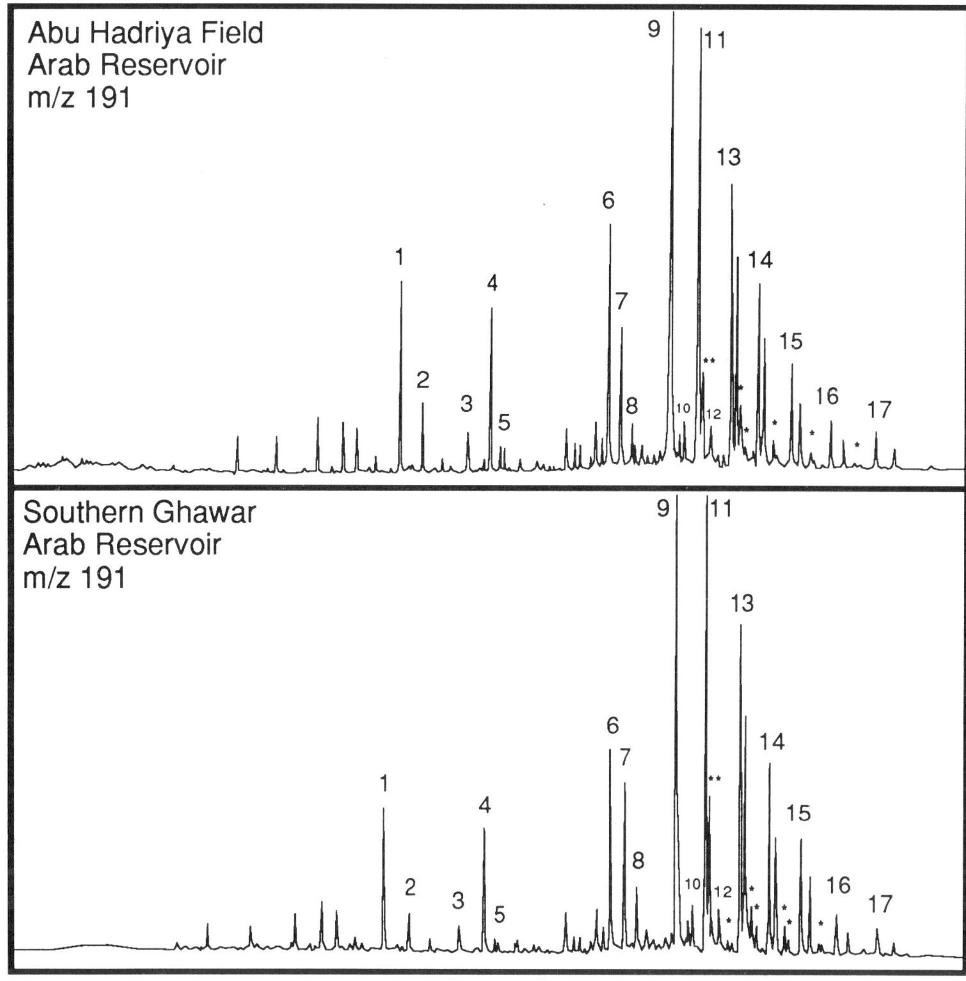

Fig. 19. Mass 191 biomarker fragmentograms for representative Arab-reservoired oils (see Table 2 for identities of peaks)

Burial History/Thermal Modeling

Burial history/thermal models were constructed for more than 50 wells in eastern Saudi Arabia using the BasinMod (version 2.9) software package from Platte River Associates. This modeling software has been successfully applied in other modeling studies (Waples et al. 1992a, b) and follows the logic described therein. The specific data required for each model are summarized below:

1. The stratigraphic section penetrated by the well is used to establish the geological burial history. The amount of burial and erosion for the observed unconformities in the well were corrected using regional seismic and geological reconstructions. The stratigraphic section below where the well penetrated was estimated using seismic lines. The lithologies modeled were also determined from the well logs.
2. The thermal history was based on the tectonic and basin reconstruction of the eastern province described in Murris (1980) and Beydoun (1991) and corrected using observed bottom-hole temperatures. Figure 21 illustrates a simplified geothermal gradient map for the eastern province. Calibration of the thermal histories was done regionally by comparison to measured maturity indicators (vitrinite reflectance, TAI, T_{max}). The most plausible thermal history assumes a mostly constant heat flow history from Jurassic time to present, as would be expected for the formation of a carbonate platform along

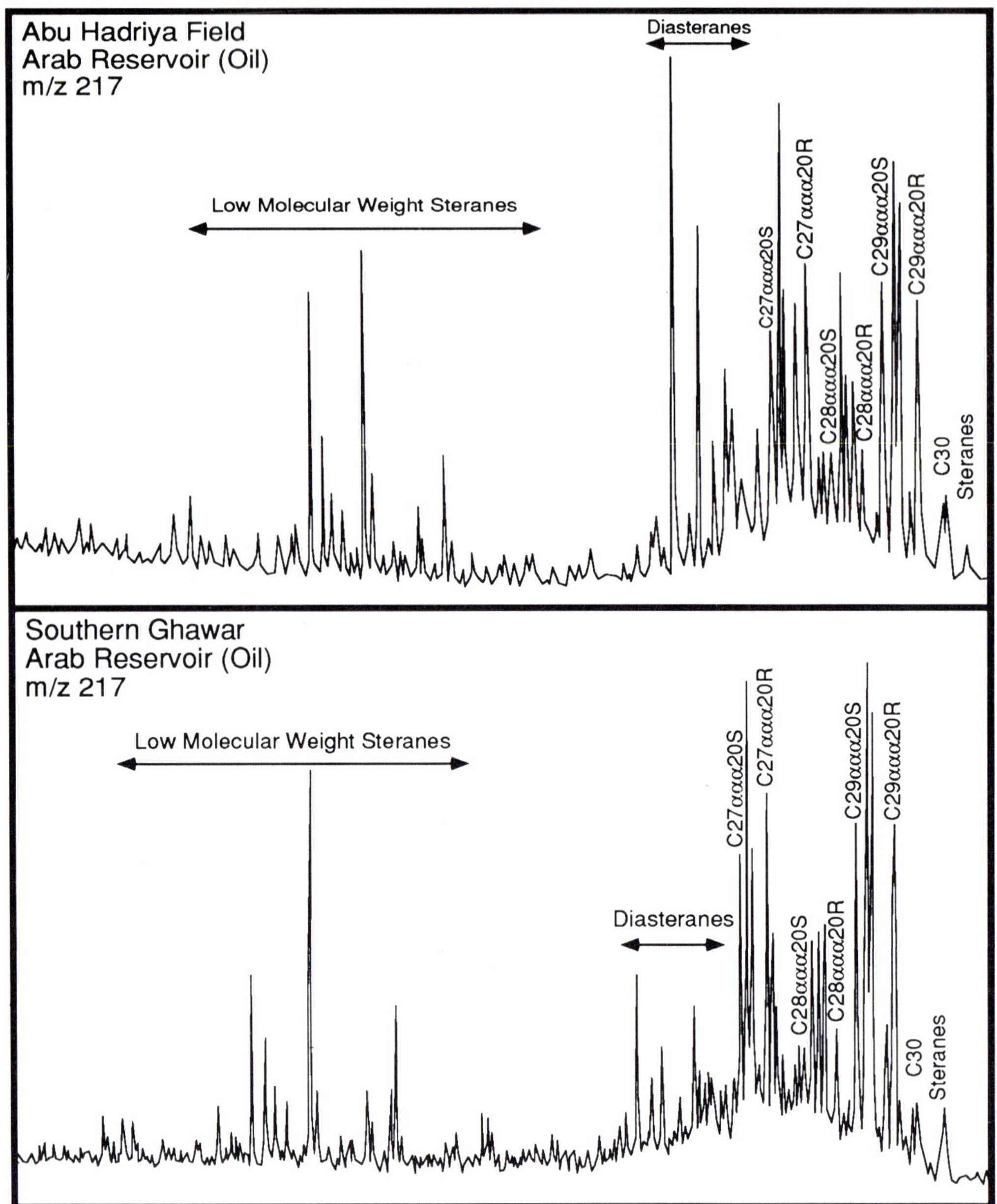

Fig. 20. Mass 217 biomarker fragmentograms for representative Arab-reservoired oils

a passive margin. However, we did assume a slightly higher heat flow during the late Jurassic/early Cretaceous caused by the reactivation of faulting.

From the modeling results, the maturity at the base Hanifa/top Tuwaiq Mountain (source rock interval) at four time periods was mapped (present-day, 25 Ma B.P., 50 Ma B.P. and 75 Ma B.P.) (Figs. 22 to 25) to determine when the principal source rocks generated and expelled their respective hydrocarbons. Table 1 lists the maturation windows used, their respective equivalent % vitrinite reflectance (VRe) ranges, and any important comments related to these maturation windows.

The present-day level of thermal maturity (Fig. 25) generally increases from the exposed western rim of the basin in central Saudi Arabia (immature;

< 0.5% VRe) towards the northeast (Safaniya area) and southeast (east of the Ghawar structure) where post-oil expulsion maturities are attained (0.8 to 0.9% VRe). As a specific note regarding the oil expulsion window, empirical observations have shown that the Hanifa and Tuwaiq Mountain source rocks probably attain peak expulsion maturity at about 0.75% VRe (Cole et al. 1994). A detailed study in the Ghawar area indicated that the source rock intervals have little residual pyrolytic yield at a measured vitrinite reflectance of 0.8–0.85%. As is readily observed, the present-day maturity contours follow the geothermal gradient trends which is expected, since this is the primary calibration tool of the models. However, it is interesting that there are distinct and quite variable present-day thermal regimes across the Eastern Province. The lowest gradients (< 20 °C/km) occur southwest of the Khurais area and along the western margin of the basin. Gradients are still 'cool' (between 20 and 25 °C/km) north of the Khurais area and increase progressively towards the east. The highest gradients (> 35 °C/km) are in the northeastern offshore area (Safaniya area) and southeast of the Ghawar structure.

Results from the thermal maturity modeling are predictable and follow the geothermal gradient trends. The least mature source rocks occur along the western margin of the basin and progressively increase in maturity toward the east. By taking slices back through time, we can determine when generation and expulsion occurred. At 75 Ma B.P. most of the Hanifa and Tuwaiq Mountain source rocks were attaining early maturity (≥ 0.45% VRe) (Fig. 22). At this time oil kitchens in the Safaniya area to the northeast and to the east of the Ghawar structure were just beginning to attain peak oil expulsion maturity (~ 0.75% VRe range). At 50 Ma B.P., the kitchens expanded outward from these two areas (Fig. 23). Several areas had now attained peak expulsion maturity and began filling the broad, gentle structures that had started forming during the late Cretaceous. By 25 Ma B.P., most of the Hanifa and Tuwaiq Mountain Formations north and east of the basin center had attained expulsion maturity (Fig. 24). Expelled oils by this time had probably filled most of the structures in the kitchens and migrating hydrocarbons began filling the structures along the basin margins. Today, the kitchens in the Ghawar area and eastward, and those in the northeast have passed through the oil expulsion window (Fig. 25). The source rocks exit the oil window toward Qatar where maturity is in the gas generation phase. The remaining areas east of the basin center are still within the oil generation window. In the western part of the

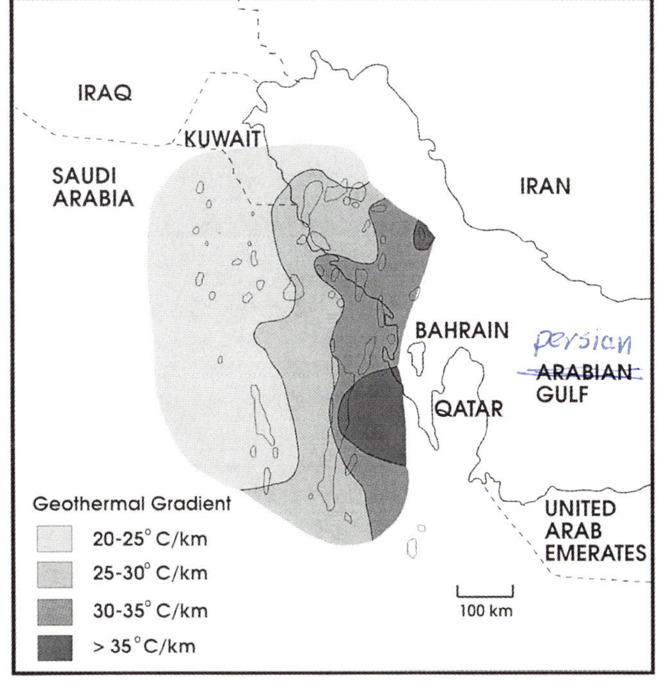

Fig. 21. Simplified geothermal gradient map of eastern Saudi Arabia. Gradients were calculated from corrected bottom hole temperatures measured at more than 50 wells across the region. Geothermal gradients show an increase from the exposed western margin of the Arabian basin (Khurais and El Haba areas) towards the northeast (Safaniya area) and southeast (east of the Ghawar structure)

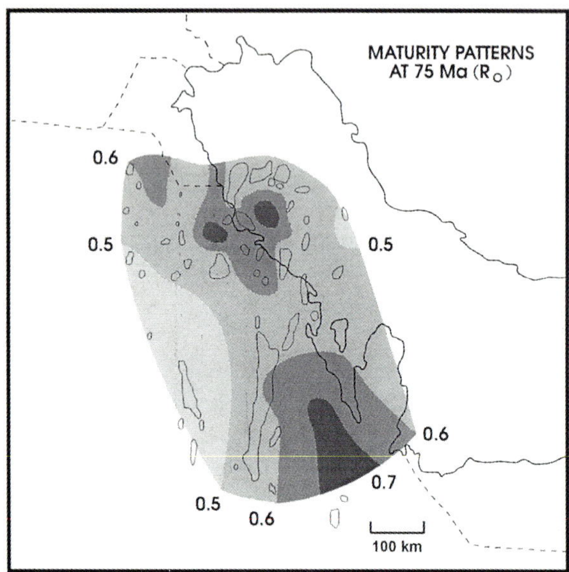

Fig. 22. Map showing thermal maturity trends for the top of the Hanifa Formation at 75 Ma ago. Contours in % vitrinite reflectance equivalent

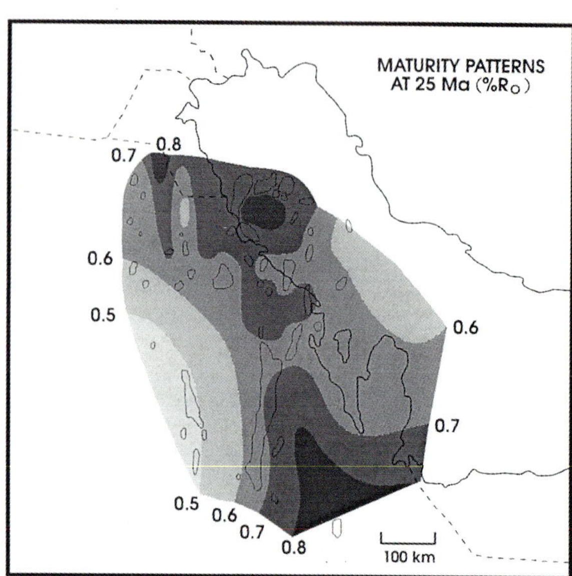

Fig. 24. Map showing thermal maturity trends for the top of the Hanifa Formation at 25 Ma ago. Contours in % vitrinite reflectance equivalent

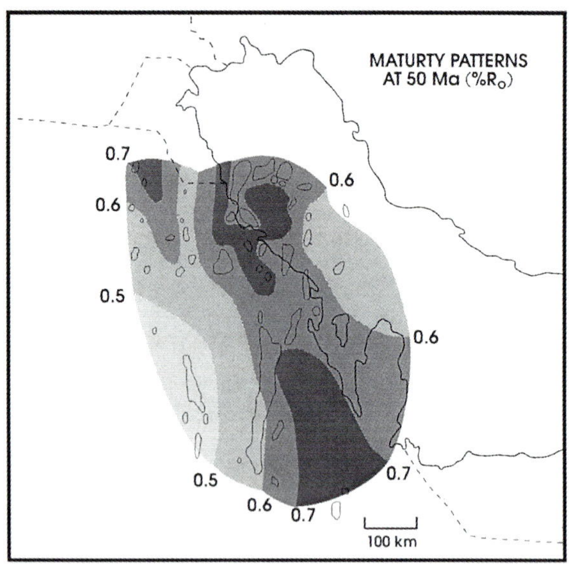

Fig. 23. Map showing thermal maturity trends for the top of the Hanifa Formation at 50 Ma ago. Contours in % vitrinite reflectance equivalent

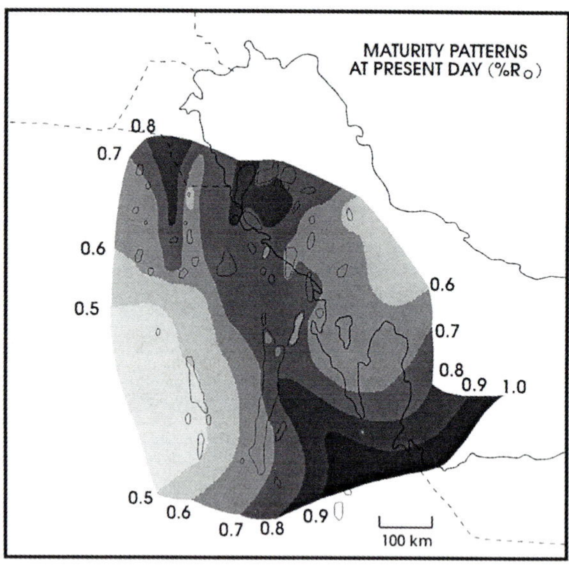

Fig. 25. Map showing presentday thermal maturity trends for the top of the Hanifa Formation. Contours in % vitrinite reflectance equivalent

basin, from El Haba south to Khurais, the source rock is either immature or only starting to enter the oil generation and expulsion window.

Based on the present-day maturity trends, most of the oil fields in eastern Saudi Arabia do not require long distance migration to be filled. The 400+ foot thick source rock units have attained sufficient maturity to have charged these structures. The only fields filled through longer distance migration are those along the western margin such as Khurais and El Haba, and these have large enough fetch areas to easily fill most of the structures.

Table 1. Maturation windows and corresponding % vitrinite reflectance equivalence (%VRE). (After Dow and O'Conner 1982; van Gijzel 1982; Robert 1988)

Maturation windows	% VRE range	Comments
Immature	0.2 to 0.45	No significant generation
Early mature	0.45 to 0.6	Significant oil generation, minor expulsion
Expulsion mature	0.6 to 0.9	Oil expulsion window, expulsion 90% complete by 0.8–0.85%, peak expulsion about 0.7–0.75%
Post-oil expulsion	0.9 to 1.3	Oil preservation to 1.3%, oil cracking to lighter oil and gases begins about 1.0%, cracking of residual kerogens to gas begins
Wet gases/condensate	1.3 to 1.8	Oil cracking continues, condensate range at $>1.3\%$, wet gases by 1.6 to 1.8%
Dry gases	>1.8	Gases become progressively drier at $>1.8\%$, kerogems essentially spent by 2.0–2.2%, gas preservation to 3.2% in carbonates and $>4.0\%$ in clastics

Table 2. Identities of peaks in the m/z 191 mass fragmentogram

Peak no./symbol[a]	Name
1	C_{23} Tricyclic
2	C_{24} Tricyclic
3	C_{25} Tricyclic
4	C_{24} Tetracyclic
5	C_{26} Tricyclic
6	22,29,30-Trisnorhopane-II (Ts)
7	22,29,30-Trisnorhopane (Tm)
8	17α(H)-29,30-Bisnorhopane
9	17α,21β(H)-30-Norhopane
10	17β,21α(H)-30-Norhopane (normoretane)
11	17α,21β(H)-Hopane
12	17β,21α(H)-30-Hopane (moretane)
13	17α,21β(H)-29-Homohopanes (C_{31} 22R and 22S)
14	17α,21β(H)-29-Bishomohopanes (C_{32} 22R and 22S)
15	17α,21β(H)-29-Trishomohopanes (C_{33} 22R and 22S)
16	17α,21β(H)-29-Tetrakishomohopanes (C_{34}22R and 22S)
17	17α,21β(H)-29-Pentakishomohopanes (C_{35} 22R and 22S)
**	17α(H)-30-nor-29-Homohopane and 2α-methyl-17α(H),21β(H)
(H)-	hopane co-elution
*	30-Norhopane (17αH) and/or hexahydrobenzohopane series

[a] In Figs. 14 and 19.

Summary

The Hanifa and Tuwaiq Mountain Formations in Saudi Arabia contain thick, regionally extensive laminated organic-rich lime mudstone units. Immature potential source rocks from the basin margin have an average TOC content of about 3%, although values as high as 13% have been observed. The average Rock-Eval total pyrolitic yield ($S_1 + S_2$) is about 25 mg HC/g rock, with a maximum of 88 mg HC/g rock. Hydrogen indices average between 600 and 800 mg HC/g TOC. Organic petrography indicates a kerogen assemblage dominated by fluorescent lamalginite with subordinate amounts of vitrinite and inertinite. H/C and O/C ratios of kerogen plot along the type II pathway on a van Krevelen diagram. Therefore, the Hanifa and Tuwaiq Mountain Formations contain organic-rich units having excellent, oil-prone source potential. Comparison of the immature Hanifa and Tuwaiq Mountain source rocks with samples from the mature parts of the basin suggest that the mature source rocks probably had similar source rock quality and hydrocarbon generation potential.

Thermal modeling of the source rocks indicate that oil generation and expulsion began about 75 Ma B.P. in the eastern part of the basin. By 50 Ma B.P. the oil kitchens had expanded westward and oil had started filling the broad, gentle structures which had started forming during the late Cretaceous. Today, the Hanifa and Tuwaiq Mountain source rocks east of the Ghawar structure have passed through the oil generation window. The source rocks in the basin center are still within the oil generation window. In the western margin of the basin, the source rocks are either immature or just starting to enter the oil window.

Ayres et al. (1982) consider the Hanifa and Tuwaiq Mountain source rocks to be the primary source of oil for the Jurassic reservoirs of Saudi Arabia. We agree with their assessment and we also recognize that other source rocks are present, both within Saudi Arabia and in adjacent areas, which may have contributed oil. However, the most likely source for the oil contained in the large Jurassic carbonate reservoirs in the Ghawar-Khurais region is the Hanifa and Tuwaiq Mountain source rocks. These oils have very similar bulk composition and isotopic, chromatographic, and biomarker fingerprints to the Hanifa and Tuwaiq Mountain source rock extracts and kerogen. The timing of oil generation and expulsion from these source rocks coincides with the development of the large structures which now form these major reservoirs. Furthermore, these source rocks underlie and are interlayered with good quality reservoir rocks, which allows for very efficient expulsion and migration. The Jurassic source rocks and reservoirs are capped by regionally extensive evaporites of the Hith Formation. All of the necessary components to form an efficient and productive petroleum system are present in the Arabian basin of eastern Saudi Arabia.

Acknowledgments. The authors would like to thank the staff of the Geochemistry Unit of the Lab R&D Center of Saudi Aramco for carrying out the analyses provided in this chapter. We would also like to acknowledge Kieraville Konsultants of Kieraville, Australia for providing a significant portion of the organic petrography. The authors also thank H.I. Halpern, M.H. Tobey, H.H. Chen, A. Rahman, A.A. Al-Haji, and M.A. Abu Ali for valuable discussions and contributions during the preparation of this chapter. The interpretation of the petroleum geology of Saudi Arabia is the result of the contributions of numerous Saudi Aramco and US company (Texaco, Exxon, Chevron, and Mobil) affiliated geochemists and geologists, both past and present. Appreciation is given to the Saudi Arabian Ministry of Petroleum and Mineral Resources and to Saudi Aramco for permission to publish this chapter.

References

Abu-Ali MA, Franz UA, Shen V, Monnier F, Mahmoud MD, Chambers TM (1991) Hydrocarbon generation and migration in the paleozoic sequence of Saudia Arabia. SPE Prof Pap 21376, Proc 7th Ann SPE Middle East Oil Show, Bahrain, 345–355

Ala MA, Kinghorn RRF, Rahman M (1980) Organic geochemistry and source-rock characterization of the Zagros petroleum province, SW Iran. J Petrol Geol 3: 61–89

Alsharhan AS, Kendall CGStC (1986) Precambrian to Jurassic rocks of Arabian Gulf and adjacent areas: their facies, depositional setting, and hydrocarbon habitat. Am Assoc Pet Geol Bull 70: 977–1002

Ayres MG, Bilal M, Jones RW, Slentz LW, Tartir M, Wilson AO (1982) Hydrocarbon habitat in main producing areas, Saudi Arabia. Am Assoc Pet Geol Bull 66: 1–9

Beydoun ZR, (1991) Arabian Plate hydrocarbon geology and potential – a plate tectonic approach. Am Assoc Pet Geol, Tulsa, Stud Geol 33: 76

Clark JP, Philp RP (1989) Geochemical characterization of evaporite and carbonate depositional environments and correlation of associated crude oils in the Black Creek basin, Alberta, Bull Can Petrol Geol 37: 401–416

Clayton CJ (1991a) Effect of maturity on carbon isotope ratios of oils and condensates. Org Geochem 17: 887–900

Clayton CJ (1991b) Carbon isotope fractionation during natural gas generation from kerogen. Mar Petrol Geol 8: 232–240

Cole GA, Carrigan WJ, Colling EL, Halpern HI, Al-Khadrawi MR, Jones PJ (1994) The organic geochemistry of the Jurassic petroleum system in Eastern Saudi Arabia. In: B. BeauchampB, Embry AF, Glass D (eds). Carboniferous to Jurassic Pangea, Can Soc Petrol Geol, Calgary, Mem 17 (in press)

Connan J, Dessort D (1987) Novel family of hexacyclic hopanoid alkanes (C_{32}–C_{35}) occurring in sediments and oils from anoxic paleoenvironments. Org Geochem 11: 103–113

Connan J, Bouroullec J, Dessort D, Albrecht P (1986) The microbial input in carbonate-anhydrite facies of a sabkha paleoenvironment from Guatemala: a molecular approach, Org Geochem 10: 29–50

Cook AC, Sherwood NR (1991) Classification of oil shales, coals and other organic-rich rocks. Org Geochem 17: 211–222

Dow WG, O'Connor DI (1982) Kerogen maturity and type by reflected light microscopy applied to petroleum exploration, In: How to assess maturation and paleotemperatures. Soc Econ Paleontol Mineral, Tulsa, Short Course Number 7, pp 133–158

Droste H (1990) Depositional cycles and source rock development in an epeiric intra-platform basin: the Hanifa Formation of the Arabian peninsula. Sediment Geol 69: 281–296

Droste H (1993) Source rock development and relative sea-level changes in an intra-platform basin: the upper Jurassic Hanifa Formation of the Arabian Peninsula, Am Assoc Pet Geol Bull 77: 533.

Huang WY, Meinschein WG (1979) Sterols as ecological indicators. Geochem Cosmochem Acta 43: 739–745

Katz BJ (1983) Limitations of Rock-Eval pyrolysis for typing organic matter. Org Geochem 4: 195–199

Langdon GS, Malecek SJ (1987) Seismic stratigraphic study of two Oxfordian carbonate sequences, Eastern Saudi Arabia. Am Assoc Pet Geol Bull 71: 403–418

Langford FF, Blanc-Valleron MM (1990) Interpreting Rock-Eval pyrolysis data using graphs of pyrolizable hydrocarbons vs. total organic carbon. Am Assoc Pet Geol Bull 74: 799–804

Mahmoud MD, Vaslet D, Husseini MI (1992) The lower Silurian Qalibah Formation of Saudi Arabia: an important hydrocarbon source rock. Am Assoc Pet Geol Bull 76: 1491–1506

McGillivray JG, Husseini MI (1992) The Paleozoic petroleum geology of Central Arabia. Am Assoc Pet Geol Bull 76: 1473–1490

McGuire MD, Koepnick RB, Markello JR, Stockton ML, Waite LE, Kompanik GS, Al-Shammery MJ, Al-Amoudi MO (1993) Importance of sequence stratigraphic concepts in development of reservoir architecture in upper Jurassic grainstones, Hadriya and Hanifa reservoirs, Saudi Arabia. SPE Prof Pap 25578, Proc 8th Ann SPE Middle East Oil Tech Conf & Exhibition, Manama, Bahrain, pp 489–499

Mello MR, Telnaes N, Gaglianone PC, Chicarelli MI, Brassell SC, Maxwell JR (1988) Organic geochemical characterization of depositional paleoenvironments in Brazilian marginal basins, Org Geochem 13: 31–46

Moldowan JM, Seifert WK, Gallegos EJ (1985) Relationship between petroleum composition and depositional environment of petroleum source rocks. Am Assoc Pet Geol Bull 69: 1255–1268

Murris RJ (1980) Middle East: stratigraphic evolution and oil habitat. Am Assoc Pet Geol Bull 64: 597–618

Oil & Gas Journal (1992) Worldwide production report 90 (52): 39–85

Pelet R (1983) A model for organic sedimentation on present-day continental margins. In: Brooks J, Fleet AJ (ed) Marine Petroleum Source Rocks, Geol Soc, London, Spec Publ 26: 167–180

Peters KE (1986) Guidelines for evaluating petroleum source rocks using programmed pyrolysis. Am Assoc Pet Geol Bull 70: 318–329

Peters KE, Moldowan JM (1991) Effects of source, thermal maturity, and biodegradation on the distribution and isomerization of homohopanes in petroleum, Org Geochem 17: 47–61

Peters KE, Moldowan JM (1993), The biomarker guide – interpreting molecular fossils in petroleum and ancient sediments. Prentice Hall, Englewood Cliffs, NY, 585 pp

Powers RW, Ramirez LR, Redmond CD, Elberg EL (1966) Sedimentary geology of Saudi Arabia, In: Geology of the Arabian Peninsula. USGS Prof Pap 560–D, 150 pp

Robert P (1988) Organic metamorphism and geothermal history: microscopic study of organic matter and thermal evolution of sedimentary basins. Elf-Acquitaine and Reidel, Boston, 311 pp

Seifert WK, Moldowan JM (1978) Application of steranes, terpanes, and monoaromatics to the maturation and migration, and source of crude oils, Geochim Cosmochim Acta 42: 77–95

Steineke M, Bramkamp RA, Sander NJ (1958) Stratigraphic relations of Arabian Jurassic oil. In: Weeks LG (ed) Habitat of Oil, Am Assoc Pet Geol symp 40th ann meeting New York, March 28–30, 1955: 1294–1329

Stoneley R (1990) The Middle East basin: a summary overview. In: Brooks J (ed) Classic petroleum provinces, Geol Soc, London, Spec Publ 50: 293–298

Suess E (1980) Particulate organic flux in the oceans – surface productivity and oxygen utilization. Nature 288: 260–262

Summons RE, Walter MR (1990) Molecular fossils and microfossils of procaryotes and protists from Proterozoic sediments. Am J Sci 290–A: 212–244

Tissot BP, Welte DH (1984) Petroleum formation and occurrence. Springer, Berlin Heidelberg New York, 699 pp

Van Gijzel P (1982) Characterization and identification of kerogen and bitumen and determination of thermal maturation by means of qualitative and quantitative microscopical techniques. In: How to assess maturation and paleotemperatures. Soc Econ Paleontol Mineral, Tulsa, Short Course Number 7, pp 159–216

Volkman JK (1986) A review of sterol markers for marine and terrigenous organic matter. Org Geochem 9: 84–99

Waples DW, Machihara T (1991) Biomarkers for geologists – A practical guide to the application of steranes and triterpanes in petroleum geology. Am Assoc Pet Geol, Tulsa, Methods 9, 91 pp

Waples DW, Kamata H, Suiza M (1992a) The art of maturity modeling. Part 1: Finding a satisfactory geologic model. Am Assoc Pet Geol Bull 76: 31–46

Waples DW, Suiza M, Kamata H (1992b) The art of maturity modeling. Part 2: Alternative models and sensitivity analysis. Am Assoc Pet Geol Bull 76: 47–66

The Kimmeridge Clay Formation of The North Sea

B.S. Cooper[1], P.C. Barnard[2], and N. Telnaes[3]

Abstract

The Kimmeridge Clay Formation and its equivalents of Callovian to early Ryazanian age are by far the most important oil source rocks of the northern North Sea area. The source rocks comprise marine claystones which were deposited in waters of increased salinity with short episodes of high salinity, during which uranium-enriched shales accumulated. Much of the sedimentary sequence in the inner parts of the basin comprises mass-flow deposits and storm-activated reworked deposits. The mineralogy and quantity and quality of the organic content is extremely varied, both regionally and in vertical sequence. A comparison of the biomarker and other geochemical characteristics leads to a conclusion that at least eight definable kerogen types contribute to the organic content, and a knowledge of their distributions is necessary before oil to source rock correlation can be performed with confidence. However, such interpretations require a large database and computer-based multivariate analysis.

It is necessary for the petroleum geologist to understand how these variations can be explained by palaeogeographical factors so that the quantity and quality of oil-generating organic matter can be mapped with confidence. Thus, more precise estimates can be made of the quantities of oil and gas which have been generated in the un-drilled kitchen areas and, where migrated oil is discovered in the adjoining structures, its provenance can be correlated with the kitchen source rock or shown to be from an alternative source.

[1]B S Cooper and Associates, Northern Cottage, Appleton Le Moors, York, YO6 6TF, UK

[2]Simon Petroleum Technology Limited, Llandudno, Gwynedd, LL30 1SA, UK

[3]Norsk Hydro Petroleum Research Centre, Bergen, Norway

Introduction

This chapter sets out to describe the palaeogeographic setting and geochemical characteristics of the Upper Jurassic-Lower Cretaceous Kimmeridge Clay Formation of the North Sea basin (Fig. 1). For the purposes of this chapter the northern North Sea is defined as the area north of 55°N (the Mid North Sea High) to about 62°N, and includes all basins within the area both in the UK and Norwegian sectors of the North Sea. The southern North Sea is generally restricted to the marine areas of subcrop of the Jurassic sediments south of the Mid North Sea High. Carbonaceous claystones of Middle Jurassic to Lower Cretaceous marine formations are the prime source rocks of the northern North Sea basin hydrocarbons. Similar, but immature source rocks within this age range are also present in the southern North Sea basin and outcrop from the east coast of England to the south coast in the Wessex basin. Northwards, source rocks of similar age subcrop along the Norwegian shelf into the Barents Sea and appear again in Svalbard, on the Russian Platform and in the West Siberian basin (Fig. 3). Throughout these contiguous areas the lithologies and organic contents of the source rocks vary considerably.

In spite of nearly 30 years of North Sea exploration with associated comprehensive geochemical studies, there is no general agreement on the processes which control the spatial distribution of kerogens in these source rocks. In the most recently published review, Miller (1990) describes persuasively how petrophysical, chemical and palaeoenvironmental modelling can explain many features of the Upper Jurassic source rock of the North Sea. Other models which have aided in the development of ideas have been described by Baird (1986), Barnard et al. (1981), Cooper and Barnard (1984), Cornford (1984), Cornford et al. (1986), Demaison et al. (1983), Reeder and Scotchman (1985), Thomas et al. (1985) and Williams (1986).

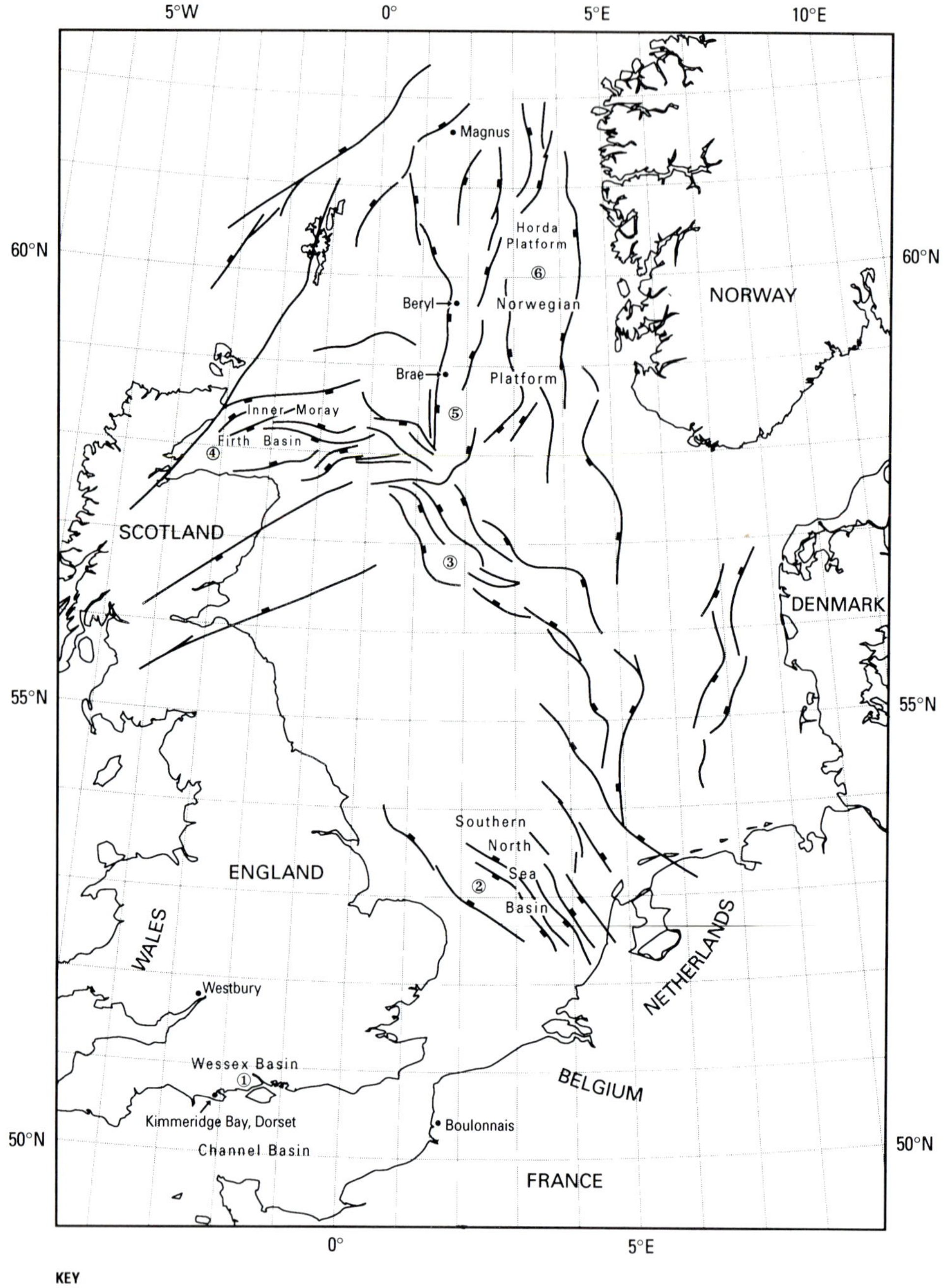

Fig. 1. Location map for the study area

Stratigraphy

Lower and Middle Jurassic sediments were deposited during a period of thermal subsidence after a Triassic rifting event. During the Middle Jurassic a further phase of rift faulting began with uplift and vulcanism at the junction of the Viking, Central and Moray Firth Graben systems (Thorne and Watts 1989). Elsewhere, the Middle Jurassic is mostly dominated by deltaic and pro-deltaic sediments passing into marine shales in the extreme

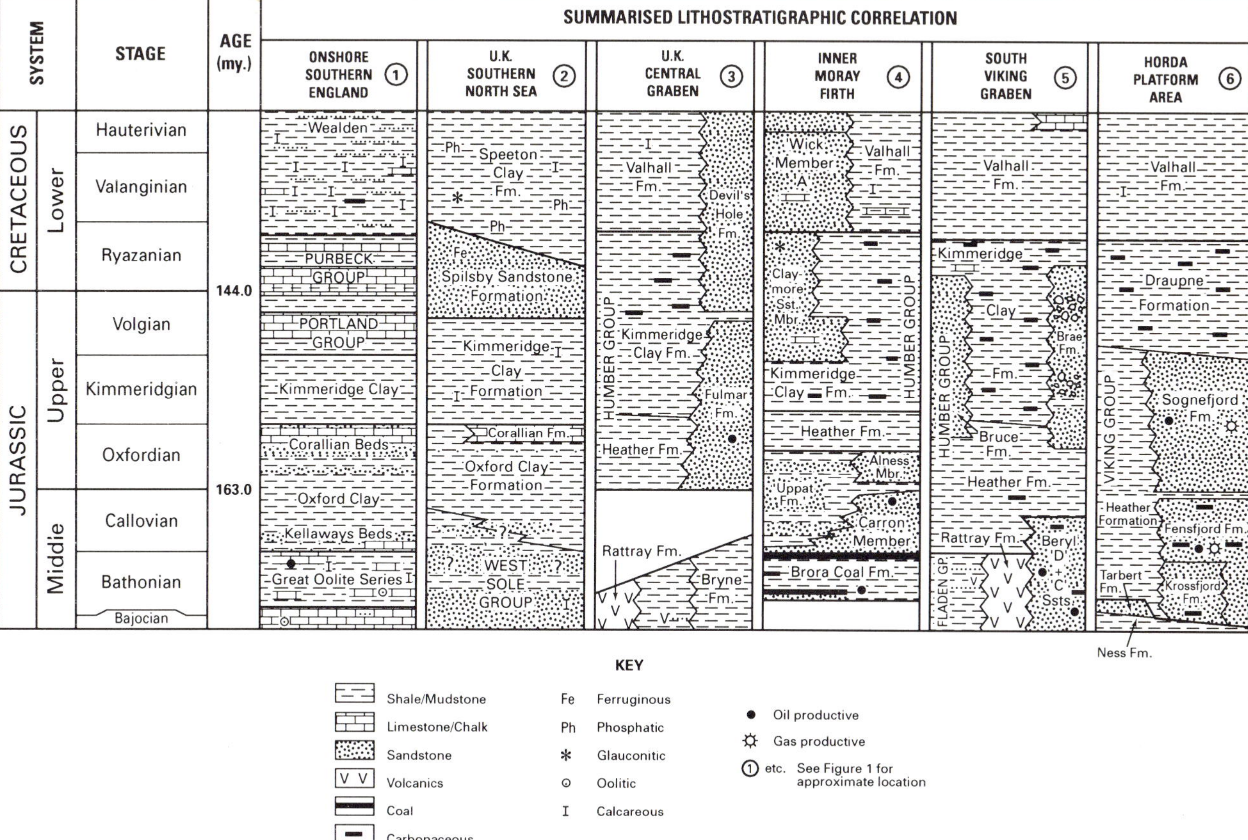

Fig. 2. Summary stratigraphy of the Upper Jurassic-Lower Cretaceous in the study area

north and shales and limestones in central and southern England.

Sediments of the Upper Jurassic are dominantly carbonaceous shales with turbidite sands accumulating against rotating fault blocks and shallow water marine sands on the margins of the basin. On platformal areas at greatest distance from sediment source, limestones were formed.

There are several source rock and non-source rock formations within the marine section of late Middle Jurassic to Early Cretaceous age and they are shown diagrammatically in Fig. 2. In the British onshore and offshore areas the source rock unit is designated the Kimmeridge Clay Formation with various equivalents in the Norwegian and other sectors. Typically the Kimmeridge Clay Formation is dark grey, black or dark brown claystone with infrequent carbonate bands, while the underlying and older Heather Formation is dark grey, often silty, mudstone with frequent carbonate bands. Formational boundaries are based on lithological parameters and are not necessarily isochronous, but for the purpose of this chapter, the Upper Jurassic marine carbonaceous claystones are divided as follows:

1. middle Volgian to Ryazanian — Upper Kimmeridge Clay Formation
2. middle Kimmeridgian to early Volgian — Lower Kimmeridge Clay Formation
3. middle Oxfordian to early Kimmeridgian — Upper Heather Formation
4. middle Callovian to early Oxfordian — Middle Heather Formation
5. Bathonian to early Callovian — Lower Heather Formation

These boundaries are often marked by a hiatus or minor unconformity which are probably tectonically related. Contemporaneous non-argillaceous formations are variously of deltaic/lagoonal origin (e.g. Hugin Formation), mass flow deposits (e.g. Brae Formation), deep water marine sands (e.g.

Magnus Formation) and shallow water marine sands (e.g. Sognefjord and Fensfjord Formations).

The Kimmeridge Clay Formation is the main hydrocarbon source rock unit in the North Sea area but the Heather Formation can be as important in certain restricted areas and possibly of greater potential in the northern extreme of the Viking Graben and in the Sogn Graben. The thickness of the Kimmeridge Clay Formation varies from less than 50 m over platformal areas to greater than 1250 m in grabenal depocentres (Barnard and Bastow 1992), but the Heather Formation is thicker and the combined thicknesses reach maximum values of about 1000 m in the Viking Graben and over 2000 m in the Central Graben.

Average organic carbon contents in the northern North Sea for the Kimmeridge Clay Formation are 6%, and for the Heather Formation 3%, but show extreme variability; kerogen contents vary from purely sapropelic, usually type II, to mixtures of types III and IV (Thomas et al. 1985). The few detailed studies of the lithology and geochemistry of the Kimmeridge Clay Formation of the northern North Sea are amplified by those carried out on outcrop samples from the type section at Kimmeridge Bay, Dorset.

Lithologies

The cores of offshore Kimmeridge Clay Formation which have been described show some of the variability in lithology which is seen in core and outcrop from the English and French onshore sections. At outcrop dark grey mudstones predominate, but they can be texturally distinguished as banded, calcareous, blocky, rubbly and flinty, with oil shales being distinctive as brown, leathery bands up to 1 m thick; fissility in fresh rock is rare. Light-coloured, carbonate-rich zones comprise about 5% of the section, and are either coccolith-rich limestones or cementstones of diagenetic origin. At least four cycles of an upwards succession of claystone, bituminous shale, oil shale and coccolithic limestone are present in cored boreholes of the onshore section near outcrop, and in the southern North Sea (Cox and Gallois 1981). This cyclicity is not readily recognised from electric logs and may be widespread over platformal areas, but is not discernable in cuttings samples, although Barnard et al. (1981) did note cyclicity in kerogen facies in a section on the northern margin of the Outer Moray Firth basin.

Aigner (1980) examined the faunal contents of the mudstones and bituminous shales of the Dorset section (Kimmeridgian to Volgian) and concluded that they were essentially similar, differing only in their mode of occurrence. The mudstones represent low oxygen conditions but with a burrowing fauna, and the bituminous shales represent periods of anoxia at sediment surface, alternating with oxygenating current events which produced graded bedding containing shelly debris reworked from the mudstones and followed by short periods of colonisation by bivalves. These characteristics support the prevalence of shallow water conditions and the occurrence of vertically embedded ammonites indicates a semi-fluid nature for the sediment surface.

A study of the Dorset section concentrating on the section containing the richest oil shale, the Blackstone Band (late Volgian) (Wignall and Myers 1988; Wignall 1989) includes measurements of species diversity, dominance and degree of fragmentation, and concludes that the environment of deposition was generally dysaerobic but interrupted by both short periods of anaerobic conditions and by storm-induced oxygenation events. Storm-dominated sandstones of early Kimmeridgian age in the Corallian Group occur to the northeast (Sun 1992).

The lithofacies and benthic faunal contents of the Dorset and Boulonnais (northern France) sections, representing far-offshore and near-shore environments of deposition, are described in detail by Oschmann (1991). In this study it is postulated that sediment surface conditions were generally dysaerobic but underwent seasonal periods of anaerobic bottom water which persisted for 1 to 3 months. Throughout the section frequent shell beds represent winnowing by storm events, and in the French section, the interdigitation of sands, of high energy depositional environments and carbonaceous mudstones, also indicates shallow water deposition. Aigner (1980) has suggested that water depths were in the order of 50-80 m.

Petrographic study of two of the oil shales in the Dorset (Kimmeridge Bay) section (O'Brien and Slatt 1990) shows that "shale" is a misnomer, their compositions being organic content 40 and 55%, carbonate 10 and 30%, clay minerals 5 and 15%, quartz and feldspar 10 and 15%, and pyrite 5 and 10%. The organic content is described as 'hash', that is, it is composed of discrete but very small particles. Further studies using SEM (Largeau et al. 1990) show that a substantial part of the organic content is composed of ultralaminae, with affinities to the resistant cell walls of Chlorophycean algae, particularly *Scenedesmus*. There has been no report of filamentous organic structures which might indicate the previous presence of mat-forming bacteria.

In the Dorset section, Mann and Myers (1989) demonstrate an upwards change from kaolinite-rich to illite-smectite-dominated clay mineralogies at the Kimmeridgian-Volgian transition which correlates with the palynoflora changing from humid sub-tropical to arid or semi-arid conditions. These changes coincide with the emergence of land areas around the Wessex and Channel basins culminating in deposition, in the mid-Volgian, of carbonates and then evaporites in a restricted basin (Cope et al. 1992).

Soft, possibly semi-fluid sediments are indicated for upper Callovian organic-rich mudstones (lower Oxford Clay Formation) which outcrop in eastern England (Hudson and Martill 1991). Organic contents are between 2 and 10%, the kerogen is dominantly sapropelic, and the clays contain a rich and diverse fauna of vertebrates and invertebrates. The internal fabric and contents of the clays demonstrate that they accumulated rapidly in shallow water and near a coastline. Interruptions of accumulation by periods of winnowing are marked by shell beds and less obvious diastems, the shell beds signifying a large removal of fine-grained material.

In the Kimmeridge Clay Formation of the Outer Moray Firth basin, Stow and Atkin (1987) recognised distal claystone equivalents of turbidite sands and breccias of the Piper Formation. The lithologies are silt-laminated claystone representing fine-grained turbidites, fissile laminated claystone possibly of turbidite origin but distal and deposited in deeper water, and silty bioturbated claystone, more common on highs.

In a well section in the South Viking Graben (Pearson and Small 1983), chlorite dominates in shales within the Sleipner Sand Formation (Callovian-Bathonian) but changes to a preponderance of illite-smectite in shales within the Hugin Shale Formation (lower Oxfordian), Heather Formation (middle Oxfordian) and Kimmeridge Clay Formation (upper Oxfordian to Volgian) with the exception of shales within the Brae Sand Formation (upper Oxfordian) which are kaolinitic. A little further north, in the Brae area, the Kimmeridge Clay Formation in the UK 16/7a-22 well averages 25% quartz, 30% calcite and 40% clay, but in well UK 16/7a-19, 50% quartz, 5% calcite and 45% clay (Shaw and Primmer 1991). Illite-smectite is a conspicuous component of the clay minerals, but SEM shows that, with illite and kaolinite, it is often a neoformation after burial. The carbonate content appears to be largely provided by microfossils.

Geochemistry and Palaeoenvironmental Indicators

The depositional model for the Kimmeridge Clay Formation, or particularly for that section which is of Kimmeridgian to Ryazanian age, has been described by Miller (1990), and puts the North Sea basin at 35° to 45°N at that time. Global palinspastic reconstructions (SPT 1992) as part of a tectonic and lithofacies study of actual and potential hydrocarbon source rocks on a global scale, have confirmed Miller's (1990) interpretation of palaeogeography and have considerably enhanced, on a globally consistent basis, the lithofacies and climate predictions (Fig. 3). The climate in the hinterland and to the north is likely to have been arid, although locally the climate may have been humid, as round the Wessex basin, where river flows were sufficient to maintain peat swamps in coastal plains.

It is argued persuasively by Miller (1990) that seawater flowed southwards from the Boreal Ocean to the Tethyan Ocean but, in the shallow waters of the archipelago of islands and shoals (Fig. 3), evaporation created waters of increased salinity (from 34 to 42‰) which sank and flowed as bottom currents into deeper water. Areas of deeper waters were most often in the more rapidly subsiding grabens and separated by sills, and the denser waters accumulated until they were able to spill over the sills. In this way the more saline waters migrated along the seabottom depressions, their paths being determined by local tectonic features but mostly from north to south. A stable water stratification ensued, the halocline being at about 50 m water depth and periods of widespread waters of increased salinity are marked by "hot shales".

Organic-rich, sapropelic mudstones of Callovian-Oxfordian age become increasingly frequent passing northwards through the North Viking Graben and are of Kimmeridge Clay Formation aspect although on an age basis they may be ascribed to the Heather Formation. Occasionally, as well as being organic-rich, they also display enhanced uranium to carbon contents, indicative of increased salinity during deposition.

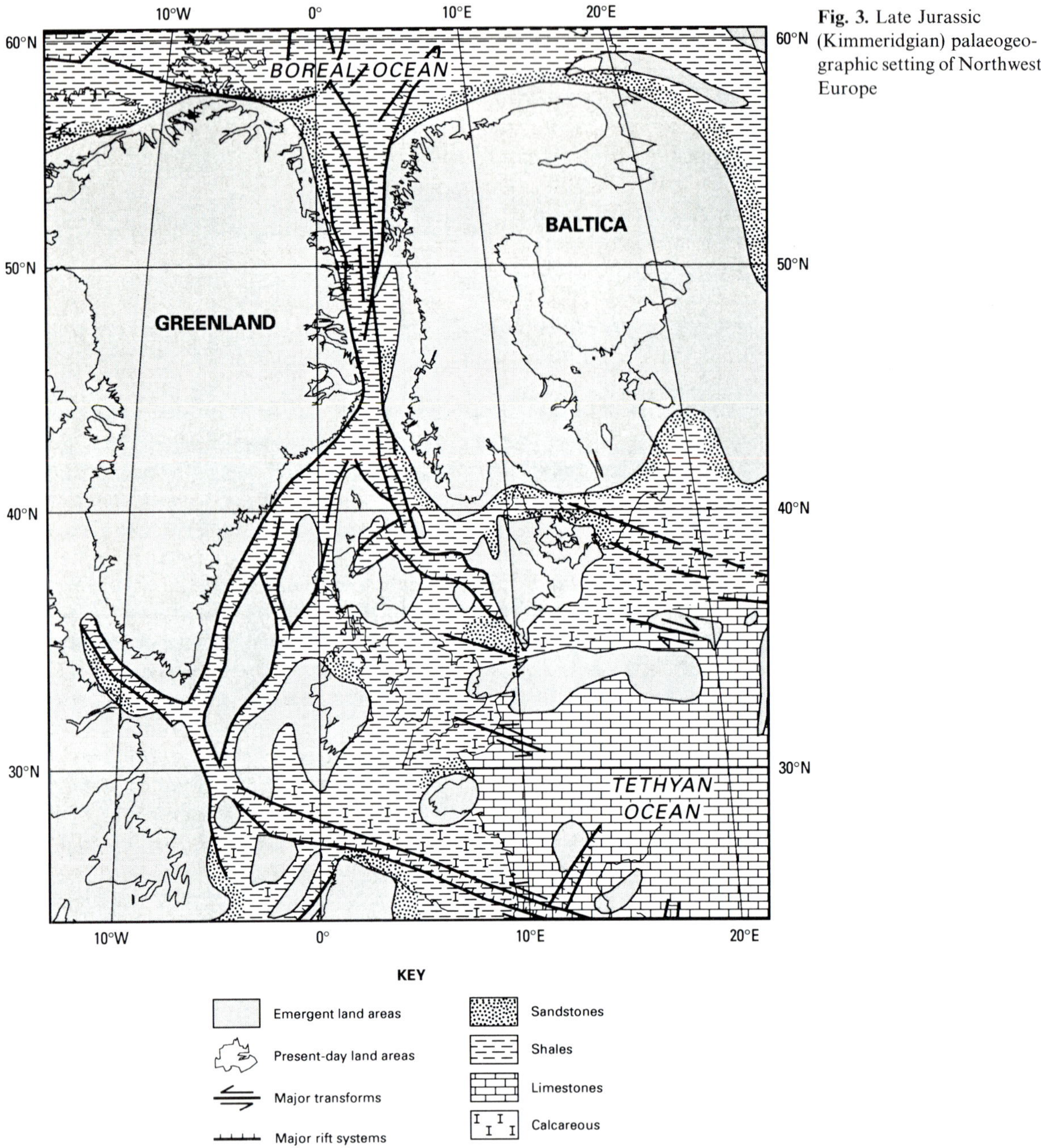

Fig. 3. Late Jurassic (Kimmeridgian) palaeogeographic setting of Northwest Europe

Observed Kerogen Facies

Studies of the organic content of the Kimmeridge Clay Formation comprise both microscopic analysis of kerogen concentrates and chemical analysis of whole rocks. Optical studies of palynofloras give valuable information on planktonic algae and transported spores, and the relative abundances of the different kerogen types are used for mapping kerogen facies. The dominant kerogen type in the low maturity source rocks is light-coloured yellow to orange, platey or granular material, without internal or peripheral structure, i.e. amorphous organic material (AOM), with darker degraded woody material (vitrinite), black granular material (inertinite) and palynomorphs (exinite) being subordinate in amount. Difficulties are encountered in total recovery of kerogen from the rock by removal of the mineral component without preferential loss of amorphous material, and in the actual identification and determination of abundance of kerogen components. For example, cutan and suberan would be hard to recognise if in a partially degraded form. Although sapropelic amorphous kerogen can often be distinguished from humic kerogen by its fluorescence, chemical analysis shows that non-fluorescent amorphous kerogen yielding high values of hydrogen index is fairly common; also, small amounts of humic amorphous kerogen admixed with sapropelic amorphous kerogen suppress fluorescence. The amorphous kerogen is variably granular, spongey or platey in texture, which may be related to different but unknown origins.

In southern England, the Middle and Upper Kimmeridgian at Westbury (Nohr-Hansen 1989) and the lower Volgian at Kimmeridge Bay, Dorset (Tyson 1989), have been studied in detail optically, with parallel chemical studies. The Kimmeridgian, with TOC values mostly in the range from 1.5 to 5.0%, shows variation from 10 to 90% AOM, the remainder being equally distributed between phytoclasts (vitrinite with inertinite) and exinites. The Volgian samples, bituminous mudstones and oil shales, with TOC from 2 to 57% contain 80 to 90% AOM. A comparison with samples from the northern North Sea, at the same level of maturity but different environments of deposition, shows a range of compositions from 90% AOM to 90% phytoclast for near-shore sediments and graben turbidites respectively. However, for deeper water platformal mudstones, the trend fits better with the Nohr-Hansen data plot, from 90 to 10% AOM with equal amounts of phytoclast and exinite. It is apparent that sorting processes during transport control the relative amounts of phytoclast and exinite, although the cause of the highly variable ratio of brown wood to black wood (vitrinite to inertinite) is not explained by this data set. Overall the palynofacies data give the same information as Rock-Eval pyrolysis data, but in addition the abundances of minor components, such as chlorophycean algae or changes in non-age-specific spore assemblages, give information on climate and water quality.

Calculated Kerogen Facies

Rock-Eval pyrolysis of carbonaceous sediment gives the parameters, Hydrogen Index (HI), Oxygen Index (OI), and temperature of maximum rate of evolution of hydrocarbons ($T_{max.}$) which are interpreted to indicate dominant kerogen type and maturity respectively (Espitalié et al. 1977). The kerogen types were initially classified as I (torbanitic kerogen), II (marine sapropelic kerogen), III (vitrinite-rich kerogen) and IV (inertinite-rich kerogen) with further sub-types defined by variation in $T_{max.}$ or other features of the chemistry of the kerogen (Barnard et al. 1981; Horsfield 1984; Cooper 1990). Precision of analysis can be high, but HI has been shown to vary with the nature and amount of mineral matrix (Katz 1983), OI is affected by

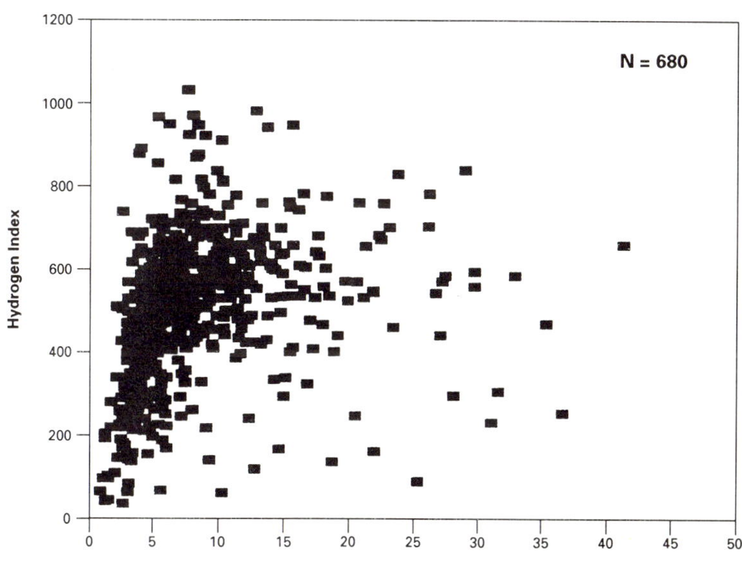

Fig. 4. Cross plot of hydrogen index versus TOC (%) for a representative set of cored boreholes in the immature Kimmeridge Clay Formation of central and southern England

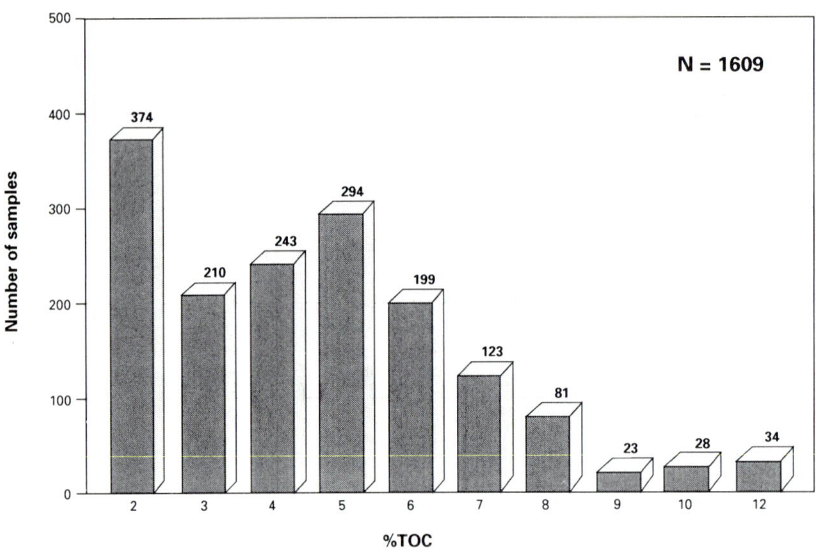

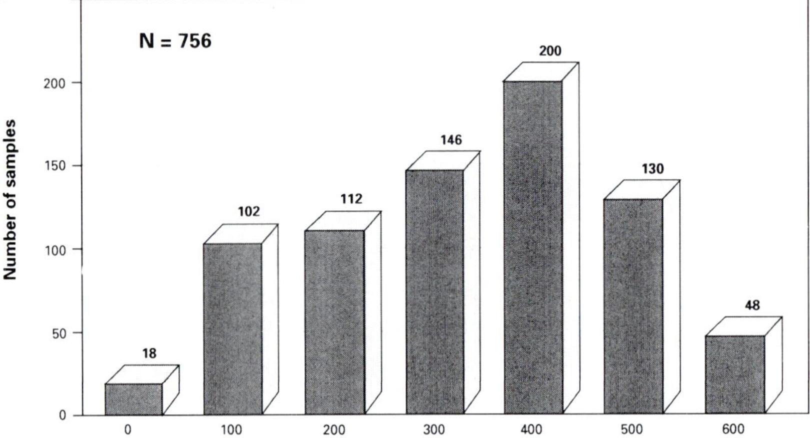

Fig. 5. Organic carbon and hydrogen index statistics for the Kimmeridge Clay Formation and its equivalents in the Norwegian North Sea

occluded CO_2 in the mineral matrix when TOC is less than 1% and HI is less than 200, and $T_{max.}$ is not always replicable to within 3 °C between laboratories. However, within the constraints of these limitations, it is possible to calculate the abundances of kerogen components (Barnard et al. 1981).

The organically richest samples for the Upper Jurassic have been recorded for the Dorset section, with a maximum TOC of 56% (Tyson 1989) for an oil shale with hydrocarbon yields of upto 250 kg/t. Maximum HI values for this section approach 1000 (our data), at a level of maturity indicated by Ro 0.5%. The Westbury section at a similar level of maturity yields maximum values of TOC 12.9% and HI 676, (Nohr-Hansen 1989). However, in both sections TOC and HI values can also be below 1% and 100 respectively. The most complete published data set on organic richness and pyrolysis data for onshore boreholes penetrating the Kimmeridge Clay Formation and based on our studies is provided by Gallois (1979). A cross plot of a representative selection of these data (Fig. 4) shows an average TOC of around 8.0% and hydrogen indices typically around 500-600, richer than the equivalent data for the Norwegian North Sea shown in Fig. 5. Among North Sea samples close to the immature-mature transition, TOC values extend to 15% and HI values to 700. Data showing organic richness (TOC) and source potential (HI) for a large database of samples from the Kimmeridge Clay Formation and its equivalents in the Norwegian sector of the North Sea are shown in Fig. 5.

Samples from onshore and wells offshore comprised the database for facies and maturity studies by Barnard et al. (1981), Cooper and Barnard (1984) and Cooper (1990). In the majority of samples, amorphous kerogen contributes more than 90% of the kerogen and is usually of a type II composition, but in addition, two types of type I kerogen are recognised, with low (403 °C) and medium (430°C) values of $T_{max.}$ in contrast to the high $T_{max.}$ Type I of Espitalie et al. (1977). A small number of samples containing amorphous kerogen as the predominant component, are of type III composition. It was assumed that these were distinct and different kerogen types with separate modes of origin which become admixed during sedimentation. However, it seems possible that these different kerogens formed under changing environmental conditions of salinity and oxicity but from the same substrate, so that intermediate compositions are not due to mixing of separate kerogen types but gradations in the balance of different biological activities (Khorasani and Michelsen 1992). Nevertheless, kerogen composition can still be treated as a mixture of end members whether the mixing is at molecular or subparticle level. On the basis of association with alkanes from solvent extracts which showed $\delta^{13}C$ of -31‰ and marked carbon preference index (CPI), type II and type I (medium $T_{max.}$) kerogens were adduced to be of land plant cuticle origin, with type I (low $T_{max.}$, $\delta^{13}C$ of -26‰) of algal origin, by Cooper and Barnard (1984). However, this interpretation is subject to review.

Although outcrop samples show that there is rapid vertical variability in TOC and kerogen content over short sections, this is much less obvious in offshore sections except for the hot shale (gamma ray log response >120 units API) intervals; some well sections appear, in their electric log response and chemical analyses, to be remarkably uniform over long intervals. To achieve an average value for the kerogen contents for a section represented by cuttings samples, the electric and caliper logs are used to determine how bulk and hand-picked samples represent different parts of the section. The kerogen contents can then be calculated, first at the present level of maturity, and then to a common level of maturity, usually taken at the immature-mature transition (Ro 0.4%, SCI 3.5). Usually the dispersion of values about the average is no more than the analytical and calculational sensitivity (about 20%), but occasionally a section will show extreme oscillations between sapropel-rich and humic-rich, and this can be designated a particular kerogen facies.

In this way, kerogen facies can be mapped for each interval from the upper Bathonian to the lower Ryazanian. In the northern North Sea, such maps demonstrate the progress of the Upper Jurassic transgression with kerogen facies zones migrating outwards from the basin centre. An example of kerogen facies for the late Kimmeridgian-middle Volgian period in the northern North Sea is shown in Fig. 6. In general, from the palaeo-shoreline towards the grabenal areas, the kerogen facies change progressively through enrichments in type IV, type III to type II, with type I occurring in patches in the grabenal areas. High-standing areas during deposition within the basin are marked by type III or IV. Down-slope from active fault systems, which created sub-aerial uplifted blocks as sources of mass-flow deposits, sequences of sands and humic clays also contain remnant quiet water deposits containing type I kerogen. Shallow water muds were also activated by either storms or tectonism to pour over the graben edge so that, for example, carbonaceous claystones containing high contents of types II and IV occupy large areas of the South Viking Graben and appear to have been supplied from a steep slope on the graben margin between the Brae and Beryl structures. At other times, well-sorted sands spilled over faulted platformal edges to interdigitate with sapropelic muds as in the Magnus area. It may be concluded that the distribution of kerogens is controlled by a combination of the hydraulic properties (particle size, density) of the land-derived inertinite, vitrinite, cuticle, etc. and by the nature of the environmental conditions in which algal kerogen is preserved and in which bacterially derived kerogens are formed and transmuted (Farrimond et al. 1984).

Biomarkers and Carbon Isotopes

The alkane and aromatic fractions of solvent extracts from Kimmeridge Clay Formation samples yield a plethora of data. Gas chromatography of the alkanes of sample extracts near the immature/early mature transition show *n*-alkane envelopes which slowly diminish in concentration towards C_{31}, large amounts of pristane and phytane, and perceptible amounts of biomarkers in the C_{26-32} region. Analyses of outcrop samples from southern and eastern England (Douglas and Williams 1981; Farrimond et al. 1984; Ebukanson and Kinghorn

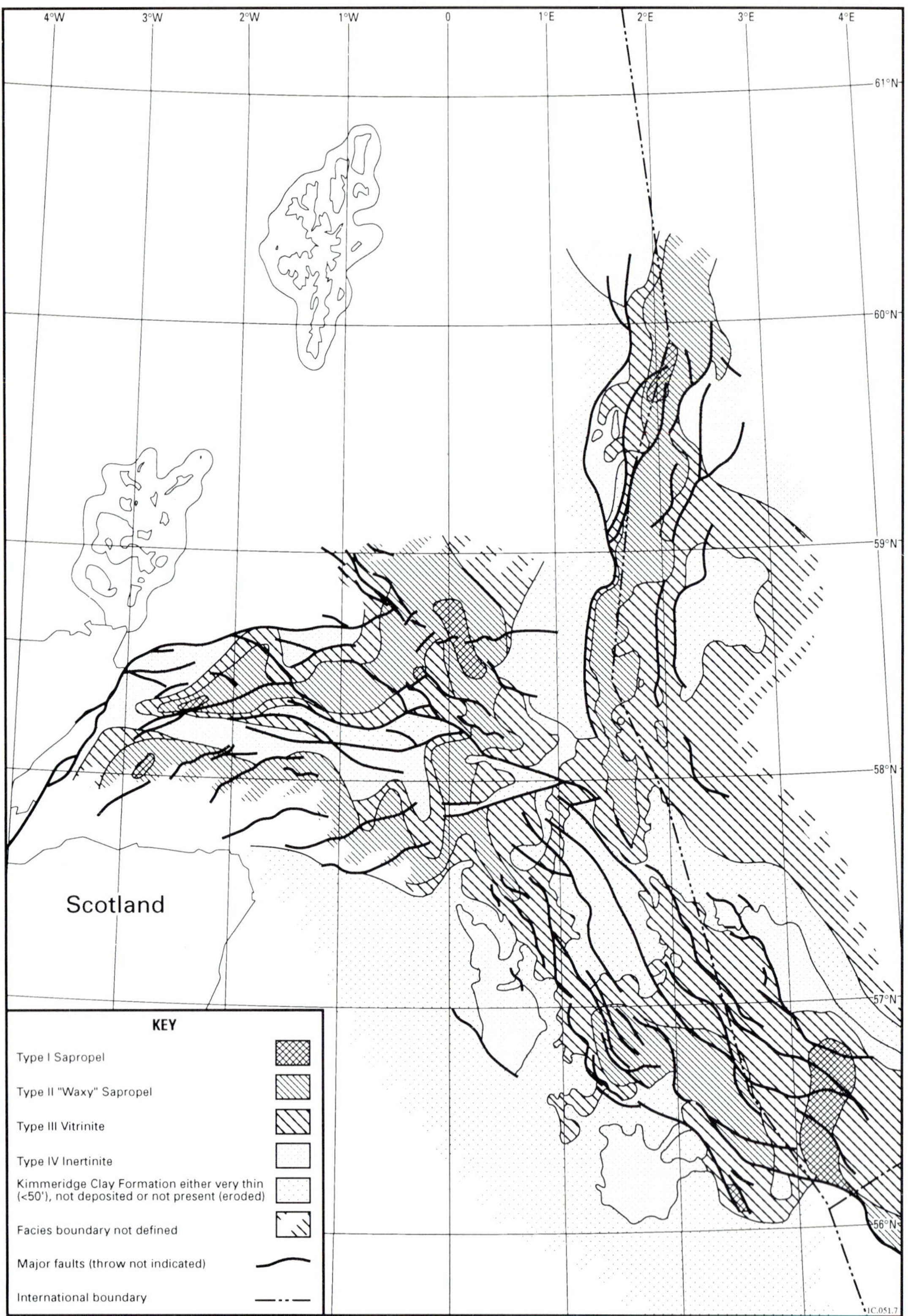

Fig. 6. Kerogen facies of the late Kimmeridgian-middle Volgian period in the central and northern North Sea

Fig. 7. Terpane (m/z 191) and sterane (m/z 217) fragmentograms for a representative Kimmeridge Clay Formation equivalent source rock extract

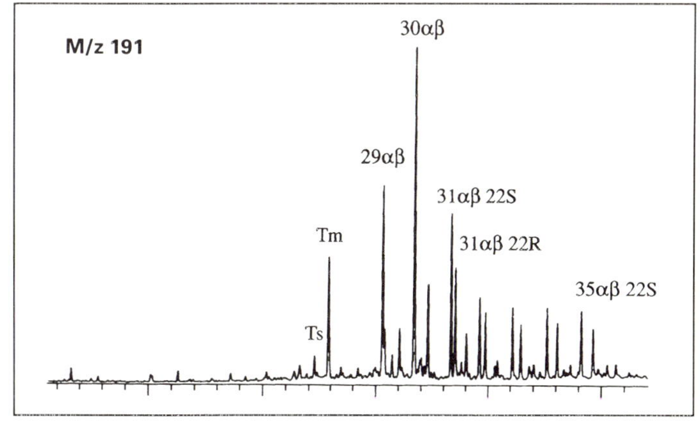

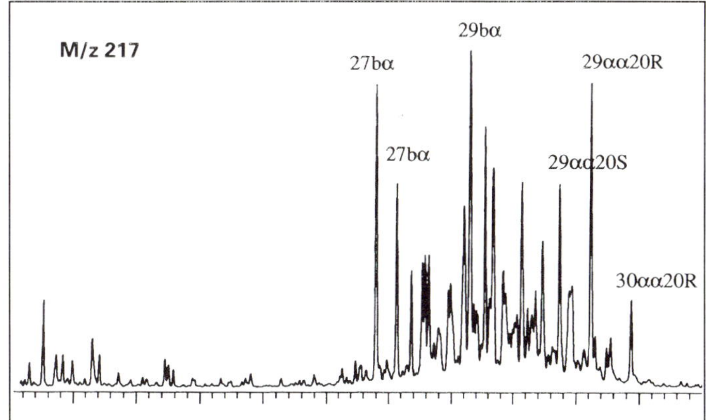

1990) surprisingly also show pronounced odd over even carbon preference (CPI) in the *n*-alkanes in many samples including oil shales. While some immature/low mature kerogens, such as vitrinite and exinite, contribute proportionally much greater amounts of hydrocarbon to the extract than others, such as alginite and amorphous kerogen, in this case we conclude that the high values of CPI must relate either to an unknown marine source or to the unrecognised presence of cuticle.

A large amount of information has been gathered on biomarkers in sediments and oils of the North Sea, principally on relative concentrations within the hopane and sterane groups. Ratios based on peak heights, such as hopane/sterane, may show inter-laboratory variation for the same sample due to the use of different equipment. Absolute measurements of concentrations are possible and our experience shows that they are extremely valuable but are not often published. In such detailed geochemical studies it is also usual to measure $\delta^{13}C$ for the alkane and aromatic fractions as discussed later.

The biomarker data are most often used in visual comparisons of fragmentograms (Fig. 7) during oil-source rock correlation, although, if sufficient data are available, multivariate analysis can also provide useful additional information (Telnaes and Cooper 1991). Cross plots of alkane and aromatic $\delta^{13}C$ values, and ternary diagrams of % abundance of the $C_{27,28}$ and C_{29} steranes are usually interpreted in terms of the relative marine and terrestrial inputs. A representative dataset of sterane composition for a large group of North Sea oils and Kimmeridge Clay Formation source rocks from the same areas is shown in Fig. 8. The broad similarity in composition of the oils is particularly notable. The spread of data for the source rock extracts is perhaps significantly broader, indicating some greater variation in the source material inputs to the sediment.

The main problems in oil-source rock correlation, however, are that analysed source rock sections are usually on the periphery of the kitchen and are therefore less mature and contain a different kerogen facies. The analysed source rock

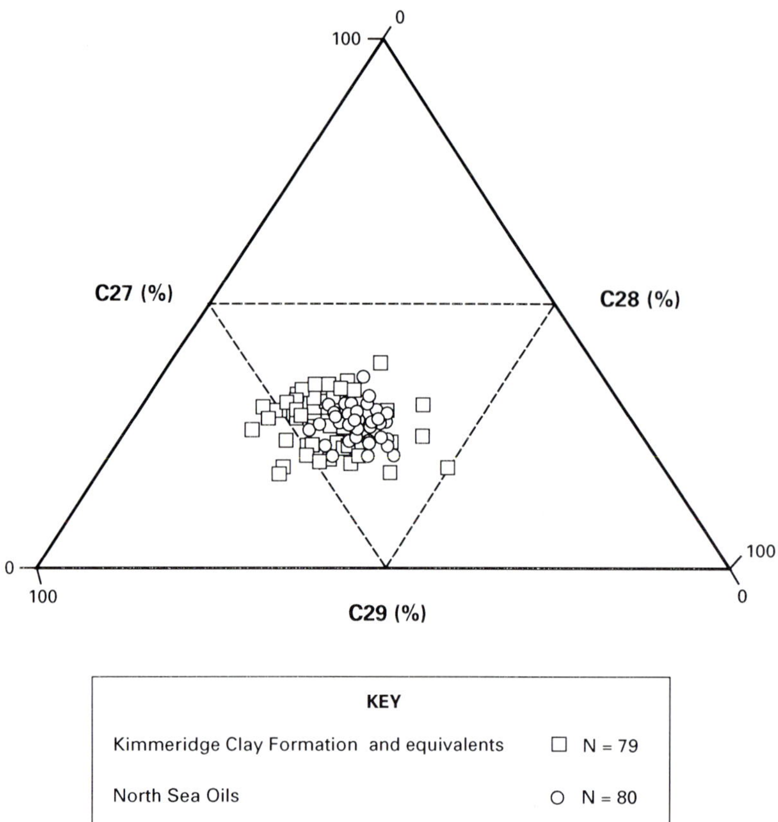

Fig. 8. Sterane carbon number distribution for a representative selection of North Sea oils and Kimmeridge Clay Formation source rock extracts

kerogen may contain greater or lesser amounts of type III kerogen which contributes significantly to the alkanes extracted from the source rock but is not itself a significant source of oil. Although kerogen mapping can predict the kerogen facies in the kitchen, the biomarker and isotope signatures of the oil derived from that kitchen are much less reliably predicted. An approach to solving this problem is the topic of the next section of this chapter.

The unusual feature of the Upper Jurassic sediments and particularly their derived oils is the presence, often in high amounts, of 28, 30 bisnorhopane (Grantham et al. 1980); its origin is uncertain (Waples and Machihara 1991) but hydrous pyrolysis of extracted kerogen shows that it is not a component of the kerogen structure. It is present in high abundances in both platformal and grabenal source rocks (and in lagoonal coal) which also contain a high humic component and are sulphur-rich. A further frequent feature is a considerable enrichment in the C_{33-35} extended hopanes, which are a result of preferential preservation in the presence of sulphur (de Leeuw and Sinninghe Damste 1990), and this is therefore frequently seen in sediments with limited iron availability, such as carbonates and evaporites. The relative abundances of C_{27-30} steranes show great variability as do the $\delta^{13}C$ values of alkane and aromatic fractions which range from -24 to -32‰.

The large variance in relative concentrations of biomarkers can best be handled by statistical methods. The results of multivariate analysis show that this approach can be fine-tuned to extract valuable additional information by using it on large datasets and subsequently refined to abstract further information from data subsets. Results from a large dataset show how Kimmeridge Clay Formation samples compare with contemporaneous deposits beyond the North Sea. The examples chosen (Fig. 9) are for molecular ratios often used as maturity parameters (Waples and Machihara 1991). Each of the ratios shows a trend of increase with maturity (Ro or SCI), but shows a wide dispersion in plotted points, a dispersion also noted by Waples and Machihara (1991) for other data sets, which is demonstrated indirectly when two of the ratios are cross plotted (Fig. 9). The amount of dispersion in the data suggests that this is not due to differing burial histories, but to

KIMMERIDGE CLAY FORMATION EQUIVALENTS

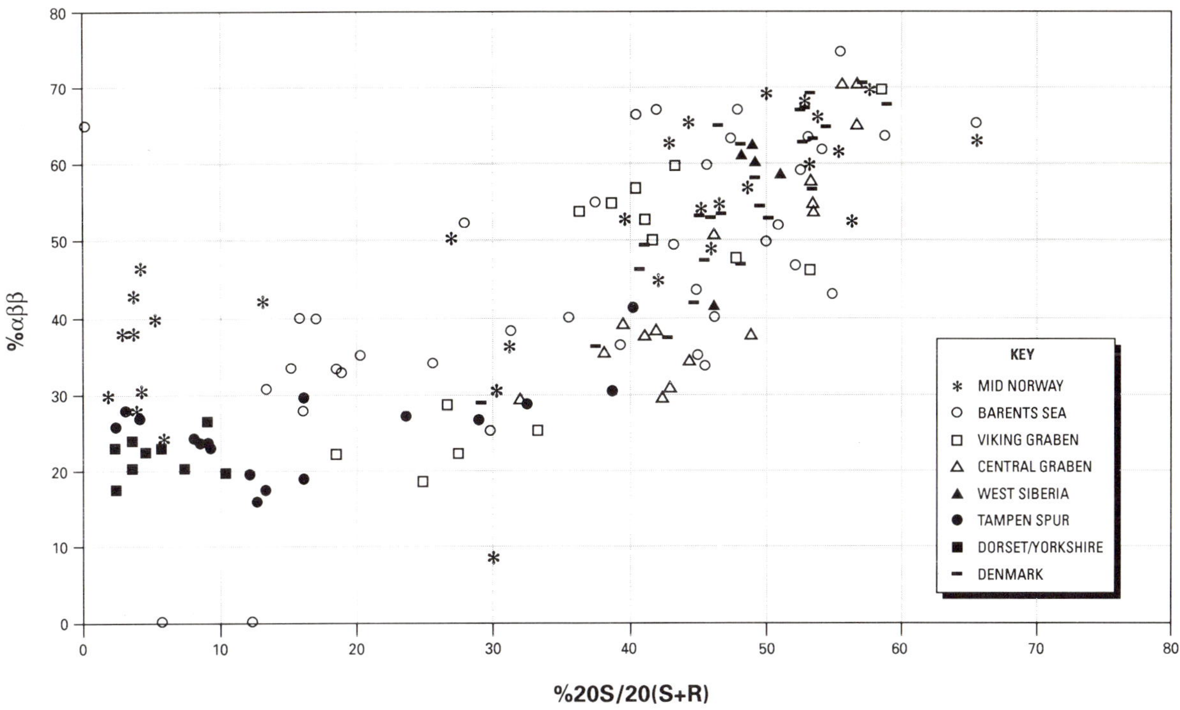

LEGEND:

%αββ = Abundance of C_{29} 5α(H)14β(H) 17β(H) 20(R+S) sterane as % of total C_{29} sterane abundances

%20S/20(S+R) = Abundance of C_{29} 5α(H)14α(H)17α(H) 20S sterane as % of total C_{29} 5α(H)14α(H)17α(H) 20(S+R) steranes

Fig. 9. Cross plot of supposedly maturity dependent sterane ratios from the Kimmeridge Clay Formation and its equivalents

stronger facies influence on these ratios than has been previously suspected.

Kerogen Composition

The above discussion has identified the stratigraphic and tectonic setting for the Kimmeridge Clay Formation and the information which may be derived from the biomarkers and isotopes as to the environment of deposition. A model to accommodate these observations is now required.

The organic input into typical marine sediments comprises:

1. terrestrial sourced material, comprising fresh plant tissues and reworked fragments of peat from coastal plains;
2. phytoplankton remains, which may be in the form of zooplankton fecal pellets;
3. microbial tissue, which may form at the surface as mats, but is also generated within the sediment in the sulphate-reducing and methanogenesis zones.

Each of these sources provides resistant tissue, such as lignin and lipid biopolymers, and other materials, mostly carbohydrates, which act as a substrate for initial or further bacterial metabolism. Typical coals, which have no potential to generate oil, form from freshwater peat in which, in the absence of a significant sulphate-reducing zone, proto-kerogen derived from mirobial exudates and residues is formed during methanogenesis. In contrast, lagoonal coals with oil potential form in a saline, usually tidal, environment and, with a plentiful supply of sulphate, sulphate reduction is prolonged (Thompson et al. 1994). It may be concluded that bacterial activity in the sulphate-reducing zone can add substantially to the lipid biopolymer content of kerogen, that is as types I and II, but not in the methanogenesis zone, where type III is formed.

Hopane may be assumed to be a general indicator of total microbial activity in the sediment when compared with the total amounts of steranes, which are a measure of algal and land plant contribution (Waples and Machihara 1991). Although there may be algal and bacterial sources for enhanced amounts of C_{29} sterane, generally, the relative amounts of C_{27} and C_{29} steranes can be used to distinguish algal from landplant inputs.

The initial organic contributions to a sediment which will eventually become a high quality source rock, such as the Kimmeridge Clay Formation, are algal and land plant lipid biopolymers (algenan, cutan and suberan), and their associated carbohydrates, the latter being transmuted to residual lipid-rich microbial tissue after several phases of biodegradation. The $\delta^{13}C$ values for the carbohydrates can be expected to be 3 to 5‰ more positive than their lipid counterparts, and, during biodegradation of carbohydrate to microbial biopolymer, it is anticipated that there is little isotopic fractionation (Waples and Machihara 1991). Possible values for hydrocarbons (+/−2‰) in source rocks deposited under temperate to tropical climatic conditions including the Kimmeridge Clay Formation, which we have observed to be associated with or generated from these sources, are:

1. Land plant cuticle −30‰
2. Microbial lipid from land plant substrate −26‰
3. Algal biopolymer −26‰
4. Microbial lipid from algal substrate −22‰

In addition, hydrocarbons derived from contemporaneously reworked peat (to be preserved as particles of vitrinite) may also be present and take values similar to those for alkanes from Upper Jurassic coals which typically have values of −28 to −29‰. However, this simplest of models will be shown to be insufficient for describing the geochemistry of the Kimmeridge Clay Formation.

A Revised Classification of Kerogen Types

If we are to optimise the potential application of geochemial interpretation to petroleum exploration we must be able to:

1. map kerogen facies such that other palaeoenvironmental factors can help in interpolating values between sample points, and
2. examine the composition of oil, principally its isotopic and biomarker signatures, and state, with more precision than currently obtained, from which kerogen facies it was derived.

To do this, it must be accepted that there are a greater number of kerogen types deposited into or generated within sediments than usually envisaged. We define a kerogen type as being sedimentary organic matter which has been generated from a distinctive progenitor and by a distinctive process, so that its chemical and isotopic composition lies within definite limits. Pure kerogen types are rarely found; kerogen usually occurs as mixtures which may be arbitrarily divided into kerogen facies. We anticipate that the number of kerogen types which may be available to a source rock within a discrete sedimentary basin may be in the region of 10 to 20, on the basis of the variety of organic materials which may be supplied to the sedimentary environment and the numerous processes of organic matter generation taking place within the sediment.

Various data for a carefully selected set of samples from offshore wells, within a narrow maturity range of Ro 0.45–0.55%, SCI 3.5–5.5, and analysed in one batch, are presented in Table 1. The samples mostly represent mudstones and turbidites deposited in water depths of 50-100 m with a smaller representation from shallow water sediments. This small dataset can be used as the basis for a preliminary exercise using multivariate analysis, so that a possible model of interaction can be established and the number of variables estimated. From this study it is determined that at least eight individual kerogen types or end members are present, comprising types Im (microbial), IIa (alginite), IIb (microbial), IIc (cuticle), IIIv (vitrinite), IIIa (microbial), IIIb (microbial), and IV (inertinite). Much of the calculation is by iterative approximation after inserting likely values for kerogen properties.

The amounts of each of the kerogen group types have been calculated from Rock-Eval pyrolysis data by the method of Barnard et al. (1981). A plot of CPI, corrected for maturity, against abundance of type III (Fig. 10a) shows values between 0.91 and 1.88, the maximum possible being 2.15 with a general increase from low to high values of % type III. Features to be noted are the general increase in CPI with increase in % of type III, that significant values of CPI occur at low concentrations of type III, and not all high concentrations of type III show high values of CPI. This means, there is a type II

Table 1. Geochemical parameters for selected source rocks

Sample		TOC		$\delta^{13}C$‰		C_{27} Ster	Hop/	Abundance of kerogen types: From Rock-Eval (*)				From all data (†)							
no.	SCI	(%)	HI	Alks	Arom	(%)	ster	I	II	III	IV	Im	IIa	IIc	IIb	IIIv	IIIa	IIIb	IV
1 (S)	5.0	3.3	204	27.7	24.7	21	3.92	0	15	76	9	0	1	1	12	9	13	54	9
2	4.5	3.3	139	29.2	28.3	38	2.08	0	7	18	75	0	2	0	4	1	3	14	75
3 (S)	4.0	7.2	218	28.4	27.3	34	1.20	0	24	43	33	0	8	2	14	1	21	21	33
4 (S)	4.5	12.5	214	27.0	24.7	23	5.61	0	19	68	13	0	1	1	18	4	19	44	13
5	3.5	23.9	184	28.6	26.5	30	1.36	0	15	15	70	0	3	1	11	0	8	7	70
6	5.0	4.8	250	29.3	28.4	34	1.61	0	35	34	31	0	6	2	26	1	11	22	31
7 (S)	4.5	8.4	306	29.4	28.9	43	1.63	0	42	37	21	0	10	1	30	3	12	22	21
8	5.5	5.7	269	27.6	26.8	43	2.07	0	33	41	26	0	8	1	24	4	20	17	26
9 (S)	4.0	10.8	249	27.8	25.3	14	3.40	1	22	62	15	1	1	2	19	8	11	43	15
10	4.0	5.4	284	29.7	28.8	32	1.24	2	30	48	20	2	8	3	18	0	11	37	20
11 (S)	4.0	3.8	263	28.8	27.6	40	1.48	12	18	41	29	12	9	2	7	2	13	27	29
12	5.0	4.5	227	28.9	28.7	34	1.40	12	27	29	32	12	7	2	18	5	7	17	32
13	4.5	3.5	309	27.0	25.9	27	1.44	26	18	42	13	26	6	4	8	6	20	16	13
14	3.5	7.3	613	30.2	28.7	38	1.08	36	31	31	2	36	14	3	13	0	1	30	2
15	5.0	12.9	568	29.0	28.2	41	1.84	47	14	34	5	47	10	2	2	3	1	30	5
16	3.5	6.0	629	28.0	28.4	48	1.29	51	19	19	11	51	16	1	2	0	19	0	11
17 (S)	4.5	58.1	617	27.3	26.2	33	2.89	56	10	24	10	56	5	2	4	0	7	17	10
18	4.0	5.6	639	26.6	26.8	38	1.99	62	11	22	5	62	9	2	0	2	17	4	5

SCI: Spore Colour Index of thermal maturity (1–10 scale).
TOC: Total organic carbon (%).
C_{27}ster(%): C_{27} steranes as % of $C_{27} + C_{29}$ steranes.
Hop/ster: ratio of hopane to total C_{27}–C_{29} steranes.
$\delta^{13}C$‰: Values quoted are all negative relative to PDB.
(*): Data derived from Rock-Eval pyrolysis data after method of Barnard et al. (1981)
†: Interpretation of kerogen type follows method described in this chapter
(S): After sample number indicated sample deposited in water depth of less than 50 m.

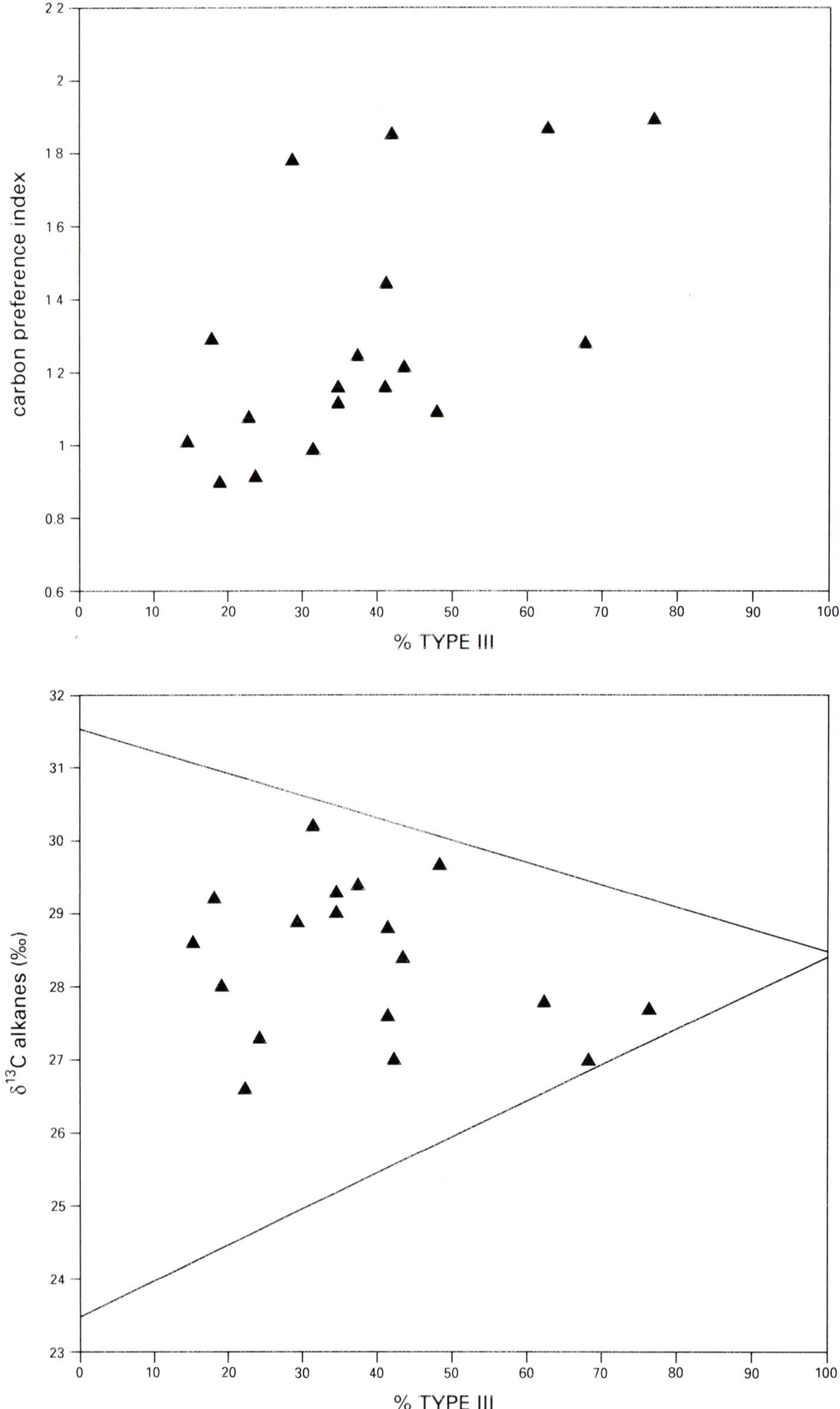

Fig. 10. Cross plots of carbon preference index (CPI) and alkane carbon isotope enrichment against percentage of vitrinitic kerogen for selected Kimmeridge Clay Formation source rocks

Table 2. Amounts of hydrocarbons (per unit of TOC) present in different kerogen type relative to the maximum (= 100) for any hydrocarbon class

Kerogen sub-type	C_{15+} alkanes	C_{15+} aromatics	C_{25-31} alkanes	Biomarkers; steranes + hopane
I (microbial)	43	1	14	6
IIa (alginite)	20	100	14	100
IIb (microbial)	67	9	14	4
IIc (cuticle)	77	10	100	4
IIIv (vitrinite)	67	15	100	33
IIIa (microbial)	67	71	14	8
IIIb (microbial)	100	32	14	30
IV (inertinite)	0	0	0	0

kerogen present which exhibits pronounced CPI, there is a type III component present other than vitrinite, and a common component in the mixtures carries a CPI of less than 0.9. A plot of alkane $\delta^{13}C$ against % type III (Fig. 10b) shows that the $\delta^{13}C$ for the samples enriched in this type II kerogen are more negative than − 29‰. Assuming that 100% type III is likely to be detrital vitrinite with $\delta^{13}C$ of − 28 to − 29‰, an envelope around the plots suggests that end members containing 0% type III have approximate values at about − 24.5 and − 31‰.

Each kerogen type can be expected to contain solvent extractable material, including alkanes, aromatics, biomarkers etc. in concentrations relative to TOC which are individual to that kerogen type. The concentrations of hydrocarbons associated with each kerogen type also vary with maturity. For this sample set, the relative contributions of four hydrocarbon fractions, measured against the maximum contribution of any one kerogen type, have been determined from our database and adjusted by iteration (Table 2).

The steranes/(steranes + hopane) ratio takes values between 0.10 and 0.65 allowing for the adventitious presence of hopanes and steranes in purely microbial or purely plant tissue respectively; the steranes ratio, st_{27}/st_{27+29} takes values between 0.1 and 0.6 based on variation within a larger data set. These ratios have not been adjusted for maturity. CPI which is very sensitive to maturity has been adjusted to a scale of maximum values which change from 2.15 to 1.05 over the maturity range of the analysed samples.

A plot of the sterane-hopane function against $\delta^{13}C$ (Fig. 11) shows the samples fall within a band, and the C_{27}-rich members tend to be more negative in carbon isotopc. From our larger data set of oil and source rock data for North Sea oils and Kimmeridge Clay Formation rock extracts we note that the highest and lowest values of alkanes $\delta^{13}C$ are − 23 and − 32‰, and conclude that values close to these (− 21 to − 23‰ and − 32 to − 34‰) correspond to two of the kerogen types. Perusal of the relative concentrations of C_{27} and C_{29} steranes and the ratio of steranes to hopane indicates that an algal component takes a value between − 32 and − 34‰, a microbial component between − 21 and − 23‰, a land plant component from − 28 to − 30 and a second microbial component from − 26 to − 28‰. After iteration it is possible to allocate values of about − 33.5, − 30.0, − 28.5 and − 22.0‰ to the alkanes of the algal, terrestrial and two microbial components respectively. The microbial components are, however, each themselves mixtures of types II and III.

The range of adjusted CPI values is from 0.91 to 1.88, and kerogen sub-types IIc, IIIv and IIa are allocated values of 2.15, 2.15 and 1.0 with other microbial kerogens 0.7, to give the proportion of IIc, IIIv and, by comparison with total type III from Rock-Eval, microbial types IIIa and IIIb. Inspection of the data accumulated to this point shows that the abundance of type I, from Rock-Eval, correlates with a microbial component characterised by alkane $\delta^{13}C$ of − 22.0‰. The abundances of other components follows by allocating and adjusting $\delta^{13}C$ values.

By these procedures it has been demonstrated that there are up to eight kerogen components present in this sample set, although the calculated relative amounts are very sensitive to the assumptions and approximations made during the calculation, and the amounts of alginite and cutinite may be underestimated. Nevertheless, the results show that, to achieve greater rigidity in the interpretation

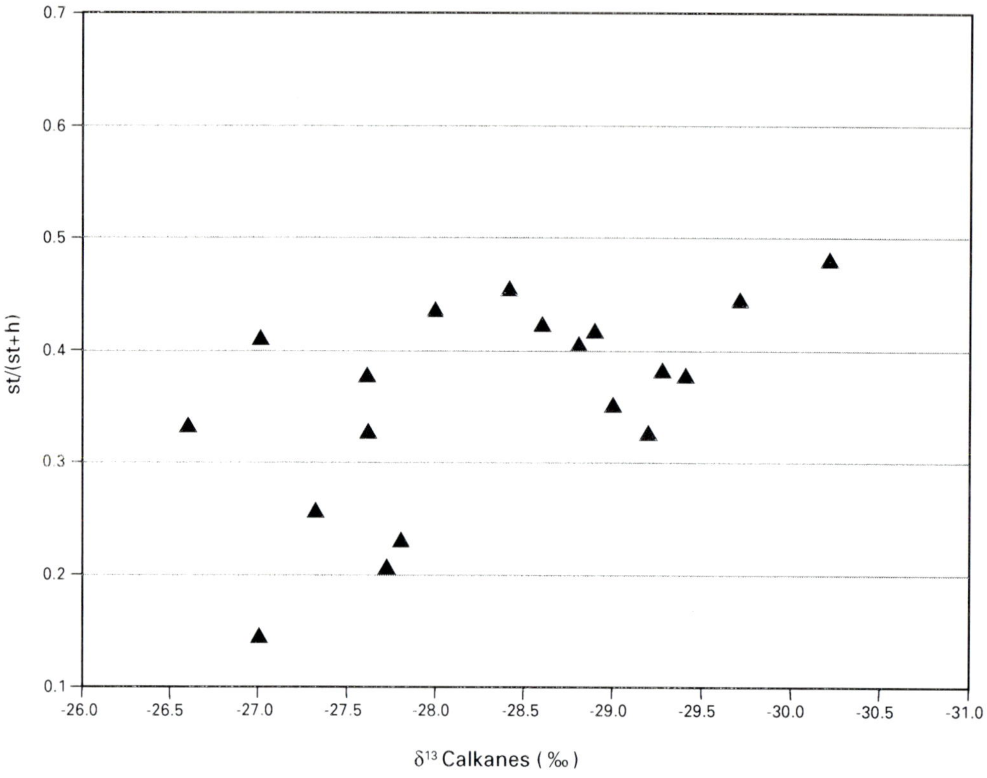

Fig. 11. Cross plot of total steranes over steranes plus hopane abundance against alkane fraction carbon isotope enrichment

of geochemical data, multivariate analysis is highly beneficial when carried out on datasubsets which lie in narrowly defined maturity classes and possibly in classes which are typified by particular biomarker profiles. From such numerical analysis it may be expected that a system of kerogen typing will emerge which will use different parameters from those used previously and be more comprehensive that systems currently in use.

Implications of the Model

The results of the foregoing numerical interpretation show that the distribution of kerogen types is related to the various environments of deposition in the basin. The calculated abundances of cutinite from the above are negligible although they appear to be significant elsewhere in platformal deposits where high CPI values coincide with high HI. The amount of particulate vitrinite (type IIIv) is at a level expected from abundances observed in kerogen concentrates. Alginite (IIa) is fairly uniformly distributed at 10-20%, but low in shallow water mudstones. Type IIb kerogen is prevalent in the source rocks of the Kimmeridge Clay Formation and is a frequent major component of these samples but it is to be noted that it is not a detrital component as previously envisaged (Barnard et al. 1981) but a microbial contribution. The high abundance of type Im kerogen, also of microbial origin, presumably reflects accumulation in areas of greatest water depth and in a chemical environment distinctly different from elsewhere. A possible cause of the change in microbial activity to give type Im is a lack of available iron for sulphide precipitation during prolonged sulphate reduction arising from enhanced sulphate concentrations. There is a possibility that methanotrophy plays a part, although higher concentrations of type I kerogen do not obviously overlie deeper sections of enhanced carbon content in which methanogenesis may have occurred. The relative amounts of IIIa and IIIb are the least precise in the estimates and erratic in their distribution, although it is possible to see a descending trend in IIIa paralleling a decline in IIb. It is also possible that a small but significant amount of a further type III component is present, the three types variously representing (a) original organic matter derived with the clay minerals from a land

Table 3. Predicted isotopic composition of extracts and oils sourced by facies mixtures of newly defined kerogen types

Source rock example	1	2	3	4
Kerogen type	Kerogen contents (%)			
Ia	10	20	40	60
IIa	0	20	10	0
IIb	30	30	30	30
IIIa	20	10	5	0
IIIb	30	10	5	0
IV	10	10	10	10
Kerogen facies	III	II	II/I	I
Expected extract				
$\delta^{13}C$ alkanes (‰)	− 26.7	− 30.8	− 30.3	− 28.0
$\delta^{13}C$ aromatics (‰)	− 24.4	− 29.7	− 29.6	− 28.0
Expected oil generated				
$\delta^{13}C$ alkanes (‰)	− 28.0	− 30.2	− 29.4	− 28.0
$\delta^{13}C$ aromatics (‰)	− 28.0	− 32.8	− 32.2	− 28.0

source and, (b) by products of methanogenesis after formation of types Im and IIb.

The overall result shows that in a study of kerogen compositions, as many as eight or more sub-types may be present which have an effect on the geochemical characteristics used in kerogen facies discrimination and in oil to source rock correlation. Additional parameters, such as pristane, phytane, norhopane and C_{35} hopane contents which partially describe the microbial input, and other environmental indicators, TOC, sulphur, uranium and carbonate content which reflect sulphate availability, are needed. The sensitivity of many of the variables to maturity, demands that data sets are divided into rather more constricted maturity bands, and the measurement of biomarker concentrations should be absolute rather than relative.

When source rocks are kerogen-rich but the kerogen comprises both sapropelic and humic components, the geochemical characteristics of the source rock do not exactly match those of its generated oil, as shown in the synthetic examples of Table 3.

In the calculation of aromatic $\delta^{13}C$ values, it is assumed that both alkanes and aromatics in each kerogen sub-type are generated from the same lipid molecular structures and have closely similar $\delta^{13}C$ values so that only when mixtures of kerogen subtypes are dominated by one or two components, such as in some lacustrine and deltaic source rocks, are alkane and aromatic $\delta^{13}C$ values coincident.

Source Rock Variation in a Single Section

A second example of source rock variation is from a small fault-bounded basin within the Norwegian Platform. The analysed section, including 50 m of core, is close to the immature-mature transition (Ro 0.45%, SCI 4.5) and comprises 11 m of black fissile shales of the Draupne Formation (middle Volgian to Ryazanian) lying unconformably above 51 m of dark grey silty mudstones of the Heather Formation (Oxfordian). Occasional thin limestones occur in the Draupne Formation and thin sandstones in the Heather Formation. A wide range of analyses of the organic, mineral, major element and trace element contents have been described in outline (Telnaes and Cooper 1991); TOC and HI values are 5.5–7.2%, 493–591 (Draupne) and 1.1–4.8%, 175–295 (Heather).

A cross plot of HI and OI shows that the kerogen contents in the Draupne Formation are dominantly mixtures of type I and type II, with minor (< 5%) amounts of types II and III. However, the ranges of OI (10-19) and $T_{max.}$ (417–426°C) in the Draupne Formation samples indicate that the type I is represented by at least two sub-types, one of which has OI about 20 and $T_{max.}$ about 428°C and probably is the same as that identified by Cooper and Barnard (1984) as a major contributor in the Outer Moray Firth basin. The CPI values of these samples, 1.3 to 1.8, are unlikely to have been provided by the small amounts of types II and III, but by a sub-type I. This latter sub-type is probably derived from land-plant cuticle, although an algal/bacterial source has been suggested (Farrimond et al. 1984) and kerogens with similar characteristics occur in Silurian and older source rocks. The other sub-type, with OI and $T_{max.}$ at about 11 and 415°C respectively, and associated with higher contents of C_{35} homohopane, is probably the same as found elsewhere in grabenal deposits.

A second feature of this sequence is a 1 m-thick, micritic dolomite layer, apparently of diagenetic origins which marks the unconformity at the top of the Heather Formation. The unconformity represents an episode of erosion, probably submarine rather than subaerial, during a period of sea level lowering, which removed an estimated 10 to 20 m of section. The dolomite has TOC of 3.4% and HI of 256 but yields a very high content of C_{35} homohopanes relative to hopane, and is enriched in uranium and molybdenum. It may be envisaged that the erosional event removed the sulphate-reducing zone and brought fresh sulphate-rich seawater into contact with the methane-generating Heather Formation. The diffusion zone for sulphate in the top few metres of Heather Formation is marked by a transition from typical Heather Formation distributions of hopanoid biomarkers to characteristics more similar to Draupne Formation. Microbial activity would therefore be sustained by upwards diffusion of methane and the processes of methane oxidation, sulphate reduction and microbial carbon dioxide assimilation caused dolomite precipitation at the sediment surface. The low availability of iron in the sediment, having been removed as sulphide in the previous phase of sulphate reduction, allowed free sulphur species to be present and ensure the preservation of the homohopanes. It may be expected that diagenetic formation of carbonate-rich layers may be expected whenever there is an abrupt mobilisation of sediment from shallow water and its redeposition in deeper water. The precipitation of carbonate crusts is coincidentally also a present-day feature of methane seepage at sea bottom in the North Sea.

Summary

Although it may be that the accumulation of source rocks in the North Sea during the Upper Jurassic was governed by more complex processes than for other source rocks, this complexity as described above, may reflect the greater amount of data and interpretation which has been accumulated; most other source rocks are probably formed by a similar complexity of processes.

From the above discussion, it can be seen that none of the generally accepted factors giving rise to source rocks in the global context is operative on its own in the Northwest European Jurassic source rock system of which the Kimmeridge Clay Formation of the North Sea is part. The oceanic anoxic event concept whereby high sea level stand results in impingement of the oxygen minimum zone on the continental shelf is clearly not appropriate given the diversity of depositional environments which can be recognised and the relatively shallow waters (never much greater than 80 m). The configuration of the basins within developing continental rift systems is also not conducive to widespread upwelling and we believe that nutrient supply may be derived dominantly from erosion of the many emergent land masses surrounding the rifts and their associated drainage systems rather than from deep ocean circulation (SPT 1992).

This highly important source rock system then seems to be dependent upon factors intimately related to the system of rifts with submarine sills constantly evolving and shifting as the rifts developed. These basins existed in large mid- to high-latitude open oceans with major oceanic water movement through the system. In this complex tectonic setting these organic-rich, smectite-rich and probably thixotropic Jurassic claystones were deposited at shallow depths in seawater of increased salinity and density. At slow rates of accumulation and with a halocline stabilising the water column, dysaerobic to anoxic conditions at the sediment surface were created and an open texture in the sediment allowed enhanced sulphate diffusion into the sediment.

In places, the sediment readily flowed, when disturbed by storms or tectonism, down the barely perceptible slopes of the basin. When this happened, the newly exposed sediment surface and the redeposited sediment, in both of which the organic matter had already undergone partial microbial diagenesis, came into contact with a replenished source of sulphate and extended the activities of the sulphate-reducing biota to form the high amounts of sapropel which are characteristic of the Kimmeridge Clay Formation. The availability of iron controlled the concentration of sulphur and sulphides, and when the supply of iron was exhausted, sulphur species reacted with and preserved particular organic structures and may also have caused a shift in kerogen composition from type II to type I. The variety of input and microbial processes creates a broad assemblage of possible kerogen sub-types and these must be identifed and measured before further progress can be made in predicting source rock qualities in undrilled areas and in carrying out reliable oil to source rock correlations.

Acknowledgments. The authors acknowledge the permission of Simon Petroleum Technology (P.C.B.) and Norsk Hydro (N.T.) to publish this contribution. We would also like to thank our many colleagues for their contributions to discussions which have assisted in the development of the ideas presented in this chapter.

References

Aigner T (1980) Biofabrics and stratinomy of the Lower Kimmeridge Clay (Upper Jurassic, Dorset, England). Neues Jahrb Geol Palaontol Abh 159: 324–338

Baird RA (1986) Maturation and source rock evaluation of Kimmeridge Clay, Norwegian North Sea. Am Assoc Petrol Geol Bull: 1–11

Barnard PC, Collins AG, Cooper BS (1981) Identification and distribution of kerogen facies in a source rock horizon–examples from the North Sea basin. In: Brooks J (ed) Organic maturation studies and fossil fuel exploration. Academic Press, London, pp 271–282

Barnard PC, Bastow MA (1992) Hydrocarbon generation, migration, alteration, entrapment and mixing in the Central and Northern North Sea. In: England WA, Fleet AJ (eds) Petroleum migration. Geol Soc, London, Spec Publ 59: 167–190

Cooper BS, Barnard PC (1984) Source rocks and oils of the central and northern North Sea. In: Demaison G, Murris R J (eds) Petroleum geochemistry and basin evaluation. Am Assoc Petrol Geol, Tulsa, Mem 35: 303–314

Cooper BS (1990) Practical petroleum geochemistry. Robertson Sci Publ, London, 174 pp

Cope JCW, Ingham JK , Rawson PF (eds) (1992) Atlas of palaeogeography and lithofacies. Geol Soc, London, Mem 13, 153 pp

Cornford C (1984) Source rocks and hydrocarbons of the North Sea. In: Glennie KW (ed) Introduction to the petroleum geology of the North Sea. Blackwell Sci Oxford, pp 171–204

Cornford C, Needham CEJ, de Walque L (1986) Geochemical habitat of North Sea oils and gases. In: Spencer, AM (ed) Habitat of hydrocarbons on the Norwegian continental shelf. Graham and Trotman, London, pp 39–54

Cox BM, Gallois RW (1981) The stratigraphy of the Kimmeridge Clay of the Dorset type area and its correlation with some other Kimmeridgian sequences. Rep Inst Geol Sci United Kingdom 80/4

de Leeuw JW, Sinninghe Damste JS (1990) Organic sulfur compounds and other biomarkers as indicators of palaeosalinity. In: Orr WL, White CM (eds) Geochemistry of sulfur in fossil fuels. Am Chem Soc, Washington DC, Symp Ser 429: 417–443

Demaison G, Holk AJJ, Jones RW, Moore GT (1983) Predictive source bed stratigraphy: Guide to regional petroleum occurrence, North Sea basin and eastern North American continental margins. In: 11th World Petrol Congr, Sect PD1(2), Wiley, New York, pp 1–13

Douglas AG, Williams PFV (1981) Kimmeridge oil shale – a study of organic maturation. In: Brooks J (ed) Organic maturation studies and fossil fuel exploration. Academic Press, London, pp 256–269

Ebukanson EJ, Kinghorn RF (1990) Jurassic mudrock formations of southern England: lithology, sedimentation rates and organic carbon content. J Petrol Geol 1: 221–228

Espitalie J, Laporte JL, Madec M, Marquis F, Leplat P, Paulet T (1977) Méthode rapide de charactérisation des roches–meres de leur potentiel pétrolier et leur degré d'evolution. Rev Inst Français du Pétrole 32: 23–43

Farrimond P, Comet P, Eglinton, G Evershed MA, Hall MA, Park DW, Wardroper AMK (1984) Organic geochemical study of the upper Kimmeridge Clay of the Dorset type area. Mar Petrol Geol 1: 340–354

Gallois RW (1979) Oil shale resources in Great Britain. Ins Geol Sci Nottingham, 2 vols

Grantham PJ, Posthuma J, De Groot K (1980) Variation and significance of the C27 and C28 triterpane content of a North Sea core and various North Sea crude oils. In: Douglas AG, Maxwell JR (eds) Advances in organic geochemistry 1979. Pergamon Press, Oxford, pp 29–38

Horsfield B (1984) Pyrolysis studies and petroleum exploration. In: Brooks J, Welte D (eds) Advances in petroleum geochemistry, Vol 1, Academic Press, London, pp 247–298

Hudson JD, Martill DM (1991) The Lower Oxford Clay: production and preservation of organic matter in the Callovian (Jurassic) of central England. In: Tyson RV, Pearson TH (eds) Modern and ancient continental shelf anoxia. Geol Soc, London, Spec Publ 58: 363–379

Katz B (1983) Limitations of "Rock-Eval" pyrolysis for typing organic matter. Org Geochem 4: 195–199

Khorasani GK, Michelsen JK (1992) Primary alteration-oxidation of marine algal organic matter from oil source rocks of the North Sea and Norwegian Arctic: new findings. In: Eckardt CB, Maxwell JR, Larter SR and Manning DA (eds) Advances in organic geochemistry 1991. Pergamon Press, London, pp 327–344

Largeau C, Derenne S, Clarray C, Casadevall E, Raynaud JF, Lugardou B, Berkaloff C, Corolleur M, Posseau B (1990) Characterisation of various kerogens by Scanning Electron Microscopy (SEM) and Transmission Electron Microscopy (TEM)–morphological relationships with resistant outer walls in extant microorganisms. Meded Rijks Geol Dienst 45: 91–101

Mann AL, Myers KJ (1989) The effect of climate on the geochemistry of the Kimmeridge Clay Formation. In: Biomarkers in petroleum–Memorial Symposium for Wolfgang K Seifert. Am Chem Soc, pp 139–142

Miller RG (1990) A palaeoceanographic approach to Kimmeridge Clay formation. In: Huc AY (ed) Deposition of organic facies. Am Assoc Petrol Geol, Tulsa, Stud Geol, 30: 13–26

Nohr-Hansen H (1989) Visual and chemical analyses of the Lower Kimmeridge Clay, Westbury, England. In: Batten DJ, and Keen MC (eds) North West European micropalaeontology and palynology Ellis Horwood Ltd, Chichester, pp 118–134

O'Brien NR, Slatt RM (1990) Argillaceous rock atlas. Springer, Berlin Heidelberg New York, 141 pp

Oschmann W (1991) Distribution, dynamics and palaeoecology of Kimmeridgian (Upper Jurassic) shelf anoxia in western Europe. In: Tyson RV, Pearson TH (eds) Modern and ancient continental shelf anoxia. Geol Soc, London, Spec Publ 58: 381–396

Pearson MJ, Small JS (1983) Illite-smectite diagenesis and palaeotemperature in northern North Sea Quaternary to Mesozoic shale sequences. Clay Miner 23: 109–132

Reeder ML, Scotchman IC (1985) Hydrocarbon generation – central and northern North Sea. Oil Gas J: 137–144

Shaw HF, Primmer TJ (1991) Diagenesis of mudrocks from the Kimmeridge Clay Formation of the Brae area, UK North Sea. Mar Petrol Geol 8: 270–277

Simon Petroleum Technology (1992) Phanerozoic palaeogeographic reconstructions, palaeoenvironments and the major source rocks of the World. Non-proprietary report

Stow DAV, Atkin BP (1987) Sediment facies and geochemistry of Upper Jurassic mudrocks in the central North Sea area. In: Brooks J, Glennie K (eds) Petroleum geology of North West Europe. Graham and Trotman, London, pp 797–808

Sun SQ (1992) A storm-dominated offshore sandstone interval from the Corallian Group (Upper Jurassic), Weald basin, southern England. Mar Petrol Geol 9: 274–286

Telnaes N, Cooper BS (1991). Oil-source rock correlation using biological markers, Norwegian continental shelf. Mar Petrol Geol 8: 302–310

Thomas BM, Moller-Pedersen P, Whitaker M , Shaw ND (1985) Organic facies and hydrocarbon distribution in the Norwegian North Sea. In: Thomas BM et al. (eds) Petroleum geochemistry in exploration of the Norwegian Shelf. Graham and Trotman, London, pp 3-26

Thompson S, Cooper BS, Barnard PC (1994) Some examples and possible explanations for oil generation from coals and coaly sequences. In: Scott AC, Fleet AJ (eds). Coal and coal-bearing strata as oil-prone source rocks? Geol Soc, London, Spec Publ 77: 119–138.

Thorne JA, Watts AB (1989) Quantitative analysis of North Sea subsidence. Bull Am Assoc Petrol Geol 73: 88–116

Tyson RV (1989) Late Jurassic palynofacies trends, Piper and Kimmeridge Clay Formations, UK onshore and northern North Sea. In: Butler DJ, Keen MC (eds) North West European Micropalaeontology and palynology. Ellis Horwood, Chichester, pp 135–172

Waples DW, Machihara T (1991) Biomarkers for geologists. Am Assoc Petrol Geol, Tulsa, Methods 9, 91 pp

Wignall PB (1989) Sedimentary dynamics of the Kimmeridge Clay tempests and earthquakes. J Geol Soc Lond 146: 273–282

Wignall PB, Myers KJ (1988) Interpreting benthic oxygen levels in mudrocks: a new approach. Geology 16: 452–455

Williams PFV (1986) Petroleum geochemistry of the Kimmeridge Clay of southern and eastern England. Mar Petrol Geol 3: 258–281

The Egret Member, a Prolific Kimmeridgian Source Rock from Offshore Eastern Canada

M.G. Fowler[1] and K.D. McAlpine[2]

Abstract

The Jeanne d'Arc basin is situated in the northeastern part of the Grand Banks, off the east coast of Newfoundland. Several large oil and gas fields have been discovered in this basin, the largest of which is the giant Hibernia field. The principle source rock for most of the oils in the Jeanne d'Arc basin is the Kimmeridgian-aged Egret Member of the Rankin Formation. This interval is easily identified on downhole and geochemical logs, principally because of its higher TOC content compared to surrounding units. The Egret Member ranges in thickness from 55 m to in excess of 200 m. Based on the analysis of cutting samples (no core has yet been taken of this unit), it shows average TOC contents up to 4.58%, and average hydrogen index values of immature samples in the 500-700 range. The organic matter is predominately amorphous in appearance, with terrestrial-derived material significant only in some wells on the eastern side of the basin.

Most Egret Member extracts have hydrocarbon distributions that show the characteristics of marine-derived organic matter deposited under reducing conditions. These include pristane/phytane ratios of usually less than one, high abundance of 4-methylsteranes derived from dinosterol, a predominance of C_{27} over C_{29} regular steranes, a low abundance of tricyclic terpanes relative to hopanes, and a smooth distribution of C_{31}-C_{35} 17α(H)-hopanes. Some wells on the eastern side of the basin (e.g., Archer K-19 and Fortune G-57) show biomarker characteristics that suggest a greater contribution from terrestrially-derived organic matter in this area. The depositional environment for the Egret Member is interpreted to be a silled marine basin with high (periodic?) primary productivity, especially from dinoflagellates. The bottom waters were suboxic to anoxic. The primary organic matter was extensively reworked by anaerobic bacteria.

Quantitative geochemical techniques show the Egret Member to be a prolific oil source rock and suggest that perhaps as little as 15% of Jeanne d'Arc basin recoverable oil reserves may presently be discovered.

Introduction

The Jeanne d'Arc basin is situated in the northeastern Grand Banks area, off the east coast of Newfoundland, eastern Canada (Fig. 1). The Grand Banks of Newfoundland comprise a broad late Cretaceous-Tertiary continental margin on a basement of Paleozoic and Precambrian crust at the northern end of the Appalachian Orogen of eastern North America. The Jeanne d'Arc basin is one of a series of interconnected sedimentary basins that formed during the Mesozoic Era. These basins and the intervening basement highs are truncated by a prominent peneplain, the Avalon Unconformity, and are buried beneath a relatively thin cover of undeformed Upper Cretaceous and Tertiary strata. None of the Mesozoic-Cenozoic sedimentary section outcrops onshore Newfoundland. The Jeanne d'Arc basin is the only one of the Grand Banks basins in which significant accumulations of hydrocarbons have been found.

Drilling in the Jeanne d'Arc basin began in 1971, but it was not until the 1979 discovery of the Hibernia field that large volumes of hydrocarbons were shown to be present within this basin. A recent estimate suggested that 1.5×10^9 bbls (0.24×10^9 m^3) of oil, 109×10^6 bbls (0.02×10^9 m^3) of gas liquids and 101.9×10^9 m^3 (3.6×10^{12} ft^3) of gas have been discovered in the basin with an

[1]Institute of Sedimentary and Petroleum Geology, 3303-33rd Street N.W., Calgary, Alberta T2L 2A7, Canada

[2]Atlantic Geoscience Centre, Bedford Institute of Oceanography, P.O. Box 1006, Dartmouth, Nova Scotia, B2Y 4A2, Canada

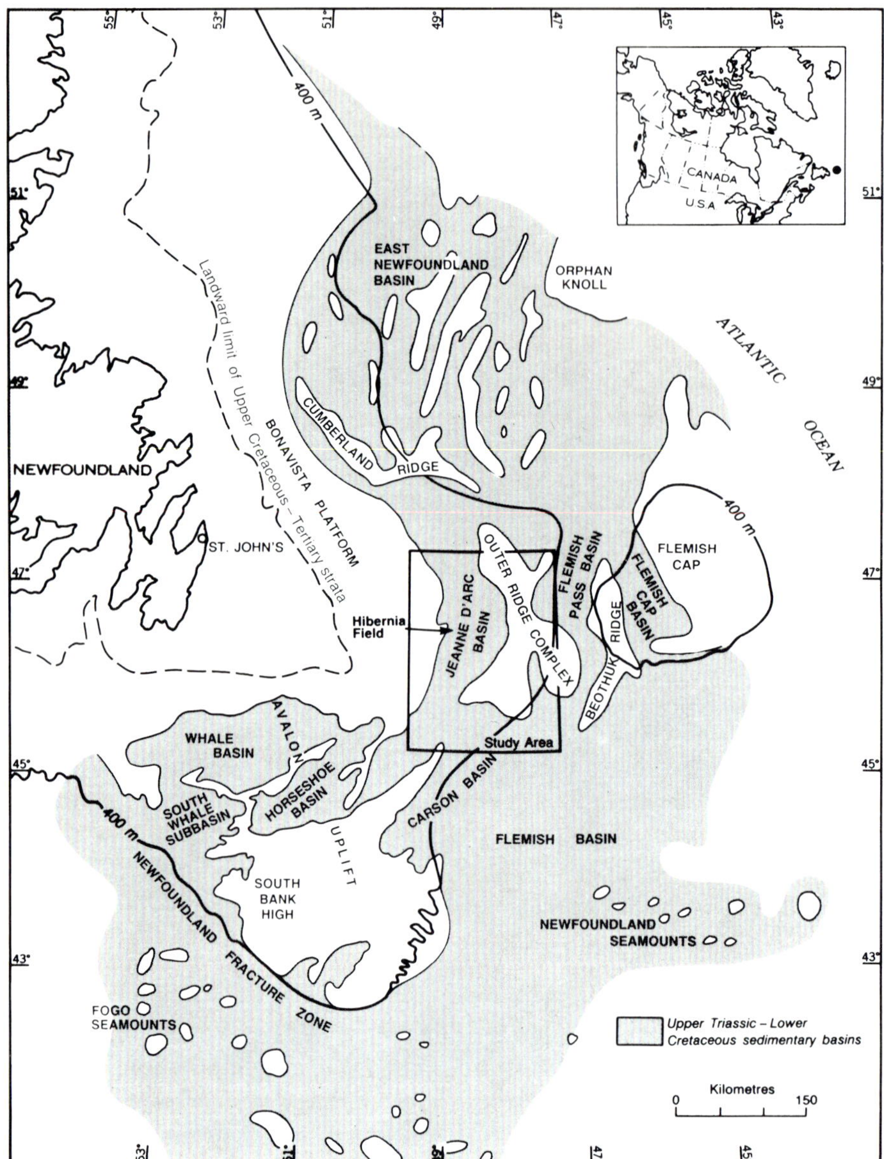

Fig. 1. Simplified tectonic element map of offshore Newfoundland (Sinclair et al. 1992) showing principal positive structural elements and Mesozoic basins underlying the continental margin around Newfoundland. *Rectangle* outlines area of Fig. 2

average expectation that 4.7×10^9 bbls (0.75×10^9 m^3) of technically recoverable oil will be eventually discovered (Taylor et al. 1991; Sinclair et al. 1992). The most significant discoveries to date have been the giant Hibernia oil field and the large Terra Nova, Whiterose, and Hebron discoveries (Fig. 2). No hydrocarbons have yet been produced from this basin but the Hibernia Field is presently being developed and production is expected to start around 1996.

Regional Setting

The Grand Banks of Newfoundland form a broad continental shelf, the easternmost promontory of the North American continental plate. As a consequence of sequential south to north continental breakup and opening of the North Atlantic Ocean and Labrador Sea (first Africa from the southern Grand Banks, then Iberia from the eastern Grand

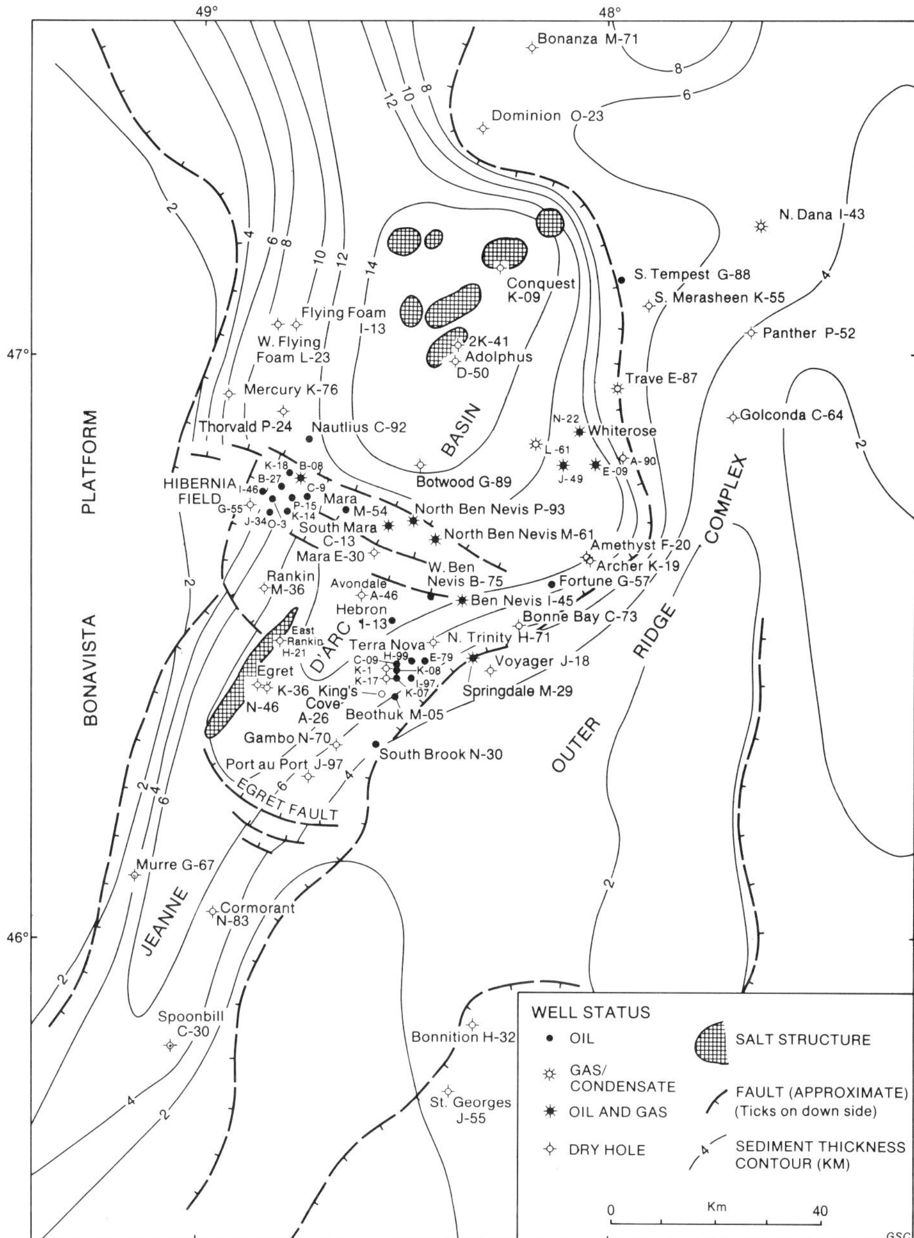

Fig. 2. Diagrammatic tectonic element and sediment thickness map of the Jeanne d'Arc basin. (After Grant and McAlpine, 1990)

Banks, then the British Isles from the northern Grand Banks and finally Greenland from both North America and Northern Europe), the Grand Banks area was exposed to a succession of stress regimes, recorded in the subsurface geology of the area. The most conspicuous subsurface elements of this region, revealed by over 25 years of petroleum exploration, are a series of five interconnected sedimentary basins containing deformed rocks of Late Triassic through Early Cretaceous age, namely the South Whale subbasin of the Scotian basin and the Whale, Horseshoe, Carson and Jeanne d'Arc basins (Fig. 1). These Mesozoic basins and the Paleozoic and Precambrian rocks that underlie and surround them were subject to periods of deformation and erosion during the Late Jurassic and Early Cretaceous that produced a prominent peneplain, the Avalon Unconformity, over the central Grand Banks and thick clastic deposits on its flanks (McAlpine 1990a). Slow regional subsidence during the Late Cretaceous and Tertiary resulted in a relatively undisturbed and thin cover of fine-grained marine shelf deposits.

The Jeanne d'Arc basin is the deepest of the failed-rift basins, containing in excess of 22 km of sedimentary fill (Enachescu 1986; Keen et al. 1987). The lithostratigraphy of the Jeanne d'Arc basin is described in detail in McAlpine (1990a, b); (Fig. 3). Although the basin lies well inboard of the continent-ocean boundary, the sedimentary fill is composed of distinctive depositional sequences that are each interpreted to record a unique tectonic episode of continental separation. In brief, six primary Mesozoic-Cenozoic sequences record two periods of rift subsidence, separated by a Middle Jurassic quiet period and followed by a two-phase transition to passive continental margin subsidence. The six sequences are, from oldest to youngest:

1. Aborted rift – Late Triassic to Early Jurassic continental redbeds and restricted marine evaporites and carbonates of the Eurydice through Iroquois Formations that were deposited in a vast rift system between the North America-Greenland and Europe-Africa plates.
2. Epeiric basin – Early to Late Jurassic normal marine shales and carbonates and minor, fine-grained deltaic Sediments of the Downing through Rankin Formations that were deposited over the Grand Banks as Africa decoupled from North America and plate movement was accommodated along the southern Grand Banks transform margin.
3. Late rift – Latest Jurassic and Neocomian sandstone-dominated deltaic and estuarine sediments of the Jeanne d'Arc through Eastern Shoals Formations that were deposited in newly developing graben on the flanks of the Grand Banks.
4. Transition to drift: Phase I – Barremian to Cenomanian shallow to deep estuarine sandstones and shales of the Avalon through Eider Formations that were deposited as Iberia moved away from the eastern Grand Banks.
5. Transition to drift: Phase II – Late Cretaceous and Paleocene open-marine shelf, overlapping deltaic deposits and distal turbidites and chalky limestones of the Dawson Canyon Formation and South Mara unit of the Banquereau Formation that were deposited during separation of the British Isles from the northern Grand Banks. Late Cretaceous and Tertiary unconformities relate respectively to the breakup between Labrador and Greenland and between Greenland and northern Europe.
6. Passive margin – Tertiary neritic shales, siliceous mudstones and minor chalks of the Banquereau Formation that were deposited during thermal subsidence of the continental margin.

All known Jeanne d'Arc basin source rocks were deposited during the Late Jurassic portion of Sequence 2 and Sequence 3 (Fig. 3). A potential source rock of Kimmeridgian age was first reported by Swift and Williams (1980). This was later defined as the Egret Member of the Rankin Formation (McAlpine 1990a) and it has been shown to be the major source for hydrocarbons within the basin (Powell 1985; Creaney and Allison, 1987; von der Dick 1989; von der Dick et al. 1989; Fowler and Brooks 1990). More recent work has suggested that some oils in the Ben Nevis area have a contribution from a second source rock whose organic matter contains a greater amount of higher land plant material (Fowler et al. 1988, 1989; von der Dick et al. 1989; Fowler and Brooks, 1990). Potential source rocks, other than the Egret Member, that have been identified within the basin include an interval of Callovian-Oxfordian age within the underlying Voyager Formation (Fowler et al. 1988) and a Tithonian age interval within the overlying Jeanne d'Arc Formation or Fortune Bay Shale (von der Dick et al. 1989; Fowler et al. 1990; McAlpine and Fowler, 1990) (Fig. 3). In addition, occasional beds within the Rankin Formation, above and below the Egret Member, show source potential in some wells. Based on the results of drilling undertaken to date, the Egret Member is the most widespread of these units. The Voyager Formation has been identified in only two wells (Rankin M-36 and Panther P-52). As these two wells are on opposite sides of the basin (Fig. 2), there is a possibility that this is because of the limited penetration of this formation and the Voyager Formation source interval may be more widespread. The Tithonian-aged and Lower Rankin intervals appear to be present only in wells on the eastern side of the basin.

Virtually all oil and gas discoveries have been made in clastic units of Sequence 3 and Sequence 4 with minor accumulations in Sequences 2 and 5 (Fig. 3). Several wells have encountered hydrocarbons in stacked pools; the cause of this has been shown to be episodic vertical migration along

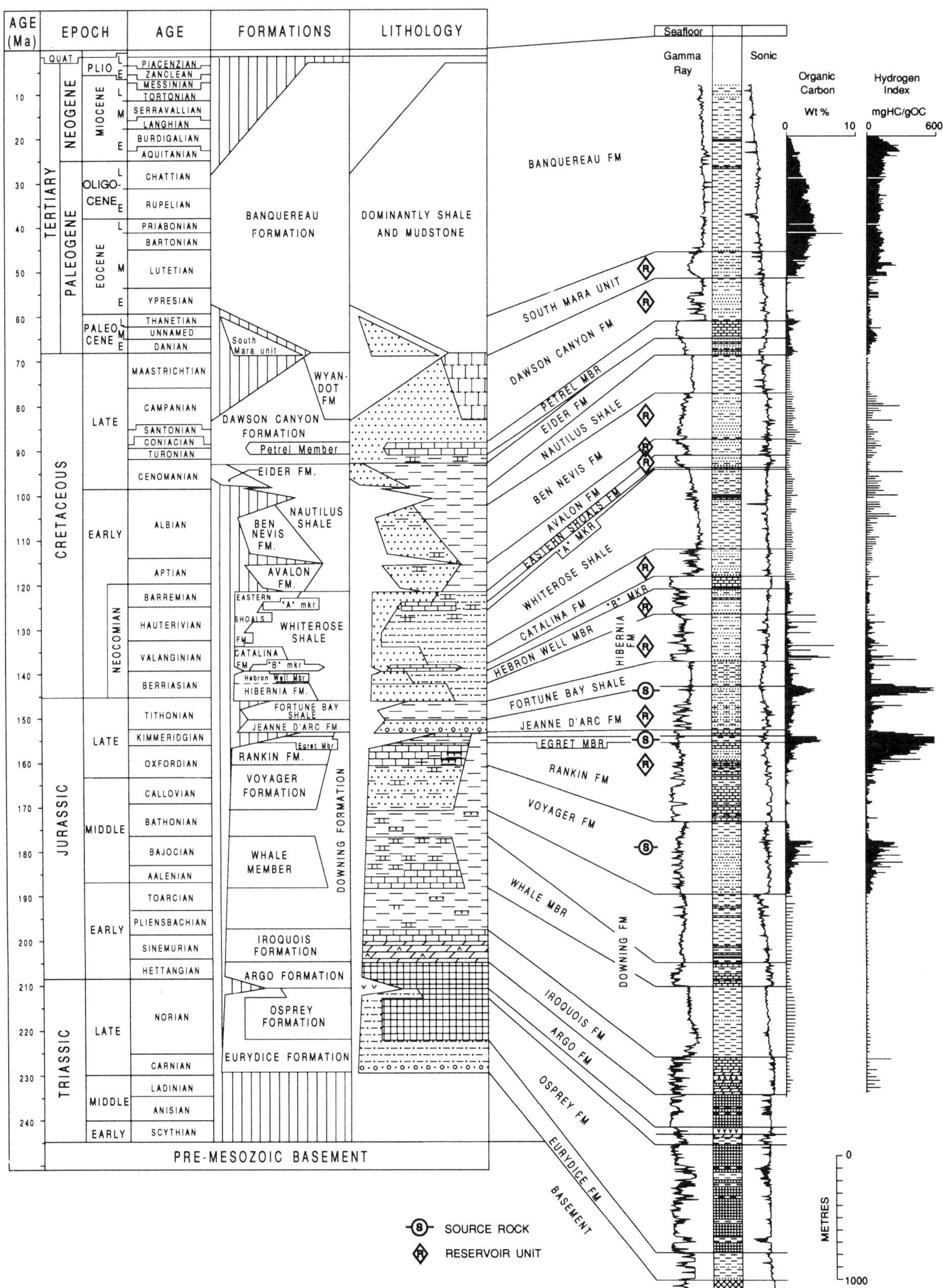

Fig. 3. Lithostratigraphic column and composite lithological/ geochemical log for the Jeanne d'Arc basin highlighting source rocks and reservoir units

faults, driven by pressure differential between a pervasive deep overpressured regime sealed by the Fortune Bay Shale (Fig. 3) and the overlying hydropressured regime (Grant and McAlpine 1990; McAlpine 1990a).

Source Rock Characteristics of Egret Member

The source rock characteristics of the Egret Member described here are based largely on Rock-Eval/TOC analyses from 30 wells (Snowdon and Fowler, 1986; Fowler and Snowdon, 1988, 1989; Fowler et al. 1990, 1991), and extract and organic petrological studies on 41 selected samples (Tables 1 and 2). Almost all these data were obtained from the analyses of cutting samples taken usually over 10-m intervals, although there is less coverage in some wells due to the unavailability of samples. Unfortunately, there has been no coring of the Egret Member within the Jeanne d'Arc basin. This restricts our detailed understanding of how the organic-rich facies is distributed relative to other lithologies within the Egret Member and leads to "smearing" of its characteristics. There is further sampling bias because most wells that penetrate the Egret Member tend to be near the margins of the basin where this unit is immature to marginally mature (Fig. 4d, Table 1).

Another problem is that some wells were drilled using oil-based muds and other organic additives (Table 1). Where heavy hydrocarbon materials were used (e.g. gilsonite at Lancaster F-70, Sarnia 'A' DMO at Baccalieu I-78, both wells in the Flemish Pass basin), no useful data on the Egret Member can be obtained (Fowler, 1993). Other wells were contaminated by light oil-based muds such as "Biovert". Allowing such samples to dry, and most of the oil to evaporite, prior to Rock-Eval analysis allowed the Egret Member to be easily distinguished on geochemical logs (e.g., well Whiterose A-90; Fowler et al. 1990). However, the extract and hydrocarbon yields of these contaminated samples are still very high (Table 2) and their C_{15+} saturated fraction gas chromatograms are dominated by residual C_{13}-C_{16} compounds originating from the oil additives (Fowler 1993). Biomarker analysis of these samples still provides useful data because of the limited molecular weight range of the additives.

The Egret Member may often be identified by its distinctive downhole log characteristics (McAlpine 1990a). It has a lower sonic velocity, lower bulk density, and higher resistivity compared to adjacent shaly units (Grant and McAlpine, 1990). The anomalous log characteristics result from its high organic carbon content coupled with oil generation where it is mature. Gamma ray readings are not diagnostic of the member. This unit is known to occur in wells (Table 1, Fig. 4) throughout the basin north of the Egret Fault until somewhere south of Bonanza M-71, where it is not present (Fowler and Snowdon, 1989; von der Dick et al. 1989). It is also present in wells drilled on the Outer Ridge Complex such as Panther P-52 (Fowler and Snowdon, 1988) and in the Flemish Pass basin (e.g., Baccalieu I-78) (Foster and Robinson, 1993). Creaney and Allison (1987) suggested a Kimmeridgian source rock was present in the Murre G-67 well which was drilled south of the Egret Fault (Fig. 2). However, more recent work has indicated that the Kimmeridgian source interval is not present in this well (von der Dick, 1989; Fowler et al. 1990). The Egret Member was probably deposited south of its present distribution but eroded during the development of the Avalon Unconformity (McAlpine, 1990a). Sinclair (1988) has suggested that Kimmeridgian source rocks may have been deposited in areas beyond the Triassic-Lower Jurassic rift basins such as in the East Newfoundland basin located north of the Jeanne d'Arc basin (Fig. 1). However, no Upper Jurassic-aged rocks have been encountered by wells drilled in the East Newfoundland basin, and seismic data suggests that Upper Jurassic rocks may not be present northwest of the Jeanne d'Arc basin (A. Edwards, pers. comm. 1990).

The Egret Member shows considerable variation in its thickness, lithology, and in the characteristics of its organic matter (Fig. 4). As evident from Table 1, the thickness of the Egret Member as defined by McAlpine (1990a, b) varies from 55 m at Rankin M-36 to 226 m at Fortune G-57 and possibly 255 m at Lancaster G-70 in the Flemish Pass basin. These thicknesses compare to a range from 30 m in South Mara C-13 to in excess of 500 m in the east reported by von der Dick (1989) and Tankard et al (1989) for the "Egret Formation", and between 60 m at Flying Foam I-13 to 200 m at Hibernia K-18 reported by Creaney and Allison (1987) for the "Kimmeridgian source rock". Some of the discrepancies regarding the thickness of the source unit between the different groups can be attributed to different definitions of what

Table 1. Depth range of Egret Member in different wells as indicated by down hole logs or Rock-Eval data on cutting samples, number of samples over this interval (N), average Rock-Eval data for the given intervals and sample problems

Well	Downhole log[a] interval thickness			Rock-Eval[b]									
	Top (m)	Bottom (m)	Thickness (m)	Top (m)	Bottom (m)	N[c]	TOC range %	TOC average	HI range	HI average	T_{max} range	T_{max} average	Sample problems
Archer K-19	3279	3449	170	3290	3440	14	2.13–6.62	2.92	210–519	273	424–436	429	
Baccalieu I-78	4208	4334	126										Heavy oil-based mud
Beothuk M-05	3387	3495	108	3385	3495	12	1.98–4.36	3.1	444–584	508	432–442	437	
Egret K-36	2642	2703	61	2661	2707	4	3.3–4.11	3.66	425–681	586	421–426	423	
Egret N-46	1970	2047	77										Poor sample coverage
Flying Foam I-13	3281	3353	72	3310	3328	3	1.66–4.34	2.57	284–697	474	427–432	429	
Fortune G-57	4163	4389	226 +	4170	4380	22	1.57–3.99	2.64	252–429	332	435–448	442	
Hibernia K-18	4819	4907	88	4790									Oil staining
Lancaster G-70	4595	4850	255										Heavy oil-based mud
North Dana I-43	3883	4000	117	3890	4000	11	1.22–1.93	1.58	157–219	184	439–442	441	
Panther P-52	3218	3291	73	3250	3320	8	1.6–8.22	3.61	406–659	520	421–434	428	
Port au Port J-97	1833	1912	79	1830	1890	7	1.12–3.28	2.15	353–504	423	410–417	414	
Rankin M-36	2436	2491	55	2440	2490	6	3.7–5.23	4.58	624–744	697	412–416	414	
South Brook N-30	1701	1789	88	1705	1785	9	1.24–3.38	2.29	456–573	517	409–421	413	Light oil-based mud
South Merasheen K-55	2627	2746	119	2630	2720	10	1.03–6.22	3.1	185–587	354	425–432	429	Light oil-based mud
South Tempest G-88	3730	3814	84	3800	3840	4	1.71–3.27	2.49	287–358	327	445–449	447	
Springdale M-29	3002	3134	132										? Oil-based mud
Terra Nova K-08	3641	3790	149	3655	3795	14	2.1–5.01	3.4	330–648	518	429–439	434	
Terra Nova K-18	3635	3790	155										
Trave E-87	3046	3120	74	3050	3120	8	2.35–4.24	3.05	680–810	746	426–431	428	
Voyager J-18	2712	2812	100	2720	2810	10	1.78–4.87	3.14	423–595	508	427–432	430	
Whiterose A-90	2882	3020	138	2890	2960	8	2.33–3.3	2.88	450–604	545	441–450	445	Light oil-based mud

[a]Range and thickness of Egret Member assessed using downhole logs.
[b]Range of Egret Member indicated by Rock-Eval analysis of cutting samples collected over 10-m sections.
[c]N is number of Egret Member cutting samples used to assess range and average of Rock-Eval parameters.

Table 2. Rock-Eval, extract and biomarker data for 41 Egret Member cutting samples that have been extracted. Biomarker data is from CAD-GC-MS-MS analysis. Only GC-MS-SIM data is available for the Fortune G-57 4290 m and 4370 m, and the Trave 3060 m samples. This data was not included in the table because the values obtained from this form of GC-MS analysis are somewhat different to those from CAD-GC-MS-MS (Fowler and Brooks, 1990) and could be misleading when compared to the rest of the data

										Steranes						Terpanes			
Well	Depth (m)	TOC (%)	T_{max}	HI	Ext. yield mg/g orgC	HC yield mg/g OrgC	pr/ph	pr/nC_{17}	nC_{17}/nC_{27}	%C_{27}	%C_{28}	%C_{29}	C_{29} ααα	C_{29} αββ/ αββ + ααα	C_{27} dia/ αααR	Ts/Tm	29/X	29/30	C_{29} ster/ hop
Archer K-19	3290	1.55	435	235	81.4	31.1	1.59	0.81	4.14	36	22	42	0.27	0.28	0.63	0.73	1.35	0.55	0.70
	3350	2.94	441	506	91.8	49.4	1.62	0.46	1.38	37	17	46	0.34	0.29	0.34	0.87	1.62	0.37	0.13
	3400	2.95	440	491	109.1	61.6	2.15	0.64	1.64	52	17	31	0.42	0.34	0.61	0.90	1.71	0.41	0.12
	3450	3.00	435	421	87.7	46.1	1.80	0.91	1.92	41	18	41	0.47	0.32	0.62	0.79	1.88	0.40	0.21
Beothuk M-05	3405	3.59	427	551	94.5	49.0	1.45	1.07	3.50	62	9	29	0.47	0.36	1.54	0.57	2.61	0.25	0.32
	3445	3.77	428	471	146.9	70.3	0.72	0.64	1.91	48	13	39	0.91	0.60	3.67	0.78	2.52	0.22	0.13
	3475	4.89	429	513	372.3	111.5	0.42	0.41	4.54	65	12	23	0.98	0.62	1.45	0.82	3.02	0.43	0.08
	3495	2.79	429	395	251.1	123.6	0.70	0.51	2.73	59	14	27	1.00	0.63	3.85	0.98	2.33	0.37	0.10
Egret K-36	2670	4.13	413	571	64.7	14.4	0.56	2.31	1.77	72	11	17	0.09	0.27	0.13	0.29	3.71	0.13	0.78
	2706	4.11	421	631	95.8	29.6	0.48	1.45	2.82	68	12	20	0.18	0.25	0.19	0.29	4.83	0.18	0.48
Flying Foam I-13	3310	4.66	436	642	123.4	71.1	0.82	0.99	5.74	58	11	31	0.85	0.53	1.59	0.70	3.02	0.25	0.31
Fortune G-57	4190	2.74	437	295	226.1	127.1	1.55	1.14	4.57	42	11	47	0.98	0.59	4.34	1.54	1.64	0.17	0.09
	4220	2.58	437	424	219.0	121.9	1.31	0.77	5.03	39	11	50	0.84	0.60	4.63	1.43	1.40	0.18	0.09
	4290	3.26	439	404	218.1	117.4	1.91	0.65	3.32										
	4370	3.33	441	317	231.8	128.0	1.47	0.65	3.72										
Hibernia K-18	4840	6.62	448	125	228.2	132.2	1.34	0.50	12.05	43	12	45	0.80	0.58	9.50	3.67	1.57	0.52	0.19
	4890	5.56	453	90	513.3	354.9	1.16	0.47	23.90	21	11	68	0.95	0.58	9.20	2.69	2.15	0.25	0.08
North Dana I-43	3950	2.13	436	127	154.1	64.4	1.56	0.69	2.96	32	12	56	1.14	0.67	8.00	3.30	1.05	0.20	0.09
Panther P-52	3270	8.10	435	604	99.5	44.7	1.90	2.11	2.79	49	14	37	0.68	0.36	0.83	0.54	3.67	0.44	0.23
	3300	3.39	428	516	146.2	82.0	1.30	1.53	3.82	62	14	24	0.73	0.37	0.59	0.77	2.96	0.37	0.19
Port au Port J-97	1850	2.78	423	498	56.8	18.0	0.90	1.06	3.72	65	12	23	0.04	< 0.01	0.01	0.05	10.70	0.57	1.85
	1900	1.98	423	402	73.7	22.4	0.99	1.07	5.16	67	9	24	0.05	< 0.01	0.02	0.07	6.76	0.22	1.90
Rankin M-36	2440	4.43	414	744	45.8	14.9	0.54	2.51	0.03	86	5	9	0.04	< 0.01	0.04	0.12	9.55	0.16	1.45
	2470	5.23	413	740	46.6	13.9	0.46	2.35	0.04	70	10	20	0.03	< 0.01	0.03	0.15	8.75	0.24	3.53
South Brook N-30[a]	1705	3.50	413	358	698.6	484.7	1.53	0.27		72	8	20	0.50	0.52	0.50	0.89	2.91	0.41	0.23
	1735	4.58	419	472	456.7	308.7	1.50	0.32		65	12	23	0.26	0.47	0.20	0.36	3.58	0.41	0.46
South Merasheen K-55*	2660	5.09	428	574	560.7	373.1	1.73	0.44		52	18	30	0.26	0.38	0.65	0.88	2.30	0.49	0.42
SouthTempest G-88	3820	3.21	446	404	204.6	116.3	1.59	0.47	4.22	50	19	31	0.92	0.64	5.27	2.50	0.86	0.33	0.03
Terra Nova K-08	3665	3.11	429	574	115.6	61.0	1.01	0.86	3.96	57	16	27	0.66	0.42	2.03	0.78	1.91	0.51	0.27
	3725	2.99	434	533	132.9	70.9	0.97	0.68	4.88	56	16	28	0.84	0.57	3.35	1.01	2.11	0.34	0.16
	3785	3.98	437	555	210.0	108.0	1.00	0.72	5.26	68	12	20	0.78	0.63	3.26	1.49	1.58	0.35	0.06
Trave E-87	3060	3.50	420	768	224.7	163.4	0.77	1.25	4.27										
	3080	6.58	436	631	119.3	68.3	0.95	1.31	3.88	52	16	32	0.72	0.33	0.50	0.75	3.35	0.24	0.87
	3100	2.14	428	707	194.3	122.9	0.90	0.75	4.29	50	19	31	0.63	0.34	0.43	0.74	2.94	0.37	0.64
Voyager J-18	2730	3.98	423	583	95.3	37.1	0.92	1.30	3.38	62	11	27	0.23	0.33	0.67	0.58	2.04	0.20	0.57
	2770	3.63	421	676	104.7	36.0	1.20	1.44	2.67	69	13	18	0.20	0.35	0.87	0.37	2.88	0.41	0.79
	2790	3.13	427	645	173.2	90.8	0.53	0.84	3.38	69	10	21	0.67	0.54	1.93	0.52	2.52	0.20	0.22
	2800	2.81	431	593	196.4	105.4	0.49	0.69	2.88	62	14	24	0.88	0.60	2.33	0.57	2.31	0.30	0.16
Whiterose A-30[a]	2890	3.38	421	496	350.5	191.5	1.85	0.52		51	15	34	0.39	0.34	0.40	0.84	2.39	0.18	0.98
	2900	2.30	434	532	321.8	166.1	1.83	0.55		57	13	30	0.36	0.30	0.35	0.75	2.88	0.33	0.44
	2950	3.29	440	518	372.4	167.1	2.27	0.55		51	9	40	0.42	0.44	0.75	0.64	3.44	0.34	0.14

[a]Drilled with oil-based muds. Hence extract and hydrocarbon yields, pr/ph, pr/nC_{17}, and nC_{17}/C_{27} are not true values. This process severely affects the following parameters, extract and hydrocarbon yields, pr/ph, pr/nC_{17} and nC_{17}/nC_{27} (this last ratio is extremely large for these samples and is not given). TOC, T_{max} and HI are from Rock-Eval analysis. Extract yield and hydrocarbon yield are the amounts of extract obtained using azeotropic chloroform-methanol and hydrocarbons after subsequent fractionation normalized to organic carbon. Other ratios are: pr/ph is pristane/phytane, pr/nC_{17} is pristane/*n*-heptadecane, nC_{17}/nC_{27} is *n*-heptadecane/*n*-heptacosane, %C_{27}, %C_{28}, and %C_{29} are the normalised proportions of C_{27}:C_{28}:C_{29} 5α(H),14α(H),17α(H) 20R steranes, C_{29} ααα S/R is the ratio of the C_{29} 5α(H),14α(H),17α(H) 20S and 20R steranes, C_{29} αββ/ααα + αββ is the ratio of C_{29} 5α(H),14β(H),17β(H) and 5α(H),14α(H),17α(H) steranes, C_{27}dia/αααR is C_{27} 13β(H),17α(H) 20S diasterane/5α(H),14α(H),17α(H) 20R sterane, Ts/Tm is 18α(H)-trisnorhopane/17α(H)-trisnorhopane, 29/X is 17α(H)-norhopane/18α(H)-norhopane, 29/30 is 17α(H)-norhopane/17α(H)-hopane, and C_{29} ster/hop is the ratio of C_{29} 5α(H),14α(H),17α(H) 20R sterane and 17α(H)-hopane.

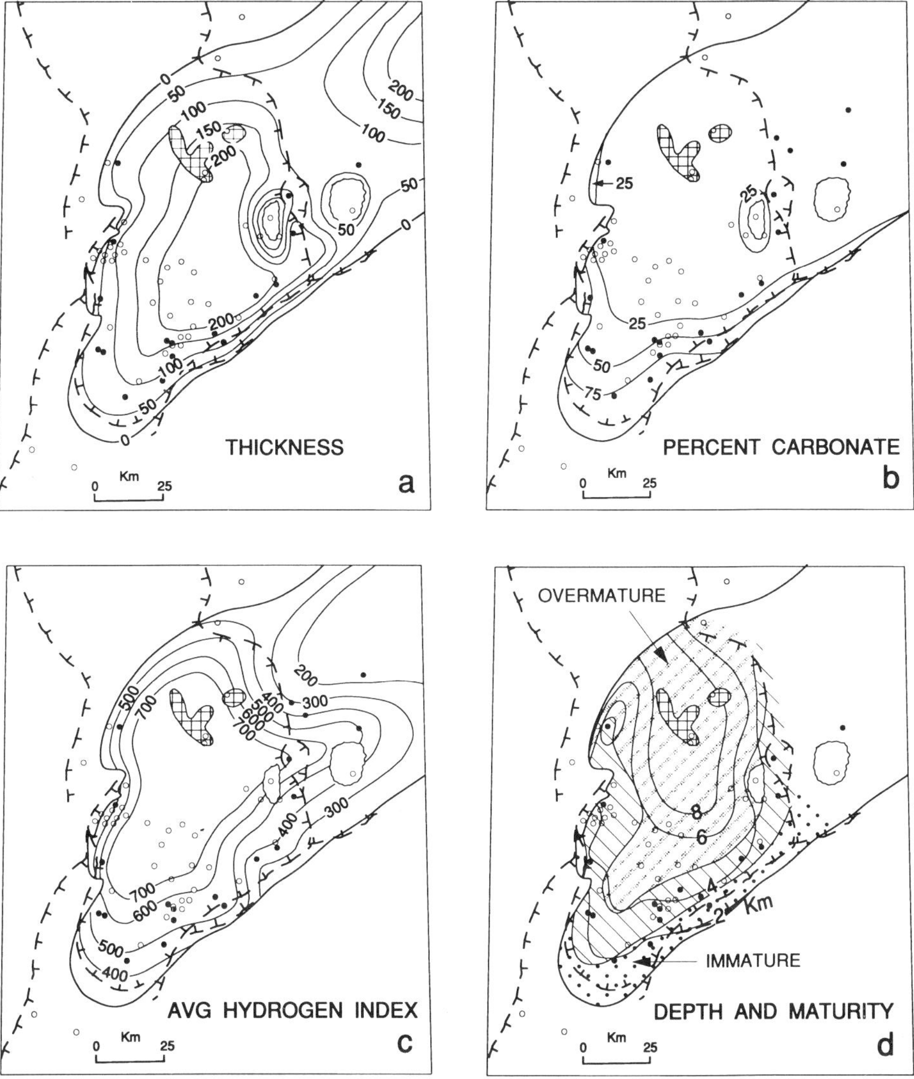

Fig. 4a-d. Sketch maps of the Egret Member source rock in the Jeanne d'Arc basin showing **a** thickness, **b** percent carbonate, **c** estimated, original, average hydrogen index and **d** depth and present-day maturity. *Small circles* are well locations (see Fig. 2 for names of wells). *Black circles* are wells that penetrated the Egret Member

constitutes the source interval. In other cases our data show that neither the Egret Member nor any other potential source rock is present, for example at South Mara C-13 (Snowdon and Fowler, 1986). Here we consider just the Egret Member, but as is evident from the geochemical log for Panther P-52 (Fig. 5) other intervals within the Rankin Formation above and/or below the Egret Member, particularly in the northeastern part of the basin, can have significant hydrocarbon potential. Some previous authors have included these intervals in their Kimmeridgian source rock sections. The extended interval reported by Creaney and Allison (1987) for Hibernia K-18 is probably related to the effects of bitumen saturation which affects the Rock-Eval analyses of samples from this well.

Around the southern perimeter of the basin, north of the Egret Fault (e.g., Rankin M-36), the Egret Member consists of thinly interbedded and laminated brown marls and calcareous mudstones (Fig. 4b). Basinward and to the northeast, as in Trave E-87, slightly calcareous gray shale and siltstone progressively become the dominant lithofacies. As can be seen from Fig. 3, gamma ray and sonic log responses are distinctly serrated. This may be caused either by interbedded carbonate,

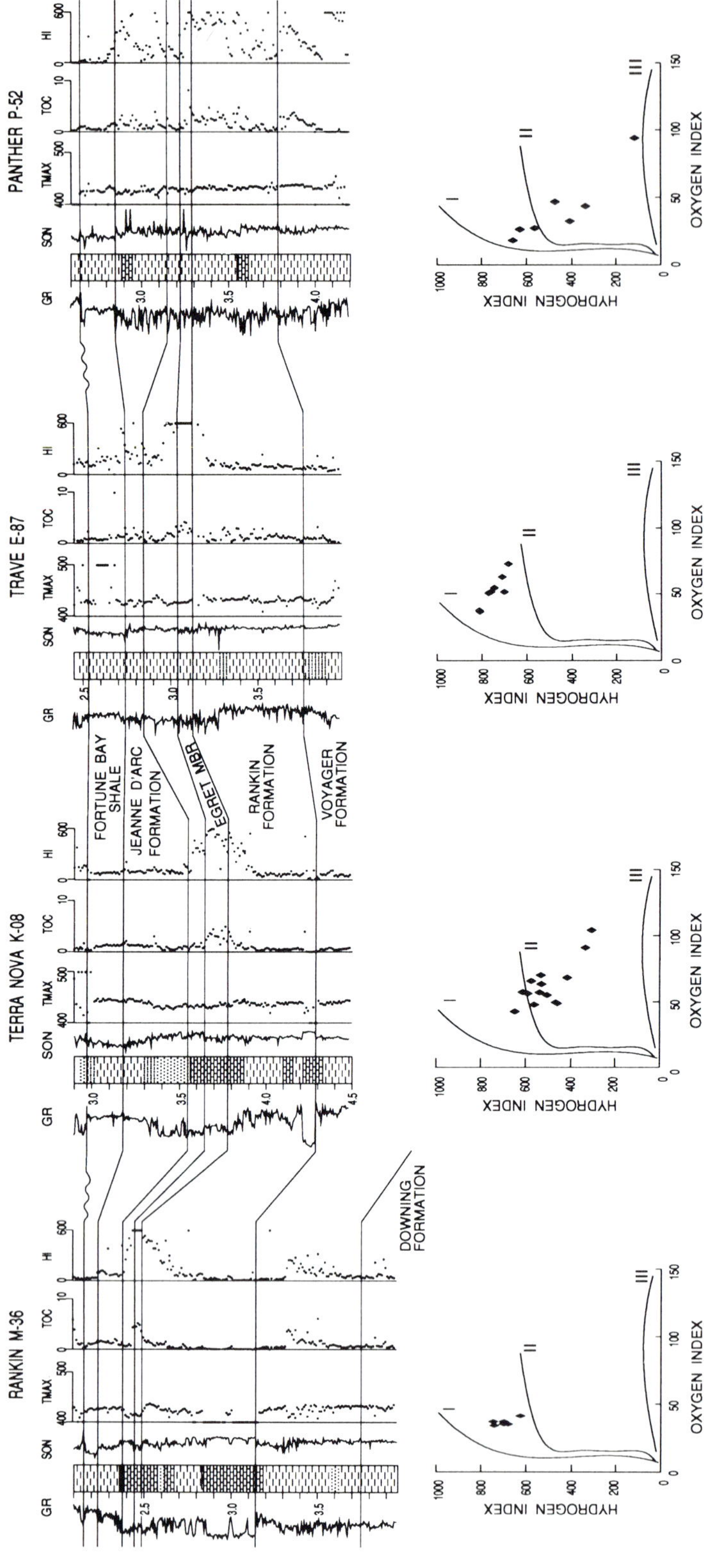

Fig. 5. Geochemical (Rock-Eval) and petrophysical log correlation of the Upper Jurassic source rocks in the Jeanne d'Arc basin. The oxygen index versus hydrogen index plot beneath each well is representative of the Egret Member only. The *numbers to the left* of the lithology columns indicate depth in kilometres below rotary table

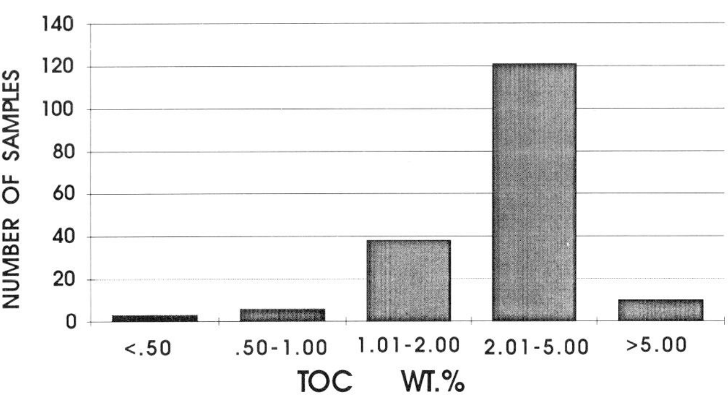

Fig. 6. Histogram of TOC content of Egret Member cutting samples

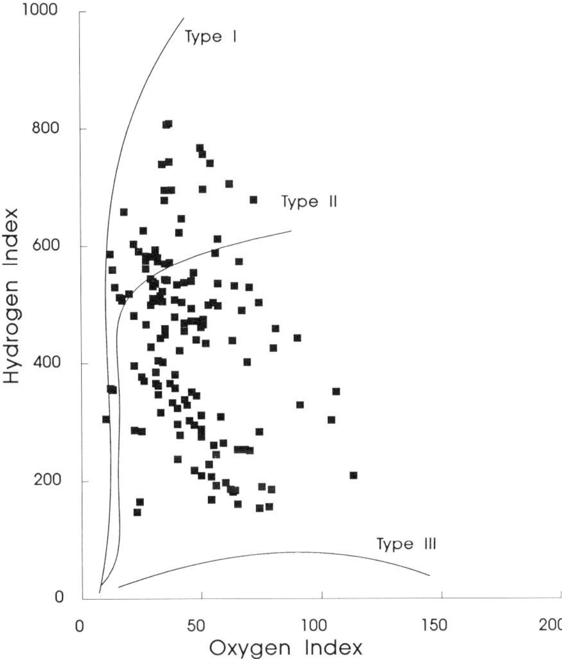

Fig. 7. Oxygen index versus hydrogen index for Egret Member cutting samples from 17 wells in the Jeanne d'Arc basin. The wells are those given in Table 1 with Rock-Eval data

shale, and sand lithologies or by interbedded source and nonsource beds. This alternation of source and non-source intervals cannot be confirmed from geochemical analyses of the cutting samples that are an average of the lithologies over usually a 10-m interval. Recently, however, a detailed correlation of petrophysical data and Rock-Eval analyses for the Egret Member determined that, where the source rock is immature, non-source layers comprise at least 25% of the gross source interval (Huang et al. 1994). This study further indicates that the major oil source is in the more shaly layers and the nonsource layers are limestone/marl in the south and siltstone/sandstone in the north.

The results of Rock-Eval/TOC analyses on cutting samples indicate that the Egret Member shows TOC contents up to approximately 8% with the majority of the samples in the 2–5% range (Fig. 6). The unit as a whole has average values of 1.6–4.6% TOC (Table 1). North Dana I-43, the most northerly well drilled on the Outer Ridge Complex, shows lower TOC contents than other wells. Average HI values where the Egret Member is immature are in the 500–700 range (Fig. 4c). Significantly higher average HI values are found at Rankin M-36 and Trave E-87 (Table 1). On an OI versus HI plot (a pseudo-van Krevelen diagram) Egret Member organic matter usually plots in the type II organic matter field (Figs. 5 and 7). Because these results are obtained from cutting samples collected over 10-m intervals, it is probable that they are an average of units with higher or lower TOC. Hand-picking samples from the organic-rich intervals does give TOC contents up to 9%, but HI values are not significantly higher. The spread of values in Fig. 7 into the type III area of the plot is because of wells such as North Dana I-13 where more oxidising conditions occurred during deposition of the Egret Member.

Microscopical examination of the organic matter of the Egret Member indicates that it is predominately amorphous (lamalginite) throughout the basin. The lamalginite, which fluoresces in immature samples, often occurs as thin layers alternating with organically lean dolomite layers (T. Gentzis, pers. comm. 1992). Very little structured material is observed such as occasional algae (usually *Tasmanites*) and dinoflagellate fragments. Terrestrially derived material is rare in samples from most wells. Mature samples are usually heavily bitumen-stained.

A cross section that illustrates the change in character of the downhole and geochemical logs in the Egret Member across the basin is shown in

Fig. 5. In Rankin M-36 the interval from 2436 to 2491 m corresponds to the Egret Member. These samples have TOC contents from 3.7 to 5.23% and contain immature type II organic matter. The immaturity of this organic matter is also indicated by the low extract and hydrocarbon yields, and by biomarker maturation parameters (Table 2). This interval therefore has excellent source potential. The Rankin Formation below the Egret Member down to 2550 m may also have some source potential. The change in the Rock-Eval parameters between the Egret Member and the underlying Rankin Formation is more gradual than indicated by the log characteristics. This may partly be caused by cavings in the cuttings samples rather than a gradual change in depositional conditions.

In Terra Nova K-08 the Egret Member shows TOC contents up to 5%, but lower HI values than Rankin M-36. It still plots as type II organic matter. The lower HI values are because of the greater maturity of organic matter in this well which is in the early part of the oil window (Avery, 1988). This is evident from the higher extract and hydrocarbon yields for these samples compared to those from Rankin M-36 (Table 2), and from the similarity of the biomarker distributions of the Egret Member extracts and oils from this well (Fig. 12). The Rock-Eval results suggest that the Rankin Formation both below and above the Egret Member has minor source potential at Terra Nova K-08.

In Trave E-87, the Egret Member occurs at depths of 3046 to 3120 m (Fig. 5). It is less calcareous and more shaly than at Rankin M-36. It has TOC values up to 4.24% and HI values up to 810 (Table 1). The organic matter is therefore type I-II. Rock-Eval results suggest that the Rankin Formation above and below the Egret Member has minor source potential. The organic matter in this section is early mature (Tables 1 and 2) and hence cannot be responsible for the highly mature condensates found in the Hibernia Formation reservoirs in this well. These hydrocarbons presumably migrated from more mature sediments further west in the basin centre (Avery et al. 1986).

The Egret Member in the Panther P-52 well is defined by log characteristics as occurring between 3218 and 3291 m. This interval shows TOC values up to 8.22% and HI values up to 660 mg HC/g TOC. The organic matter is marginally mature type II. As is evident from Fig. 5, this interval corresponds to the richest part of an extensive section within the Rankin Formation with a considerable potential to generate hydrocarbons.

Hydrocarbon Characteristics

The C_{15+}-saturated fraction gas chromatograms of most Egret Member extracts show the characteristics expected of marine-derived organic matter deposited under reducing conditions (Fig. 8). The less mature samples (e.g., Fig. 8a) usually show a bimodal distribution of *n*-alkanes with the C_{15}-C_{17} members in highest abundance and a high abundance of acyclic isoprenoids and polycyclic alkanes. The C_{23}-C_{31} *n*-alkanes can show an odd or even carbon number predominance. With increasing maturity, the abundance of *n*-alkanes relative to the branched and cyclic compounds increases, and they show a unimodal distribution around nC_{17} (Fig. 8b, c). The ratio of pristane to phytane (pr/ph) is less than one for many samples (Table 2) suggesting anoxic conditions of deposition (Didyk et al. 1978). Higher pr/ph ratios (1.3-2.1) are observed for wells in the eastern side of the basin (North Dana I-43, Panther P-52, South Tempest G-88, Archer K-19, and Fortune G-57), possibly suggesting more oxic conditions of deposition in that area.

The distribution of biomarkers in Egret Member extracts is relatively uniform throughout the basin, although as discussed later, there are some local differences. Most samples show a strong predominance of C_{27} over C_{29} steranes with C_{28} steranes in much lower concentrations (Figs. 9, 10; Table 2). The ratio of diasteranes relative to regular steranes in the extracts examined to date shows a closer relationship to the maturity rather than the lithology of the Egret Member at that location. C_{28}-C_{30} 4-methylsteranes are present in these extracts and in Jeanne d'Arc basin oils in high abundance (Fowler and Brooks 1990). The C_{30} 4-methylsteranes are those derived from dinosterol and hence indicate an important contribution by dinoflagellates to the organic matter of these samples (Summons et al. 1987; Goodwin et al. 1988). Although the Egret Member was deposited in a marine environment, C_{30} 4-desmethylsteranes which have been suggested to be indicators for marine-derived organic matter (Moldowan et al. 1985) are present in very low concentrations relative to other steranes. Terpane distributions are similar throughout the basin. Characteristics of m/z 191 fragmentograms (Fig. 11) include tricyclic terpanes in low concentrations relative to the hopanes in samples of all maturity levels, C_{30} hopanes in much higher concentrations than C_{29} hopanes, and a smooth distribution of C_{31}-C_{35} hopanes. This latter feature

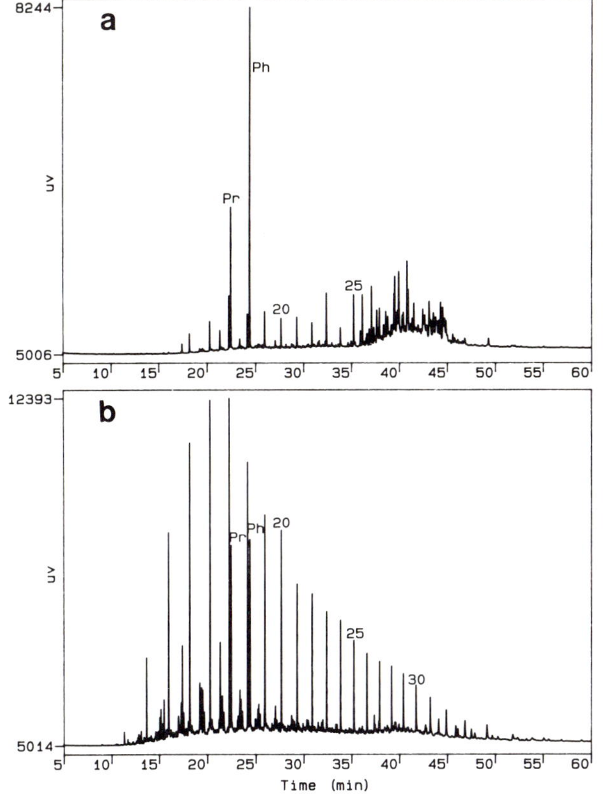

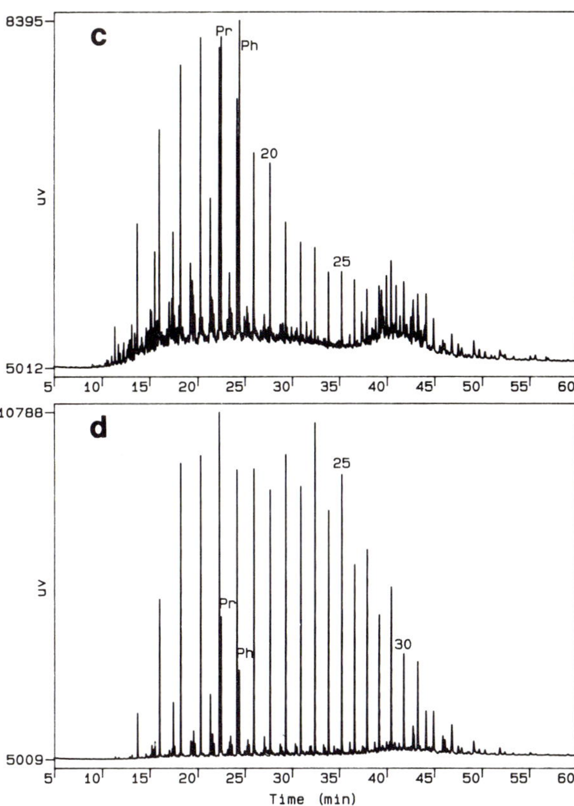

Fig. 8a-d. Representative saturate fraction gas chromatograms of extracts of Egret Member cutting samples showing the effects of increasing maturity (**a-c**) and differing depositional environment (**d**). Samples are: **a** Rankin M-36, 2470–2480 m; **b** Trave E-87, 3080–3090 m; **c** Terra Nova K-08, 3785–3795 m; and **d** Archer K-19, 3350–3360 m. *20*, *25* and *30* are the C_{20}, C_{25} and C_{30} *n*-alkanes; *Pr* and Ph are pristane and phytane respectively. With respect to hydrocarbon generation, the Egret Member is immature at Rankin M-36, marginally mature at Trave E-87 and Archer K-19, and mature at Terra Nova K-08

has been considered typical of source rocks deposited under marine conditions with suboxic bottom waters during deposition (Peters and Moldowan 1993). As previously reported for Jeanne d'Arc basin oils (Fowler and Brooks 1990), the ratios of Ts/Tm and of 17αH-norhopane to C_{29} 18αH-neohopane (X) in Egret Member extracts is mostly a function of maturity (Fig. 11, Table 2). A typical oil-source correlation using biomarkers is shown in Fig. 12 for an oil and an extract from the Terra Nova K-08 well. Biomarker characteristics that support the correlation include the high abundance of 4-methylsteranes, the relative abundance of the C_{27}:C_{28}:C_{29} regular steranes, a similar abundance of rearranged steranes to regular steranes and of rearranged hopanes to 17α(H)-hopanes, low abundance of tricyclic terpanes relative to 17α(H)-hopanes, and the similar C_{29}-C_{35} 17α(H)-hopane profile.

Source rock kerogens can be characterized using pyrolysis-gas chromatography (py-gc). It has also been demonstrated that the pyrogram of an asphaltene is similar to that of the kerogen from the same sample (Behar et al. 1984). Pyrograms of asphaltenes from some Egret Member extracts are shown in Fig. 13. These pyrograms are mostly dominated by alkane-alkene doublets with only minor amounts of aromatics. In the less mature samples (e.g., Fig. 13a), prist-1-ene is the dominant peak. These Egret Member asphaltene pyrograms are similar to those of kerogens classified as type IIpn by Larter and Sentfle (1985). According to these authors, this group of kerogens are from marine source beds containing dominantly amorphous organic matter from algal and bacterial sources. This is in agreement with what has been described here for the Egret Member.

When compared to the other potential source rocks in the Jeanne d'Arc basin, the Egret Member is found to most resemble the Jeanne d'Arc Formation. This is evident for example, in the ternary plot

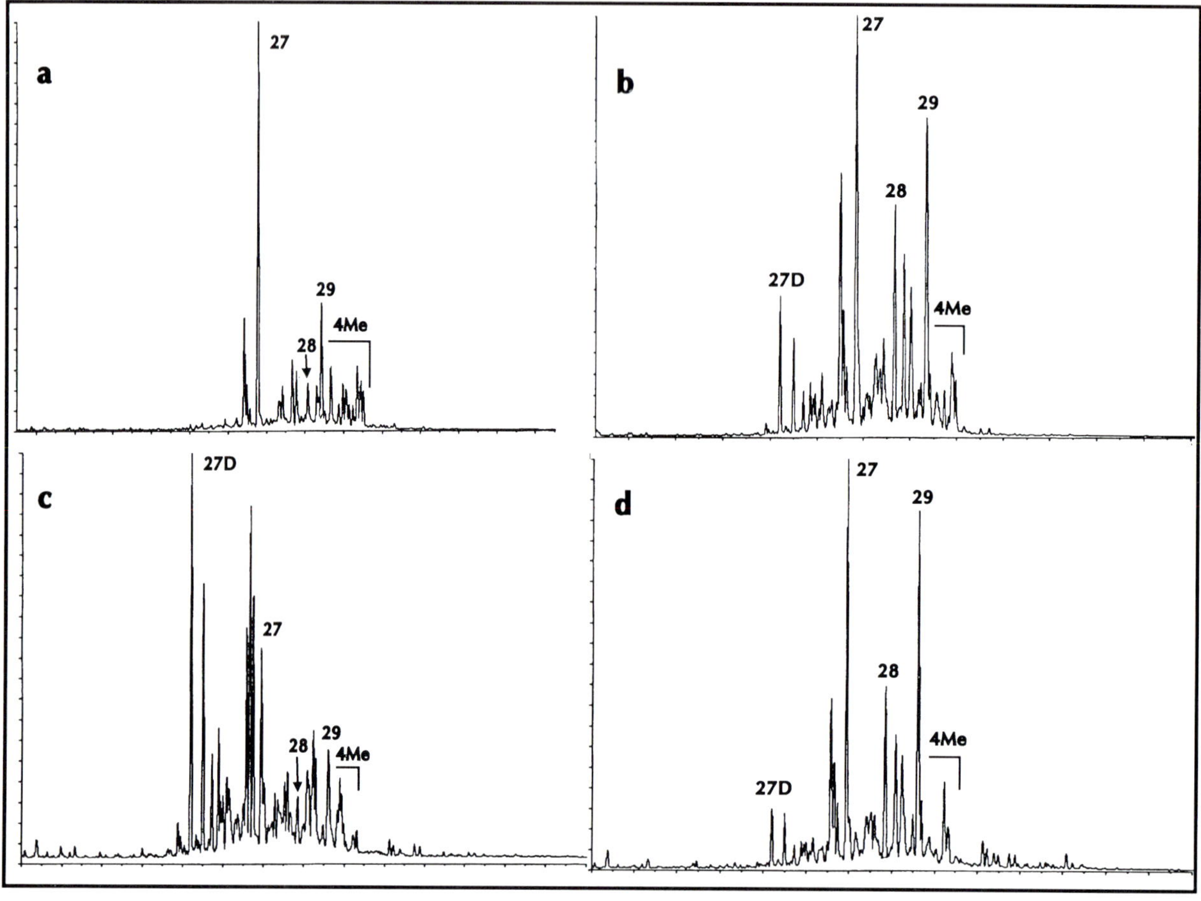

Fig. 9a-d. Representative m/z 217 mass fragmentograms showing the distribution of steranes in the extracts of Egret Member cutting samples. Samples are: **a** Rankin M-36, 2470–2480 m. **b** Trave E-87, 3080–3090 m. **c** Terra Nova K-08, 3785–3795 m, **d** Archer K-19, 3350–3360 m. *27*, *28*, and *29* are C_{27}-C_{29} 5α(H),14α(H),17α(H) 20R steranes and *27D* is the C_{27} 13β(H),17α(H) 20S diasterane. *Peaks within the area indicated with 4Me* are predominately C_{30} 4-methylsteranes

of C_{27}:C_{28}:C_{29} regular steranes (Fig. 10). Geological and geochemical evidence indicate that the Egret Member and the Jeanne d'Arc Formation source interval appear to have been deposited under similar marine conditions. At present it is difficult to distinguish migrated hydrocarbons sourced from these units. Hence, some oils in the Jeanne d'Arc basin, besides the minor oil discovery at Beothuk M-05, that is both probably sourced and reservoired within the Jeanne d'Arc Formation (based on biomarker maturity data), have a presently undetermined contribution from this unit. In the two wells where the Voyager Formation source interval has been encountered, its organic matter is mostly of terrestrial origin. This is reflected in the greater abundance of C_{29} steranes in the extracts of this formation compared to those from the Egret Member and Jeanne d'Arc Formation extracts (Fig. 10). Hydrocarbons sourced from this unit should be easily distinguishable from those of the Egret Member. This is the reason why this interval is thought to be a probable contributor to the 'Ben Nevis' type oils whose biomarkers show greater terrestrial influence than other Jeanne d'Arc basin oils (Fowler and Brooks, 1990).

Source Rock Heterogeneity

There are some localized variations to the above characteristics. Lower HI values are observed for some Egret Member samples from the Archer K-19 well than would be expected from their maturity when compared to samples from other wells (Table 1). In addition, compared to most other Egret

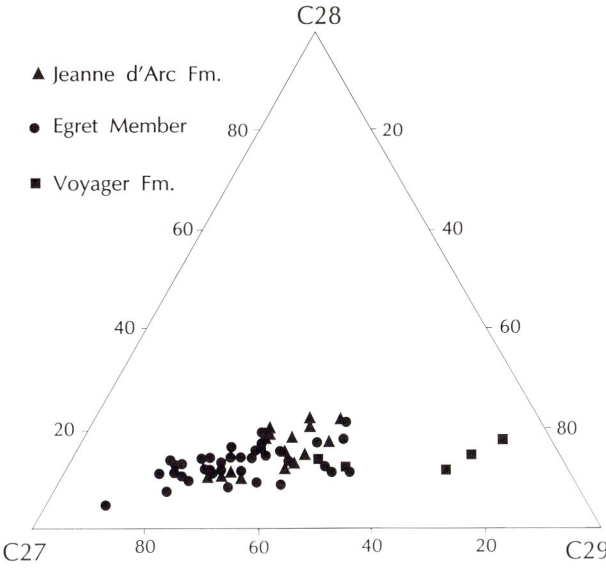

Fig. 10. Ternary plot of C_{27}:C_{28}:C_{29} sterane abundance in source rock extracts from the Jeanne d'Arc basin

Member samples, these samples from Archer K-19 contain a higher proportion of vitrinite and inertinite, have gas chromatograms that show a higher abundance of C_{20+} n-alkanes with a pronounced odd carbon number preference and a higher pr/ph ratio (Fig. 8d, Table 2), a higher abundance of C_{29} relative to C_{27} steranes (Fig. 9d, Table 2) and lesser amounts of 4-methylsteranes relative to 4-desmethylsteranes. All these characteristics suggest a greater contribution from higher land plant-derived organic matter to the Egret Member at Archer K-19. This was also noted by von der Dick (1989), who thought that this was a general trait for wells on the eastern flank because of runoff from the eastern Central Ridge. The Egret Member at Fortune G-57 has many of the same characteristics as Archer K-19 but samples from wells such as Trave E-87 and Voyager J-18 do not show evidence for a significant terrestrial contribution (Table 2). Archer K-19 is located where a late Jurassic deltaic wedge is seen on seismic data to intrude into the basin (A. Edwards, pers. comm, 1989) which would explain the greater terrestrial contribution to the organic matter in the Egret Member in this area.

Depositional Environment

The fine-grained and laminated nature of the Egret Member and its high organic carbon content suggests a low energy restricted depositional environment (McAlpine, 1990a). Isopachs (Fig. 4a) show thickest occurrences in the present-day axis of the Jeanne d'Arc basin and on the northeast Outer Ridge. A narrow zone of thin deposition between these two areas suggests a sill could have acted as a barrier to circulation. Oolitic and skeletal packstones and grainstones, often encountered below and above the unit in the south, suggests proximity to a carbonate shelf or bank (Fig. 4b). The environment was probably a shallow water (25–50 m) anoxic basin. This is supported by the absence of foraminiferal and the presence of ostracod fauna that are more tolerant of extreme marine conditions (P. Ascoli, pers. comm., 1993). Micropaleontological evidence indicates that the Rankin Formation above and below the Egret Member is normal marine. Further north, a somewhat deeper water siliciclastic environment prevailed where density currents probably played a significant depositional role. Terrestrial organic matter is low in most places, suggesting low continental runoff at the time of deposition. High planktonic productivity in the surface waters, especially from dinoflagellates, and restricted circulation in a silled basin leading to suboxic to anoxic conditions in the bottom waters resulted in a high accumulation of organic matter. The low proportion of structured compared to amorphous organic matter and the relatively high hopane to sterane ratios indicate extensive reworking of the primary organic matter during early diagenesis by anaerobic bacteria. As discussed above, there is seismic and geochemical evidence of deltaic conditions existing in the area around Archer K-19 during Egret Member deposition.

Exploration Strategy

Figure 4d is a schematic structure map on top of the Egret Member that also shows where the source rock intersects the present-day oil generative window. The depth to Egret Member (Table 1, Fig. 4) indicates the differential burial history of the Egret Member in each well. It is evident that the majority of oil discoveries lie within an area underlain by the present day mature zone and that the gas-prone wells are within or proximal to the overmature zone. Because the structural framework of the Jeanne d'Arc basin was essentially established prior to Late Cretaceous time, the general thermal

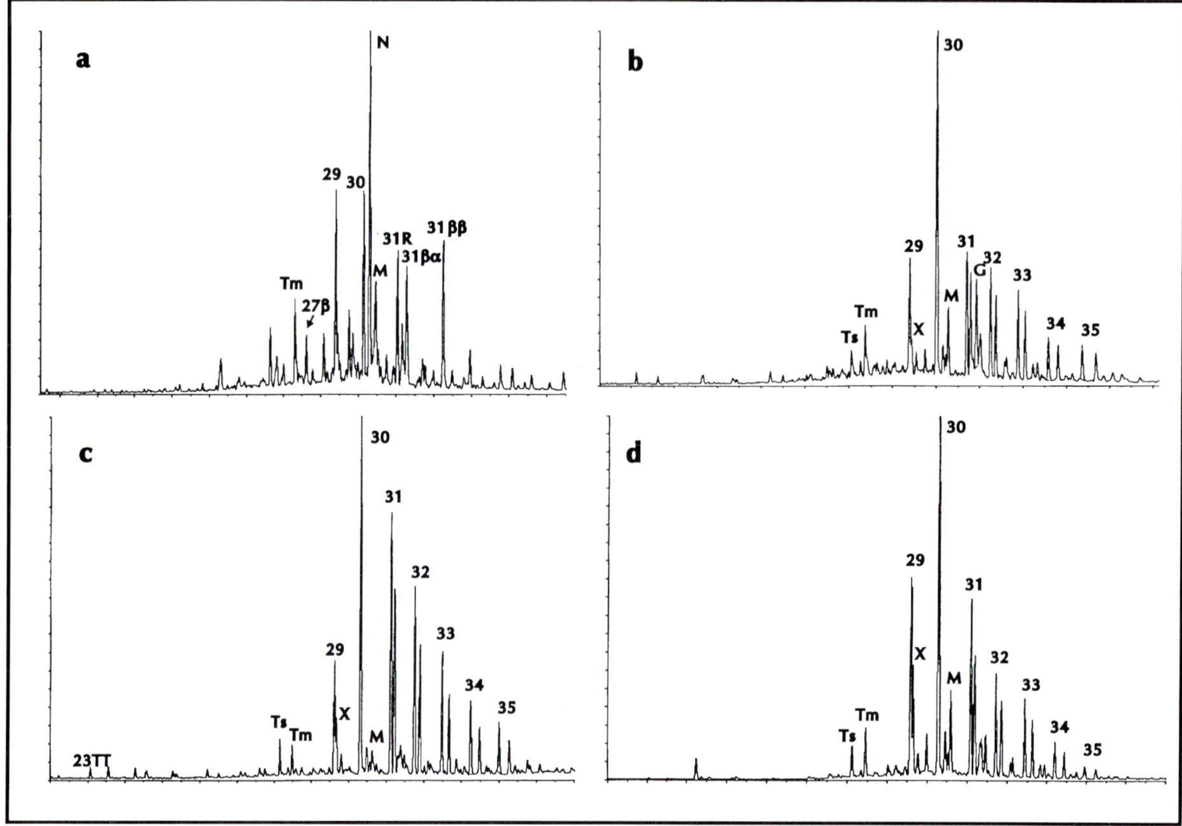

Fig. 11a-d. Representative m/z 191 mass fragmentograms showing the distributions of terpanes in the extracts of Egret Member cutting samples. Samples are: **a** Rankin M-36, 2470–2480 m. **b** Trave E-87, 3080–3090 m. **c** Terra Nova K-08, 3785–3795 m. **d** Archer K-19, 3350–3360 m. 29-35 are C_{29}-C_{35} 17α(H),21β(H)-hopanes; *23TT* is the C_{23} tricyclic terpane, *Ts* is 18α(H)-trisnorhopane; *Tm* is 17α(H)-trisnorhopane, *27β* is 17β(H)-trisnorhopane; *X* is C_{29} 18α(H)-neohopane, N is a C_{30} hopene tentatively identified as neohop- 13(18)-ene, *M* is 17β(H),21α(H)-moretane, *G* is gammacerane *31*R, *31α*, and *31ββ* are the 17α(H),21β(H) 22R, 17β(H),21α(H), and 17β(H),21β(H) homohopanes respectively

maturation configuration is likely to have existed throughout the last 100 Ma and to have ascended stratigraphically as the basin was buried beneath a relatively uniform blanket of Late Cretaceous and Tertiary strata. Modeling indicates that the creation of the maturity zones, such as the nonmature to mature boundary in the southern part of the basin, was mostly established by 100 Ma (Williamson, 1992). This model suggests that only in the east has the nonmature to mature boundary changed, where in the last 60 Ma it has moved eastward to encompass part of the Outer Ridge Complex. Peak generation appears to have been reached 50 Ma ago in the Early Tertiary, long after the structural traps had been formed (Ervine 1985; Brown et al. 1989; Williamson 1992). Therefore all drillable structures that are suitably located with respect to mature Egret Member source beds can be considered prospective for hydrocarbon entrapment (McAlpine 1990a).

Differences in the kinetics of generating bitumen from this unit may be related to lithological differences within the Egret Member. In the southern area, geochemical logs often show a suppression of T_{max} values within the Egret Member relative to the surrounding units which is not observed to the same extent in the northeast (compare the geochemical logs of Rankin M-36 and Trave E-87 in Fig. 3). This may be due to greater incorporation of sulfur into the organic matter in the southern area. This occurs to a greater extent in carbonates rather than shales because of the lower availability of iron during early diagenesis (e.g., Berner, 1985). Hence, hydrocarbons may be generated from the Egret Member at lower maturity in the southern compared to the northern part of the

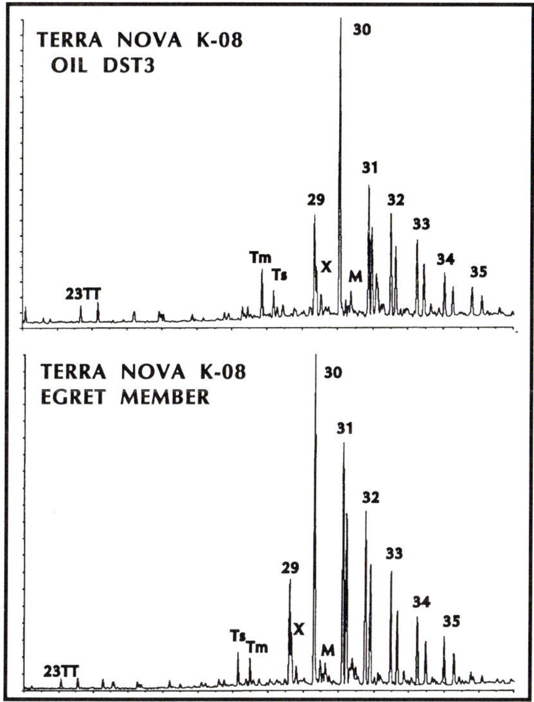

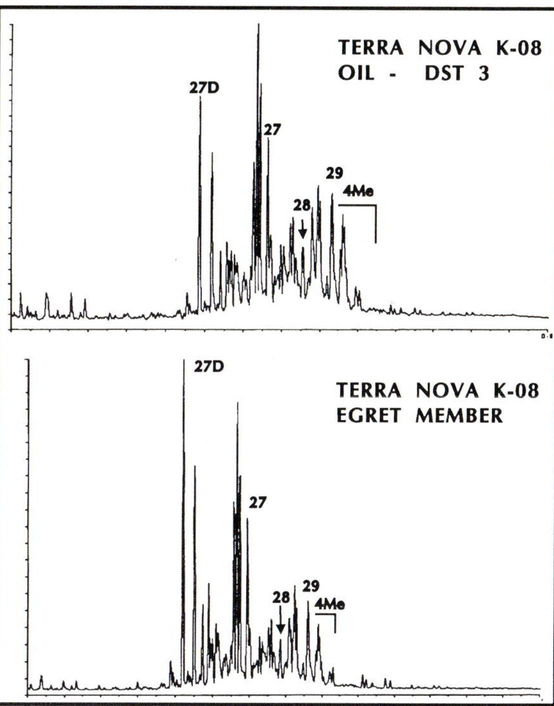

Fig. 12. Oil-source correlation for an oil from Terra Nova K-08 (DST3, 3380-3395 m.) and an Egret Member cuttings extract (3785-3795 m) from the same well. Annotation as for Figs. 10 and 11

basin. This is also observed for the Tithonian-aged source rock. Hunt et al. (1991) recently considered the kinetics of the "Egret Source Rock" from Hibernia K-18, which they assumed originally contained type II organic matter with about 7% organic sulfur. They assigned it type IIC kinetics which meant it has a "medium reaction rate" with regard to hydrocarbon generation. For reasons outlined above, hydrocarbon generation from the Egret Member in other parts of the basin may be accelerated because of a possibly higher organic sulfur content.

In much of the prospective area of the Jeanne d'Arc basin, the source rock is now overpressured while the majority of discovered hydrocarbons are in the overlying, hydrostatically pressured regime. The laterally continuous Fortune Bay Shale generally forms the cap rock for the overpressure (Grant and McAlpine 1990; McAlpine 1990a). In addition to retaining connate and diagenetic fluids in the rocks until thermal conditions are ideal for the generation of hydrocarbons, the overpressure may, in fact, promote subsequent migration. The main path for oil migration appears to be along faults and fractures that have opened sporadically in response to build-up of abnormally high fluid pressures. Pressure plots for many wells show that the rate of pressure increase with depth, within the sealed zone, is faster than can be accounted for by simple loading, probably because of thermal expansion and hydrocarbon generation. This means that shales will become mobile at depth and cause faults and fractures in the overlying section that become avenues for pressure and fluid escape and oil migration. This process is cyclic because the faults would close after pressure release. Based on the interpretation of geochemical data, an example of this was demonstrated at Hibernia K-18 by Fowler and Brooks (1990). These authors showed that stacked reservoirs over a 1577-m vertical interval were probably originally filled by a single pulse of oil that migrated via faults. Reservoirs within the lower part of the Hibernia Formation have subsequently received later generated, more mature hydrocarbons that probably "leaked" out of the overpressured zone below. Fowler and Brooks (1990) presented geochemical data that indicated that there is no longer a conduit from these sands to a reservoir at the top of the Hibernia Formation and to reservoirs within the Avalon and Catalina Formations.

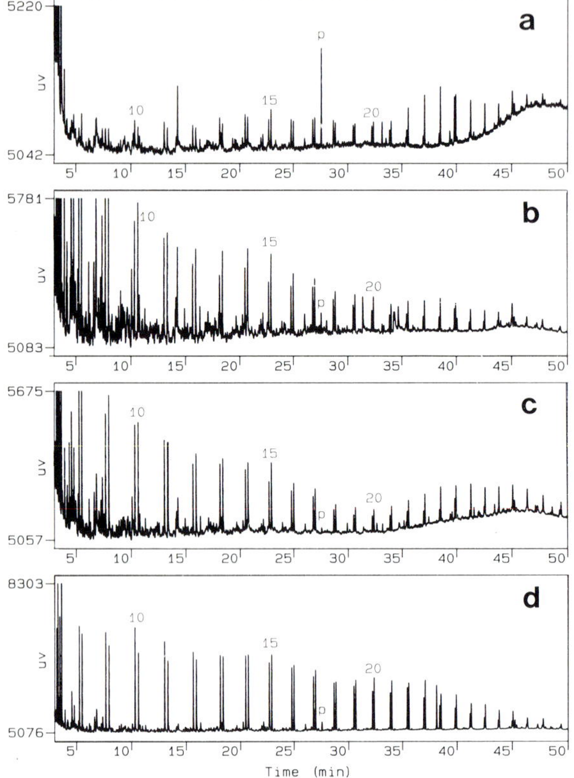

Fig. 13a-d. Pyrograms of asphaltenes from Egret Member extracts. Samples are: **a** Rankin M-36, 2470-2480 m. **b** Trave E-87, 3080-3090 m. **c** Terra Nova K-08, 3785-3795 m, and **d** Archer K-19, 3350-3360 m. *10*, *15*, and *20* are C_{10}, C_{15}, and C_{20} *n*-alkane-alkene doublets and *p* is prist-1-ene

Quantitative Estimates of Hydrocarbons Generated by Egret Member

Currently discovered, recoverable oil in the Jeanne d'Arc basin is estimated to be 1.5×10^9 bbls ($0.24 \times 10^9 m^3$) contained in 17 discoveries within 5 play groups (Taylor et al. 1991, Sinclair et al. 1992). On the other hand, past assessments of the basin's ultimate, recoverable oil potential have been 12.3×10^9 bbls (1.96×10^9 m^3) (Sheppard and Hawkins 1983), 7.1×10^9 bbls (1.13×10^9 m^3) (Procter et al. 1984), and 4.7×10^9 bbls (0.75×10^9 m^3) (Taylor et al. 1991, Sinclair et al. 1992). These assessments were based on statistical methods, Monte Carlo, probabilistic and pool size discovery modeling, respectively. An alternative resource assessment technique, applicable to frontier basins where well data are sparse, involves calculating the hydrocarbon-generating potential of likely or known source rocks and discounting the calculated volume for migration and trapping inefficiencies.

Rock-Eval analysis of cutting samples from 16 out of 20 wells that penetrated the Egret Member in the Jeanne d'Arc basin, including two on the Outer Ridge Complex, show it to be an excellent source rock (Table 1). In these wells the Egret Member ranges in thickness from 55 m to in excess of 200 m, and shows average TOC contents up to 4.58% and average hydrogen index values of immature samples in the 500–700 range (Table 1, Fig. 4). We used two geochemical techniques to calculate the amount of oil that the Egret Member may have generated, a mass balance method (Goff 1983) and another method employing hydrogen index values obtained from our Rock-Eval data. The mass balance method calculates the volume of oil generated as the mathematical product of the bulk volume of the source rock, the percent by volume of organic material, the genetic potential (about 70% for type II kerogen), the fraction of oil in the hydrocarbon yield (about 80%), the transformation ratio (about 0.5 from Rock-Eval data), and the volume increase on oil generation (1.15 using specific gravities of 1.0 for kerogen and 0.87 for the oil). For the Hydrogen index method the volume of oil generated is the mathematical product of the bulk volume of the source rock, the density of the source rock, the percent by weight TOC, the hydrogen index, the fraction of oil in the hydrocarbon yield, and the volume increase on oil generation. Parametric data for the equations were obtained by measuring the areas between contours on isopach and geochemical facies maps and calculating weighted averages. The following results are updated from McAlpine (1990a) by incorporating more recent well data and geochemical analyses. Additionally, the results apply to only the mature and overmature area of the Egret Member within the Jeanne d'Arc basin, not including the Outer Ridge Complex (Fig. 4d). Within this area of 6350 km^2 the Egret Member averages 150 m thick, 3.4% by weight TOC or 10.2% by volume organic matter, and has an average hydrogen index of 630 mg hydrocarbon/g organic carbon.

The mass balance and hydrogen index techniques of calculating oil generation yielded similar quantities of 226×10^9 bbls (36×10^9 m^3) and 265×10^9 bbls (42×10^9 m^3). The average estimate of 245×10^9 bbls (39×10^9 m^3) does not take into account expulsion, migration, and trapping inefficiencies and, of course, does not include potential contributions from other source intervals.

Estimating conservatively that 15% of the oil has ended up in reservoirs, and using a recovery factor of 30% gives estimated recoverable reserves of about 11×10^9bbls (1.8×10^9 m^3). A comparison with current estimated recoverable reserves of 1.5×10^9 bbls (0.24×10^9 m^3) (Taylor et al. 1991; Sinclair et al. 1992) suggests that less than 15% of potential recoverable oil reserves in the Jeanne d'Arc basin may presently be discovered. Hence, further exploration of this frontier area may show the Egret Member to be an even more prolific source rock than is presently suggested based on known reserves and previous total resource estimates.

Summary

The Kimmeridgian-aged Egret Member of the Rankin Formation is the most important hydrocarbon source rock within the Jeanne d'Arc basin. It ranges in thickness from 55 m to in excess of 200 m and shows average TOC contents up to 4.58%. The organic matter is mostly amorphous type II with hydrogen index values of immature samples in the 500–700 range. The Egret Member is easily identified on downhole and geochemical logs, principally because of its higher TOC content compared to surrounding units. It was probably deposited under relatively shallow water conditions in a semi-silled basin where periodically high planktonic productivity occurred in surface waters, especially from dinoflagellates. Because of the restricted circulation, bottom waters were reducing enough to allow reworking of this primary organic matter by only anaerobic bacteria. The lithology of the Egret Member is interbedded shales and carbonates with the shales becoming more dominant toward the northeast of the basin. As the Egret Member has not been cored, detailed sedimentological studies have not been possible.

Modeling suggests that the peak generation of hydrocarbons from the Egret Member in the Jeanne d'Arc basin occurred about 50 Ma ago. The Egret Member is presently overpressured over most of the area where it is mature. Oil migration is thought to have occurred primarily via faults into reservoirs and carrier beds. Estimates of the quantity of hydrocarbons that could have been generated from the Egret Member indicate considerably more oil may be present in this frontier basin than has been discovered to date.

Acknowledgments. We wish to acknowledge our colleagues at the ISPG (Sneh Achal, Paul Brooks, Ron Fanjoy and Marg Northcott) plus Ross Stewart (Arctic Geochem) for their technical assistance, Mike Avery (AGC) for allowing us to use his vitrinite reflectance data and for his assistance in computer drafting. We thank Lloyd Snowdon, Don Stachiw, Hans von der Dick, Trevor Powell, and Barry Katz for their critical comments on previous versions of this chapter. This is GSC contribution no. 19192.

References

Avery MP (1988) Vitrinite reflectance (Ro) of dispersed organics from Petro-Canada Terra Nova K-08. Geol Surv Can Open File Rep 1805, 15 pp

Avery MP, Bell JS, McAlpine KD (1986) Vitrinite reflectance measurements and their implications for oil and gas exploration in the Jeanne d'Arc basin, Grand Banks, eastern Canada. In Current research, Part A. Geol Surv Can Pap 86-1A: 489–498

Behar F, Pelet R, Roucache J (1984) Geochemistry of asphaltenes. Org Geochem 6: 587–595

Berner RA (1985) Sulphate reduction, organic matter decomposition and pyrite formation. Philos Trans R Soc Lond A315: 25–38

Brown DM, McAlpine KD, Yole RW (1989) Sedimentology and sandstone diagenesis of Hibernia Formation in Hibernia oil field, Grand Banks of Newfoundland. Am Assoc Petrol Geol Bull 73: 557–575

Creaney S, Allison BH (1987) An organic geochemical model of oil generation in the Avalon/Flemish Pass subbasins, east coast Canada. Bull Can Petrol Geol 35: 12–23

Didyk BM, Simoneit BRT, Brassell SC, Eglinton G (1978) Organic geochemical indicators of palaeoenvironmental conditions of sedimentation. Nature 272: 216–222

Enachescu ME (1986) Integrated geophysical study of Newfoundland continental margin (east coast Canada). In: Expanded abstracts of the Society of Exploration Geophysicists, 56th Annu Meet Exposition, Houston, November, 1986 pp 488–492

Ervine WB (1985) A synthesis of maturation data for the East Newfoundland basin. Geol Surv Can Open File Rep 1178, 105 pp

Foster DG, Robinson AG (1993) Geological history of the Flemish Pass basin, offshore Newfoundland. Am Assoc Petrol Geol Bull 77: 588–609

Fowler MG (1993) The effects of the use of oil-based drilling muds and other organic additives on organic geochemical analyses of samples from the Jeanne d'Arc basin, offshore eastern Canada. In (eds). Oygard, K. et al. Organic geochemistry, poster sessions from the 16th International Meeting on Organic Geochemistry, Stavanger, 1993', Falch Hurtigtrykk Oslo pp 6–9

Fowler MG, Brooks PW (1990) Organic geochemistry as aid in the interpretation of the history of oil migration into different reservoirs at the Hibernia K-18 and Ben Nevis I-45 wells, Jeanne d'Arc basin, offshore eastern Canada. Org Geochem 16: 461–475

Fowler MG, Snowdon LR (1988) Rock-Eval/TOC data from an additional seven wells located within the Jeanne d'Arc basin, offshore Newfoundland. Geol Surv Can Open File Rep 1735, 47 pp

Fowler MG, Snowdon LR (1989) Rock-Eval/TOC data from wells located in the southern Grand Banks and the Jeanne d'Arc basin, offshore Newfoundland. Geol Surv Can, Open File Rep 2025, 49 pp

Fowler MG, Brooks PW, Snowdon LR, McAlpine KD (1988) Petroleum geochemistry of the Jeanne d'Arc basin. In: Program with abstracts. Geol. Assoc Can, Mineral Assoc Can, Can Soc Petrol Geol, Joint Annu Meet, St. John's, May 23–25, 1988, 13: A40

Fowler MG, Brooks PW, Snowdon LR (1989) Gas chromatography and gas chromatography-mass spectrometry data of some Jeanne d'Arc basin oil saturate fractions. Geol Surv Can Open File Rep 2074, 96 pp

Fowler MG, Snowdon LR, Stewart KR, McAlpine KD (1990) Rock- Eval/TOC data from nine wells located offshore Newfoundland. Geol, Surv Can Open File Rep 2271, 72 pp

Fowler MG, Snowdon LR, Stewart KR, McAlpine KD (1991) Rock- Eval/TOC data from five wells located within the Jeanne d'Arc basin, offshore Newfoundland. Geol Surv Can Open File Rep 2392, 41 pp

Goff JC (1983) Hydrocarbon generation and migration from Jurassic source rocks in the East Shetland basin and Viking Graben of the North Sea. J Geol Soc Lond 140: 445–474

Goodwin NS, Mann AL, Patience RL (1988) Structure and significance of C_{30} 4-methyl steranes in lacustrine shales and oils. Org Geochem 12: 495–506

Grant AC, McAlpine KD (1990) The continental margin around Newfoundland. In: Keen MJ, Williams GL (eds) Geology of the continental margin off Eastern Canada. Geol Surv Can Geol Can 2: [also Geol Soc Am Geol N Am 1] 239–292

Huang Z, Williamson M, Fowler MG, McAlpine KD (1994) Predicted and measured petrophysical and geochemical characteristics of the Egret Member oil source rock, Jeanne d'Arc basin, offshore eastern Canada. Mar Petrol Geol 11: 294–306

Hunt JM, Lewan MD, Hennet RJ-C (1991) Modelling oil generation with time-temperature index graphs based on the Arrhenius Equation. Am Assoc Petrol Geol Bull 75: 795–807

Keen CE, Boutilier R, de Voogd B, Mudford BS, Enachescu ME (1987) Crustal geometry and models of the evolution of the rift basin on the Grand Banks of eastern Canada: constraints from deep seismic data. In: Beaumont C, Tankard AJ (eds) Sedimentary basins and basin-forming mechanisms. Can Soc Petrol Geol, Calgary, Mem 12: 101–115

Larter SR, Sentfle JT (1985) Improved kerogen typing for petroleum source rock analysis. Nature 318: 277–280

McAlpine KD (1990a) Mesozoic stratigraphy, sedimentary evolution, and petroleum potential of the Jeanne d'Arc basin, Grand Banks of Newfoundland. Geol Surv Can Pap 89–17, 50 pp

McAlpine KD (1990b) Lithostratigraphy of fifty-nine wells, Jeanne d'Arc basin. Geol Surv Can Open File Rep 2201, 97 pp

McAlpine KD, Fowler MG (1990) Quantitative assessment of hydrocarbon potential of Jeanne d'Arc basin source rocks using geological and geochemical data. In: Program with abstracts. Can Soc Petrol Geol Annu Conv Basin Perspectives, Calgary, May, 1990, p 93

Moldowan JM, Seifert WE, Gallegos EJ (1985) Relationship between petroleum composition and depositional environment of petroleum source rocks. Am Assoc Petrol Geol Bull 69: 1255–1268

Peters KE, Moldowan JM (1993) The biomarker guide. Prentice Hall, Englewood Cliffs, 363pp

Powell TG (1985) Paleogeographic implications for the distribution of Upper Jurassic source beds: offshore eastern Canada. Bull Can Petrol Geol 33: 116–119

Procter RM, Taylor GC, Wade JA (1984) Oil and natural gas resources of Canada – 1983. Geol Surv Can Pap 83–31, 59 pp

Sheppard MG, Hawkins D (1983) Petroleum resource potential of offshore Newfoundland and Labrador. Resource Assessment Division, Petroleum Directorate of Newfoundland and Labrador, Spec Rep PD 83-1, 14 pp

Sinclair IK (1988) Evolution of Mesozoic-Cenozoic sedimentary basins in the Grand Banks area of Newfoundland and comparisons with Falvy's (1974) rift model. Bull Can Petrol Geol 36: 255–273

Sinclair IK, McAlpine KD, Sherwin DF, McMillan NJ, Taylor GC, Best ME, Campbell GR, Hea JP, Henao D, Proctor RM (1992) Petroleum resources of the Jeanne d'Arc basin and environs, Grand Banks, Newfoundland. Geol Surv Can Pap 92-8, 48 pp

Snowdon LR, Fowler MG (1986) Rock-Eval/TOC data from seven wells located within the Jeanne d'Arc basin, offshore Newfoundland. Geol Surv Can Open File Rep 1382, 40 pp

Summons RE, Volkman JK, Boreham CJ (1987) Dinosterane and other steroidal hydrocarbons of dinoflagellate origin in sediments and petroleum. Geochim Cosmochim Acta 51: 3075–3082

Swift JH, Williams JA (1980) Petroleum source rocks, Grand Banks area. In: Miall AD (ed) Facts and principles of world petroleum occurrence. Can Soc Petrol Geol, Calgary, Mem 6: 567–587

Tankard AJ, Welsink HJ, Jenkins WAM (1989) Structural styles and stratigraphy of the Jeanne d'Arc basin, Grand Banks of Newfoundland. In: Tankard AJ, Balkwill HR (eds) Extensional tectonics and stratigraphy of the North Atlantic margins. Am Assoc Petrol Geol, Tulsa, Mem 46: 265–281

Taylor GC, Best ME, Campbell GR, Hea JP, Henao D, Procter RM (1991) Petroleum Resources of the Jeanne d'Arc basin, Grand Banks of Newfoundland. Geol Surv Can Open File Rep 2150, 13 pp

von der Dick H (1989) Environment of petroleum source rock deposition in the Jeanne d'Arc basin off Newfoundland. In: Tankard AJ, Balkwill HR (eds) Extensional tectonics and stratigraphy of the North Atlantic margins. Am Assoc Petrol Geol, Calgary, Mem 46: 295–303

von der Dick H, Meloche JD, Dwyer J, Gunther P (1989) Source-rock geochemistry and hydrocarbon generation in the Jeanne d'Arc basin, Grand Banks, offshore eastern Canada. J Petrol Geol 12: 51–68

Williamson MA (1992) Subsidence, compaction, thermal and maturation history of the Egret Member source rock of the Jeanne d'Arc basin, offshore Newfoundland. Bull Can Petrol Geol 40: 135–150

Petroleum Generation in the Nonmarine Qingshankou Formation (Lower Cretaceous), Songliao Basin, China

Li Desheng[1], Jiang Renqi[1] and Barry Jay Katz[2]

Abstract

The Songliao basin, covering an area of 260 000 km^2 in northeast China, is known as one of the world's most prolific oil- and gas-producing basins. The super giant Daqing oil field is located in the center of the basin with major oil plays in the Lower Createaceous Yaojia and Qingshankou Formations.

The nonmarine Qingshankou Formation also acts a primary source for the basin's hydrocarbon reserves. Dark lacustrine mudstones with a thickness of between 300 and 400 m are widely distributed in the basin. The Qingshankou Fm. kerogens range from sapropelic to gas-prone. The average organic carbon and total hydrocarbon contents are 2.3 wt.% and 1540 ppm, respectively. The generated crude oils display many typically lacustrine oil characteristics including high wax contents, high nickel/vanadium ratios, low sulfur contents, and light carbon isotopic compositions.

The Gulong depression, west of the Daqing anticlinal belt, was a primary depocenter for the Qingshankou Fm. Here, the unit covers approximately 50 000 km^2 and is currently buried to depths of 1900 to 2400 m. The hydrocarbon reserves associated with Qingshankou Fm. can be attributed to the interrelationship among the source-reservoir-seal system which includes encapsulating lacustrine mudstones, and lenticular delta-front and channel filled sand bodies.

Recent drilling has indicated that substantial Qingshankou-derived oil remains to be produced.

[1] Research Institute of Petroleum Exploration and Development, P.O. Box 910, Beijing 100083, Peoples' Republic of China

[2] Texaco Inc., 3901 Briarpark, Houston, Texas 77042, USA

Introduction

The Lower Cretaceous Qingshankou Formation comprises the main source rock of the Songliao basin in northeast China. Another important source rock, the Nenjiang Formation, is less thermally mature. The basin covers an area of ~260 000 km^2 (Fig. 1) and has been an area of active research for the past 30 years. Numerous aspects of the basin have been investigated, including

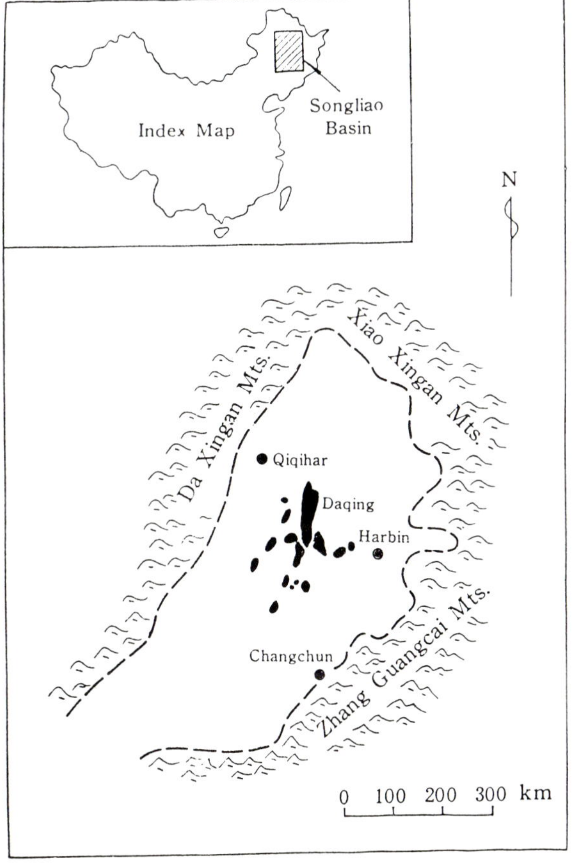

Fig. 1. The location of the Songliao basin

its tectonic evolution (Li Desheng 1982; Zhang Kai et al. 1983; Liu Jiaqi 1989; Tian Zaiyi 1990; and Liu Bin and Zhao Chunman 1991), sedimentation history (Wang Hengjian and Chao Wenfu 1981; Du Bomin et al. 1984; Qiu Yinan et al. 1988; Zhang Jinliang and Yu Huilong 1989), biostratigraphy (Chin Chen 1980; Gao Ruiqi 1980; Ye Dequan 1988), and hydrocarbon occurrence (Yang Wanli et al. 1985). There have also been various attempts to integrate these diverse aspects into a complete synthesis (Yang Wanli 1985; Lee 1986; Li Desheng 1987).

The present chapter focuses on petroleum generation in the Qingshankou Formation by providing a comprehensive picture of the unit's depositional setting, sedimentary facies, source rock characteristics, and hydrocarbon occurrences. Of particular interest is the cause for the numerous excellent source rock horizons within the Qingshankou Fm. This chapter combines a review of previous studies with new observations in an attempt to link depositional environments and the development of these prolific source beds.

Geological Setting

The Songliao basin is a Meso-Cenozoic intraplate, rift-depression, with a basement of Paleozoic metamorphic rocks and Hercynian granites (Li Desheng 1980, 1984). Crustal thickness varies in the

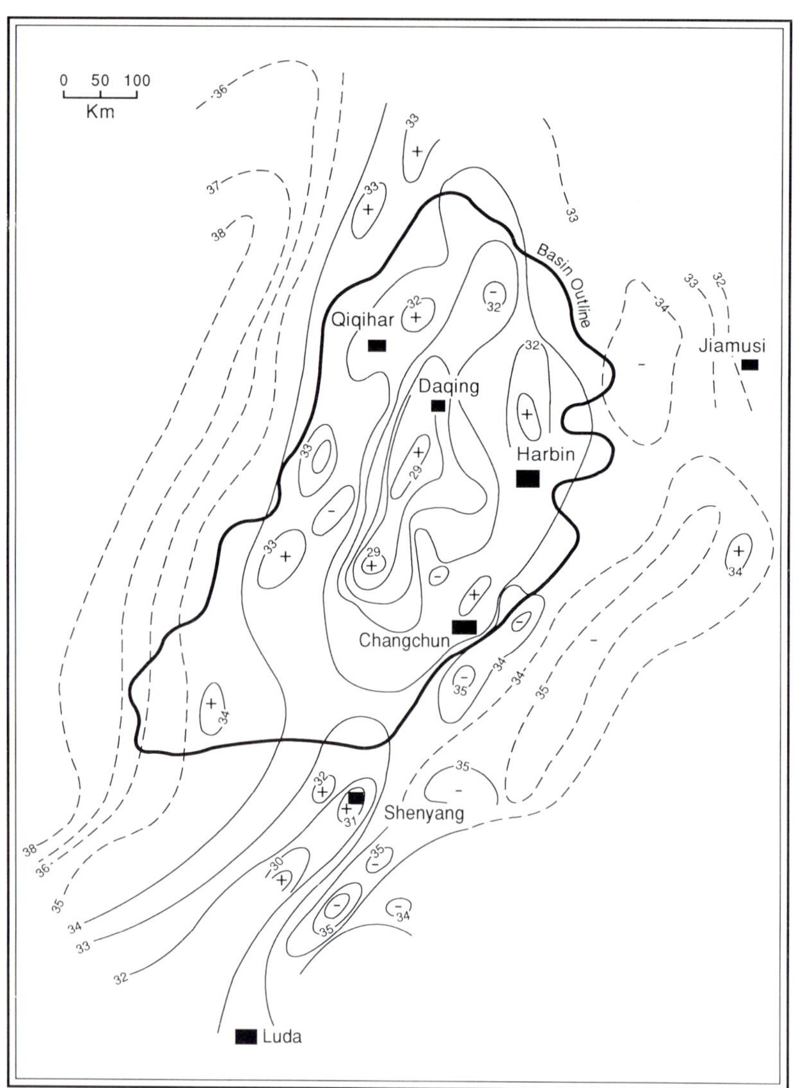

Fig. 2. Depth to Moho in kilometers in the Songliao basin area

vicinity of the basin, the Mohorovicic Discontinuity (Moho) ranges from ~ 27 km in the depressions to greater than 35 km in the surrounding mountains (Fig. 2). Basin fill ranges in age from middle Jurassic to Recent, with thickness between 3000 and 7000 m (Fig. 3).

In the Paleozoic, large-scale crustal uplift was caused by Hercynian movement. This uplift was associated with magmatic activity and the emplacement of granite. Extensional stresses associated with middle and late Jurassic block faulting formed a series of discrete rift valleys (Fig. 4). Jurassic sediments, composed primarily of sandstones and gray mudstones interbedded with coal and volcanics, were deposited in rifts, achieving in some cases thicknesses between 3 and 5 kms.

During the early Cretacous, the various separated rift depressions evolved into one large depression covering an initial area of ~ 42 000 km^2. The Albian to Aptian, Quantou, Qingshankou, Yaojia, and Nenjiang Formations were deposited within this rapidly subsiding single basin which incorporated several rivers and a large lake system. Total thicknesses of these sediments can be in excess of 3000 m. As the basin evolved its areal extent expanded. By Qingshankou time, deep lacustrine facies covered an area of ~ 87 000 km^2 and by Nenjiang-1 time more than 100 000 km^2 of deep

System	Series	Stage	Formation	Member	Lithology	Thickness (m)	Age (M.Y.)
Q	Pleis.					0 - 143	2.0
Tertiary	Plio.		Taikang			0 - 165	
	Mio.		Daan			0 - 123	
	Oli-Eo		Yian			0 - 256	
	Pales.	Danian	Mingshui	M2		0 - 381	
				M1		0 - 243	67.0
Cretaceous	Upper	Ceno-manian	Sifangtai			0 - 413	92.0
	Lower	Albian	Nenjiang	N5		0 - 355	
				N4		0 - 300	
				N3		47 - 118	
				N2		50 - 252	106.8
				N1		21 - 222	108.5
			Yaojia	Y2+3		0 -150	
				Y1		0 - 60	112.0
		Aptian	Qingshan-kou	Qn2		53 - 552	119.2
				Qn1		25 - 112	120.5
			Quantou	Q4		0 - 128	
				Q3		0 - 529	
				Q2		0 - 479	
				Q1		0 - 885	131.0
		Upper Neo-comian	Denglouku	D4		0 - 212	
				D3		0 - 562	
				D2		0 - 700	
				D1		0 - 119	144.0
Ju						>1000	

Fig. 3. Stratigraphic column for the Songliao basin

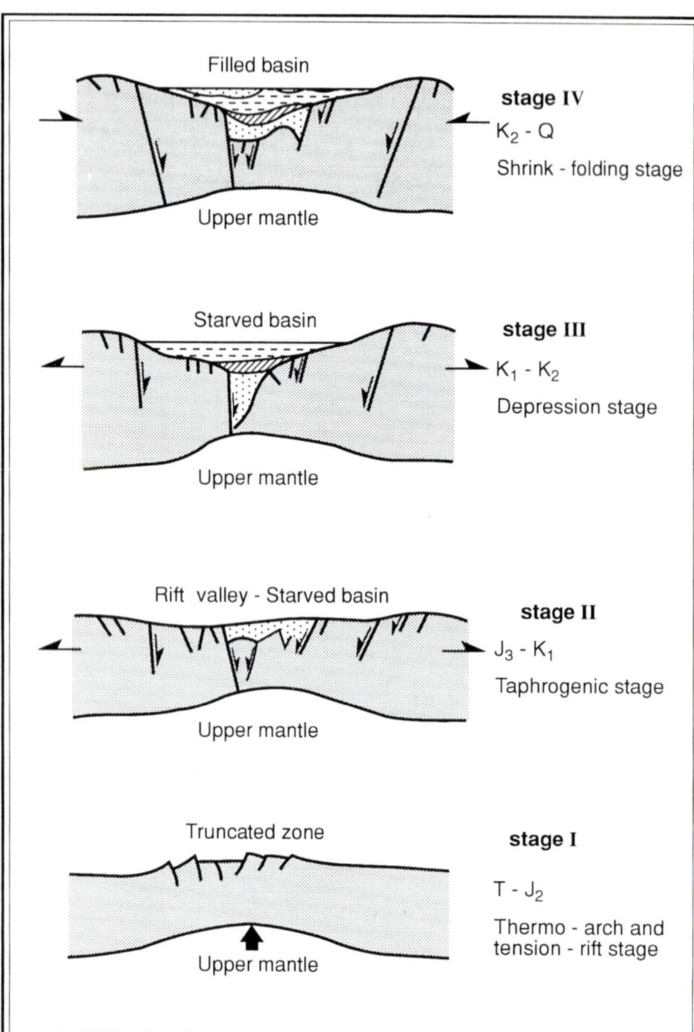

Fig. 4. Tectonic evolution of the Songliao basin

lake deposition was occurring. It is this deep lake facies in both the Qingshankou and Nenjiang Formations that contains the main source rocks of the basin.

With the onset of spreading in the Japan Sea, near the end of the Cretaceous, the tectonic framework of the basin was altered and became dominated by westward compressional stresses. This resulted in a series of anticlinal and synclinal regions and formed the current tectonic framework of the Songliao basin (Fig. 5).

Paleolake Geographic Conditions and Lithofacies

The Qingshankou lacustrine system of the Songliao basin was among the largest lacustrine systems on the Asian continent, covering 87 000 km^2 during deposition of Member 1 (Qn1) and 68 000 km^2 during deposition of Member 2 (Qn2).

The strata contain abundant fossils. Such diverse groups as dinosaurs, insects, lizards, crocodiles, turtles, and plants are found within the paleolake (Fig. 6). The fossils found include 26 genera of conchostraca, 27 genera of chara, 15 genera of gastropods, 19 genera of ostracodes, and several genera of lamellibranchs, fish, and algae. Many of these genera are endemic to the Songliao basin. for example, 98% of the conchostraca, 67% of the chara, 93% of the gastropods, 90% of the ostracodes, and 100% of the fish are limited to the basin. Of the more than 800 species of Cretaceous fossils identified in the basin, over 96% were limited to freshwater with the remaining fossils having been able to survive in either fresh or brackish water.

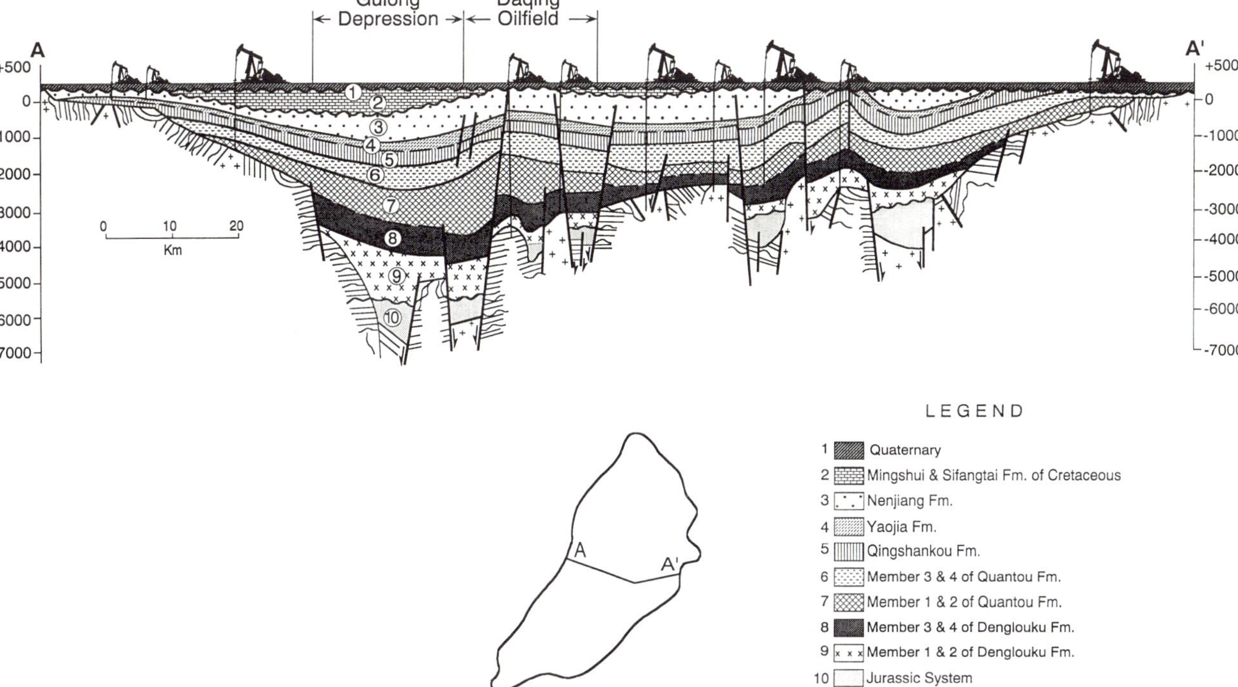

Fig. 5. Structural cross section of the Songliao basin

There are no marine fossils such as foraminifera, corals, ammonites, or belemnoidea present. Thus, the fossil evidence suggests that the Qingshankou lake was freshwater.

A nonmarine freshwater lake system is also supported by the lithogeochemical character of the sediment (Ye Dequan et al. 1980) and the geochemical characteristics of the produced crudes: high wax content, high nickel/vanadium ratio, and low sulfur content.

The size of the lake and the nature of its flora and faunal population during the Cretaceous indicates a subtropical, humid climate for the Songliao basin. This led during the Cretaceous to lush plant growth along the lake shores. These plants were capable of supplying large quantities of continental material into the lake basin proper.

The Qingshankou Fm. may exceed 600 m in thickness, and has been divided into two members. Qn1, the older member, obtains a maximum stratigraphic thickness of about 120 m (Fig. 7). Qn2, the younger member, obtains a maximum stratigraphic thickness of about 550 m (Fig. 7). The black mudstones, which represent the source intervals within this unit, obtain maximum thicknesses of $\sim$ 100 m in Qn1 and $\sim$ 400 m in Qn2 (Fig. 8). In the axial portion of the basin the Qingshankou Fm. obtains its maximum depth of burial of about 1800 m (Fig. 9).

During Qingshankou time, a poly-directional drainage system existed within the Songliao basin, with flow towards the basin center. Five to seven major drainage systems existed along the basin margins (Fig. 10). The coalescing of these drainage patterns resulted in various depositional facies, including alluvial fans, fluvial flood plains, deltas, and lacustrine settings. These depositional facies form a ring pattern around the lake basin proper (Fig. 10).

The alluvial facies are composed primarily of poorly sorted sand and gravel. They do not display any internal bedding structure. The flood plain facies are characterized by gray sandstone and massive red and green mudstones. These sediments display a generally finding-upward character. The delta plain with its fan shape was cut by a series of distributory streams. The sediments are characterized by gray, green, and lesser amounts of red mudstones, and interbedded silts and sandstones. Within both the flood plain and the delta plain facies, channels, splays, and natural levee deposits can all be recognized. Delta front facies consist of river mouth bar sands, delta front sheet sands, and channel and deposits with interbedded muds. The

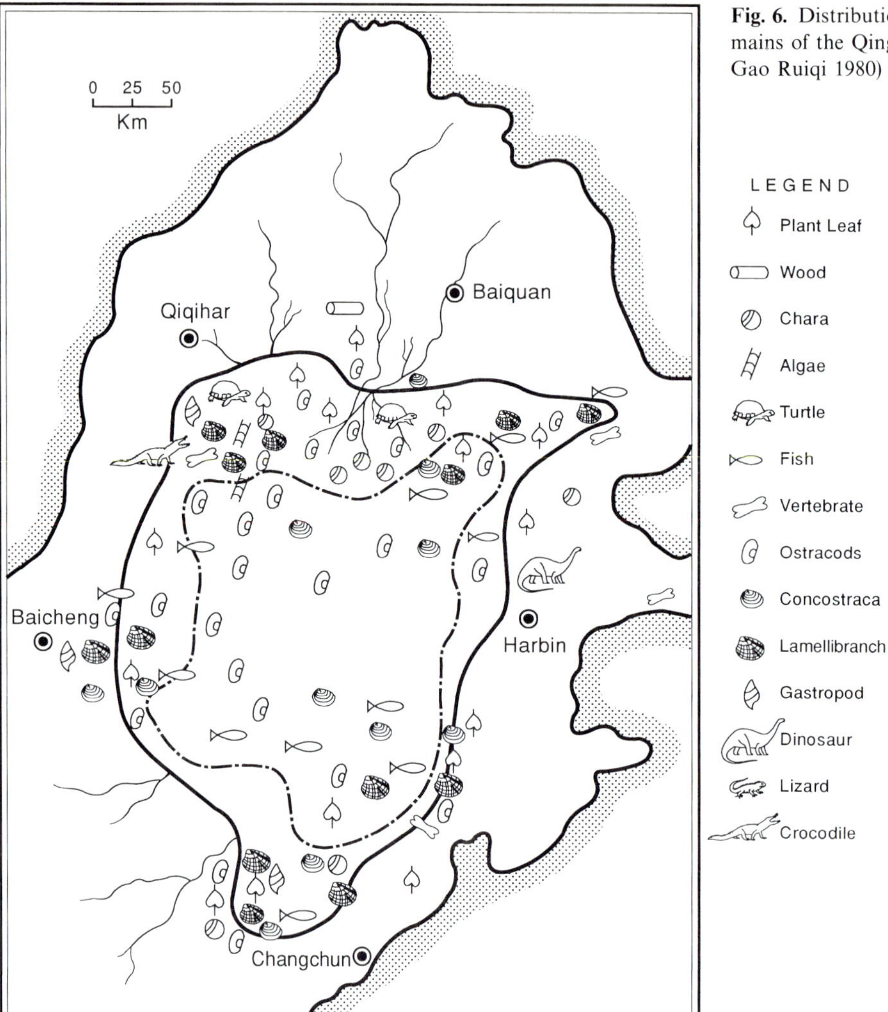

Fig. 6. Distribution of plant and animal remains of the Qingshankou Formation. (After Gao Ruiqi 1980)

shallow lake facies consist of thin beds of mudstone, bioclastic limestone, oolitic limestone, and siltstones. Deeper-water lacustrine deposits consists mainly of black shales, oil shales, and argillaceous limestones. These deeper-water rocks were deposited below wave-base. Also observed is a near-shore silting facies which consists mainly of red and green mudstones interbedded with small amounts of siltstones.

It is as a result of these multiple drainage systems and changes in lake level that sand bodies within the Qingshankou Fm. are largely encapsulated by source rocks. This proximal relationship has favorably influenced the system's hydrocarbon migration and accumulation history.

Source Rock Characteristics

Organic Enrichment

The distribution of organic matter within the Qingshankou Fm. is very strongly influenced by depositional setting. As observed with the facies patterns, the organic carbon isopleths form a roughly concentric pattern. The organic carbon content increases toward the basin center in the deep lake (Fig. 11). Elevated levels of organic carbon were also observed within swamp facies, and to a lesser degree in a small lake bay and within the delta complex.

An examination of the Qingshankou Fm. reveals that this pattern is best developed in Qn1,

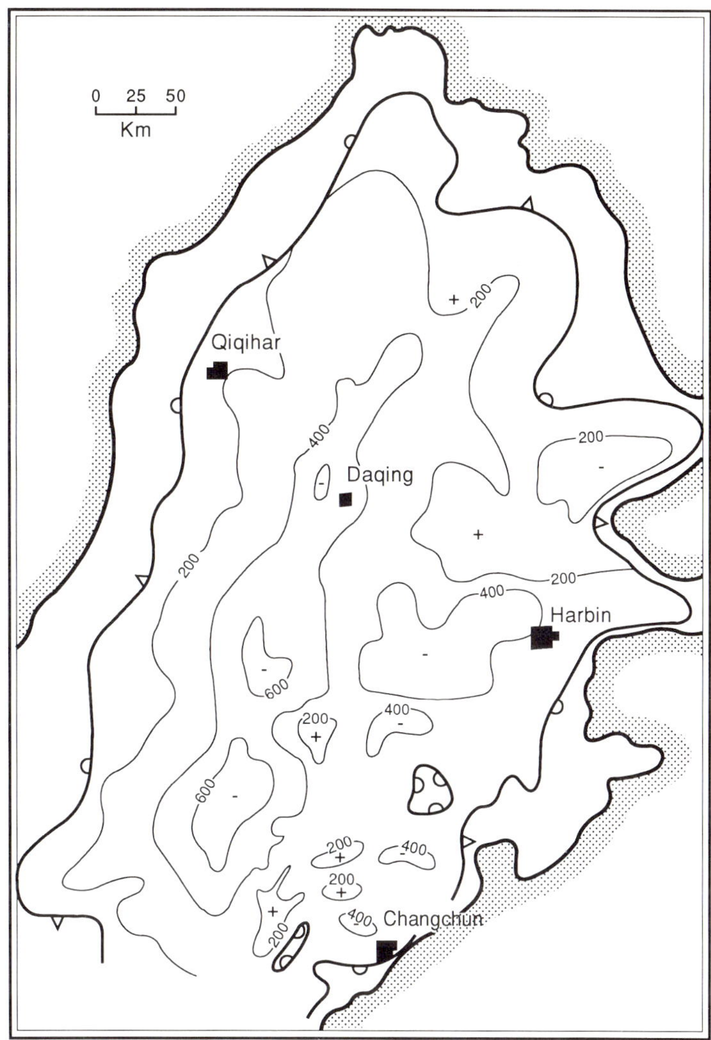

Fig. 7. Isopach map (in m) of the Qingshankou Formation

where organic carbon contents of greater than 4.0 wt.% are observed in paleolake deeps. Maximum individual organic carbon values as high as 8.4 wt.% have been reported within the deep lake facies (Yang Wanli et al. 1983).

Qn2 displays lower levels of organic enrichment (Fig. 11). In Qn2 the higher levels of organic enrichment appear to be associated with the delta complexes. This appears to be the result of the poorer development of the deep lake basin during this later phase of basin evolution (Fig. 10). In the deltas, organic matter was mainly accumulating in the bottom- and top-set beds, where values as high as 2.0 wt.% are observed. On average, within the delta complex, the level of organic enrichment is about 0.7 wt.%.

Another measure of hydrocarbon source rock potential is the total pyrolytic hydrocarbon yield ($S_1 + S_2$). Yang Wanli et al. (1983) have also noted a strong dependence between average hydrocarbon yield and depositional environment. The deep lake facies display an average generation potential of 22.65 mg HC/g rock. The sediments of the shallow lake facies have average hydrocarbon yields of 2.17 mg HC/g rock. The noncoaly portions of the lake margins display an average generation potential of less than 1 (0.56) mg HC/g rock. As will be noted below, this increase in hydrocarbon yield is greater than would be expected if due solely to an increase in organic content without a change in kerogen quality.

Total pyrolytic hydrocarbon yields greater than 2.5 mg HC/g rock are typically considered necessary for a rock to be considered representative of a commercial hydrocarbon source (Bissada 1982).

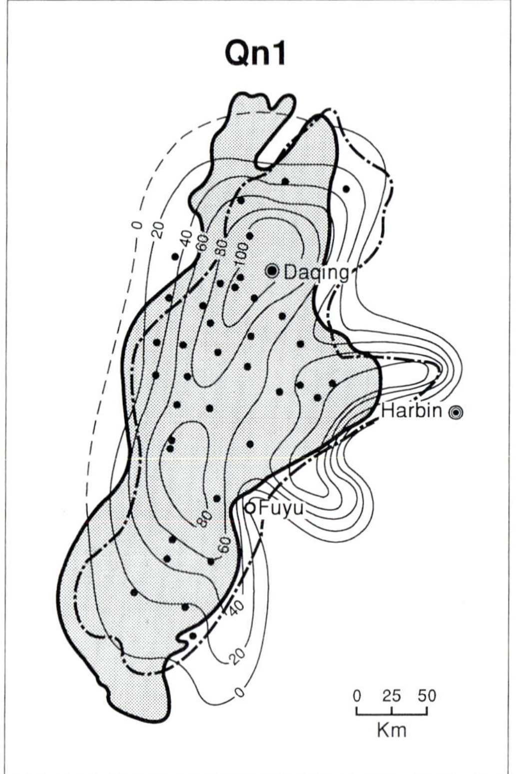

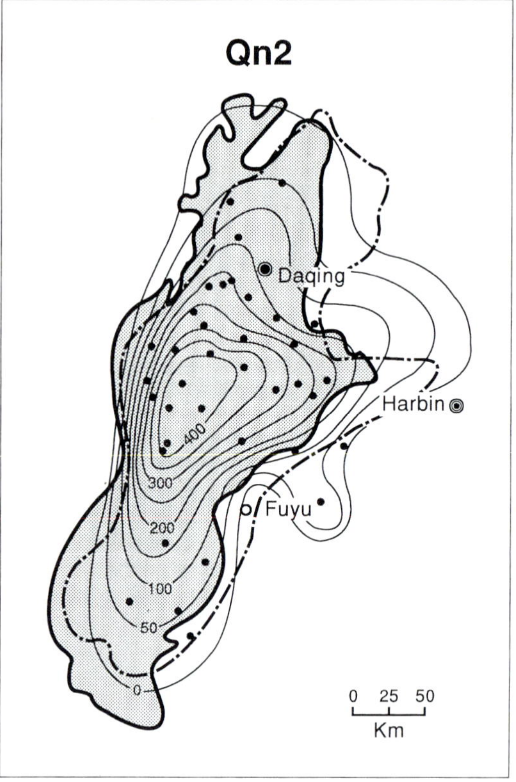

Fig. 8. Thickness (in m) of the Qn1 and Qn2 members of the Qingshankou Formation. *Shaded area* represents those areas which have entered into the oil window. *Solid circles* represent exploration wells used in the construction of the isopachs

Kerogen Character

Kerogen within the Qingshankou Fm. has been characterized through several means. The relationship between the hydrogen and oxygen indices (Fig. 12) reveals a broad spectrum of kerogen types, ranging from oil-prone, type I organic matter to gas-prone, type III organic matter. The type I material, with hydrogen index values >650 mg HC/g TOC, are restricted to the deep lake facies. The type III material, with hydrogen index values < 200 mg HC/g TOC, is largely restricted to the shallow lake and lake bog facies. Intermediate values appear to be associated with more marginal lacustrine facies.

The elemental analysis of isolated kerogen reveals a pattern similar to that of the pyrolysis data (Fig. 13). Type I kerogen, with atomic H/C ratios >1.4 are restricted to the deep lake facies. This kerogen type also appears largely restricted to Qn1, which is consistent with the better definition of the deep lake facies during this period (Fig. 10). Type II, mixed kerogens dominate in the "deeper" lacustrine facies of Qn2. Type III kerogen, with atomic H/C ratios < 0.8, are largely limited to the lake-shore and lake bog facies.

Visual kerogen analysis reveals that the deep-lake facies are dominated by liptinitic material (>90%). Much of this material appears to have been derived from freshwater algae and, to a lesser extent, bacterially reworked terrestrial plant material. The lake-shore and lake bog facies, in contrast, are dominated by vitrinite (>90%). The mixed kerogens contain varying proportions of the liptinic and vitrinitic end-members. Within these mixed kerogens bacterially reworked terrestrial material rather than freshwater algal material appears to dominate in the liptinite fraction.

Bitumen Content and Character

The total chloroform "A" bitumen yields (Fig. 14) of the sediments display a pattern similar to those previously noted for depositional facies, organic richness, and kerogen composition. As would be expected, higher average extract yields were noted

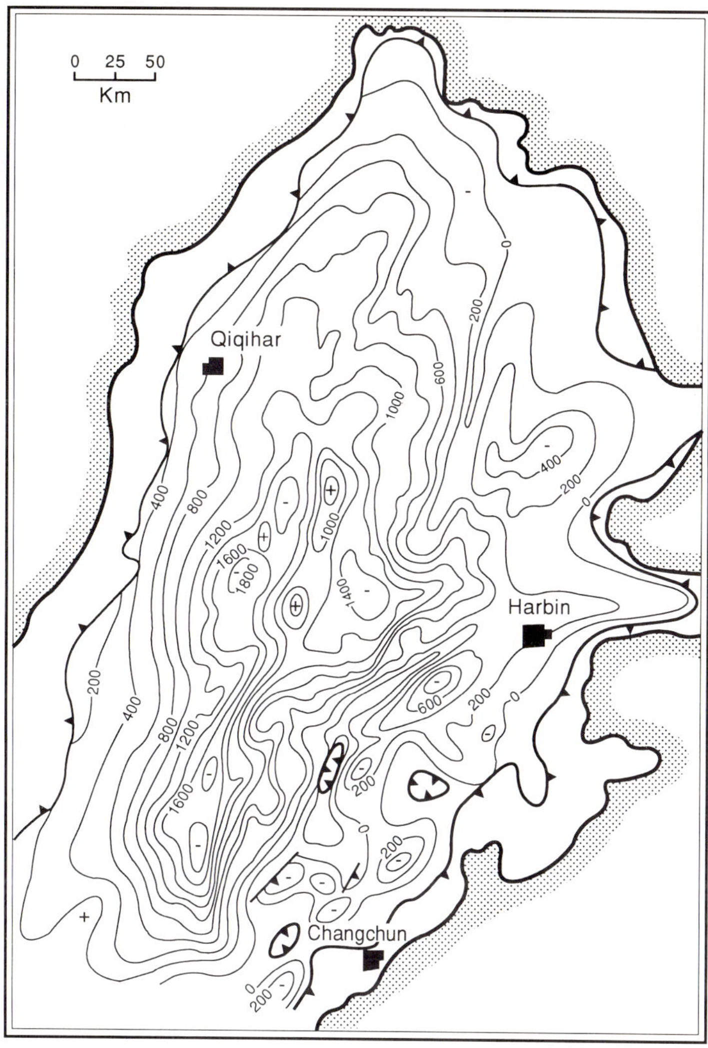

Fig. 9. Depth of burial (in m) to the top of the Qingshankou Formation

in the Qn1 member than in the Qn2 member, 5300 ppm and 510 ppm, respectively. Maximum bitumen yields in excess of 12 000 ppm are present within the deep lake facies of Qn1. In contrast, in Qn2 maximum yields in excess of 4000 ppm were observed.

Total hydrocarbon content (saturated + aromatic hydrocarbons) of the bitumens displays the influence of thermal maturity (Fig. 15). This is best observed in the Qn1 member. Within the Qn1 member, the highest yields were associated with a depocenter located on the eastern margin of the deep basin located near Harbin. In contrast, the highest hydrocarbon contents within the Qn2 bitumens were observed in a depocenter on the northwest margin of the lake. The lower hydrocarbon concentrations in Qn2 reflect both the poorer quality of the organic matter and the lower levels of thermal maturity.

The HC/TOC ratios expressed as mg/g display a general relationship with depositional facies. Within the deep lake setting, values are typically greater than 200. Along the lake margin and in the bog facies an average ratio of ~50 is observed. The intermediate lake facies display an average ratio of ~90.

In addition to absolute yield, there are clear differences in the character of the bitumens associated with each of the different facies. The saturate/aromatic hydrocarbon ratios in the deep lake facies are typically 2 to 3, but may range upward to greater than 8. The dominant *n*-alkanes have carbon numbers of 19, 21, and 23. The dominance by compounds in this molecular weight range is

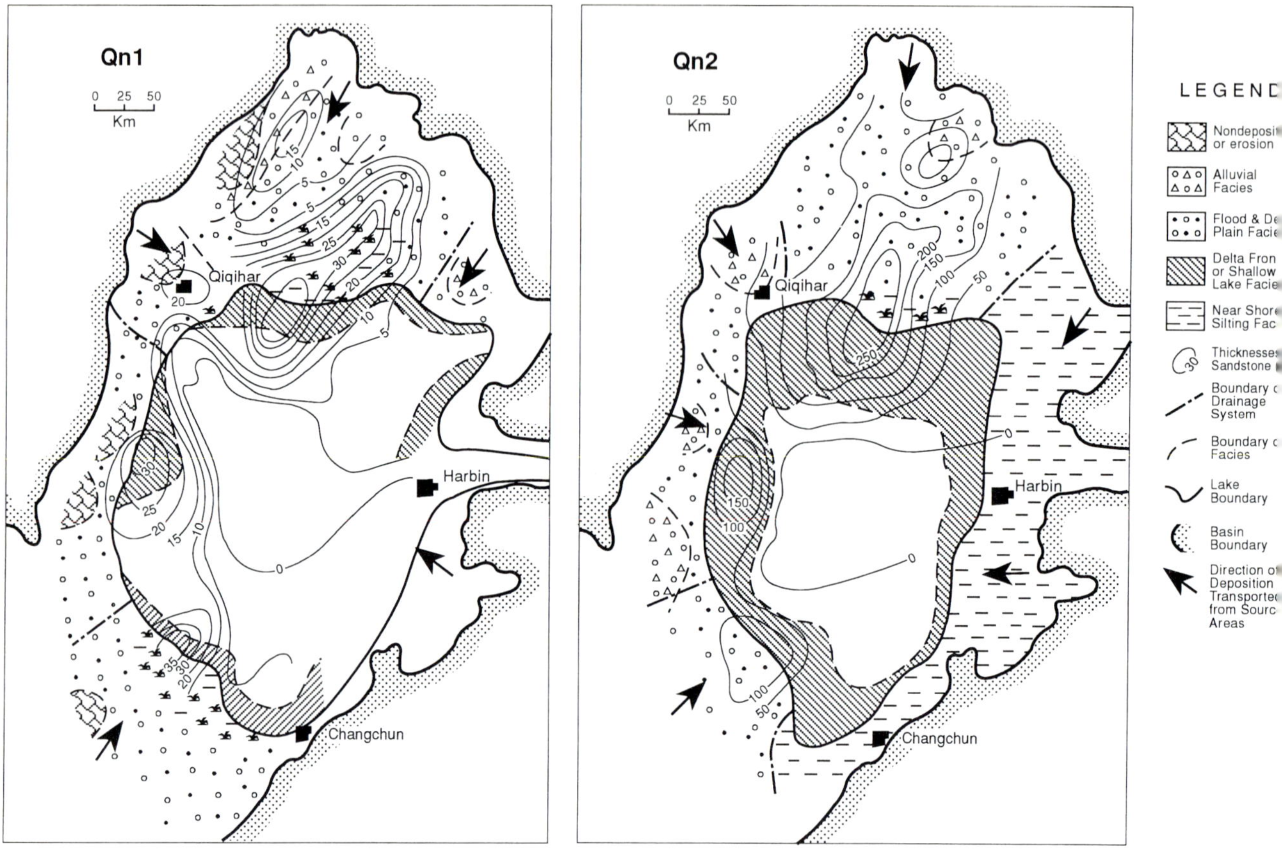

Fig. 10. Facies distribution of the Qn1 and Qn2 members of the Qingshankou Formation. (After Wang Hengjian and Chao Wenfu 1981)

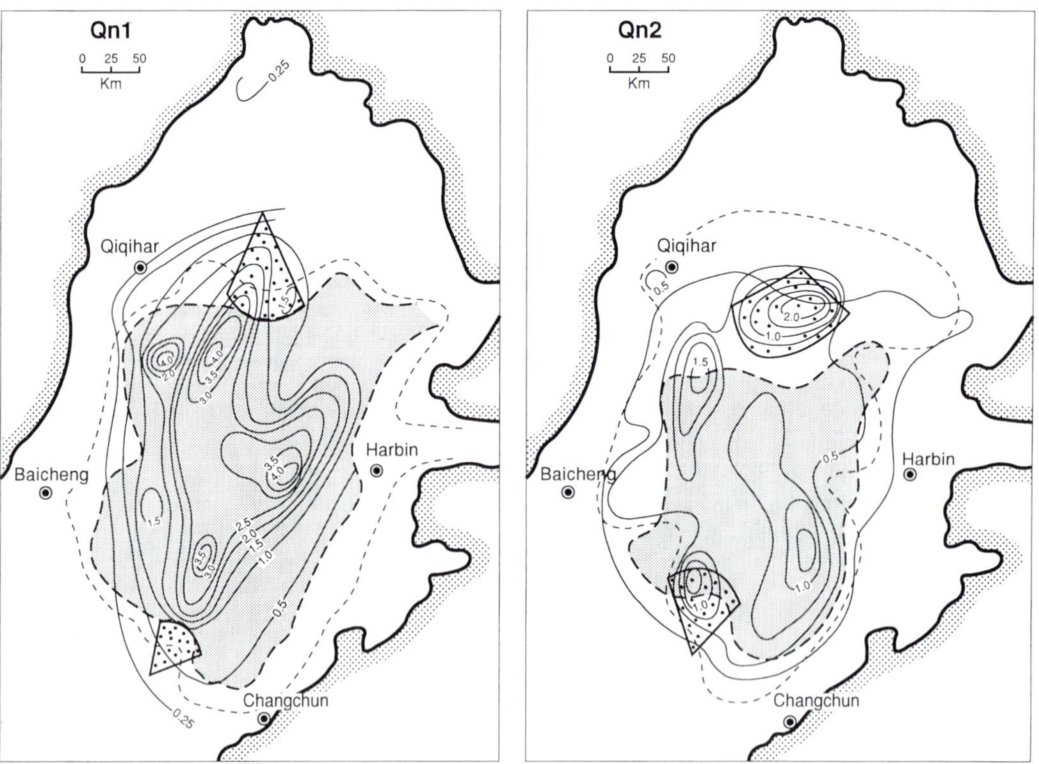

Fig. 11. Distribution of organic carbon in the Qn1 and Qn2 members of the Qingshankou Formation. *Shaded areas* represent deeper lake facies

consistent with an algal and/or bacterial precursor (Limjback 1975). The $(nC_{21} + nC_{22})/(nC_{28} + nC_{29})$ ratios, an indicator of waxiness in samples of comparable maturity (i.e., lower values indicate a greater degree of waxiness), range between 2 and ~ 10. Pristane/phytane (Pr/Ph) ratios typically range between 1.2 and 1.8. The isoprenoid/n-alkane ratios, Pr/nC_{17} and Ph/nC_{18}, which range from 0.3 to 0.5 and 0.24 and 0.59, respectively (Yang Wanli et al. 1983) are also consistent with a largely algal source (Fig. 16). These bitumens also display an average whole-bitumen carbon isotopic composition of ± − 26‰ PDB.

In contrast, the lake margin and bog facies typically display a dominance of the longer-chain, higher molecular weight n-alkanes. There is a dominance by nC_{25}, nC_{27}, and nC_{29}. The $(nC_{21} + nC_{22})/(nC_{28} + nC_{29})$ ratios are less than 2, typically between 1.1 and 1.4. Such a ratio is consistent with a high plant cuticle precursor where waxes tend to dominate. The sediments within this facies also display saturate to aromatic hydrocarbon ratios less than 2. These lower ratios are result of a higher proportion of vitrinite within the precursor kerogen. The Pr/nC_{17} and Ph/nC_{18} ratios range from 2.1 to 2.4 and 2.3 to 5.7, respectively (Fig. 16). The whole-bitumen carbon isotopic compositions are lighter than those of the deep lake facies, with $\delta^{13}C$ values < − 27‰ PDB.

The intermediate water depth facies display intermediate bitumen compositions. For example, the n-alkane distribution displays a clear bimodal pattern, with nC_{20} and nC_{27} being the dominant peaks.

Thermal Maturity and Hydrocarbon Generation

No large-scale uplift and erosion has occurred in the central depression of the Songliao basin since Qingshankou time. Therefore, the sediments display a regular increase in thermal maturity with depth (Fig. 17). This increase in thermal maturity occurs at a relatively rapid rate as a consequence of the rather elevated average geothermal gradient, ~ 37 °C/km. Interval gradients within the Qingshankou Fm. as high as 78 °C/km have been noted in the P1 well as a consequence of the low thermal conductivity of the organic-rich shale.

Within the basin, four stages of thermal maturity have been defined by Yang Wanli et al. (1983).

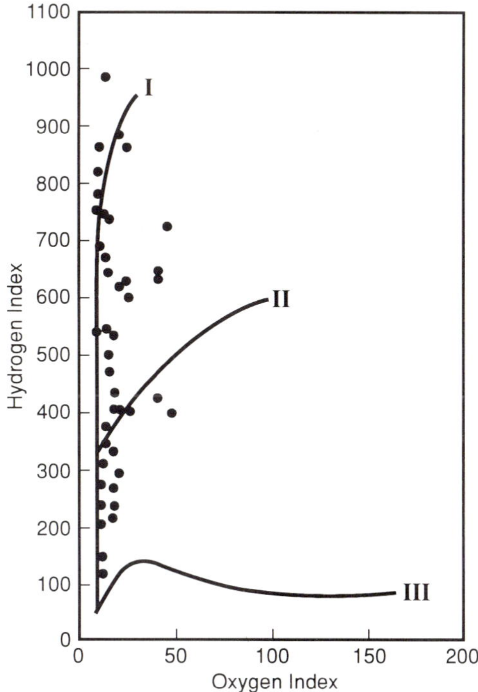

Fig. 12. Modified van Krevelen diagram based on whole-rock pyrolysis results. (After Yang Wanli et al. 1983)

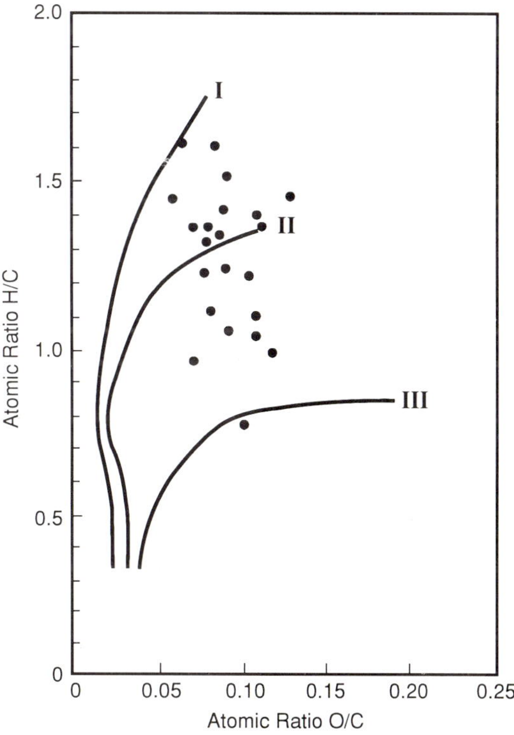

Fig. 13. Van Krevelen diagram based on the elemental analysis of isolated kerogens. (After Yang Wanli et al. 1983)

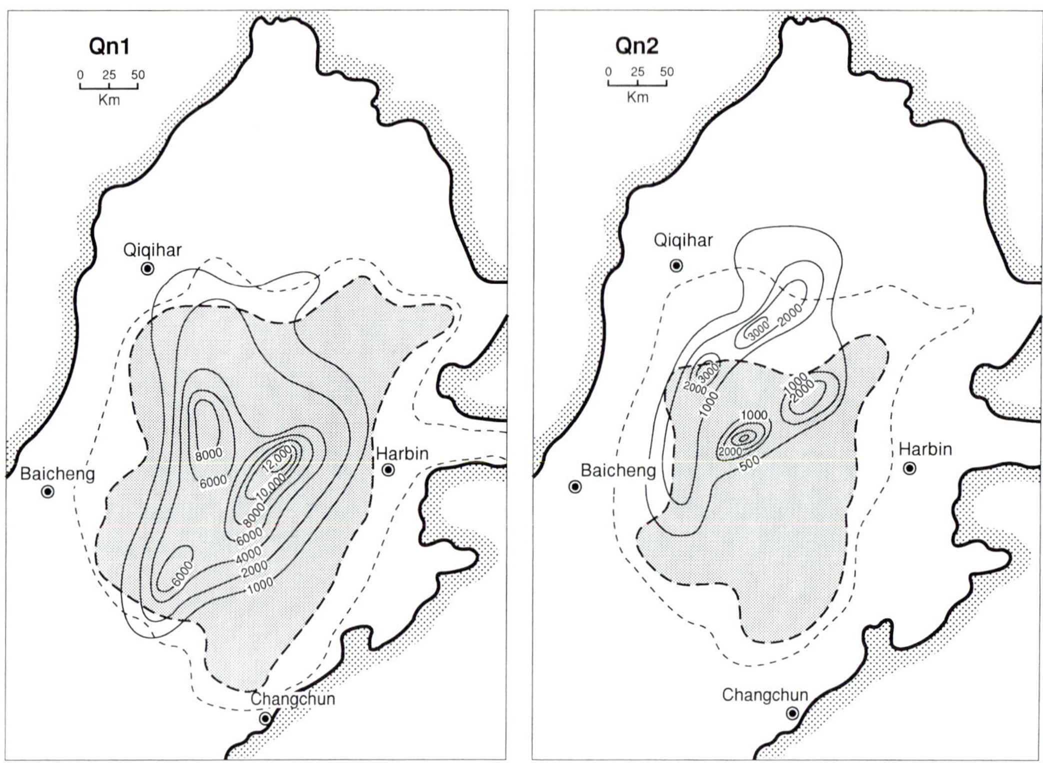

Fig. 14. Concentration of chloroform extract A (in ppm) in the Qn1 and Qn2 members of the Qingshankou Formation. *Shaded areas* represent deeper lakes facies

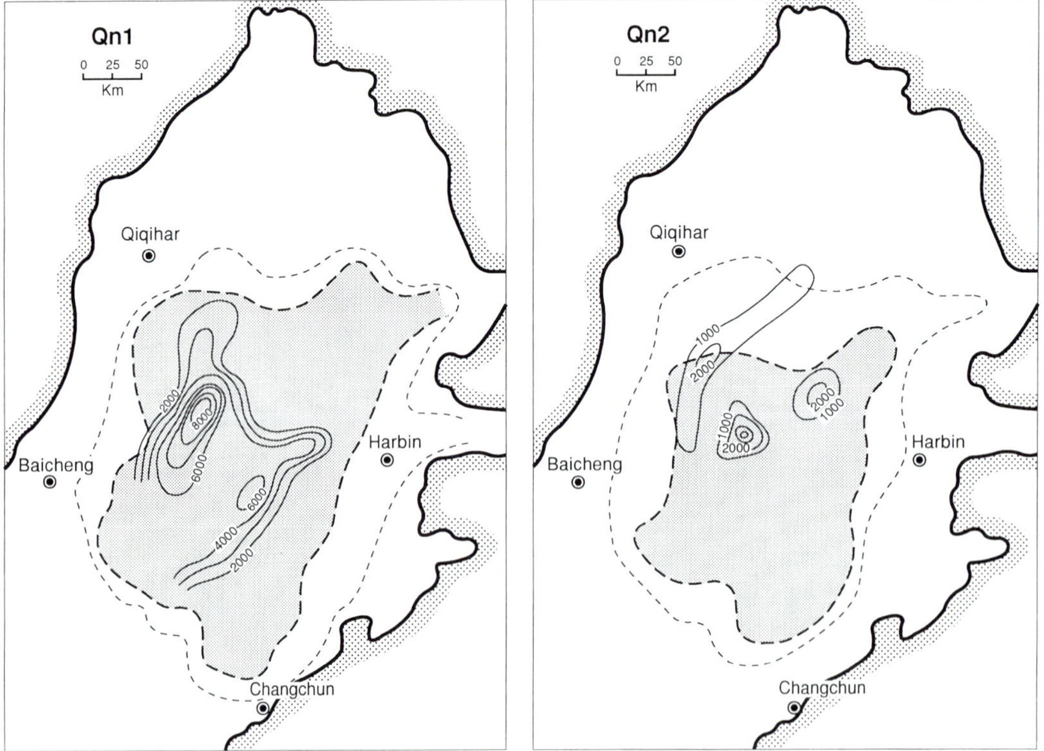

Fig. 15. Concentration of extracted hydrocarbons (in ppm) in the Qn1 and Qn2 members of the Qingshankou Formation. *Shaded areas* represent deeper lake facies

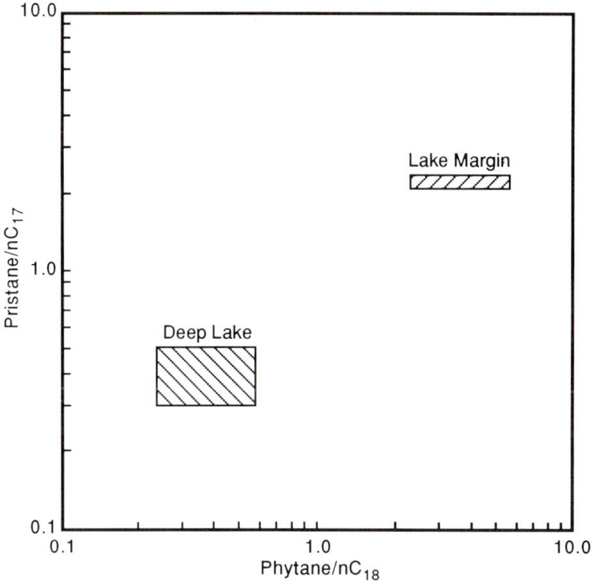

Fig. 16. Differences in the relationship between the pristane/nC_{17} and phytane/nC_{18} ratios in lake margin and deep lake facies

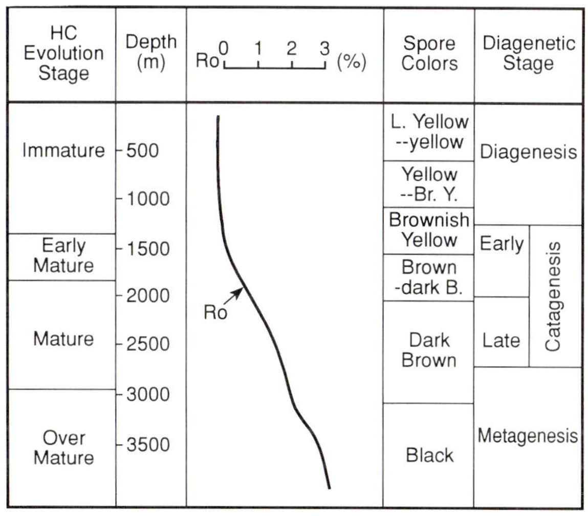

Fig. 17. Thermal maturity profile for the Songliao basin

The first stage generally extends to depths of ~1300 m where temperatures of ~70 °C. are achieved. Spore colors typically are light yellow to brownish yellow, and vitrinite reflectance values are less than 0.5%. Hydrocarbon generation is limited to biogenic gas generation.

An early mature stage occurs at depths between 1300 and 1900 m, with temperatures between 70 and 90 °C. Spores and pollen grains are brownish yellow to brown, and vitrinite reflectance values are between 0.5 and 1.0%. Mainly heavy oil is generated during this burial stage.

A mature stage exists between 1900 and 2900 m, with subsurface temperatures between 90 and 120 °C. The vitrinite reflectance values range between 1.0 and 2.0%. Spore colors typically are brown to brownish black. This stage is associated with the generation of light oils and condensates.

The overmature stage exists at depths greater than 2900 m, with formation temperatures in excess of 120 °C. Spores are generally black and vitrinite reflectance values are greater than 2.0%. This stage is associated with dry gas.

This interpretation of visual thermal maturity indices is also consistent with that published by Zhao Chuanben (1984), who shows only minor deviation from these reported maturity threshold depths.

This maturity evolution pathway describes the intermediate kerogen, i.e., the type II kerogen which dominates the Qn2 member and the intermediate water depth facies of the Qn1 member. The two end-member kerogens display slightly different maturation thresholds as a consequence of differences in kinetics. For example, the onset of thermal generation (i.e., early mature) of oil in the deep lake, type I facies of the Qn1 member occurs at a depth of ~1130 m. In contrast, the type III material, of the lake bog facies, typical of the Qn2 member, reaches this maturation threshold at a depth of ~1870 m. Consequently, at equivalent depths of burial, thermogenic hydrocarbon generation typically displays a decrease toward the lake margins.

An analysis of the burial history of the Gulong depression within the Songliao basin reveals that the Qingshankou Fm. entered into the main phase of hydrocarbon generation near the end of Nenjiang (late Albian) time and is still currently within the mature stage of generation (Fig. 18). A summary of the basin's generation history is presented below and in Table 1.

At the end of the Nenjiang time the Qingshankou Fm. entered into the early mature phase of generation. The mature area of the basin covered approximately 14 570 km^2. The most favorable areas for generation were the deep depressions, Gulong and Shanzhao, which surround the central Daqing structural belt (Fig. 19). Hydrocarbon generation during this period experienced an average rate of 2740 MMbbl/Ma.

At the end of the Cretaceous, the Qingshankou source rocks were mature and peak hydrocarbon

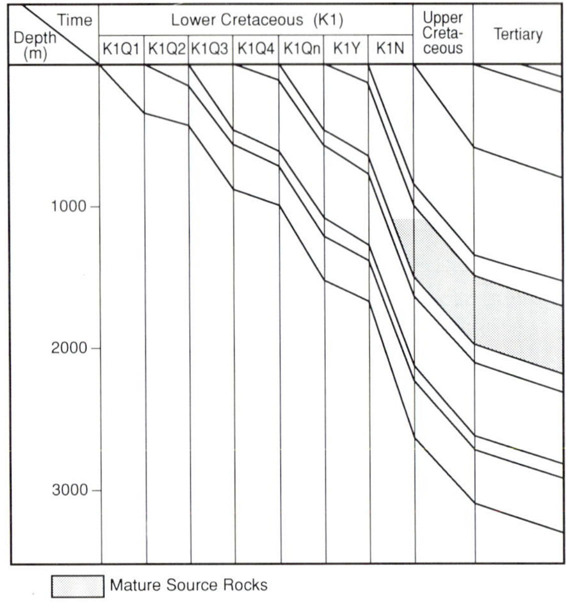

Fig. 18. Burial history of the lower Cretaceous in the Gulong Depression of the Songliao basin

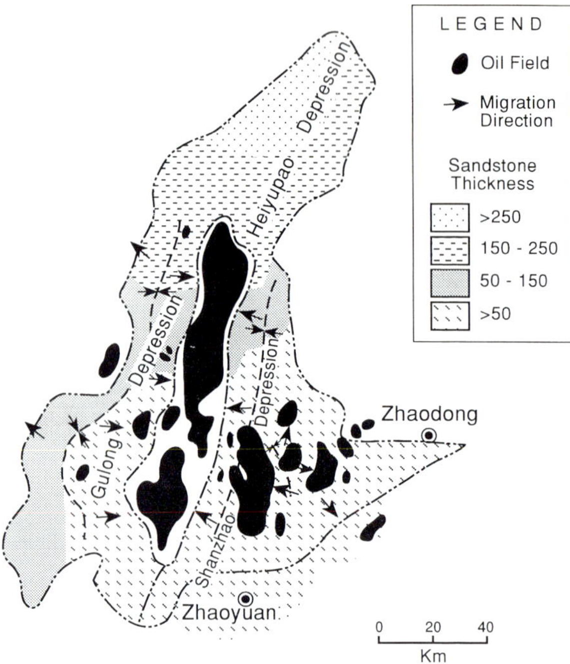

Fig. 19. General hydrocarbon migration directions in the vicinity of the Daqing oil field. Net sand thicknesses (in m) are also presented

generation was occurring. The area of mature source rocks extended to approximately 27 000 km^2. Generation achieved rates up to 16 280 MMbbl/Ma.

By the late Tertiary, the Qingshankou source was largely beyond peak generation. Hydrocarbon generation rates were reduced to about 2812 MMbbl/Ma. Much of the Qingshankou lake basin was mature to post-mature, covering an area of $\sim$33 000 km^2.

The current limits of the mature Qingshankou Fm. are presented in Fig. 8.

Oil Composition and Character

Produced oils are highly paraffinic. Saturated hydrocarbons account for greater than 50% of the nonvolatile fraction (Fig. 20). Sulfur contents are low, less than 0.2% as are both nickel ($\sim$2.3 ppm) and vanadium (0.08 ppm) (Yang Wanli et al. 1983). Nickel/vanadium ratios are high ($>$28). API gravities range from $\sim$29° to $\sim$44°. Pour points range from 25 to 35 °C.

Table 1. Generated hydrocarbons in the Qingshankou Formation

Time	Hydrocarbon generation			
	Qn1		Qn2	
	Amount (%)	Area (km^2)	Amount (%)	Area (km^2)
End of Nenjian	23.9	14 570	—	—
End of Cretaceous	63.8	26 780	67.9	22 200
End of Tertiary	84.3	33 140	91.6	27 060
Present	100	39 600	100	32 240

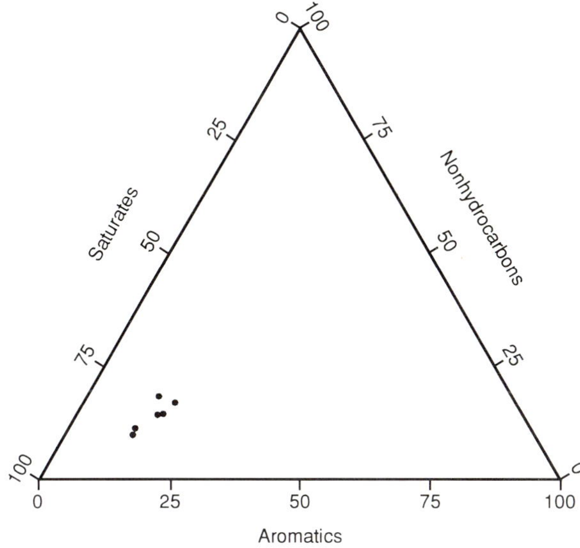

Fig. 20. Ternary diagram depicting gross composition of Qingshankou-derived oils. (Data from Li Taiming et al. 1987)

An analysis of the gas chromatographic analyses reveals an abundance of higher molecular weight *n*-alkanes (Fig. 21). The oils tend to display an odd predominance in these longer-chain *n*-alkanes (Li Taiming et al. 1987). The isoprenoid abundances are relative low compared to their associated *n*-alkanes. The pristane/nC_{17} and phytane/nC_{18} ratios are typically less than 0.25. The Pr/Ph ratios are greater than 1.0, but typically less than 1.3.

The m/z 191 ion fragmentograms are dominated by the 17α(H)-hopane series (Fig. 22). Tricyclic compounds are virtually absent. Gammacerane is also present in significant quantities. The Tm/Ts (18α(H)-22,29,30-trisnorneohopane/17α(H)-22, 29,30-trisnorhopane) ratios are less than 1. Such values are believed typical of lacustrine settings (Robinson 1987).

The m/z 217 ion fragmentograms (Fig. 23) suggest a dominance by C_{29} normal steranes (Fig. 24). The oils also appear to contain relatively low concentrations of diasteranes.

Whole-oil carbon isotopic compositions range from −29.7% to −27‰ PDB (Yang Wanli et al. 1983).

Oil Occurrence

The most significant occurrence of Qingshankou-derived oils is in the Daqing oilfield. It is located in

Fig. 21. Representative gas chromatogram of a Qingshankou-derived oil. (After Li Taiming et al. 1987)

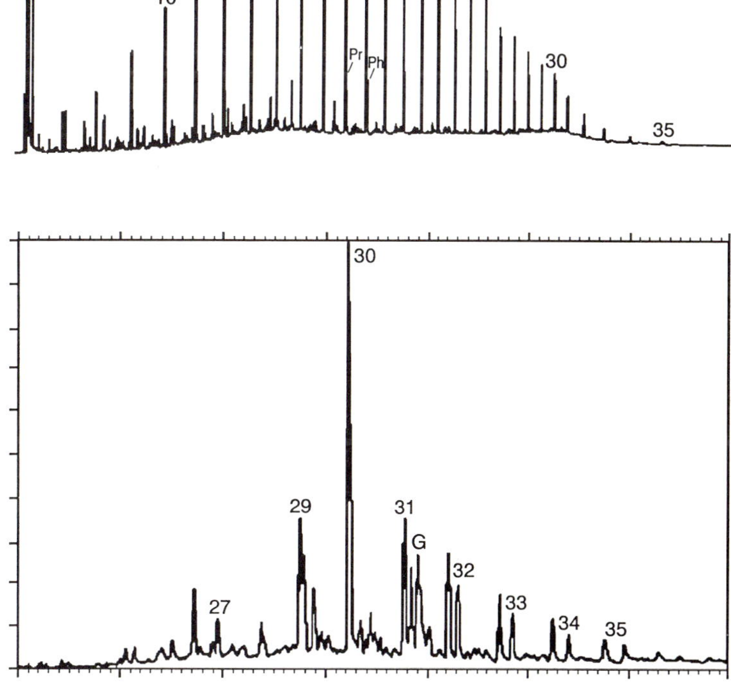

Fig. 22. Representative m/z 191 chromatogram from a Qingshankou-derived oil. (After Li Taiming et al. 1987)

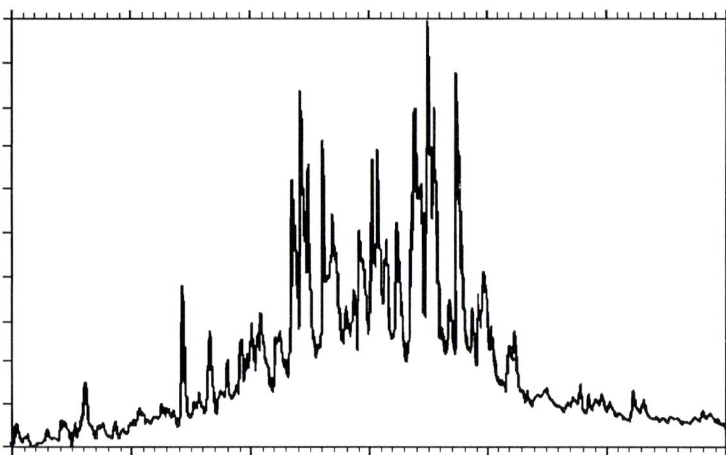

Fig. 23. Representative m/z 217 chromatogram from Qingshankou-derived oil. (After Li Taiming et al. 1987)

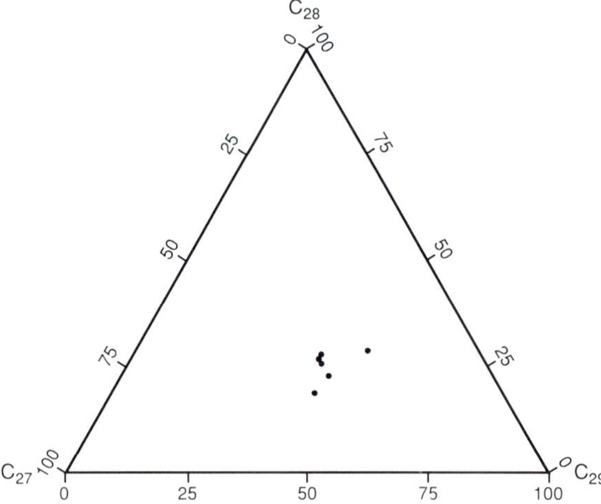

Fig. 24. Ternary diagram of normal sterane composition of Qingshankou-derived oils. (Data from Li Taiming et al. 1987)

the center of the Songliao basin in an anticlinal belt, with a north-south length of 140 km and an east-west width of 20 to 30 km. This belt consists of seven structural highs. Qingshankou (upper Qn2) and Yaojia Formation deltaic sands form primary reservoirs for these oils. Nenjiang formation mudstones provide the seal. By the end of the early Cretaceous compressional stresses resulted in better definition of the Daqing structure.

The Daqing field was discovered in 1959. The field contains approximately 10 000 wells. Since 1976 the field has produced approximately 1 million barrels/day. Cumulative production through the end of 1990 was 7.2 billion barrels.

In addition to the super giant Daqing field, Qingshankou-derived oils occur in a series of lesser deposits. Many of these reserves are found within the Qingshankou Fm. itself in turbidite sands and in fractured mudstone reservoirs. The most significant turbidate accumulation was encountered by the Jin 6 well. The field is an anticlinal trap, with 23.25 m of oil-bearing sandstone in the upper part of the Qingshankou Fm.

There are more than 20 wells which indicate the presence of oil-bearing fractured mudstones within the Qingshankou Fm. in the Gulong depression. These fractures occur in the middle and lower part of the thick black mudstones in an over-pressured interval. The fractures exist in several modes not all of which are oil-bearing. Oil occurs in horizontal fractures of between 2 and 5 mm width in ostracode-rich beds and in laminated black shales and in a fracture network typically restricted to argillaceous limestone interbeds. The latter fracture system is believed to have resulted from uplift and surface denudation.

Fracture mudstones were encountered between 1829 and 2115 m in the Ying 12 well. A zone between 2033.7 and 2083.5 m tested at 34 bbl/day with sustained rates of production of about 30 bbl/day. Lesser rates were obtained from a fractured zone in the Ha 14 well. Tests in the interval between 1995.5 and 2083.0 meters resulted in flow rates of slightly greater than 5 bbl/day and sustained rates over a 9-month period of less than 2 bbl/day.

Although the flow rates associated with these fractured reservoirs are low, it is believed that between 14 and 31 billion barrels of oil are contained within the fractured mudstone in the Gulong depression alone.

Summary

During the early Cretaceous (Aptian) rapid, regional subsidence in the Songliao basin resulted in the formation of a large lake. Within this lake, deposition of the Qingshankou Formation occurred. Floral and faunal evidence within the Qingshankou Fm. suggests that the lake was largely freshwater in nature. The Qingshankou Fm. itself has been divided into two members, Qn1 and Qn2. A series of fluvial drainage systems flowed into the lake. These drainage systems resulted in a series of coalescing deltas surrounding a deeper water lacustrine facies. The deep lake was best developed during Qn1 time. The presence of these coalescing deltas prograding into a lake with variable changes in water level resulted in a very favorable geographic relationship between potential source and reservoir facies. The deep lake facies contained the highest concentrations of organic matter and the most oil-prone kerogen. This material appears to have been derived largely from algal material with lesser amounts of bacterially reworked terrestrial material. The lake margin and lake bog facies, including the deltaic facies, contain generally less organic matter. The material present in these marginal facies is largely gas-prone and derived from higher plant material. Intermediate levels of organic enrichment and kerogen type are present at intermediate water depths. Consequently, the Qn2 member appears less oil-prone than the Qn1 member. The efficiency of generation within the Songliao basin was favored by the elevated regional geothermal gradients. A burial history analysis suggests that hydrocarbon generation began as early as late Albian time. Peak hydrocarbon generation occurred during the Late Cretaceous. Hydrocarbons generated by the Qingshankou Fm. are reservoired in several different reservoirs. Much of the oil in the Daqing oilfield is derived from this unit. Recent drilling suggests that large volumes of oil are also present in basinal turbidite sands and in fractured basinal mudstones.

Acknowledgments. The authors thank Ms. M. Walsh for her assistance in obtaining the necessary reference material. Drafting services were provided by Pauletta Kay. Earlier versions of this manuscript were read by Drs. J. Rullkötter, C. Robison, V. Robison. Permission to publish was obtained from China National Petroleum Corporation and Texaco Inc.

References

Bissada KK (1982) Geochemical constraints on petroleum generation and migration – a review. Proc 2nd ASCOPE Conf, Manila, Oct, 1981, pp. 69–87

Chin Chen (1980) Non-marine setting of petroleum in the Songliao basin of northeastern China. J Petrol Geol 2: 233–264

Du Bomin, Xing Shunquan, Zhou Shuxin (1984) Characteristics of reservoir rocks and their diagenetic evolution in the northern part of the Songliao basin. Oil Gas Geol 5: 122–131 (in Chinese)

Gao Ruiqi (1980) Characteristics of the continental Cretaceous in the Songliao basin. Acta Geol Sin 54: 9–22 (in Chinese)

Lee KY (1986) Petroleum geology of the Songliao basin, northeast China. US Geol Surv Open File Rep 86-0502, 19pp

Li Desheng (1980) Geological structure and hydrocarbon occurrence of the Bohai Gulf oil and gas basin, China. In: Mason JF (ed) Petroleum Geology in China. Pennwell Publ, Tulsa, pp 180–192

Li Desheng (1982) The structural characteristics of the eastern china petroliferous basins. Petrol Explor Dev 2: 1–14 (in Chinese)

Li Desheng (1984) Geologic evolution of petroleum basins on continental shelf of China. Am Assoc Petrol Geol Bull 68: 993–1002

Li Desheng (1987) geological characteristics of hydrocarbon generation and distribution in the Songliao basin, China. In: Kumar RK, Dwivedi P, Banerjie V, Gupta V (eds) Petroleum Geochemistry and exploration in the Afro-Asian Region. Balkema, Rotterdam, pp 191–195

Limjback GWM (1975) On the origin of petroleum. Proc 9th World Petrol Congr 2: 357–369

Li Taiming, Rullkötter J, Radke M, Schaefer RG, Welte DH (1987) Crude oil geochemistry of the southern Songliao basin. Erdöl und Kohle-Erdgas-Petrochemie vereinigt mit Brennstoff-Chemie 40: 337–346

Liu Bin and Zhao Chunman (1991) Tectonic features of the Lishu Depression, Songliao basin, Jilin. Jilin Geol 10: 44–49 (in Chinese)

Liu Jiaqi (1989) On the origin and evolution of the continental rift system. Abstr 28th Int Geol Congr 2: 314–315

Qiu Yinan, Xiao Jingsiu and Wang Zhenbiao (1988) Clastics depositional patterns within Chinese lake basins. China Earth Sci 1: 309–321

Robinson KM (1987) An overview of source rocks and oils in Indonesia. Proc 16th Annu Conv Indones Petrol Assoc, Jakarta, Oct, 1987, 1: 97–122

Tian Zaiyi (1990) The formation and distribution of Mesozoic-Cenozoic basins in China. J Petrol Geol 13: 19–34

Wang Hengjian and Chao Wenfu (1981) A model of the Cretaceous sedimentary facies in the Songliao basin. Oil Gas Geol 2: 227–242 (in Chinese)

Yang Wanli (1985) Daqing oil field, People's Republic of China: a giant field with oil of nonmarine origin. Am Assoc Petrol Geol Bull 69: 1101–1111

Yang Wanli, Li Yongkang and Gao Ruiqi (1983) Formation and evolution of nonmarine petroleum in the Songliao basin, China. Sci Res Design Inst Daqing Oil Field, China (unpubl report), 22 pp

Yang Wanli, Li Yongkang and Gao Ruiqi (1985) Formation and evolution of nonmarine petroleum in Songliao basin, China. Am Assoc Petrol Geol Bull 69: 1112–1122

Ye Dequan (1988) Subdivision of the Cretaceous of the Songliao basin based on micropaleobiotic evolution. Acta Micropalaeontol Sin 5: 111–126

Ye Dequan, Zhao Chunben, Zhang Ying (1980) The paleontological characteristics of continental fossils from the Cretaceous rocks of the Songliao basin. Acta Petrol Sin 1980 (Oct): 50–57

Zhang Jinliang, Yu Huilong (1989) Knowledge on genesis of shallow lacustrine sandbodies from hydrocarbon-bearing basins in east China. Oil Gas Geol 10: 40–44 (in Chinese)

Zhang Kai, Zhang Qing, Yao Huijun, Gao Minyuan (1983) Meso-Cenozoic tectonics and the evolution of riftogenic petroliferous basin along China's sea area and vicinity. Oil Gas Geol 4: 353–364 (in Chinese)

Zhao Chuanben (1984) Thermal alteration of spores and pollen and maturity of organic matter of the Cretaceous System, Songliao basin, northeast China. Geochemistry 3: 84–87

Sedimentological and Geochemical Characterization of the Lagoa Feia Formation, Rift Phase of the Campos Basin, Brazil

L.A.F. Trindade[1], J.L. Dias[2], and M.R. Mello[1]

Abstract

The Barremian rift-stage lacustrine sequence of the Lagoa Feia Formation comprises the source rocks f all petroleum so far discovered in the Campos basin, the most prolific oil province in Brazil. The Lagoa Feia Fm. is characterized by a complexity of terrigenous and carbonate sedimentary rocks which are subdivided into four distinct brackish to saline lacustrine sequences, with increasing upward salinity. Calcareous black shales rich in bacterial and algal-derived type I kerogen (up to 9% TOC) were deposited under anoxic conditions in brackish to saline alkaline waters. These source shales reach a net thickness of up to 200 m. Diagnostic biomarker features include low concentrations of steranes, presence of β-carotane, gammacerane, 28,30-bisnorhopane, and 25,28,30-trisnorhopane, high concentrations of hopanes and methylsteranes and high relative abundances of extended tricyclic terpanes up to C_{45}. The main source beds are identified in the Upper Barremian and are mature throughout most of the basin. A close association between source rock depocenters and oil accumulations is observed. Migration was controlled by unconformities and porous layers within the Lagoa Feia Formation and was dominantly vertical, constrained by windows in the salt layer and listric faults and unconformities in the shallower marine sequence.

Introduction

The Campos basin, the most important Brazilian hydrocarbon province, is located offshore Rio de Janeiro state, southeast Brazil, comprising an area of about 100 000 km^2 (Fig. 1). It is part of a typical passive, divergent continental margin. After 20 years of exploration, approximately $5.05 \times 10^9 m^3$ of equivalent oil in-place has been discovered (Figueiredo and Martins 1990). The Lagoa Feia Formation, considered the source rock of all hydrocarbon accumulations so far discovered in the basin, is composed of rocks deposited mainly during the Barremian rift phase. This sedimentary succession corresponds to the lake and gulf sequences (Asmus and Baisch 1983), which overlie basic volcanic rocks (120–130 Ma) and grade to middle Cretaceous strata (Fig. 2).

The stratigraphic and sedimentological attributes of the Lagoa Feia Formation have been studied by Castro et al. (1981), Bertani and Carozzi (1984), and Dias et al. (1988). Organic-rich shales of the Lagoa Feia Formation were identified as the source rocks in the Campos basin by Meister (1984), Pereira et al. (1984), Figueiredo et al. (1985), and more detailed geochemical characterization of the source intervals were performed by Dias et al. (1988), who identified four distinct depositional sequences within the Lagoa Feia Formation. This chapter focuses on the heterogeneity of the organic geochemical features of the Lagoa Feia Formation associated with facies variation.

Regional Setting and Evolution of the Campos Basin

The Campos basin is separated from adjacent basins by basement highs which strike transverse to the continental margin (Fig. 1). It is bounded from the Espírito Santo basin to the north by the Vitória High and to the south from the Santos basin by the Cabo Frio High. The origin of the basin is related to the breakup of Gondwanaland due to tensional stress during Early Cretaceous.

[1]Petrobrás/Cenpes/Divex, Ilha do Fundão, CEP 21949-900, Rio de Janerio, RJ, Brazil
[2]Petrobrás/Depex, Av. República do Chile 65, Rio de Janeiro, RJ, 20035, Brazil

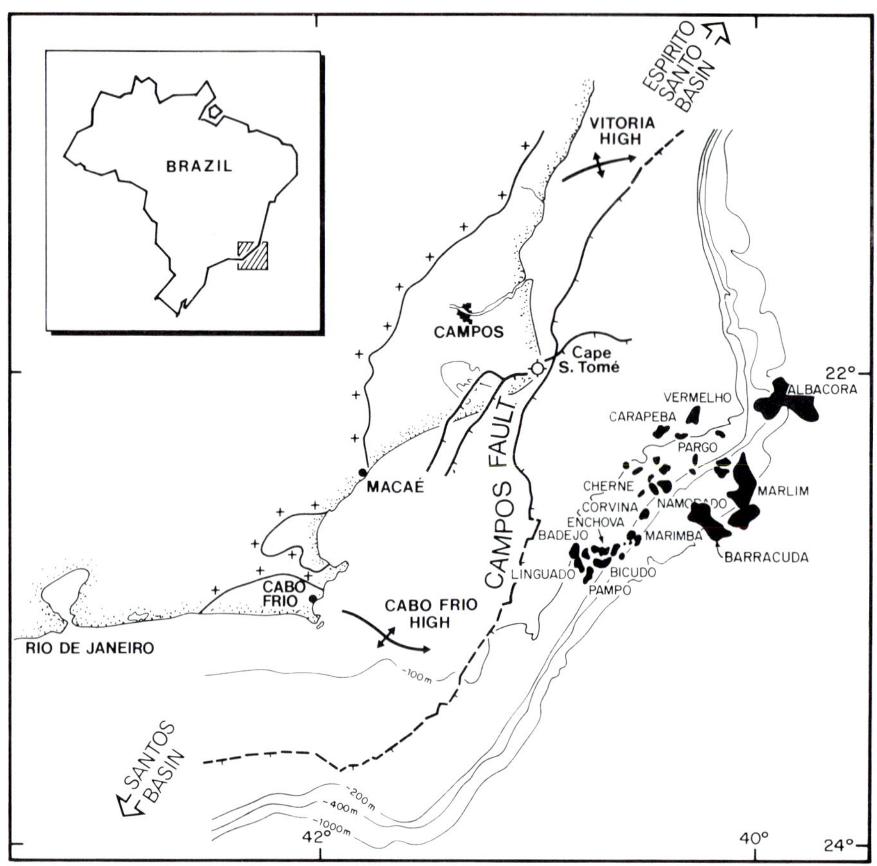

Fig. 1. Location map of the Campos basin, showing distribution of the main oil fields. (After Guardado et al. 1989)

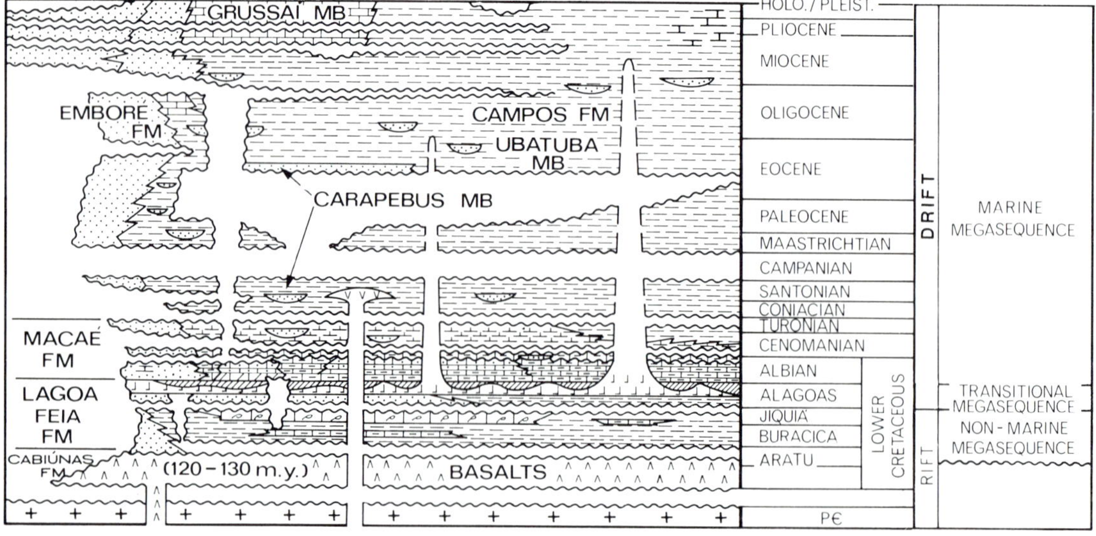

Fig. 2. Schematic stratigraphic chart of the Campos basin. (After Guardado et al. 1989)

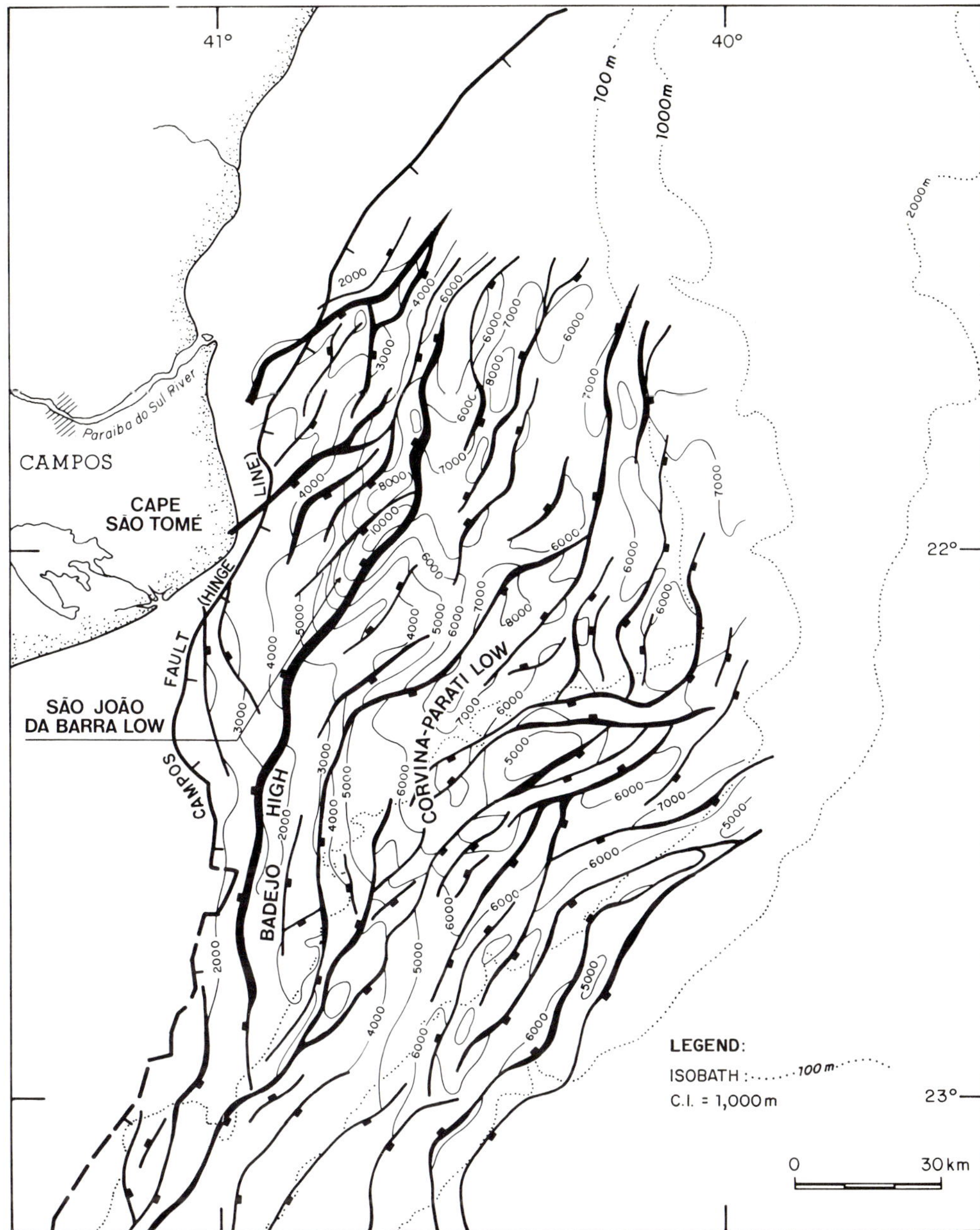

Fig. 3. Structure map at the top of Lower Cretaceous basalts, displaying the rift framework of the Campos basin. (After Guardado et al. 1989)

A northeast-trending rift valley system developed a series of horsts and half-grabens (Fig. 3). The shallow western part of the basin, where Tertiary sediments rest directly on basement rocks is separated by the Campos Fault from the thicker eastern area, which contains Lower Cretaceous sediments (Fig. 3).

The Badejo Regional High, the most prominent horst of the basin, where most of the wells which drilled the Lagoa Feia Formation are located, plunges northward and is flanked to the west by the São João da Barra Low and to the east by the Corvina-Parati Low. Large amounts of lacustrine sediments were deposited within these lows (Fig. 4).

The sedimentary fill of the Campos basin, like the other Brazilian marginal basins, can be divided into three megasequences (Ponte and Asmus 1978), each of them corresponding to a distinct set of

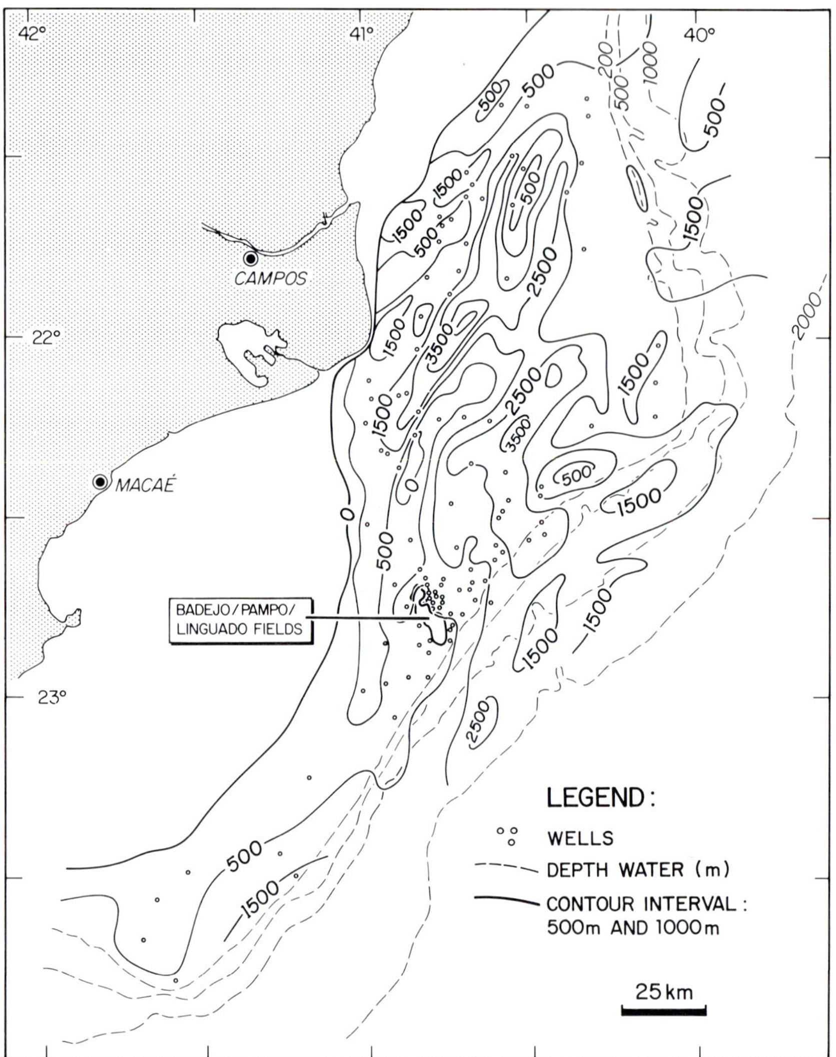

Fig. 4. Isopach map of the Lagoa Feia Formation excluding evaporites. (After Dias et al. 1988)

depositional environments and to a distinct phase of crustal rifting and basin evolution (Fig. 2):

1. Nonmarine megasequence – rift phase: the continental sequence of Barremian age overlies basalts of the Cabiúnas Formation and includes lacustrine terrigenous and carbonate sediments of the Lagoa Feia Formation affected by extensional tectonics during rifting. The Barremian tectonic activity of the rift phase was intermittent and its upper limit coincides with the Pre-Alagoas (Pre-Aptian) unconformity.

2. Transitional megasequence – drift phase: the thick Aptian salt sequence was deposited as part of the huge South Atlantic evaporitic basin (Leyden et al. 1976). These deposits originated from seawater incursions into the basin derived from the ocean to the south, when the climate was arid. The São Paulo and Walvis Ridges acted as major barriers to ocean flooding (Kumar et al. 1977; Gamboa and Rabinowitz 1981).

3. Marine megasequence – late drift phase: progressive northward opening of the South Atlantic resulted in a breakdown of topographic barriers, as reflected by the development of the Albian Macaé carbonate platform. Vertical aggradation of this platform was favored by thermal subsidence of the basin, which was accompanied by a significant eustatic sea level rise.

The end of active carbonate deposition, with deposition of calcilutites and marls, coincided with a worldwide carbonate drowning event during the late Albian/Cenomanian (Schlanger 1981). An overall transgressive event occurred in the Neo-Cretaceous, and turbidites were deposited during short-term sea level falls. During the Tertiary, the decrease in subsidence rate combined with eustatic sea level fall resulted in the progradation of a clastic wedge associated with several canyons and turbidite systems.

Depositional Sequences of the Lagoa feia Formation

The Lagoa Feia Formation, which includes the petroleum source rocks of the Campos basin, is characterized by a variety of lithologies, including siliciclastic sediments, carbonates, and evaporites. The lack of marine fossils associated with other mineralogical evidence (Bertani and Carozzi 1985) suggests a nonmarine depositional environment for this formation. The Lagoa Feia Formation can be subdivided into four depositional sequences: Basal Clastic Sequence, Talc-Stevensitic Sequence, Coquinas Sequence, and the Clastic-Evaporitic Sequence (Fig. 5). Paleoenvironmental reconstruction of these sequences was based on isopach and facies maps, controlled by cores and well cutting descriptions due to the absence of outcrops.

1. Basal Clastic Sequence (Lower Buracica/Aratu local stages): this sequence rests directly on basic volcanic rocks extruded during the Early Cretaceous (Cabiúnas Fm.). The Upper boundary of this basal sequence corresponds to an electrical log marker (LF-20) which is best identified

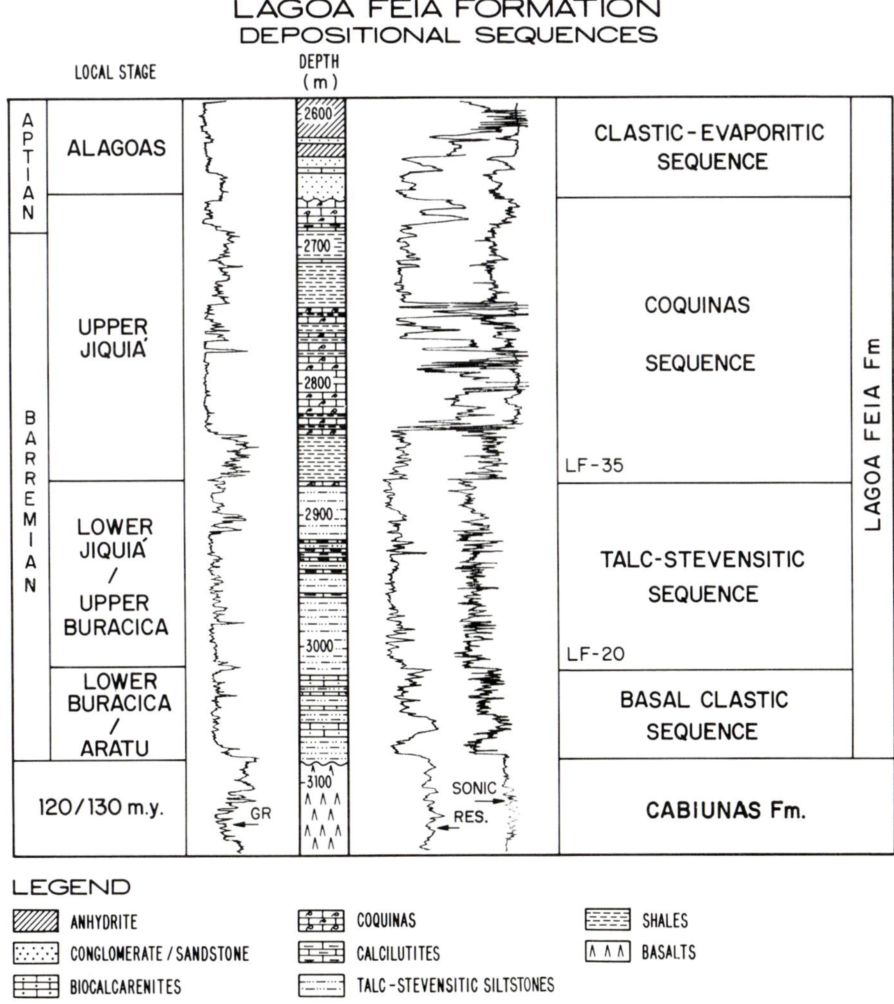

Fig. 5. Depositional sequences of the Lagoa Feia Formation as identified by electric logs. (After Dias et al. 1988)

at the Badejo, Pampo, and Linguado areas (Fig. 5).

The Basal Clastic Sequence is divided into two units: the essentially conglomeratic lower part and a terrigenous-carbonate upper part.

The lower section has been reached by only a few wells and comprises alluvial fan deposits (Fig. 6A). The conglomerates are composed mainly of basaltic fragments, quartz and feldspar grains, heavily cemented by calcite. These conglomerates grade vertically to cross-stratified sandstones and distally to brown mudstones, frequently displaying desiccation cracks.

The upper part of the Basal Clastic Sequence also comprises alluvial fan, marginal and distal lacustrine facies (Figure 6B). The marginal lacustrine facies is formed by bioclastic calcarenites, calcilutites with features of local subaerial exposition, and siltstones composed of talc peloids and stevensite peloids. The distal lacustrine facies is represented by marls, calcilutites and dark shales.

2. Talc-Stevensitic Sequence (Lower Jiquiá/ Upper Buracica local stages): the upper boundary of this sequence is defined by the LF-35 stratigraphic marker, which corresponds to a strong regional seismic reflector at the base of the Coquinas layers (Fig. 5).

The Talc-Stevensitic Sequence is characterized by a marginal lacustrine facies (Fig. 6C) which comprises siltstones and sandstones composed of talc and stevensite peloids generally dolomitized and silicified. Talc peloids present concentric structures, frequently with ostracod shell nuclei.

The origin of the peloids is controversial. Bertani and Carozzi (1984) interpret the talc peloids as kerolitic ooids which are normally associated with hyaloclastites composed of volcanic glasses and hyalotuffs, usually altered to trioctaedral smectites. However, Rehim et al. (1986) interpret the peloids as the product of the chemical precipitation from lacustrine brackish alkali water rich in Si and Mg. The stevensite peloids are formed by direct chemical precipitation, while the talc peloids are the result of diagenetic alteration of chemically precipitated magnesian silica gel.

A distal lacustrine facies, essentially composed of dark shales, also occurs. The system is bordered by alluvial fan facies, similar to those previously described.

3. Coquinas Sequence (Upper Jiquiá local stage): this is the most important sequence of the Lagoa Feia Formation because it includes both good reservoirs and the principal source rocks of the Campos basin (Fig. 7A). It is characterized by extensive limestone deposition throughout the basin, separated from the overlying depositional sequence by a regional unconformity (Figs. 2 and 5), which is mapped throughout the basin.

Several lithologies can be recognized in this sequence: calcilutites; peloidal, oolitic, or bioclastic calcarenites; calcirudites composed of pelecypods (reworked from bioaccumulated deposits) and locally of gastropods with or without calcarenite matrix; and in situ bivalves bioaccumulated limestones which are normally associated with contemporaneous depositional highs.

The coquinas are locally pure calcite, but sometimes contain abundant talc and stevensite peloids. These rocks are generally cemented by calcite, chert, or dolomite. High energy, matrix free calcirudites and calcarenites are the best reservoir rocks of the Lagoa Feia Formation. The coquinas are one of the producing reservoirs of the Badejo, Linguado, and Pampo fields in the southern area of the basin (Fig. 1).

Marls and calcareous shales rich in organic matter have been found in the syndepositional lows of the basin. These calcareous shales are the most important source rocks of the Campos basin.

The occurrence of alluvial fan deposits in this sequence is more significant than in the older ones, reflecting important tectonic event during the deposition of the Coquinas Sequence.

4. Clastic-Evaporitic Sequence (Alagoas local stage): comprises a terrigenous basal portion deposited over the 'pre-Alagoas' unconformity overlain by a thick evaporitic layer. The base of the evaporites corresponds to a very clear seismic reflector mapped throughout the basin. The top of the evaporites is taken as the upper boundary of the Lagoa Feia Formation.

The lower portion (Fig. 7B) is marked by intense progradation of the alluvial fan facies. While coarse clastic deposition was dominant near the border of the basin, synchronous deposition of muds occurred in its distal portions. The clasts are dominantly composed of basalt fragments but there is a gradual northward increase of metamorphic and igneous rock fragments. The mudflat becomes areally more extensive upward and consists dominantly of reddish brown mudstones with frequent subaerial exposition features.

The carbonate facies comprises diagenetic nodular and laminated limestones, probably algal in origin (stromatolites).

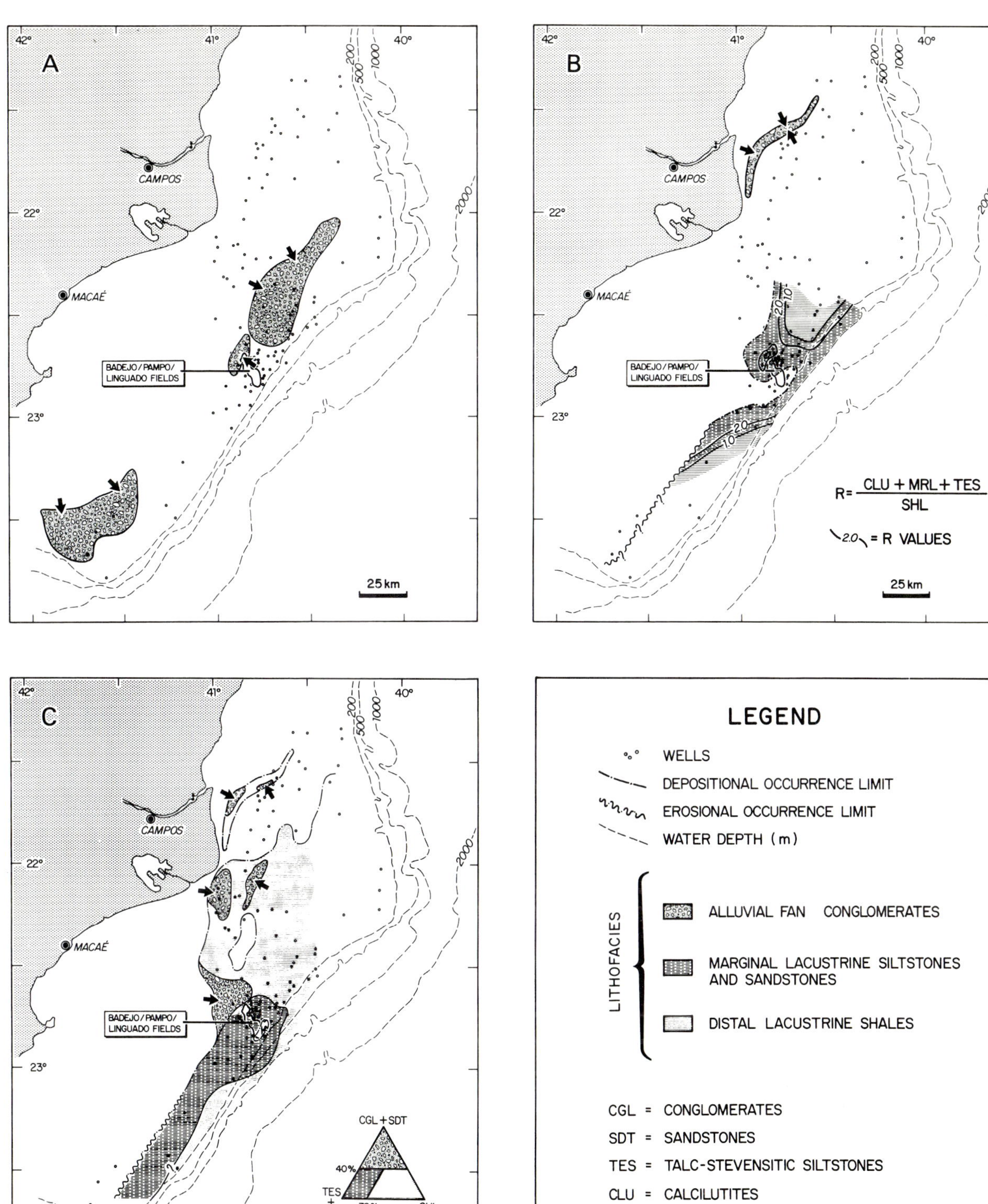

Fig. 6A-C. Facies maps of the Lagoa Feia Formation depositional sequences. **A** The Basal Clastic Sequence (lower part). **B** The Basal Clastic Sequence (upper part). **C** The Talc-Stevensitic Sequence. (After Dias et al. 1988)

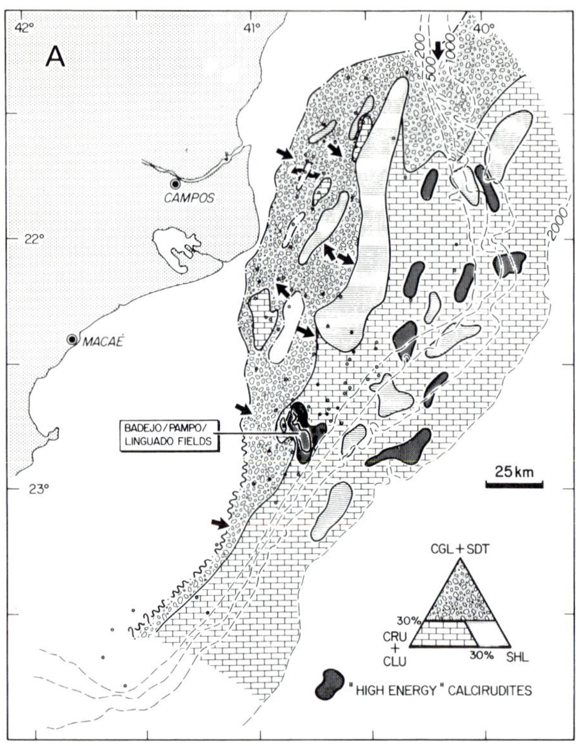

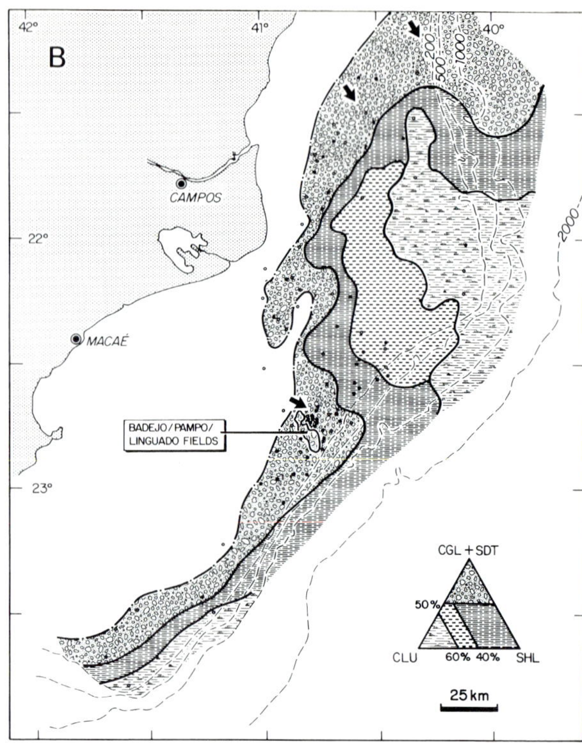

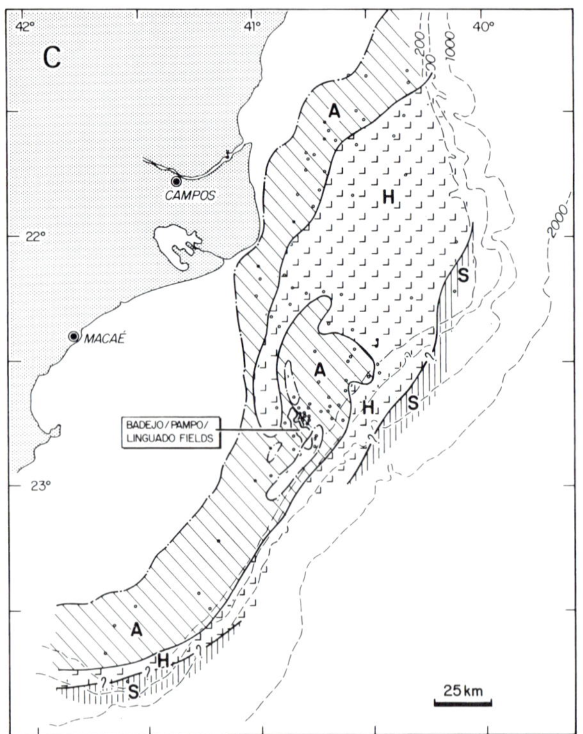

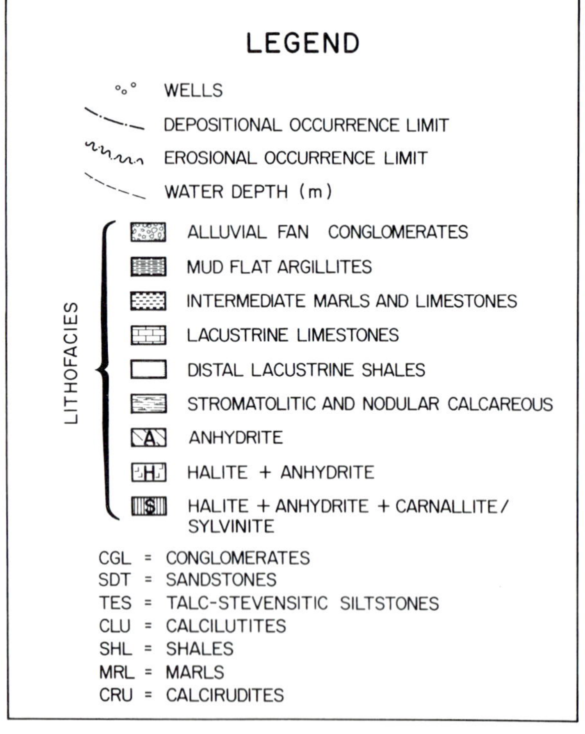

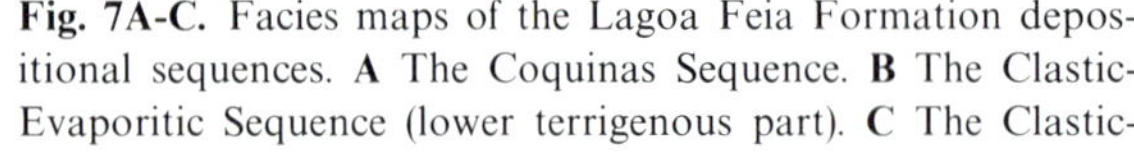

Fig. 7A-C. Facies maps of the Lagoa Feia Formation depositional sequences. **A** The Coquinas Sequence. **B** The Clastic-Evaporitic Sequence (lower terrigenous part). **C** The Clastic-Evaporitic Sequence (upper evaporitic part). (After Dias et al. 1988)

The final stage of deposition of the Lagoa Feia Formation is essentially chemical. From the proximal to the distal area, the following mineral assemblages can be recognized: anhydrite, halite + anhydrite, halite + anhydrite + carnallite + sylvinite (Fig. 7C). Limestones occur locally, associated with more proximal facies.

Organic Geochemistry of the Lagoa Feia Formation

Several studies have shown that the hydrocarbon accumulations so far discovered in the Campos basin are sourced from lacustrine calcareous black shales of the Lagoa Feia Formation (Pereira et al. 1984; Mello et al. 1984, 1988a,b; Guardado et al. 1989). Two source rock systems are identified in the Lagoa Feia Formation: lower Barremian (Talc-Stevensitic sequence) black shales and marls and an upper Barremian (Coquinas Sequence) system composed mainly of calcareous black shales and marls deposited in brackish to saline alkaline lacustrine environments (Mohriak et al. 1990). The Basal Clastic Sequence and the Clastic-Evaporitic Sequence are organically leaner and, therefore, have a less significant contribution to the petroleum accumulated in the Campos basin. The laboratory methodology employed in this chapter is described in more detail by Trindade et al. 1992.

The upper and lower Barremian organic-rich shales exemplifying the source rock systems of the Campos basin are shown in Fig. 8. These sequences include mostly well-laminated calcareous ($CaCO_3$ from 5 to 59%) black shales, very rich in organic matter (TOC up to 9%), with low sulfur content ($< 0.3\%$). The alternation of coquinas and shales in the upper Barremian sequence is shown in Fig. 8, where the shales are thicker and richer at the base of this sequence. These rocks show excellent hydrocarbon source potential (S_2), which reaches

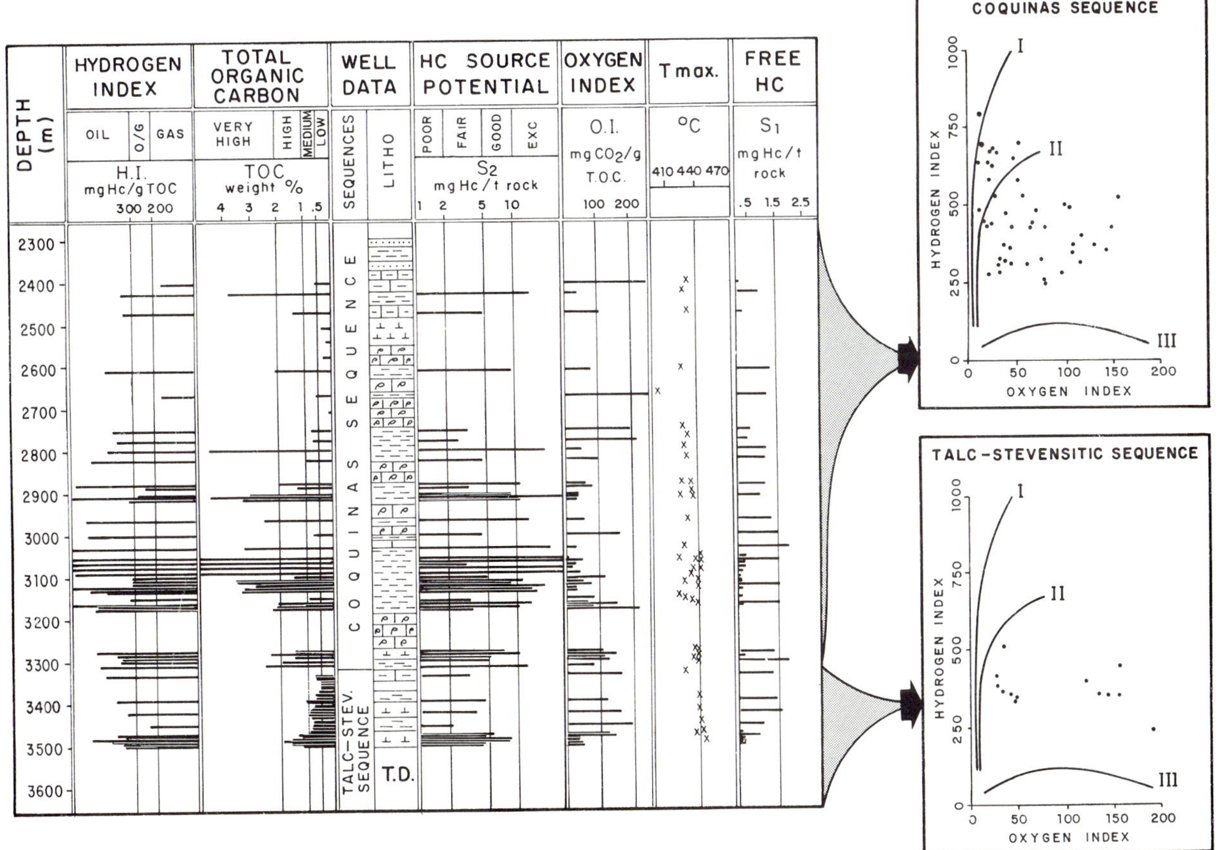

Fig. 8. Typical geochemical log for a well that drilled the Lagoa Feia Formation and van Krevelen type diagrams for the Coquinas (upper Barremian) and the Talc-Stevensitic (lower Barremian) sequences

COQUINAS SEQUENCE

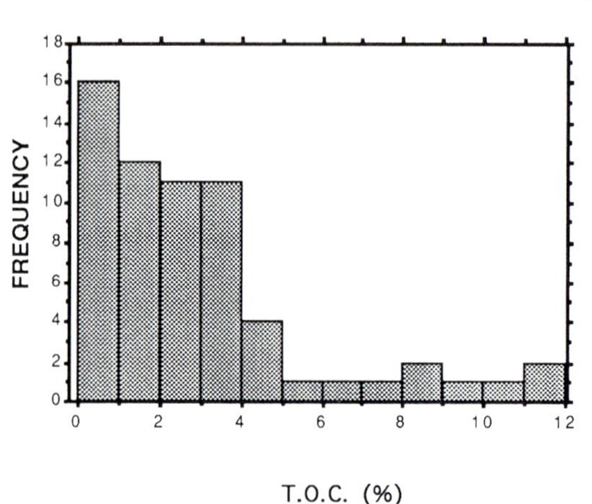

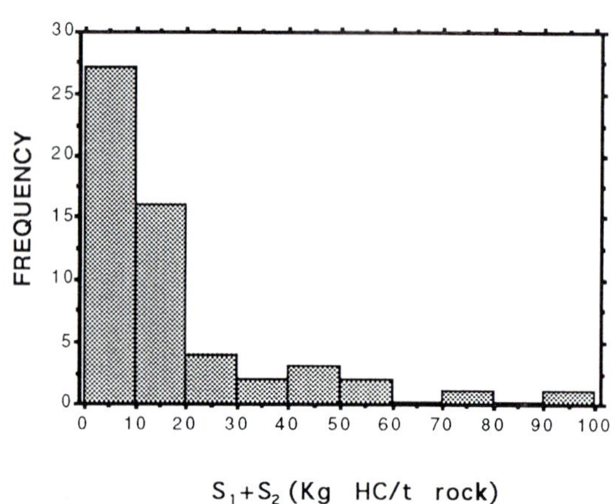

TALC-STEVENSITIC SEQUENCE

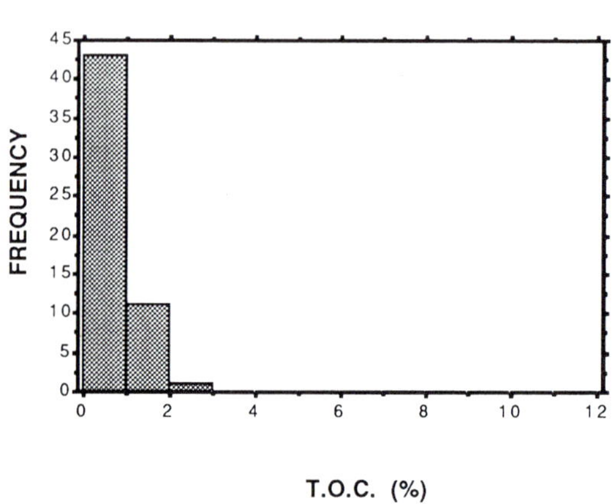

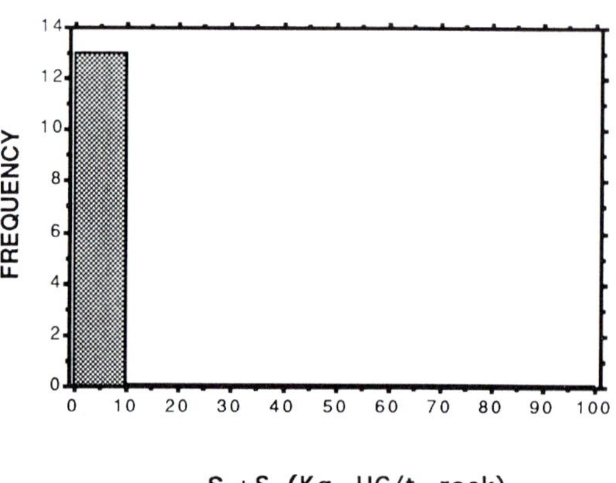

Fig. 9. Distribution of TOC (total organic carbon, wt%) and Rock-Eval $S_1 + S_2$ (kg HC/t rock) values for the Coquinas and Talc-Stevensitic sequences

92 kg HC/t rock and hydrogen index is up to 800 mg HC/g TOC. The marls and shales of the lower Barremian Talc-Stevensitic Sequence is not as rich; the TOC averages 2%, and the hydrocarbon source potential 8.5 kg HC/t rock. Comparison of van Krevelen type diagrams of both sequences shows that type I kerogen is more abundant in the upper Barremian Coquinas Sequence (Fig. 8). Visual kerogen analyses identify the organic matter as being mainly type I kerogen, composed of amorphous lipid-rich material. The better source rock attributes of the upper Barremian Sequence is evidenced by Fig. 9, where higher TOC and $S_1 + S_2$ values are associated with this sequence. There are trends of increasing values in direction to the depocenters of both sequences, especially the Corvina-Parati Low (Fig. 3), where more distal conditions allowed deposition and preservation of more algal-rich source beds. The Corvina-Parati low, as well as other deep water structural lows acted as the major oil kitchens for petroleum accumulations in the Campos basin.

Geochemical characteristics of lower Barremian and upper Barremian source intervals are shown by gas chomatograms and m/z 191 chromatograms of the alkane fraction in Fig. 10. There are some significant differences in the bulk geochemical data and biological marker distribution between the two. The lower Barremian contains higher concentrations of β-carotane (peak β),

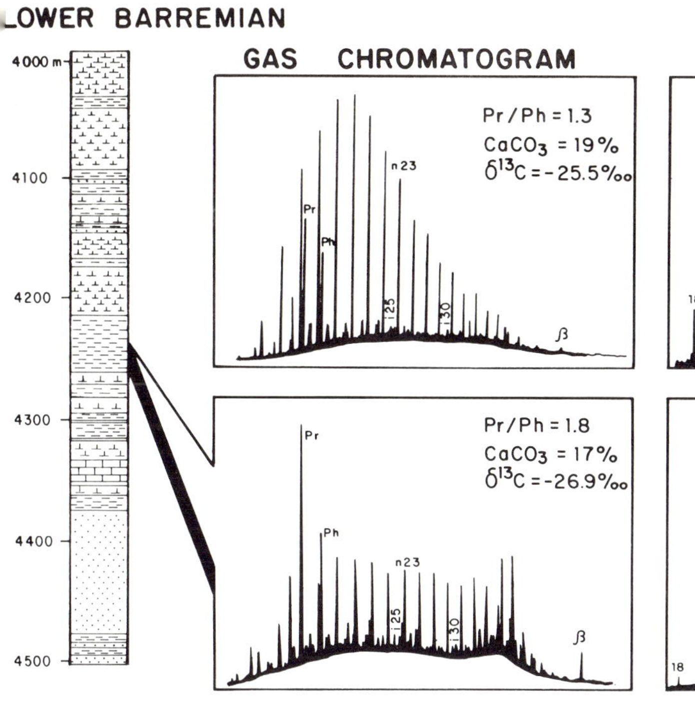

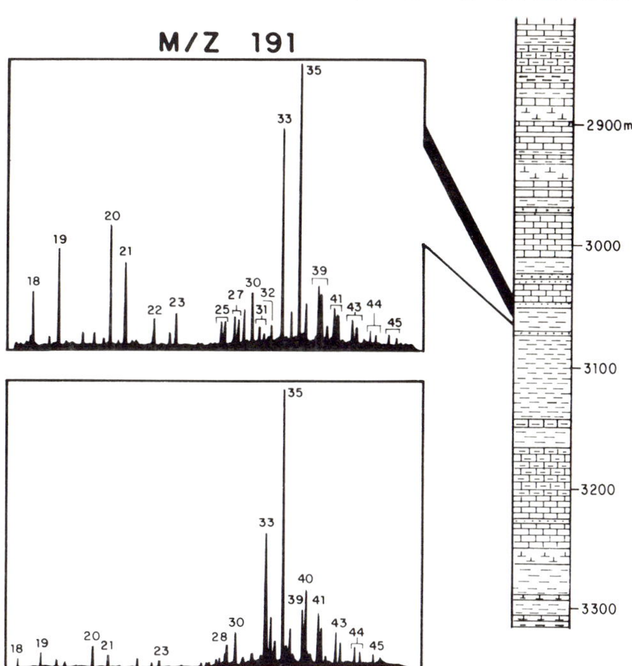

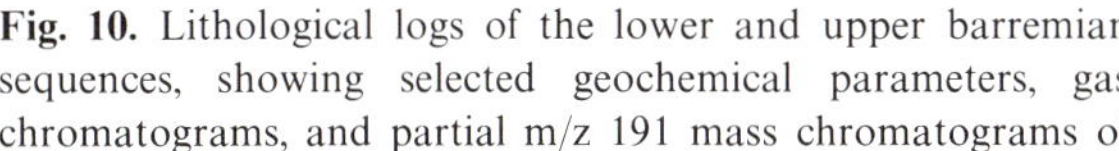

Fig. 10. Lithological logs of the lower and upper barremian sequences, showing selected geochemical parameters, gas chromatograms, and partial m/z 191 mass chromatograms of representative saturate fractions of solvent extracts from organic-rich samples are shown (see Table 1 for compound identification). (After Mohriak et al. 1990)

higher relative abundances of high molecular weight *n*-alkanes, higher abundance of gammacerane (peak 40), higher pristane/phytane ratio (around1.8), pristane/nC_{17} and phytane/n-$C_{18} > 1$, lower abundances of tricyclic terpanes (peaks 18 to 26), and lighter carbon isotopic values ($\delta^{13}C \sim -26.9‰$) than the upper Barremian. In the lower Barremian, extended tricyclics up to C_{45} were identified (de Grande 1992). The presence of C_{30} steranes and dinosterane isomers in the Coquinas Sequence, indicative of marine contribution (Moldowan et al. 1985; Summons et al. 1987; Goodwin et al. 1988; Mello et al. 1988a,b) supports the interpretation of some marine transgressions over the lake during the Neo-Barremian. Biomarker features suggest that both of the Barremian successions were deposited in a lacustrine brackish to saline environment. The differences between the upper and lower Barremian samples in Fig. 10 indicate variations within the depositional environment and specifically shallower water and enhanced higher plant input for the lower Barremian organic-rich sedimentary succession (Mohriak et al. 1990).

Biomarker distributions and concentrations of oils accumulated in reservoirs ranging from Barremian to Tertiary show a good correlation. The similarities of the oils are also expressed by carbon isotope ratios and other bulk and geochemical parameters. As a result of the large variation of reservoir depths and ages, different *n*-alkane distributions may reflect biodegradation and maturity constraints. Oils recovered from Albian and Barremian reservoirs displayed in Fig. 11 are representative of Campos basin-type oil, which show high relative proportions of tricyclic terpanes (peaks 18–26), moderate gammacerane and 28,30-bisnorhopane (peaks 40 and 32 respectively) abundances, predominance of C_{27} steranes (peaks 8–10) over their C_{29} counterparts (peaks 14–16), and presence of β-carotane (peak β in the GC trace).

Correlation of the oils with source rock organic extracts show a better match with the upper Barremian Coquinas Sequence shales (of Figs. 10 and 11). This correlation is better expressed by higher relative abundances of tricyclics, lower β-carotane concentration, lower proportions of pristane and phytane in relation to *n*-alkanes, and lighter carbon

Table 1. Compound identifications

Pr	Pristane
Ph	Phytane
i-C_{25}	C_{25} Acyclic isoprenoid
iC_{31}	C_{31} Acyclic isoprenoid
β-	β-Carotane
1	13β(H),17α(H)-Diapregnane (C_{21})
2	5α(H),14β(H),17α(H)-Pregnane (C_{21}
3	5α(H),14β(H),17α(H) + 5α(H),14α(H), 17α(H)-Pregnane (C_{21})
4	4-methyl-5α(H),14β(H),17β(H) + 4-methyl-5(H),14α(H),17α(H)-Homopregnane (C_{22})
5	5α(H),14β(H),17β(H) + 4-methyl-5α(H),14α(H), 17α(H)-Homopregnane (C_{22})
6	13β(H),17α(H)-Diacholestane, 20S
7	13β(H),17α(H)-Diacholestane, 20R
8	5α(H),14α(H),17α(H)-Cholestane, 20S
9	5α(H),14β(H),17β(H)-Cholestane, 20R + 20S
10	5α(H),14α(H),17α(H)-Cholestane, 20R
11	5α(H),14α(H),17α(H)-Methylcholestane, 20S
12	5α(H),14β(H),17β(H)-Methylcholestane, 20R + 20R
13	5α(H),14α(H),17α(H)-Methylcholestane, 20R
14	5α(H),14α(H),17α(H)-Ethylcholestane, 20S
15	5α(H),14β(H),17β(H)-Ethylcholestane, 20R + 20S
16	5α(H),14α(H),17α(H)-Ethylcholestane, 20R
17	C_{19} Tricyclic terpane
18	C_{20} Tricyclic terpane
19	C_{21} Tricyclic terpane
20	C_{23} Tricyclic terpane
21	C_{24} Tricyclic terpane
22	C_{25} Tricyclic terpane
23	C_{26} Tricyclic terpanes
24	C_{24} Tetracyclic terpane (Des-E)
Te	C_{24} Tetracyclic terpane (Des-A)
25	C_{28} Tricyclic terpanes
26	C_{29} Tricyclic terpanes
27	C_{25} Tetracyclic terpane
28	C_{27} 18α(H)-Trisnorneohopane (Ts)
29	C_{30} Tricyclic terpanes
T	25,28,30-Trisnorhopane
30	C_{27} 17α(H)-Trisnorhopane (Tm)
31	C_{31} Tricyclic terpanes
32	17α(H),18α(H),21β(H)-28,30-Bisnorhopane
N	25-Norhopane
33	17α(H),21β(H)-Norhopane
34	17β(H),21α(H)-Norhopane
35	17α(H),21β(H)-Hopane
36	C_{33} Tricyclic terpanes
37	17β(H),21α(H)-Hopane
38	C_{34} Tricyclic terpanes
39	17α(H),21β(H)-Homohopane (22S + 22R)
40	Gammacerane
41	17α(H),21β(H)-Bishomohopane (22S + 22R)
42	C_{35} Tricyclic terpanes
43	17α(H),21β(H)-Trishomohopane (22S + 22R)
44	17α(H),21β(H)-Tetrakishomohopane (22S + 22R)
45	17α(H),21β(H)-Pentakishomohopane (22S + 22R)

isotope ratios of the upper Barremian shales as compared to the lower Barremian source intervals (Fig. 10). These geochemical features indicate that the organic-rich rocks of the upper Barremian Coquinas Sequence are the major contributors to the hydrocarbon accumulations that have been discovered so far in the Campos basin. The better hydrocarbon source potential and larger thickness of the upper Barremian source rocks than the lower Barremian ones also support the hypothesis that this interval contains the main source rocks of the basin (Dias et al. 1988).

Appropriate maturity conditions of the Lagoa Feia Formation assures the source character of the upper and lower Barremian sequences (Meister 1984; Mello et al. 1984; Pereira et al. 1984; Figueiredo et al. 1985). The top of the upper Barremian shales reached the oil generation window about Eocene time in some structural lows, and did not reach the overmature zone anywhere in the basin (Guardado et al. 1989). The source rocks are still immature in the Badejo High and in the southern area of the São João da Barra Low. This relatively low source-rock maturation is a result of the low thermal flux, which was due to a reduced crustal thinning (Dias et al. 1988). The deepest occurrence of known associated gas in the area suggests that this source rock is not overmature yet, even in the deepest part of the basin. Carbon isotope ratios of methane ranging from $-45‰$ to $-52‰$, associated with C_2+ ranging from 10 to 15%, also indicate that these gases were generated within the oil window.

Recent kinetic parameters obtained through Rock Eval V and Lawrence Livermore software (Burnham et al. 1987) present broad spread in the data, ranging from 48 to 58 kcal/mol, reflecting variation in depositional conditions within the Lagoa Feia Formation (Soldan, pers. commun.). These results agree with values obtained for the Bucomazi Formation (Burwood et al. 1992), the West African equivalent of the Lagoa Feia Formation in the Lower Congo Coastal basin (Mello et al. 1991).

Variability in the molecular properties for samples occurring at different horizons of the lower and upper Barremian sequences are observed. Examples of three organic extracts from the lower Barremian organic-rich sediments with similar vitrinite reflectance values are shown in Fig. 12. There are subtle differences in the *n*-alkane distribution, β-carotane concentration (peak β), gammacerane abundance (peak 40), Pr/Ph ratio, carbonate content, and the tricyclic terpane relative abundances (peaks 18-26; cf. Mello et al. 1988a, b for details). As these samples present similar maturity conditions, variations in geochemical parameters suggest

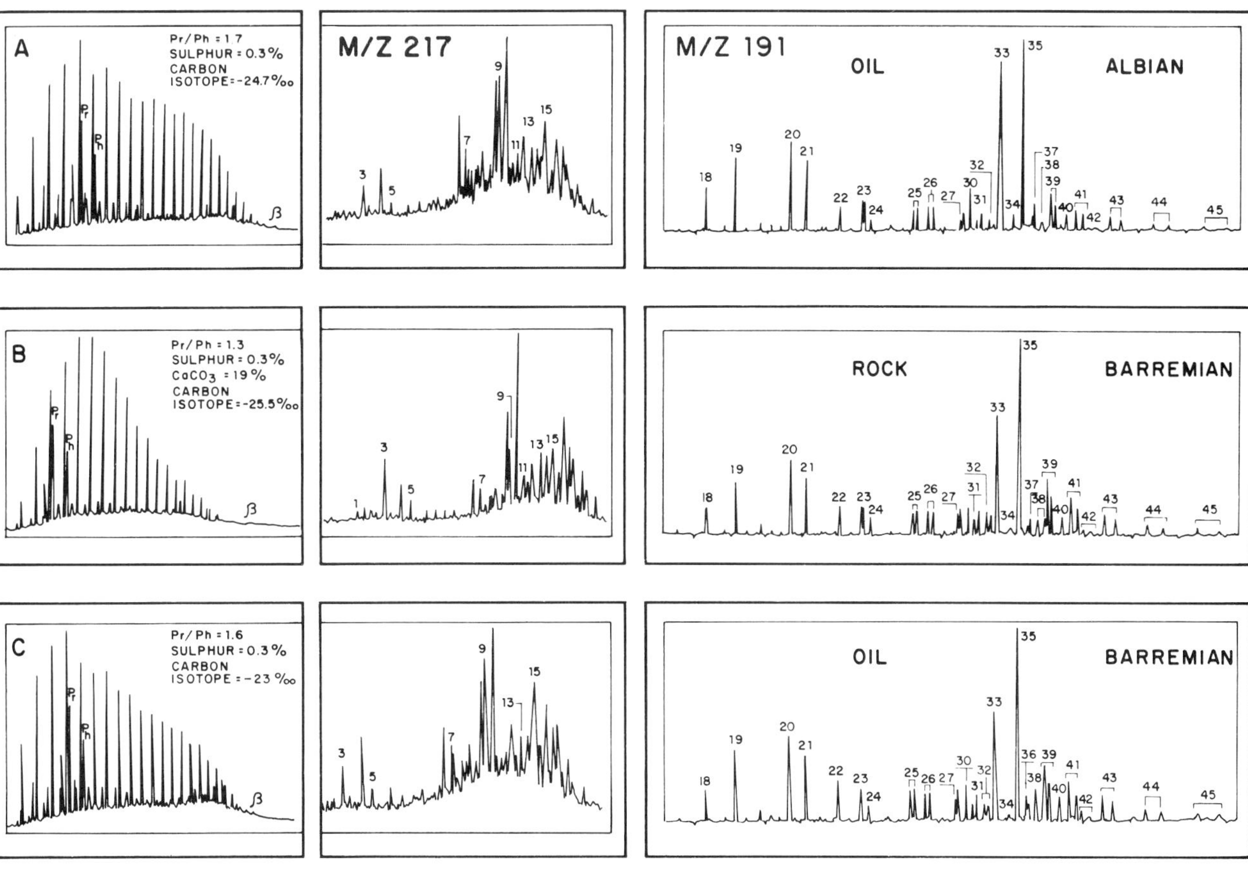

Fig. 11A-C. Oil-source rock correlation using gas chromatograms, steranes (m/z 217) and terpanes (m/z 191). Lagoa Feia organic extract is depicted in **B**, whereas typical oil samples pooled in different reservoirs are presented in **A** (Albo-Cenomanian) and **C** (Barremian). See Table 1 for compound identification. (After Mohriak et al. 1990)

fluctuations in the depositional environment, where higher gammacerane and β-carotane relative abundances are associated with higher salinities.

Depositional Environment Model

Sedimentological and mineralogical studies (e.g., Castro et al. 1981; Bertani and Carozzi 1985), as well as carbon and oxygen isotopes data (Takaki and Rodrigues, 1984) support contractions and expansions of the lake system during deposition of the upper Barremian organic-rich sediments. The block diagram in Fig. 13 shows a schematic illustration of the proposed paleoenvironment of deposition of the Campos basin during the late Barremian rift stage, where a shallow, saline alkaline lake system prevailed.

The high amount of nutrients available in the brackish to saline waters, generally associated with perenial alkaline springs, enhances the development of well-adapted, limited species. Therefore, without competition, algae and bacteria show prolific productivity within the lake. Salinity differences between an upper oxic, less saline layer and a lower anoxic, more saline, denser, and alkaline layer enhance the water column stability, leading to stratification and permanent bottom water anoxia. These conditions are lethal for microfauna and benthic organisms, but enhance anaerobic bacterial activity. Anoxic conditions in the bottom waters also enhance the preservation of organic matter, resulting in the deposition of well-laminated, organic-rich calcareous black shales (Demaison and Moore 1980; De Deckker 1988; Kelts 1988; Mello and Maxwell 1991).

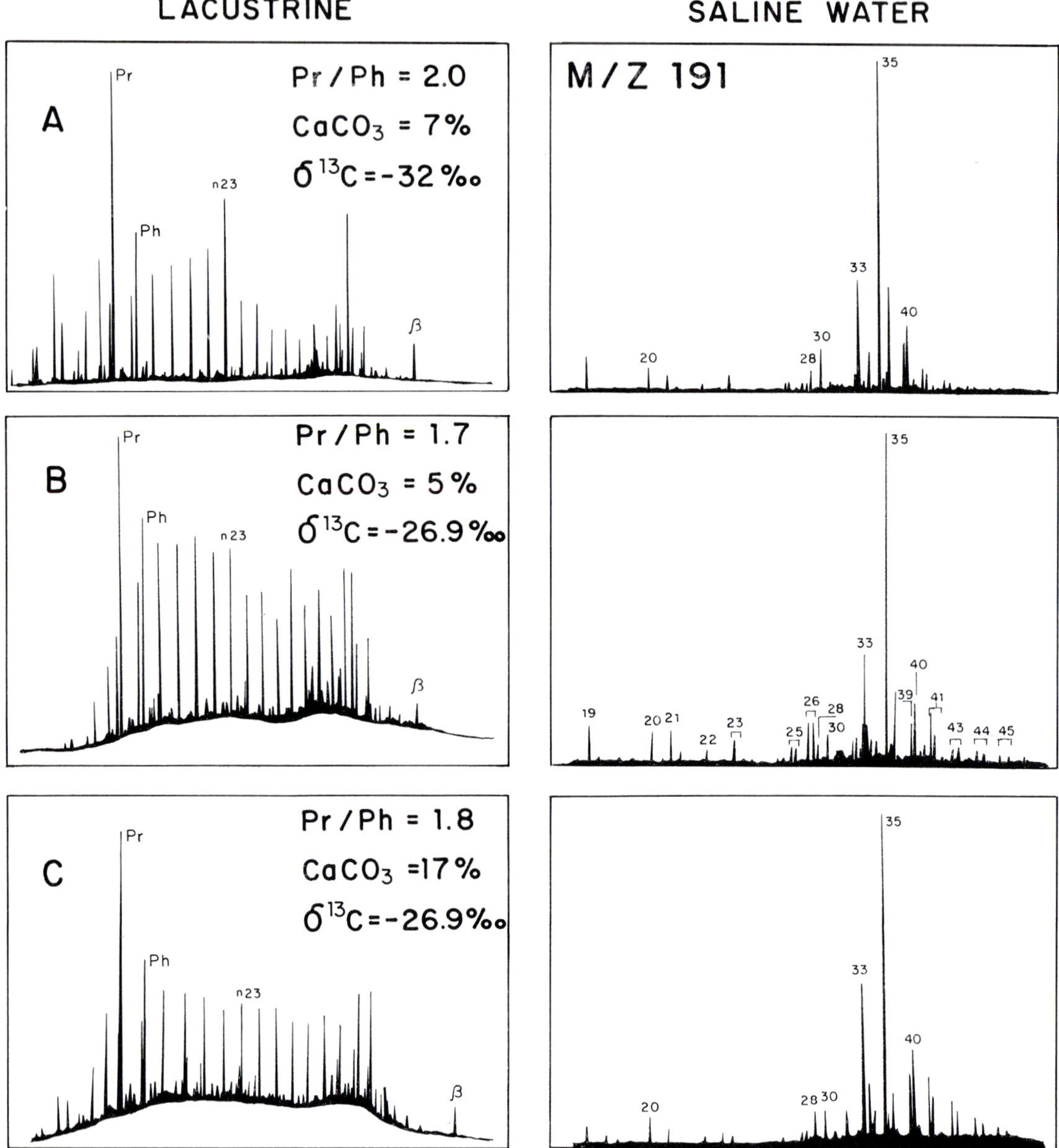

Fig. 12A-C. Gas chromatograms and terpanes (m/z 191) of upper Barremian source rocks, showing the variability in geochemical parameters observed among different samples. (After Mohriak et al. 1990)

Modern analogues generally occur in areas of high evaporation, associated with semiarid climate, including lakes Nakuru, Magadi and Bogoria in the East African rift system (Eugster 1986; Vincens et al. 1986; Talbot 1988). Few comparable examples of ancient saline lake systems have been reported in the literature. The best comparisons with the Lagoa Feia Formation analogue appear to be the well-studied Eocene Green River Formation in the Uinta basin, USA (e.g., Tissot et al. 1978; Dean and Fouch 1983; Katz this Vol.), the Chaidamu basin in China (Powell 1986) and the Junggar basin in China (Jiang and Fowler 1986; Carroll et al. 1992).

Hydrocarbon Migration

A schematic model of hydrocarbon migration and entrapment, which shows how hydrocarbons

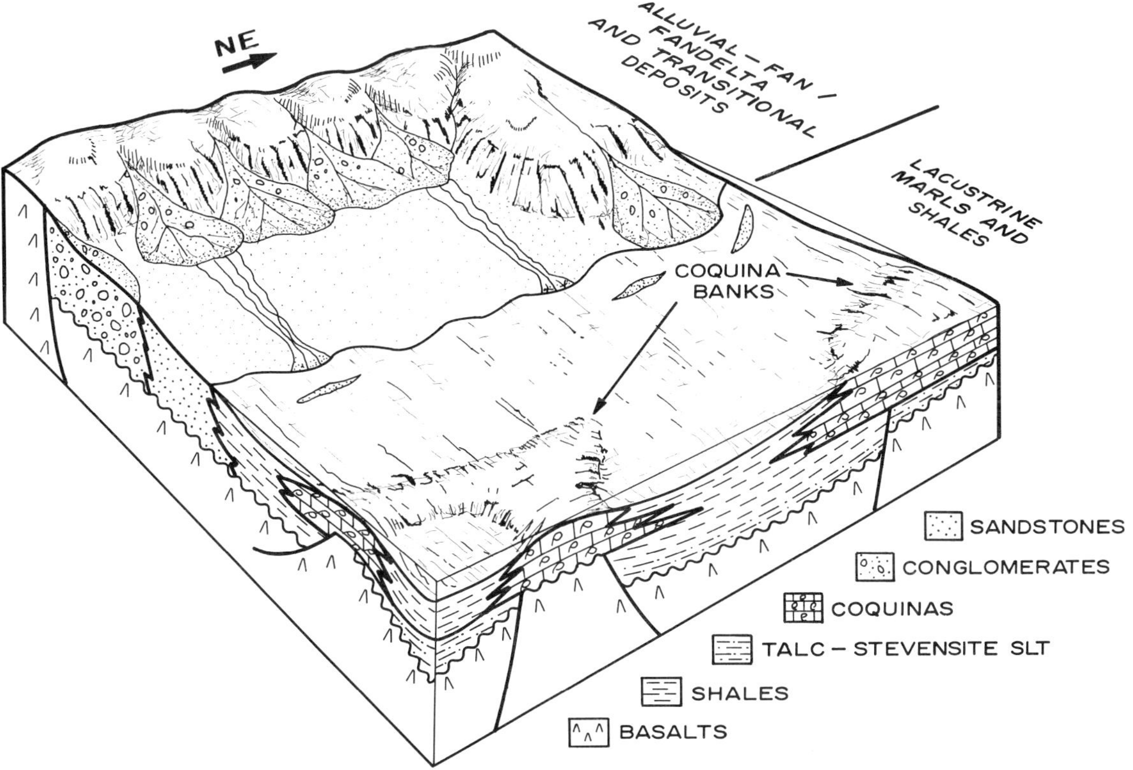

Fig. 13. Paleogeographic model for the lacustrine Lagoa Feia Formation, depicting lateral facies variations and deposition of source rocks in the distal and deeper areas. (After Guardado et al. 1989)

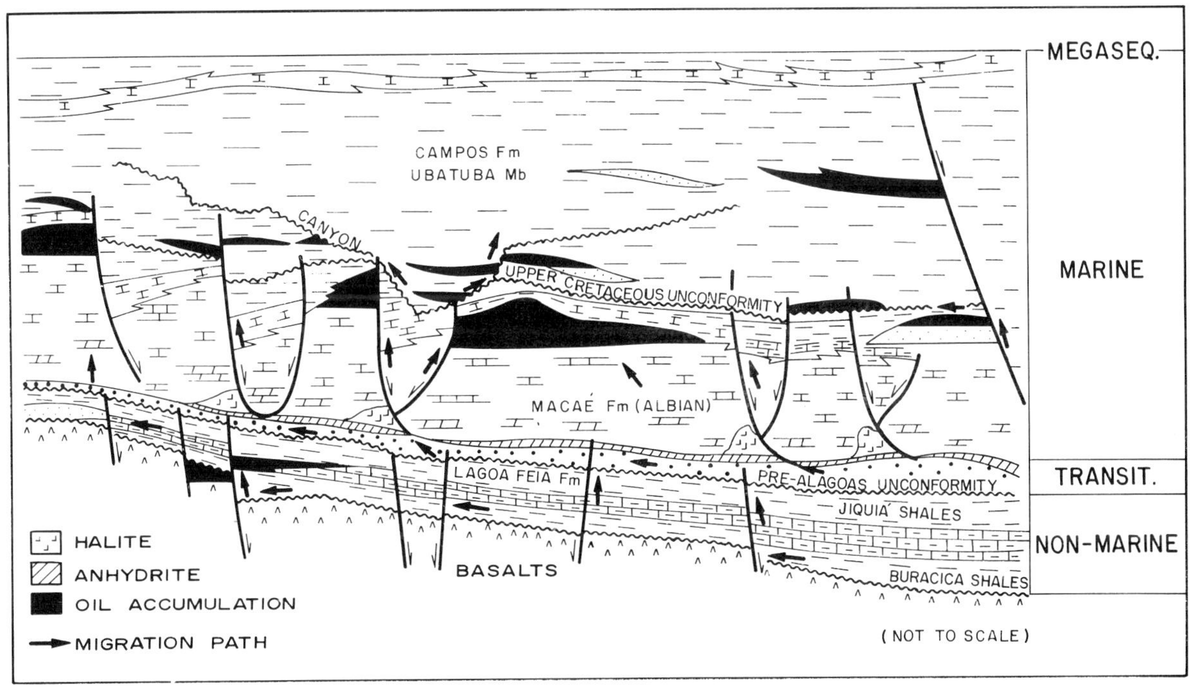

Fig. 14. Schematic petroleum migration pathways for the Campos basin oil accumulations. Petroleum generated by the Lagoa Feia source rocks migrated through windows in the salt layer and by faults and unconformities to the marine reservoirs. (After Guardado et al. 1989)

migrated from the Lagoa Feia source rocks to the upper Barremian and shallower post-salt reservoirs (Albian carbonates and Upper Cretaceous/Tertiary turbidites) is shown in Fig. 14. Hydrocarbon accumulations found in the fractured basement and coquina reservoirs of the pre-salt stage are associated with migration by direct contact or through unconformities associated with normal faults. Oils pooled in the marine sequence (Albian, Upper Cretaceous, and Tertiary) migrated through a system associated with pre-salt normal faults, windows in the evaporite layers (probably caused by halokinetic tectonism), listric faults, and regional unconformities (Figueiredo et al. 1985). The most important oil accumulations in the Campos basin are associated with deep water fans ranging from Late Cretaceous to Late Tertiary (Trindade et al. 1990). They occur as widespread sheets and are also enclosed in submarine canyons. The coalescent turbiditic sand bodies of the middle Eocene probably also acted as a hydrocarbon collector system, allowing the migration of petroleum arising from growth fault systems or unconformities.

Summary

Regional mapping of the Lagoa Feia Formation allowed the identification of four depositional sequences (Basal Clastic Sequence, Talc-Stevensitic Sequence, Coquinas Sequence and Clastic-Evaporitic Sequence), which can be related to the evolution of the basin during rift stage.

The organic rich upper Barremian calcareous black shales and marls of the Lagoa Feia Formation, deposited in anoxic brackish to saline lakes, are the petroleum source rocks of all the known oil accumulations of the Campos basin.

Variability of geochemical parameters within the Lagoa Feia Formation reflects fluctuations in the salinity of the lake and suggests that the main source rocks are associated with the upper Barremian shales included in the Coquinas Sequence.

Petroleum migration pathways involve direct contact between source and reservoir rocks within the rift sequence. Migration to the marine sequence reservoirs is related to windows in the salt layer connected to growth faults and unconformities.

Acknowledgments. The authors thank Petrobrás for permission to publish this chapter. Comments and suggestions by the reviewers A.R. Carroll and M. Kruge, were greatly appreciated. Discussions with A.L. Soldan, N.C. Azambuja Filho, and J.R. Cerqueira also improved an early version of the manuscript. We are also grateful for the analytical support provided by members of the Geochemistry Section of Petrobrás Research Center.

References

Asmus HE, Baisch PR (1983) Geological evolution of the Brazilian Continental Margin. Episodes 4: 3–9

Bertani RT Carozzi AV (1984) Microfacies, depositional models and diagenesis of Lagoa Feia Formation (Lower Cretaceous) Campos Basin, Offshore Brazil, Petrobrás/Cenpes. Ciênc Téc Petról 14: 104

Bertani RT, Carozzi AV (1985) Lagoa Feia Formation (Lower Cretaceous) Campos basin, offshore Brazil: rift-valley stage carbonate reservoirs – I and II. J Petrol Geol 8: 37–58, 199–220

Burnham AK, Braun RL, Gregg HR (1987) Comparison of methods for measuring kerogen pyrolysis rates and fitting kinetic parameters. Energy & Fuels 1 (6): 452–458

Burwood R, Leplat P, Mycke B, Paulet J (1992) Rifted margin source rock deposition: a carbon isotope and biomarker study of a West African Lower Cretaceous "lacustrine" section. In: Eckardt CB, Maxwell JR, Larter SR, Manning DAC (eds.) Advances in organic geochemistry 1991. Org Geochem 19: 41–52

Carroll AR, Brassell SC, Graham SA (1992) Upper Permian Lacustrine Oil Shales, Southern Junggar Basin, Northwest China. Am Assoc Pet Geol Bull 76: 1874–1902

Castro JC, Azambuja Filho NC, Xavier AAPG (1981) Fácies e análise estratigráfica da Formação Lagoa Feia, Cretáceo Inferior da Bacia de Campos, Brasil. 80 Congr Geol Argent 2: 567–576

Dean W, Fouch TD (1983) Lacustrine environment. In: Scholle PA et al., eds. Carbonate depositional environments, Am Assoc Pet Geol, Tulsa, Mem 33: 97–130

De Deckker P (1988) Large Australian Lakes during the last 20 million years: sites for petroleum. In: Fleet AJ, Kelts K, Talbot M (eds.) Lacustrine petroleum source rocks. Geol Soc, London, Spec Publ 40: 45–58

de Grande SMB (1992) A utilização da espectrometria de massas acoplada a espectrometria de massas em geoquímica orgânica: estudo dos terpanos tricíclicos. PhD Thesis, Univ Fed Rio de J 209 pp

Demaison GJ, Moore GT (1980) Anoxic environments and oil source bed genesis. Am Assoc Pet Geol Bull 64: 1179–1209

Dias JL, Oliveira JQ, Vieira JC (1988) Sedimentological and stratigraphic analysis of the Lagoa Feia Formation, rift phase of the Campos basin, offshore Brazil. Rev Bras Geociênc 18: 252–260

Eugster HP (1986) Lake Magadi, Kenya: A model for rift valley hydrochemistry and sedimentation. In: Frostick LE, Renant RW, Reid I, Tiercelin JJ (eds) Sedimentation in the African rifts Geol Soc, London, Spec Publ 25: 177–191

Figueiredo AMF, Martins CC (1990) 20 anos de exploração da Bacia de Campos e o sucesso nas águas profundas. Bol Geociênc Petrobrás 4(1): 105–123

Figueiredo AMF, Pereira MJ, Mohriak WU, Gaglianone PC, Trindade LAF (1985) Salt tectonics and oil accumulation in the Campos Basin, offshore Brazil (abstract). Am Assoc Pet Geol Bull 69: 255

Gamboa LAP, Rabinowitz PD (1981) The Rio Grande fracture zone in the Western South Atlantic and its tectonic implication. Earth Planet Sci Lett 52: 410–418

Goodwin NS, Mann AL, Patience RL (1988) Structure and significance of C_{30}-4-methylsteranes in lacustrine shales and oils. Org Geochem 12: 495–506

Guardado LR, Gamboa LAP, Lucchesi CF (1989) Petroleum geology of the Campos basin, Brazil; a model for producing an Atlantic type basin. In: Edwards JD, Santogrossi PA (eds) Divergent passive margin basins. Am Assoc Pet Geol, Tulsa, Mem 48: 3–79

Jiang Z, Fowler MG (1986) Carotenoid-derived alkanes in oils derived from northwestern China. Org Geochem 10: 831–839

Katz BJ (1993) The Green River shale: an Eocene carbonate lacustrine source system. In: Katz BJ (ed) this volume.

Kelts K (1988) Environments of deposition of lacustrine petroleum source rocks: an introduction. In: Fleet AJ, Kelts K, Talbot M (eds) Lacustrine petroleum source rocks. Geol Soc, London, Spec Publ 40: 3–26

Kumar N, Gamboa LAP, Schreiber B, Mascle J (1977) Geologic history and origin of the São Paulo Plateau, comparison with Angolan margin and the early evolution of the northern South Atlantic. In: Supko PR Initial Reports of the Deep Sea Drilling Project. US Government Printing Office, Washington, DC 39: 927–945

Leyden R, Asmus HE, Zembruscki S, Bryan G (1976) South Atlantic diapiric structures. Am Assoc Pet Geol Bull 60: 196–212

Meister EM (1984) Historical geology of petroleum in Campos Basin, Brazil. AAPG Annu Meet San Francisco (Abstr) 68: 506

Mello MR, Maxwell JR (1991) Organic geochemical and biological marker characterization of source rocks and oils derived from lacustrine environments in the Brazilian continental margin. In: Katz BJ (ed) Lacustrine basin exploration—case studies and modern analogs. Am Assoc Pet Geol, Tulsa, Mem 50: 77–97

Mello MR, Estrella GO, Gaglianone PC (1984) Hydrocarbon source potential in Brazilian marginal basins. Am Assoc Pet Geol Bull 68: 506 (Abstr)

Mello MR, Gaglianone PC, Brassell SC, Maxwell JR (1988a) Geochemical and biological marker assessment of depositional environments using Brazilian offshore oils. Mar Petrol Geol 5: 205–223

Mello MR, Telnaes N, Gaglianone PC, Chicarelli MI, Brassell SC, Maxwell JR (1988b) Organic geochemical characterisation of depositional palaeoenvironments of source rocks and oils in Brazilian marginal basins. In: Matavelli L, Novelli L (eds) Advances in organic geochemistry 1987. Pergamon, Oxford, pp 31–45

Mello MR, Mohriak WU, Koutsoukos EAM, Figueira JCA (1991) Brazilian and West African oils: generation, migration, accumulation and correlation. Proc 13th World Petrol Congr Wiley, New York pp 153–164

Mohriak WU, Mello MR, Dewey JF, Maxwell JR (1990) Petroleum geology of the Campos basin, offshore Brazil In: Brooks J (ed) Classic petroleum provinces. Geol Soc, London, Spec Publ 50: 119–141

Moldowan JM, Seifert WK, Gallegos EJ (1985) Relationship between petroleum composition and depositional environment of petroleum source rocks. Am Assoc Pet Geol Bull 69: 1255–1268

Pereira MJ, Trindade LAF, Gaglianone PC (1984) Origem e evolução das acumulações de hidrocarbonetos na Bacia de Campos. 330 Congr Bras Geol 10: 4763–4777

Ponte FC, Asmus HE (1978) Geological framework of the Brazilian continental margin. Geol Rundsch 67 (1): 201–235

Powell TG (1986) Petroleum geochemisty and depositional settings of lacustrine source rocks. Mar Petrol Geol 3: 200–219

Rehim HAA, Pimentel AM, Carvalho MD, Monteiro M (1986) Talco e estevensita na Formação Lagoa Feia da Bacia de Campos – Possíveis implicações no ambiente deposicional. 340 Congr Bras Geol 1: 416–422

Schlanger W (1981) The paradox of drowned reef and carbonate platforms. Geol Soc Am Bull 92: 197–211

Summons RE, Volkman JK, Boreham CJ (1987) Dinosterane and other steroidal hydrocarbon of dinnoflagelate origin in sediments and petroleum Geochim Cosmochim Acta 51: 3075–3082

Takaki T, Rodrigues R (1984) Isótopos estáveis do carbono e oxigênio dos calcários como indicadores paleoambientais – Bacias de Campos, Santos e Espírito Santo. 330 Congr Bras Geol 10: 4750–4762

Talbot MR (1988) The origins of lacustrine oil source rocks: evidence from the lakes of tropical Africa. In: Fleet AJ, Kelts K, Talbot M (eds) Lacustrine petroleum source rocks. Geol Soc, London, Spec Publ 40: 29–44

Tissot B, Derro G, Hood A (1978) Geochemical study of the Uinta Basin: formation of petroleum from the Green River Formation. Geochim Cosmochim Acta 42: 1469–1485

Trindade LAF, Soldan AL, Cerqueira JR, Ferreira JC, Mello MR, Carminatti M, Dias JL (1990) Petroleum generation, migration, accumulation in deep-water area of Campos basin, Brazil. Am Assoc Pet Geol Bull 74: 781 (Abstr)

Trindade LAF, Brassell SC, Santos Neto EV (1992) Petroleum migration and mixing in the Potiguar basin, Brazil. Am Assoc Pet Geol Bull, 76: 1903–1924

Vincens A, Casanova J, Tiercelin JJ (1986) Paleolimnology of the Lake Bogoria (Kenya) during the 4500 B.P. high lacustrine phase. In: Frostick LE, Renaut RW, Reid I, Tiercelin JJ (eds) Sedimentation in the African rifts. Geol Soc, London, Spec Pub 25: 323–333

The Napo Formation, Oriente Basin, Ecuador: Hydrocarbon Source Potential and Paleoenvironmental Assessment

M.R. Mello[1], E.A.M. Koutsoukos[1], and W.Z. Erazo[2]

Abstract

The Napo Formation of the Oriente basin, Ecuador, contains one of the most prolific hydrocarbon source rocks in South America. Its organic-rich succession, mainly composed of calcareous black shales and marlstones (TOC up to 13 wt.% and S_2 from Rock Eval up to 78 kg HC/t of rock) ranging in age from Albian to early Campanian, was deposited in dysoxic-anoxic, paralic-neritic environments with variable amounts of algal and terrestrial plant input. The facies with the highest hydrocarbon source potential were deposited during a paleobathymetric maximum at the Cenomanian-Turonian boundary, with salinity-stratified, bottom-water conditions. Such an event is also recorded in the stratigraphic sequences of most Subandean basins of South America and corresponds to a global tectonic-eustatic, sea-level rise recorded in the latest Cenomanian-earliest Turonian and to the worldwide Oceanic Anoxic Event 2 (OAE-2).

Biological marker data of the oils discovered up to now in the Oriente basin show very good correlation with the ones presented by the organic extracts from the organic-rich facies of the Napo Fm. (Cenomanian-Turonian), therefore confirming the oils' origin.

Introduction

The organic-rich layers of the Napo Formation, Oriente basin, eastern Ecuador (Fig. 1), have been considered the source rocks of almost all hydrocarbon accumulations discovered in Ecuador, although oil to source correlation is still poorly documented in the literature (Lozada et al. 1985; Rivadeneira 1986; Dashwood and Abbotts 1990). Cumulative oil production in the Oriente basin was over 1.5 billion barrels up to the end of 1990, and remaining reserves stand at around 1.8 billion barrels.

The Napo Formation ranges in age from the Albian to the early Campanian and covers an area of approximately 100 000 km^2 with a maximum known thickness of around 900 m (Figs. 1–3; Dashwood and Abbotts 1990). It was deposited in a paralic-neritic, stable-shelf environment; shales and carbonate lithofacies represent transgressive periods and the Napo sandstones all have regressive bases and trangressive tops (Fig. 2).

This paper describes an integrated geochemical and micropaleontological investigation, aiming to characterize the stratigraphy and depositional history of the organic-rich rocks of the Napo Formation and their putative oils. This multidisciplinary approach involved the ascertainment of the microbio-chronostratigraphic framework and paleoenvironmental evolution of the studied sections, the evaluation of organic carbon content, hydrocarbon source potential (Rock-Eval Pyrolysis) and thermal evolution (vitrinite reflectance, spore coloration and T_{max}), and molecular studies involving carbon isotope and biological marker investigations of alkanes using GC, GC-MS and GC-MS/MS. For this purpose, a large number of rock samples (40 outcrops and about 700 ditch-cuttings of 4 well sections) from the Napo Formation, and 20 oil samples recovered from oil accumulations in the Oriente basin were analyzed. An extensive data set of geochemical analyses of the Napo shales, published by Rivadeneira (1986) and Dashwood and Abbotts (1990), was used to complement this study.

Generalized Geological Setting

The Oriente basin is part of the sub-Andean group of foreland basins that lies east of the Andean

[1] Petrobrás/Cenpes/Divex, Ilha do Fundão, CEP 21949-900, Rio de Janeiro, RJ, Brazil

[2] Petroproduction, Filial de Petroecuador, Guayaquil, Ecuador

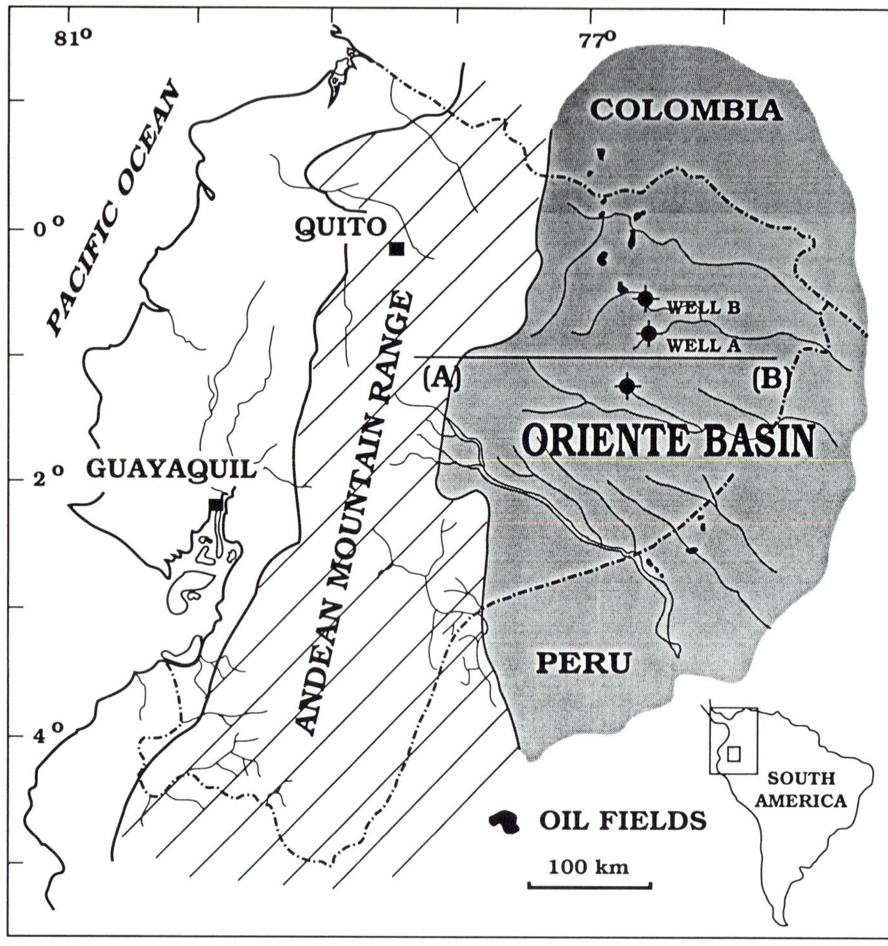

Fig. 1. Location map of the Oriente basin, Ecuador. (After Lawrence 1989). Line of section *A–B*, refers to Fig. 3

Mountain chain stretching from Venezuela to Argentina. It covers about 75 000 km^2 in eastern Ecuador, extending between the Andes on the west and the Guayana shield on the east (Fig. 1). Its sedimentary succession is over 10 000 m thick and ranges in age from the Paleozoic to the Holocene (Dashwood and Abbotts 1990; Fig. 2).

The Napo Formation lies conformably over deposits of the Aptian-Albian Hollin Formation, the latter mostly deposited in a braided fluvial system (Figs. 2 and 3; Canfield et al. 1982). It is composed of shales, limestones, and sandstones deposited in a shallow sea which transgressed over the basin from the west (Feininger 1975). It ranges in age from Albian to Campanian, achieving a maximum thickness of about 900 m in the southwest area and thinning to the northeast. The organic-rich black shales, calcareous mudstones, and marls of the Napo Formation were deposited during a time of maximum sea level rise in the mid-Cretaceous (Figs. 2 and 4–7).

Source Rock Characteristics, Stratigraphy and Paleoenvironments

Detailed geochemical and micropaleontological analyses were carried out on 740 rock samples from the Napo Formatio (40 outcrops and about 700 ditch-cuttings of 4 well sections; Table 1 and Figs. 4–7), including source rock characterization, oil-oil, and oil-source rock correlation. These results have confirmed, as reported before (e.g., Dashwood and Abbotts 1990), that all the hydrocarbons discovered so far in the Oriente basin were sourced by the shales and calcareous mudstones of the Napo Formation.

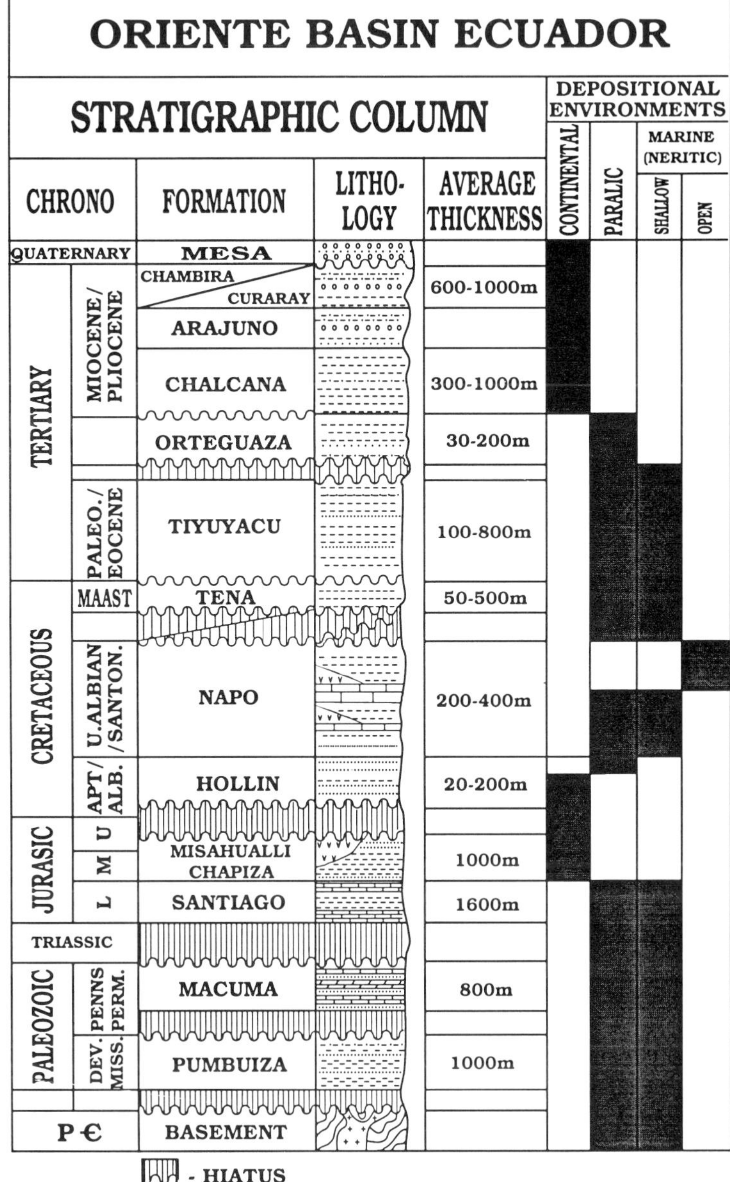

Fig. 2. Schematic stratigraphic column of the Oriente basin, Ecuador, showing the paleoenvironmental evolutionary phases

The source rock characteristics of the Napo Formation vary depending on both geographic location in the basin (east or west) and stratigraphic position (base, middle or top of the section; Figs. 2–7). The most organic-rich layers recovered so far in the studied area were deposited in the western part of the Oriente basin in a middle neritic environment, during a sea level maximum at the Cenomanian-Turonian boundary (Figs. 4 and 5).

Organic-rich rocks, coeval and similar to those of the Napo Formation, are also recorded in the stratigraphic sequences of other sub-Andean basins of South America, such as in the La Luna (Venezuela), Villeta and Gacheta (Colombia), and Chonta (Peru) Formations (Zumberge 1984; Cassani 1986; Rivadeneira 1986; Talukdar et al. 1986; Mello et al. 1989, 1990). They all correspond to a time of globally warm equable climate associated with a tectono-eustatic sea level rise recorded in the latest Cenomanian-earliest Turonian and to the worldwide Oceanic Anoxic Event 2 (OAE-2), of Schlanger and Jenkyns (1976) and Jenkyns (1980).

The planktonic microfossil content of the Cenomanian-Turonian sequence is typically composed of abundant shallow-dwelling foraminiferids (*Heterohelix globulosa*, *H. moremani*, and *H. reussi*) and

W (A)
E (B)
1000
0 (m)
– 1000
– 2000
– 3000
– 4000
– 5000
Undifferentiated Mio Pliocene
Orteguaza Fm.
Tiyuyacu Fm.
Tena Fm.
Napo Fm.
Misahualli Fm.
Undifferentiated Pre-Aptian
Misahualli Fm.
Hollin Fm.
Macuma and Pumbuiza Fm.
Macuma and Pumbuiza Fm.
Hollin Fm.
Chapiza, Sacha, Macuma and Pumbuiza Fm.

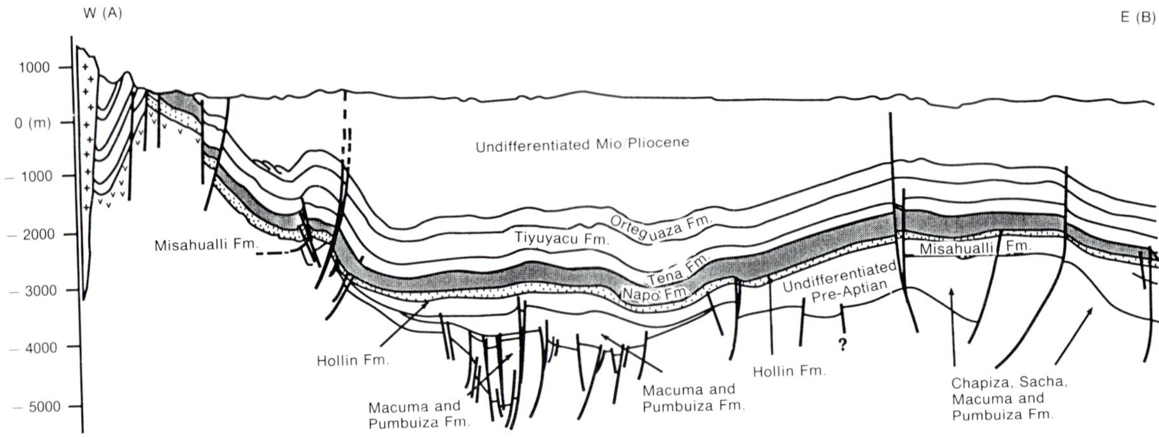

Fig. 3. Schematic structural cross section across the Andean mountain ridge and the Oriente basin, Ecuador. (After Rivadeneira and Sanchez 1989). See location in Fig. 1

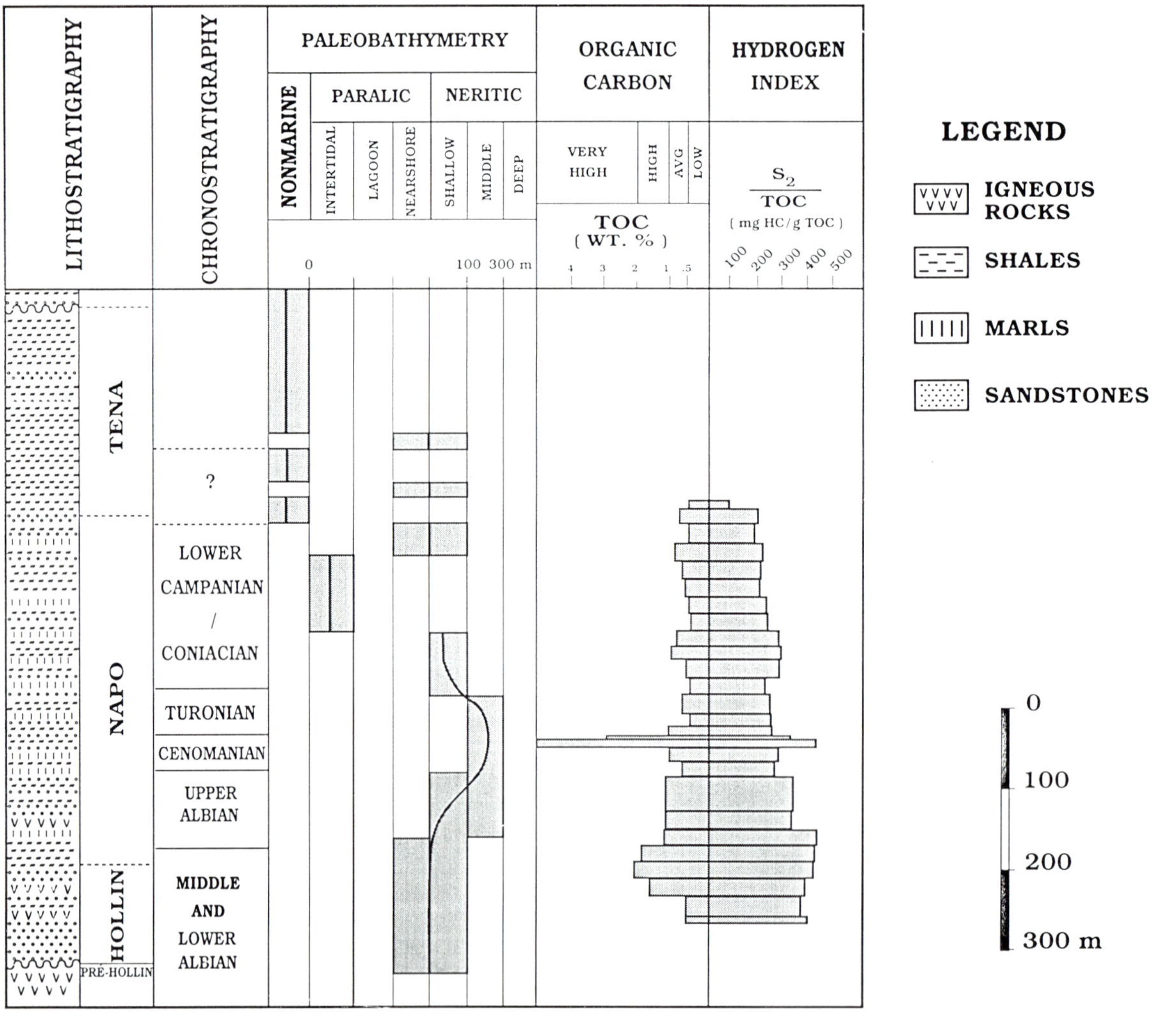

LITHOSTRATIGRAPHY
CHRONOSTRATIGRAPHY
PALEOBATHYMETRY
NONMARINE
PARALIC
NERITIC
INTERTIDAL
LAGOON
NEARSHORE
SHALLOW
MIDDLE
DEEP
0
100 300 m
ORGANIC CARBON
VERY HIGH
HIGH
AVG
LOW
TOC
(WT. %)
4 3 2 1 .5
HYDROGEN INDEX
S_2 / TOC
(mg HC/g TOC)
100 200 300 400 500
TENA
NAPO
HOLLIN
PRÉ-HOLLIN
?
LOWER CAMPANIAN / CONIACIAN
TURONIAN
CENOMANIAN
UPPER ALBIAN
MIDDLE AND LOWER ALBIAN
LEGEND
IGNEOUS ROCKS
SHALES
MARLS
SANDSTONES
0
100
200
300 m

Fig. 4. Chrono- and lithostratigraphic section of a typical well from the central part of the Oriente basin, Ecuador, showing the paleoenvironmental evolution and geochemical data (TOC, HI)

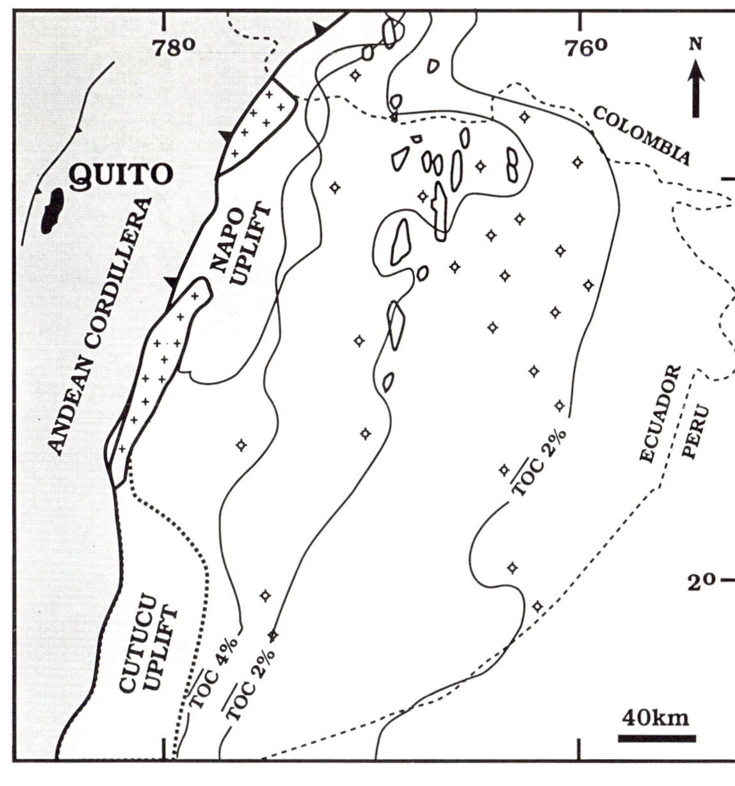

Fig. 5. Map of the areal distribution of average total organic carbon in the rocks of the Napo Formation, Oriente basin, Ecuador. (After Dashwood and Abbotts 1990)

rare, keeled, deeper-dwelling species (*Praeglobotruncana* cf. *P. praehelvetica*), together with low-diversity assemblages of dinoflagellates and calcareous nannofossils, all indicating well-oxygenated epipelagic surface waters (e.g., Koutsoukos et al. 1991a,b). On the other hand, the benthonic microfauna is highly impoverished, represented by rare agglutinated foraminiferids (*Ammomarginulina* cf. *A. colombianus*) and ostracods (*Brachycythere*? *sp.*), indicating dysaerobic to quasi-anaerobic bottom water conditions (Koutsoukos et al. 1991b). Levels with no benthonic microfauna, corresponding to the highest TOC levels, represent deposition under truly anoxic bottom water conditions. The rocks deposited under such conditions are mainly composed of well-laminated, calcareous, dark gray shales intercalated with layers of light gray shales. In general, they show medium to very high organic carbon content (up to 13 wt.%; Table 1 and Figs. 4–6), with a modal value of approximately 3 wt. %. As shown in Fig. 5, the data suggest that the richest facies were deposited primarily toward the west of the present Oriente basin. The Rock-Eval pyrolysis data show medium to excellent hydrocarbon source potential (up to 78 kg HC/t of rock), arising mainly from type II kerogen (hydrogen indices up to 740 mg HC/g of organic carbon; Table 1 and Figs. 4, 6, and 7). Most of the samples display average potential-yield values around 10 kg HC/t of rock and hydrogen indices greater than 300 mg HC/g of organic carbon (Fig. 6, Table 1). These rocks, when submitted to adequate thermal evolution conditions, generate oils classified as paraffinic-naphthenic crudes (Tissot and Welte 1984).

Petrographic analyses show an organic facies mostly composed of predominantly amorphous, sapropelic kerogen (ca. 95%), with a pale yellow fluorescence. In this facies, liptinitic and woody materials are minor contributors. The liptinite is represented mainly by cutinite with a bright yellow fluorescence, followed by sporinite with pale orange color under UV light (Mello et al. 1989, 1990).

The low thermal-evolution data obtained for all sediments studied (spore color index, vitrinite reflectance and T_{max}, e.g., Table 1 and Figs. 6–7) support the early observations of Dashwood and Abbotts (1990) that no well drilled so far in the Oriente basin has recorded fully mature (peak generation) rocks of the Napo Formation, excluding some anomalous wells associated with local igneous intrusions (Espin 1980). Indeed, the low vitrinite

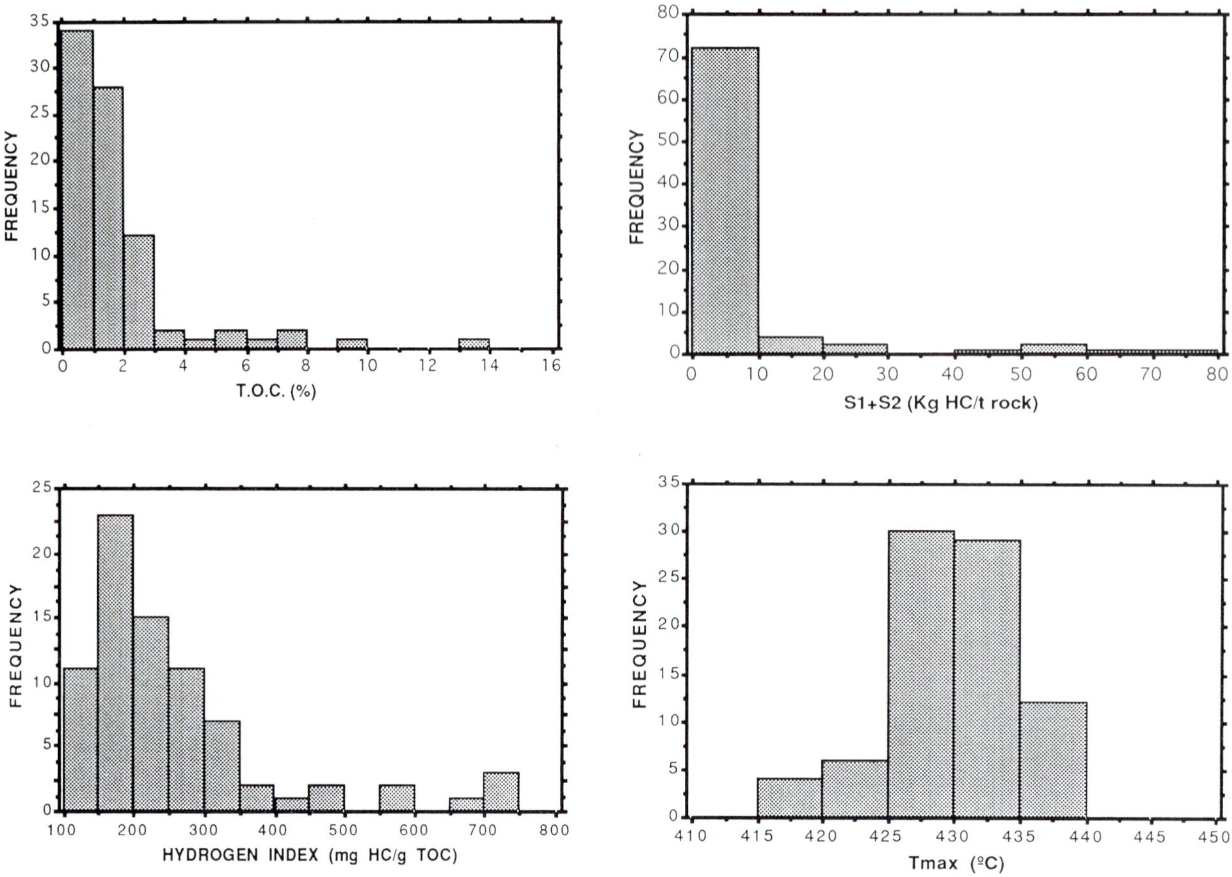

Fig. 6. Distribution of TOC (total organic carbon, weight %) and Rock Eval $S_1 + S_2$ (hydrocarbon source potential; kg HC/t rock), hydrogen index (mg HC/g TOC), and T_{max} (°C) values for the Napo sequence in the Oriente basin

reflectance values observed in the well B indicates that the Napo sedimentary section buried around 3300–3500 m, in the Oriente basin, is far from the onset of the oil window (T_{max} up to 438 °C and Ro up to 0.55%; as example see Table 1 and Figs. 6 and 7).

Since most of the Napo Formation sedimentary section in the Oriente basin is found in depth of less than 4000 m, it is likely that the Napo Formation is immature or in the onset of oil generation in most of the present Oriente basin. However, Dashwood and Abbotts (1990) showed several maps with data that suggest that areas to the southeast and northwest of the present day Oriente basin were the likely source kitchens (spent source rocks) for the oils discovered up to now in Ecuador (see also Del Solar 1982; Rivadeneira 1986).

In general, the Cenomanian-Turonian samples from the Napo Formation have medium to rich amounts of extractable organic matter (from 300 to 10 000 ppm of bitumen). As expected for thermally immature rocks, an examination of the relative abundance of saturated and aromatic hydrocarbons versus NSO compounds reveals that the NSO compounds dominate (saturates ranging from 10 to 48% of the total extractable material; Table 1).

The wide range of extractable organic material, hydrogen index (ranging from 107 to 740 mg HC/t. rock) and carbon isotope data for the total extract from the samples analyzed (ranging from − 25.0 to − 28.0‰; Table 1, Figs. 6 and 7) suggest that organic facies changes occurred during the deposition of the Napo sedimentary section. Indeed, a large number of papers have been published corroborating such an interpretation (Mello et al. 1989, 1990; Rivadeneira and Sanchez 1989; Dashwood and Abbotts 1990). It is important to note that most of the samples from the Cenomanian-Turonian sequence show carbon isotope values lighter than − 27.0‰ (Table 1).

The gas chromatography and GC-MS data for the rock extract form well A (3352 m depth;

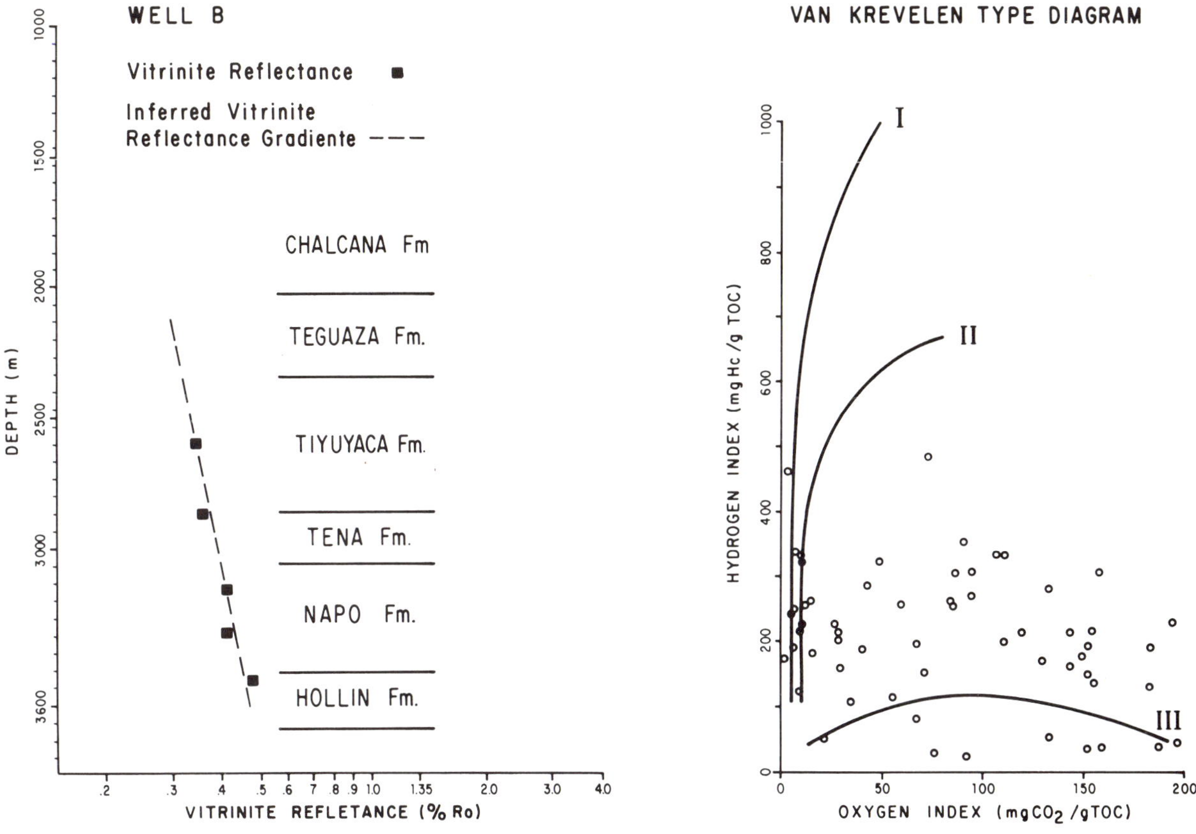

Fig. 7. Van Krevelen type diagram and vitrinite reflectance versus depth plot for well B samples (see Fig. 1 for location)

Table 1. Geochemical data for rock samples of the Napo formation

Rock samples	Napo Fm. Cen./Tur.	Napo Fm. Albian	Sample well A	Sample well B
Depth (m)	—	—	3352	3311
$CaCO_3$ (%)	< 27	< 13	13	0.8
TOC (%)	< 13	0.4–4.0	4.4	2.9
S_2 (kg HC/t ROCK)	< 78	< 10	13	7.3
HI (mg HC/g TOC)	< 740	< 350	332	249
Saturates (%)	10 to 48	10–35	32	21
$\delta^{13}C$ PDB (‰)	− 25 to − 27	− 25 to − 28	− 26.3	− 26.0
Ro (%)	0.35–0.55	0.35–0.55	0.45	0.47
$C_{29}\alpha\beta\beta/(\alpha\beta\beta + \alpha\alpha\alpha)$	0.35–0.46	0.37–0.49	0.38	0.49
$C_{29}20S/(20S + 20R)$	0.29–0.36	0.31–0.40	0.32	0.39
T_{max} (°C)	420–438	428–438	431	429

a Cenomanian-Turonian cutting sample; Figs. 1, 8 and 9; Table 2), taken as representative of the organic-rich facies of the Napo Formation, illustrate a number of geochemical and biological marker features, which characterize the Cenomanian-Turonian facies of the Napo Formation. These include: predominance of low molecular weight *n*-alkanes (around C_{15}-C_{17}); dominance of phytane over pristane (Pr/Ph < 1), low abundance of diasteranes, low hopane over sterane ratios (less than 4), high relative abundance of low molecular weight steranes, dominance of C_{27} regular steranes over their C_{28} and C_{29} counterparts, Ts/Tm ratios typically less than 1, prominence of

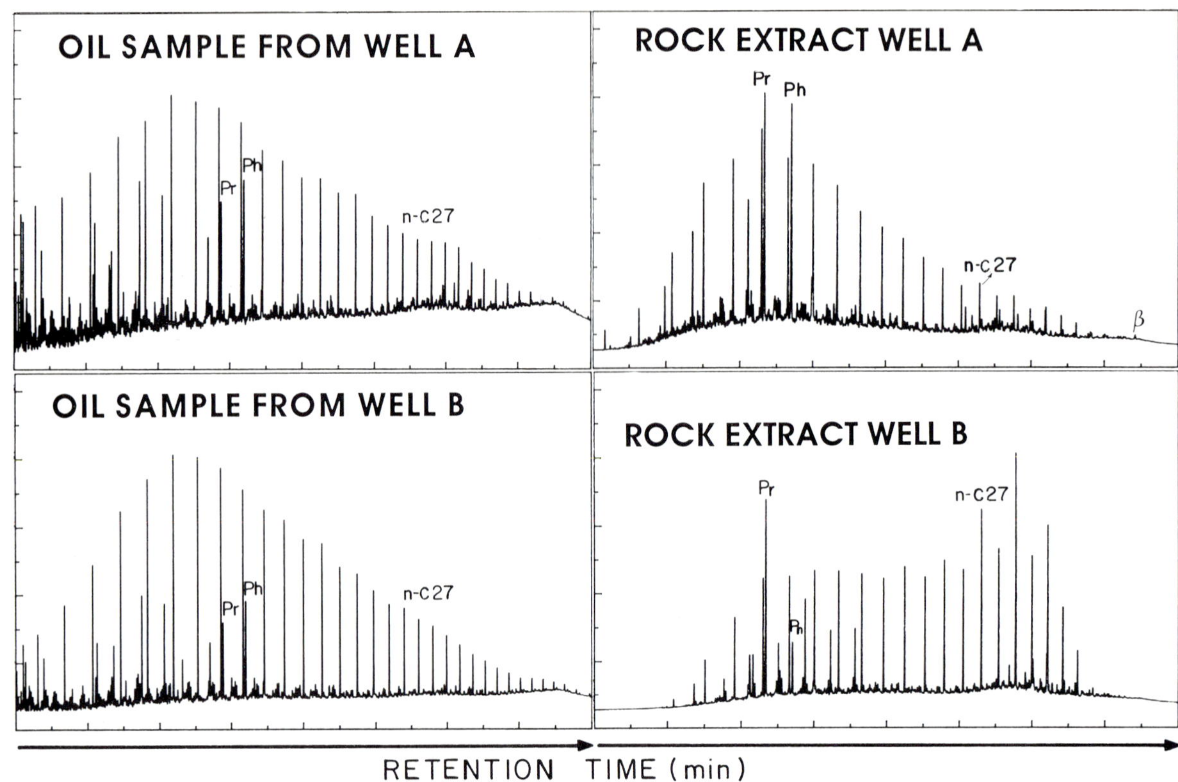

Fig. 8. Gas chromatograms of total alkanes of two typical oils recovered from Napo basal and Hollin reservoirs and two rock extracts representing: calcareous shale from the Cenomanian-Turonian sequence of the Napo Formation (extract *A*); and a dark gray shale from the upper Albian sequence of the Napo formation, Oriente basin, Ecuador (extract *B*); for peak assignments see Appendix 1

tricyclic components ranging from C_{19} to C_{39} (peaks 46 and 47 in Fig. 9). Other important features to note are the presence of gammacerane and β-carotane and, in most samples, high relative abundance of C_{35} hopanes relative to their C_{34} counterparts, and the presence, in high abundances, of C_{30} steranes and dinosteranes (originated from chrysophytacean algae and marine dinoflagellates respectively; e.g., Summons et al. 1987; Moldowan et al. 1990). Most of these features have been extensively reported as characteristic of rock extracts of Cenomanian-Turonian organic-rich facies in most of the sub-Andean basins, and have been observed previously in rocks deposited under marine dysoxic-anoxic carbonate dominated, predominantly algal, environments, with low influx of terrestrial plant debris (Palacas et al. 1984; Zumberge 1984; Cassani 1986; Cassani and Eglinton 1986; Connan et al. 1986; Talukdar et al. 1986; Mello et al. 1988a, 1989; Koutsoukos et al. 1991a).

In rock extract A (Cenomanian-Turonian), the presence of pristane $<$ phytane, gammacerane (peak 40 in Fig. 9) and β-carotane (peak β in Fig. 8) suggests an enhanced salinity (compared with normal marine) of the bottom water column in the depositional environment (e.g., Connan et al. 1986; Fu Jiamo et al. 1986; ten Haven et al. 1987, 1988; Mello 1988; Mello et al. 1988b), and suggest that salinity stratification might have played an important role in the preservation of the organic matter of the Napo Formation, for the Cenomanian-Turonian sequence.

The organic-rich layer recorded in the upper Albian interval of the Napo Formation (Figs. 2, 4), shows planktonic microfossil composed of rare foraminiferids (*Hedbergella trocoidea*, *Heterohelix globulosa*, and *H. moremani*), sparse and low-diversity assemblages of dinoflagellates and calcareous nannofossils. This impoverished biota suggests shallower and nearshore depositional environments in comparison to those present during the Cenomanian-Turonian. The benthonic microfauna is also poor, although slightly more diversified than the one recovered in the Cenomanian-Turonian strata (see above), and represented by rare calcareous (*Gavelinella berthelini-reussi* plexus) and

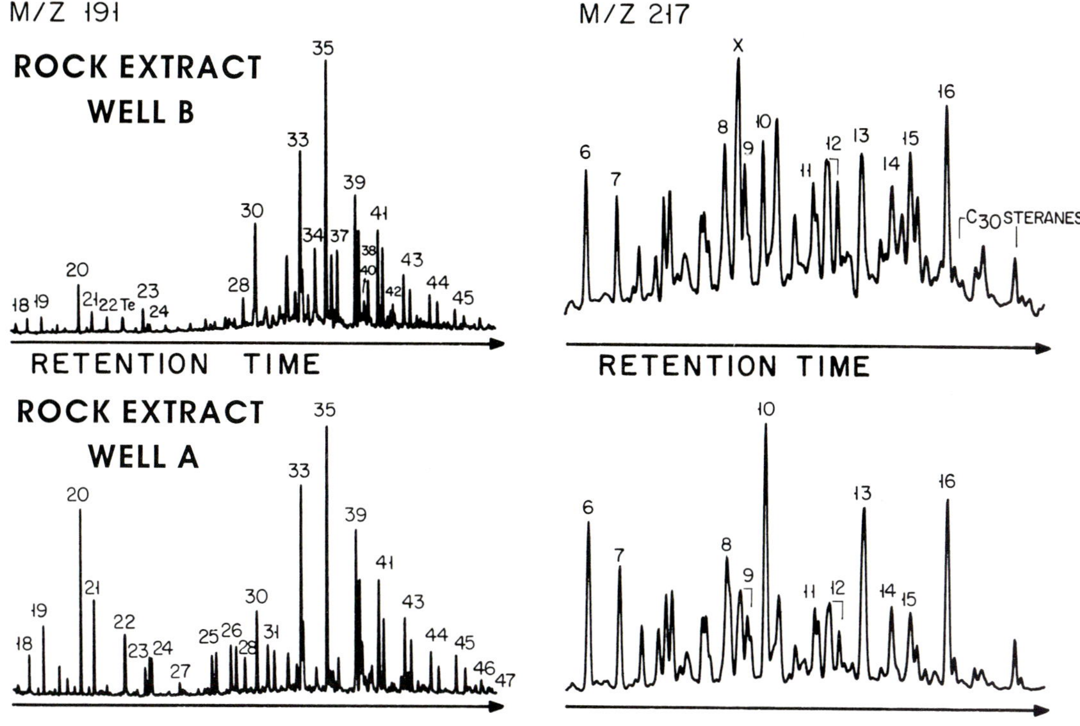

Fig. 9. Partial m/z 191 (terpanes) and 217 (steranes) chromatograms for the same rock extracts shown in Fig. 8; for peak assignments see Appendix 1

Table 2. Geochemical data for oil samples from oriente basin

Parameters	Oil samples	Oil/well A	Oil/well B
Reservoir	Napo and Hollin	Napo Basal	Hollin
API (°)	10–25	20	25
S (%)	1.4–3.0	1.9	2.2
δ^{13}C PDB (‰)	− 26.3 to − 26.8	− 26.8	− 26.7
Pr/Ph	0.7–1.2	0.86	0.79
Saturates (%)	10–50	32	40
Aromatics + NSO (%)	50–90	68	60
T_s/T_m	< 1	0.8	0.72
$C_{35}\alpha\beta/C_{34}$ hopanes	> 1	1.2	1.1
$C_{29}\alpha\beta\beta/(\alpha\beta\beta + \alpha\alpha\alpha)$	0.50–0.59	0.54	0.58
C_{29} 20S/(20S + 20R)	0.41–0.46	0.42	0.44

agglutinated (*Asanospira sp.*) foraminiferids and ostracods (*Cytheropteron* ex gr. *howelli*, *Cythereis sp. A*, *Cytherella*? *sp. A. Veenia*? *sp. A*), indicating dysaerobic bottom water conditions (Koutsoukos et al. 1991a,b). These rocks are mainly composed of dark gray shales with an organic carbon content ranging from 0.4 to 4.0 wt. % (Table 1, Fig. 4). The Rock-Eval data show medium hydrocarbon source potential (up to 10 kg HC/t of rock), mainly arising from type II/III kerogen (hydrogen indices up to 350 mg HC/g of organic carbon; Table 1, Fig. 4). The petrographical analyses show an organic facies mostly composed of mixed organic matter with dominance of liptinitic and humic macerals (vitrinite and inertinite) over the amorphous ones, suggesting higher rates of runoff with consequent increased influx of land-plant debris (Table 1, Fig. 4; Mello et al. 1989, 1990). Therefore, it may be possible that the high degree of organic-matter preservation in this interval of the Napo Formation is mostly due to a higher sedimentation rate and rapid burial, during a time (in the late Albian) of more humid climate with stronger runoff, rather than to truly widespread dysoxic-anoxic bottom water conditions.

A close examination of the GC and GC-MS data of the organic extract from well B (3311 m depth), which represents a typical example of a rock extract from the upper Albian organic-rich facies (Figs. 1, 8, 9), shows a number of biological marker features typical of rocks deposited in a shallow marine and nearshore environment (e.g., a coastal plain) with very high terrestrial plant input. These include (Table 2, Fig. 9): pristane/phytane much greater than 1, predominance of high molecular weight *n*-alkanes (C_{25} to C_{29}) with odd/even preference, dominance of C_{29} regular steranes over their C_{27} and C_{28} counterparts, very high relative abundances of diasteranes (peaks 6, 7, and X; Fig. 9), and presence of the tetracyclic terpanes des-A and des-E (peaks Te and 24, respectively, in Fig. 9). Most of these biological marker features have been considered diagnostic of higher plant input (e.g., McKirdy et al. 1984; Tissot and Welte 1984; Philp and Gilbert 1986; Seifert and Moldowan 1986; Abdullah et al. 1988; Mello 1988; Mello et al. 1983a).

Overall, the differences in the biological marker features between the two organic-rich facies of the Napo formation reflect the different depositional regimes in which the rocks were accumulated:

1. a shallow neritic nearshore environment in the upper Albian, with strong terrestrial plant input and rapid burial and preservation in probable dysoxic bottom water conditions;
2. a middle to deep neritic environment in the Cenomanian-Turonian, with high primary algal productivity and preservation in dysoxic-anoxic bottom water conditions.

The assessment of the thermal evolution, based on the biomarkers obtained from the rock extracts from the wells A (3352 m) and B (3311 m), indicates that the Napo sedimentary section buried around 3300–3500 m, in the Oriente basin, is far from the onset of the oil window (Figs. 8 and 9, Table 1). These data are in agreement with the Rock Eval and petrographic data shown in Table 1 and Fig. 6. This conclusion is based on biological marker maturity parameters (high values of nC_{17}/Pr and nC_{18}/Ph (Fig. 8), and dominance of steranes with $\alpha\alpha\alpha$ and 20R configurations (C_{29} $\alpha\beta\beta/(\alpha\beta\beta + \alpha\alpha\alpha) < 0.49$ and C_{29} $\alpha\alpha\alpha$ 20S(20S + 20R) < 0.39; see Table 1 and Fig. 9; e.g., Tissot and Welte 1984; Seifert and Moldowan 1986; Mello 1988).

Hydrocarbon Characteristics and Oil-Source Rock Correlation

Twenty oil samples comprising crude oils and impregnated oils from Napo sandstone reservoirs (oil wells and seeps) were analyzed (Table 2, Figs. 8, 10). The oils attributed to the Napo Formation usually display API gravity values from about 10° to 35° (Table 2; e.g., Dashwood and Abbots 1990). The geochemical features of the oils analyzed in this study include low to medium API gravity values (ranging from 11° to 25°), medium to high sulphur content (1.4 to 3.0%), and whole oil carbon isotopes varying within a very narrow range (from −26.3‰. to −26.8‰.; Table 2). The oils are typically dominated by aromatic and NSO compounds, with saturated contents commonly accounting for $< 50\%$ of the C_{12+} fraction (Table 2). The integration of these data, together with the lack of low molecular weight in the GC trace of the oil from well B (Fig. 8), suggests that most of the crude oils analyzed (severely biodegraded oils are not shown here) were partially/totally subjected to biodegradation processes, due to percolation of meteoric waters in the reservoirs.

As can be observed in the m/z 191 (triterpanes) and m/z 217 (steranes) mass chromatograms of two oils, taken as representative of all the samples analyzed (Fig. 10), there are a number of molecular features that characterize oils from the Oriente basin, suggested to be sourced from the Cenomanian-Turonian sequence of the Napo Formation (see below). Noteworthy are the dominance of C_{27} regular steranes over their C_{28} and C_{29} counterparts, (Table 2; see m/z 217 in Fig. 10), low hopanes/ steranes ratio (less than 2), $T_s/T_m < 1$, presence of 28, 30-bisnorhopane, high abundance of tricyclic terpanes up to C_{39} relative to pentacyclics, and dominance of C_{35} hopanes over their C_{34} homologues. In addition, the oil presents C_{30} steranes and dinosteranes (Fig. 11) in significant abundances (originated from Chrysophyta algae and marine dinoflagellates respectively; e.g., Summons et al. 1987; Moldowan et al. 1990). The biological marker features, observed in the analyzed

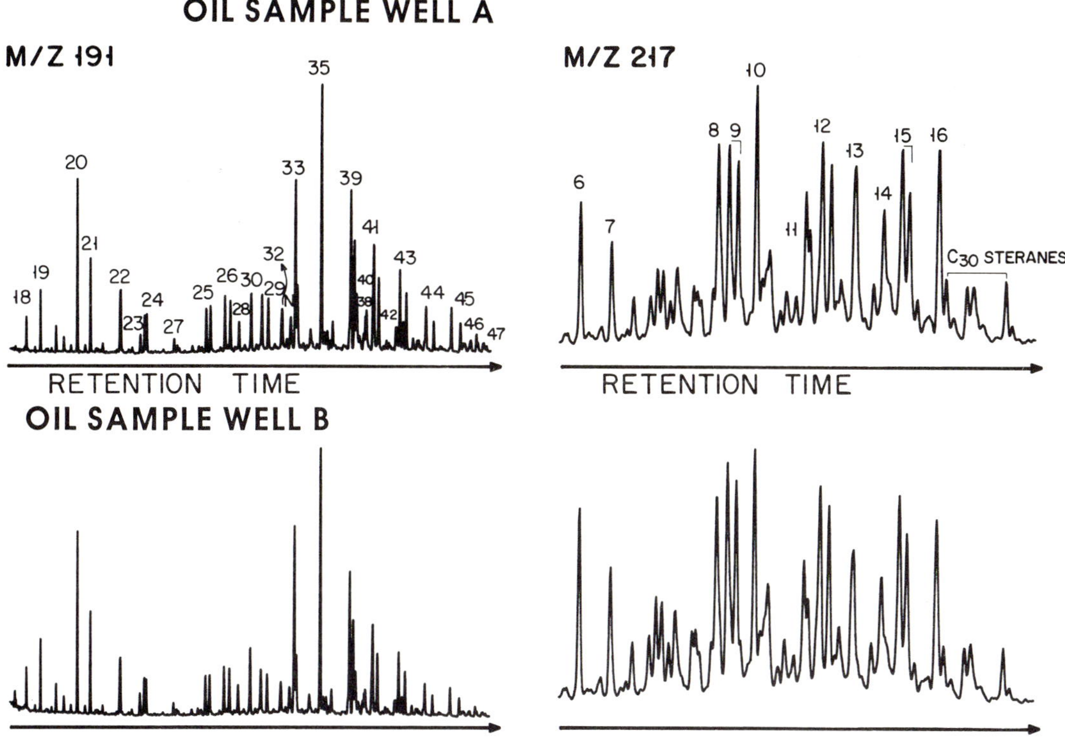

Fig. 10. Partial m/z 191 (terpanes) and 217 (steranes) chromatograms for the same oil samples shown in Fig. 8; for peak assignments see Appendix 1

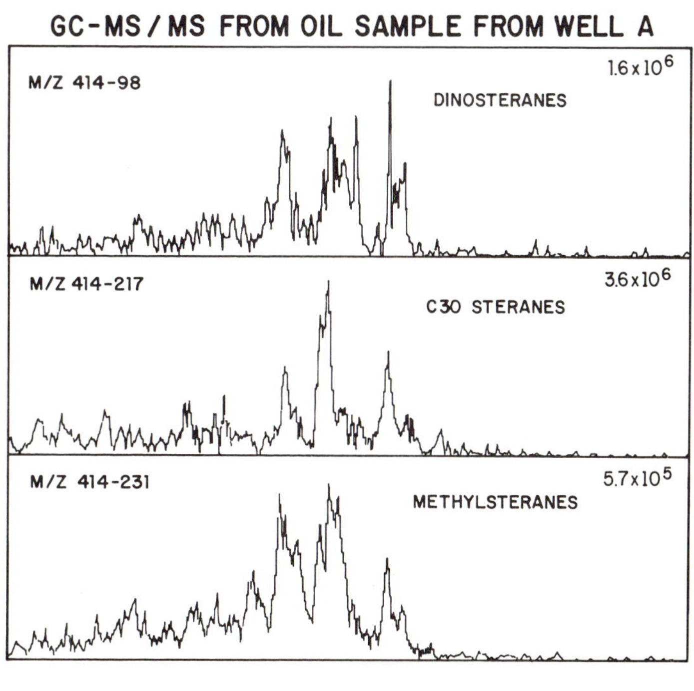

Fig. 11. GC-MS/MS data showing the C_{30} steranes (24-*n*-propylcholestanes (m/z 414-217), dinosteranes (m/z 414-98) and methylsteranes (m/z 414-231) distributions for oil sample A from Fig. 8

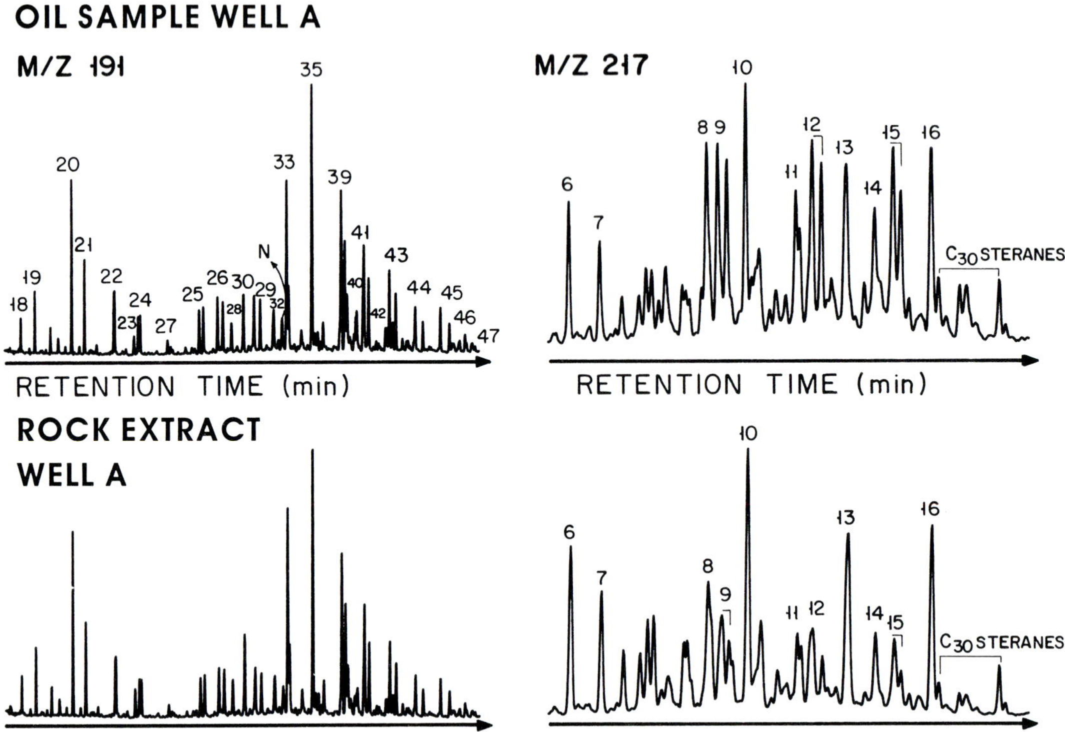

Fig. 12. Partial m/z 191 (terpanes) and 217 (steranes) chromatograms showing the oil-source rock correlation between the oil sample A and rock extract A from Fig. 8; for peak assignments see Appendix 1

samples, have been reported from marine carbonate oils derived from the La Luna and the Querecual Formations (Venezuela), Magdalena Fm. (Colombia), Sunniland Fm. (South Florida), Chonta Fm. (Peru), and Albian-Cenomanian carbonates from the Brazilian marginal basins (Palacas et al. 1984; Zumberge 1984; Cassani 1986; Talukdar et al. 1986; Mello et al. 1988b, 1989).

The presence of demethylated hopanes in the m/z 191 (peak N in Fig. 10) mass chromatogram of the oils analyzed indicates that they have undergone severe biodegradation, with the removal of all the straight chain paraffins and isoprenoids (cf. Connan et al. 1980).

Demethylated hopanes are believed to appear only after the complete loss of all straight chain compounds and isoprenoids (Connan et al. 1980; Cassani 1986). However, the presence of straight chain paraffins and isoprenoids, in these oils, together with the demethylated hopanes, suggests that there is a coexistence of nonbiodegraded and biodegraded oils in the reservoirs. Such a situation could only result form a complex generation-migration history, suggesting more than one pulse of oil migration into the reservoirs. The oil accumulation due to the first migration pulse underwent severe biodegradation with the removal of all the straight chain alkanes and isoprenoids, and formation of demethylated hopanes. After degradation ceased, other migration pulses occurred, mixing the biodegraded with the nonbiodegraded oils in the reservoirs.

Table 2 and Fig. 10 show steranes (C_{29} $\alpha\alpha\alpha$ 20S/(20S + 20R) ranging from 0.42 to 0.44 and C_{29} $\alpha\beta\beta/\alpha\beta\beta + \alpha\alpha\alpha$ (20R + 20S) ranging from 0.54 to 0.58) parameters used to assess the thermal evolution degree of the oils. These ratios suggest a moderate degree of thermal evolution (e.g., Seifert and Moldowan 1986; Mello 1988; Mello et al. 1989, 1990), indicating that all the oils studied were sourced just around the peak generation stage (Mello 1988).

Figures 8 and 12 illustrate the attempt of oil-source rock correlation using gas chromatograms, m/z 191, and m/z 217 mass chromatograms. As can be observed, there is a good match for the biological marker features arising from the oil analyzed (sample from well A) and the rock extract from the

Napo Formation at 3352.8 m depth (well A), although the rock sample is immature. On the other hand, there is no correlation whatsoever between the samples representing the paralic-neritic organic-rich facies (e.g., sample B in Figs. 8 and 9) and the oil samples shown in Figs. 8, 10 and 11. For experimental and analytical procedure see Appendix 2.

Summary

The organic-rich succession of the Napo Formation, ranging in age from Albian to early Campanian, is mainly composed of calcareous black-shales and marlstones with up to 13 wt.% of TOC. The Rock-Eval pyrolysis data indicate high hydrocarbon source-rock potential (S_2 up to 78 kg HC/t of rock), largely arising by the presence of type II kerogen (HI up to 740 mg HC/g TOC).

The highest organic-rich layers were deposited during a maximum of sea level at the Cenomanian-Turonian boundary, with salinity-stratified bottom water conditions. These rocks are characterized by a scarce and low-diversity benthonic microfauna, represented by few gavelinellids and agglutinated foraminiferida taxa, with an abundant but low-diversity planktonic assemblage mostly composed of heterohelicids. Such microfauna suggest widespread dysoxic-anoxic bottom water conditions with well-oxygenated shallow epipelagic waters.

Biological marker data of the oils discovered up to now in the Oriente basin, when compared with the organic extracts from the organic-rich facies of the Napo Fm. (Cenomanian-Turonian) show a good match, therefore confirming the oils' origin. The molecular features typical of these oils and the source rock extracts are the predominance of low molecular weight *n*-alkanes, Pr/Ph < 1, high concentrations of steranes with dominance of C_{27} compounds, $\alpha\beta$-hopanes, tricyclic terpanes up to C_{39}, low hopane/sterane ratios (< 2), Ts $<$ Tm, the presence of 28, 30-bisnorhopane, C_{30} steranes and dinosteranes, and an abundance of C_{35} hopanes over their C_{34} counterparts.

Organic-rich rocks, coeval and similar to those of the Napo Formation, are also recorded in the stratigraphic sequences along most of the sub-Andean basins of South America. They all correspond to a time of global warm equable climate associated with a tectonic-eustatic sea-level rise recorded in the latest Cenomanian-earliest Turonian, associated with the worldwide Oceanic Anoxic Event 2 (OAE-2).

Acknowledgments. The authors are indebted to the Geochemistry Group of Petrobrás for all geochemical data provided and to C.E. Souza Cruz and M.C. Barros for providing oil and rock samples from the Oriente basin. The authors are also grateful to Dr. Geoffrey B. Newton, Barry J. Katz, Luis A.F. Trindade, and Peter Szatmari (Petrobrás-Cenpes) for their helpful comments and suggestions during the review of this chapter. We also thank Petrobrás for permission to publish.

Appendix 1. Biomarker identification chart

Pr-2,6,10,14-tetramethylpentadecane (pristane)
Ph-2,6,10,14-tetramethylhexadecane (phytane)
β-β-Carotane.
6-13β(H),17α(H)-diacholestane,20S (C_{27}-diasterane)
7-13β(H),17α(H)-diacholestane,20R (C_{27}-diasterane)
8-5α(H),14α(H),17α(H),20S (C_{27}-cholestane)
9-5α(H),14β(H),17β(H),20R + 20S (C_{27}-cholestane)
10-5α(H),14α,17α(H),20R (C_{27}-cholestane)
11-5α(H),14α(H),17β(H),20S (C_{28}-methylcholestane)
12-5α(H),14β(H),17β(H),20R + 20S (C_{28}-methylcholestane)
13-5α(H),14α(H),17α(H),20R (C_{28}-methylcholestane)
14-5α(H),14α(H),17α(H),20S (C_{29}-ethylcholestane)
15-5α(H),14β(H),17β(H),20R + 20S (C_{29}-ethylcholestane)
16-5α(H),14α(H),17α(H),20R (C_{29}-ethylcolestane)
17-C_{19} tricyclic terpane
18-C_{20} tricyclic terpane
19-C_{21} tricyclic terpane
20-C_{23} tricyclic terpane
21-C_{24} tricyclic terpane
22-C_{25} tricyclic terpane
23-C_{26} tricyclic terpane
24-C_{24} tetracyclic (Des-E)
Te-C_{24} tetracyclic (DES-A)
25-C_{28} tricyclic terpanes
26-C_{29} tricyclic terpanes
27-C_{25} tetracyclic
28-C_{27} 18α(H)-trisnorneohopane (T_s)

29-C_{30} tricyclic terpanes
T-C_{27},25,28,30-trisnorhopane
30-C_{27}17α(H)-trisnorhopane (T_m)
32-17α(H),18α(H),21β(H)-28,30-bisnorhopane (C_{28})
N-25-norhopane (C_{29})
33-C_{29} 17α(H),21β(H)-norhopane
34-C_{29} 17β(H),21α(H)-norhopane
35-C_{30} 17α(H),21β(H)-hopane
37-C_{30} 17β(H),21α(H)-hopane
38-C_{34} tricyclic terpanes
39-C_{31} 17α(H),21β(H)-homohopane (22S + 22R)
40-C_{30} gammacerane
41-C_{32}17α(H),21β(H)-bishomohopane (22S + 22R)
42-C_{35} tricyclic terpanes
43-C_{33}17α(H),21β(H)-trishomohopane (22S + 22R)
44-C_{34}17α(H),21(H)-tetrakishomohopane (22S + 22R)
45-C_{35}17α(H),21β(H)-pentakishomohopane (22S + 22R)
46-C_{38} tricyclic terpanes
47-C_{39} tricyclic terpanes

Appendix 2. Experimental and analytical procedures

The oil and rock samples were submitted to bulk, elemental, liquid chromatography, and carbon isotope analysis according to routine methods (Mello 1988).

The organic carbon content was determined by combustion in a LECO carbon analyzer. A Rock-Eval with TOC was used in order to determine hydrocarbon potential yield of the rock, type of organic matter and maximum temperature of pyrolysis. Kerogen (insoluble organic matter) was separated according to the standard procedure used in palynology without oxidation step, and mounted on regular glass slides for further studies using transmitted and reflected light.

Bitumen was extracted in a Soxhlet apparatus by refluxing with dichloromethane. The organic extracts were fractionated via solid-liquid chromatography.

The GC analyses of alkanes were carried out using a Hewlett Packard 5890A equipped with a splitless injector and fitted with a 30-m DB-5 column. Hydrogen was employed as carrier gas with a temperature program of 40–80 °C at 8 °C/min and 80–320 °C at 4 °C/min. The GC-MS analyses were performed using a HP-5890 spectrometer coupled to a Hewlett Packard 5890A GC equipped with on column injector, and fitted with a 25-m SE-54 column. Helium was employed as a carrier gas with temperature program of 70–190 °C at 15 °C/min and 190 to 290 °C, at 2 °C/min. The spectrometer was operated in multiple ion detection (MID) mode, the GC-MS/MS technique was performed using a triple stage quadrupole Finnigan TSQ-70 instrument coupled to a HP-5890A gas chromatograph fitted with a DB-5 fused silica column. Helium was used as the carrier gas and the temperature program was 60–200 °C, at 15 °C/min and 200–300 °C/min at 2 °C/min.

The bitumen was oxidized to CO_2 in a continuous oxygen flow preparation line; carbon isotope ratios of bitumen were measured using a Delta E (Finnigan) mass spectrometer. The isotopic data were reported in the usual delta notation relative to the PDB standard.

References

Abdullah WH, Murchinson D, Jones JM, Telnaes N, Gjelberg J (1988) Lower Carboniferous coal depositional environments on Spitsbergen, Svalbard. In: Mattavelli L, Novelli L (eds) Advances in organic geochemistry, 1987. Pergamon, New York, pp 953–964

Canfield RW, Bonilla G, Robbins RK (1982) Sacha oil field of Ecuadorian Oriente. Am Assoc Pet Geol Bull 66: 1076–1090

Cassani F (1986) Organic geochemistry of extra-heavy crude oils from the Eastern Venezuelan Basin. PhD Thesis, Univ Bristol, 376 pp

Cassani F, Eglinton G (1986) Organic geochemistry of Venezuelan extra-heavy oils. 1. Pyrolysis of asphaltenes: a technique for the correlation and maturity evaluation of crude oils. Chem Geol 56:167–183

Connan J, Restle A, Albrecht P (1980) Biodegradation of crude oils in the Aquitaine basin. In: Douglas AG, Maxwell JR (eds) Advances in organic geochemistry, 1979. Pergamon Press, New York, pp 1–19

Connan J, Bouroullec J, Dessort D, and Albrecht P (1986) The microbial input in carbonate-anhydrite facies of a sabkha palaeoenvironment from Guatemala: a molecular approach. In: Leythaeuser D, and Rullkotter JW (eds) Advances in organic geochemistry, 1985. Pergamon Press, New York pp, 29–50

Dashwood MF, Abbotts IL (1990) Aspects of the petroleum geology of the Oriente basin, Ecuador. In: Brooks J (ed) Classic petroleum provinces. Geol Soc, London, Spec Publ 50: 89–118

Del Solar C (1982) Ocurrencia de hidrocarburos en la formacion Vivian, Cuenca Maranon, Nor-Oriente Peruano. In: Simposio Bolivariano. Exploracion Petrolera en las Cuencas Subandinas de Venezuela, Colombia, Ecuador y Peru. Asociacion Colombiana de Geologos y Geofisicos del Petroleo, Bogota

Espin P (1980) Detection of Cuerpos Igneos en el Cretacico del Oriente Ecuatoriano. In: II Congresso Ecuatoriano de Ingenieros, Geologos de Minas y Petroleos, Quito

Feininger T (1975) Origin of petroleum in the Oriente of Ecuador. Am Assoc Pet Geol Bull 59: 1166–1175

Fu Jiamo S, Guoying P, Pingan S, Brassell C, Eglinton G, Jigang J (1986) Peculiarities of salt lake sediments as potential source rock in China. In: Leythaeuser D and Rullkotter JW (eds) Advances in organic geochemistry, 1985. Pergamon Press, New York, pp 119–127

Jenkyns HC (1980) Cretaceous anoxic events: from continent to oceans. J Geol Soc Lond 137: 171–188

Koutsoukos, EAM, Mello MR, Azambuja Filho NCde, Hart MB, Maxwell JR (1991a) The Upper Aptian-Albian succession of the Sergipe Basin, Brazil: an integrated paleoenvironmental assessment. Am Assoc Pet Geol Bull 75: 479–498

Koutsoukos EAM, Mello MR, and Azambuja Filho NC (1991b) Micropalaeontological and geochemical evidence of mid-Cretaceous dysoxic-anoxic palaeoenvironments in the Sergipe Basin, northeastern Brazil. In: Tyson RV, Pearson TH (eds) Modern and ancient continental Shelf anoxia. Geol Soc, London, Spec Publ 58: 427–447

Lawrence RS (1989) Reginal variations in formation water salinity, Hollin and Napo Formations (Cretaceous), Oriente basin, Ecuador. Am Assoc Pet Geol Bull 73: 757–776

Lozada FT, Endara PB, Cordero CA (1985) Exploracíon y desarrollo del Campo Libertador. 2nd Simposio Bolivàriano de Exploracíon Petrolera en las Cuencas Subandinas Asoc Colomb Geol Geof Petr, Bogota pp 1–8

McKirdy DM, Kantsler AJ, Emmett JK, Aldridge AK (1984) Hydrocarbon genesis and organic facies in Cambrian carbonates of the Easter Officer Basin South Australia. In: Palacas JG (ed) Petroleum geochemistry and source rock potential of carbonate rocks. Am Assoc Petrol Geol, Tulsa, Stud Geol 18: 12–32

Mello MR (1988) Geochemical and molecular studies of the deposition environments of source rocks and their derived oils from the Brazilian marginal basins. PhD Thesis, Univ Bristol, 240 pp

Mello MR, Telnaes N, Gaglianone PC, Chicarelli MI, Brassell SC, Maxwell JR (1988a) Organic geochemical characterization of depositional palaeoenvironment of source rocks and oils in Brazilian margin basins. In: Mattavelli L, Novelli L (eds) Advances of organic geochemistry, 1987. Pergamon, New York, pp 31–45

Mello MR, Gaglianone PC, Brassell SC, Maxwell JR (1988b) Geochemical and biological marker assessment of depositional environment using Brazilian offshore oils. Mar Petrol Geol 5: 205–223

Mello MR, Soldan AL, Freitas LCS, Concha FJM, Triguis JA (1989) Geochemical evaluation of oil and rock samples from Oriente basin, Ecuador. Petrobrás Int Publ 1: 1–15

Mello MR, Soldan AL, Freitas LCS, Concha FJM, Triguis JA (1990) O habitat do petróleo na Bacia do Oriente, Equador. In: Resumés 2nd Latin Am Congr Org Geochem ALAGO, p 77

Moldowan JM, Seifert WK, Gallegos EJ (1985) Relationship between petroleum composition and depositional environment of petroleum source rocks. Am Assoc Pet Geol Bull 69: 1255–1268

Moldowan JM, Fago FJ, Lee CY, Jacobson SR, Watt DS, Slougui N, Jeganathan A, Young DC (1990) Sedimentary 24-n-propylcholestanes, molecular fossils diagnostic of marine algae. Science 247: 309–312

Palacas JG, Anders DE, King JD (1984) South Florida basin: A prime example of carbonate source rocks of petroleum. In: Palacas JG (ed) Petroleum geochemistry and source rock potential of carbonate rocks. Am Assoc Pet Geol, Tulsa, Stud Geol 18: 71–96

Philp RP, Gilbert TD (1986) Biomarker distributions in Australian oils predominantly derived from terrigenous source material. In: Leythaeuser D, Rullkotter JW (eds) Advances in organic geochemistry, 1985. Pergamon Press, New York, pp 73–84

Rivadeneira MV (1986) Evaluacion geochemica de rocas madres de la Cuenca Amazonica Ecuatoriana. 4th Congr Ecuatoriano Geol Min Petrol Col Ing Geol Min Petr Pinchindra, Quito, pp 31–48

Rivadeneira MV, Sanchez HF (1989) Consideraciones geologicas del preaptense en la Cuenca Oriente. 3th Simposio Bolivariano de Exploracion Petrolera en las Cuencas Subandinas Asoc Colomb Geol Geof Petr, Bogata pp 10–31

Schlanger SO, Jenkyns HC (1976) Cretaceous oceanic anoxic events: causes and consequences. Geol Mijnbouw 55: 179–184

Seifert WK, Moldowan JM (1986) Use of biological markers in petroleum exploration. In: Johns RB (ed) Biological markers in the sedimentary reocrd. Elsevier, Amsterdam pp, 261–290

Sofer Z (1984) Stable carbon isotope composition of crude oils: Application to source depositional environments and petroleum alteration. Am Assoc Pet Geol Bull 68: 31–49

Summons RE, Volkman JK, Boreham CJ (1987) Dinosterane and other steroidal hydrocarbons of dinoflagellate origin in sediments and petroleum. Geochim Cosmochim Acta 51: 3075–3082

Talukdar S, Gallango O, China-a-Lien M (1986) Generation and migration of hydrocarbons in the Maracaibo basin, Venezuela: An integrated basin study. In: Leythaeuser D, Rullkotter JW (eds) Advances in organic geochemistry, 1985. Pergamon Press, New York, pp 261–280

ten Haven HL, de Leeuw JW, Rullkotter JW, Sinninghe-Damste JS (1987) Can the pristane/phytane ratio be use as a palaeoenvironment indicator? Nature 330: 641–643

ten Haven HL, De Leeuw JW, Sinninghe-Damste JS, Schenck PA, Palmer SE, Zumberge JE (1988) Application of biological markers in the recognition of paleohypersaline environments. In: Fleet AJ, Kelts K, Talbot MR (eds) Lacustrine petroleum source rocks. Geol Soc, London, Spec Publ 40: 123–130

Tissot BP, Welte DH (1984) Petroleum formation and occurrence. Springer, Berlin, Heidelberg, New York, 699 pp

Zumberge JE (1984) Source rocks of the La Luna Formation (Upper Cretaceous) in the Middle Magdalena Valley, Colombia. In: Palacas JG (ed) Petroleum geochemistry and source rock potential of carbonate rocks. Am Assoc Petrol Geol, Tulsa, Stud Geol 18: 127–134

The Albian Kazhdumi Formation of the Dezful Embayment, Iran: One of the Most Efficient Petroleum Generating Systems

M.L. Bordenave[1] and R. Burwood[2]

Abstract

The Zagros orogenic belt of Iran is one of the most prolific petroleum provinces of the world. The Albian Kazhdumi source rock and two calcareous reservoirs of Early Miocene (Asmari) and the Cenomano-Turonian age (Bangestan) constitute by far the dominant petroleum system of the Dezful Embayment, being responsible for the accumulation of more than 7.3% of the world reserves in a surface area of only 40 000 km^2.

The Kazhdumi source rock accumulated in an intrashelf silled depression communicating with the South Tethys Ocean, under a humid equatorial climate and during a period of sea level rise. The influx of clastics provoked the building of rapidly prograding deltas, while eastwards, argillaceous material was deposited under prodelta conditions. Freshwater from rivers and deep oceanic countercurrents caused water stratification according to a 'positive water balance' mechanism, as described by Demaison and Moore (1980). Large amounts of nutrients were brought in by rivers, causing high biological activity in surface waters while euxinic conditions prevailed at depth. These conditions resulted in the deposition of up to 300 m of black low-energy marl and argillaceous limestone in the center of the depression. These sediments contain large amounts of algal type II organic matter, with TOC values as high as 11%, $S_1 + S_2$ up to 40 g HC/kg rock and source rock potential indices (SPI) in excess of 20 t/m^2.

The high productivity of the Kazhdumi source rock resulted from the coincidence of several favorable factors resulting from post-Albian geological history. Some of them, such as the existence of excellent reservoirs, the efficient caprock of the Gachsaran evaporites capping the Asmari, and large whale-back anticlines, have been known since the early days of petroleum exploration. Some were discovered more recently, i.e., the Zagros folds began to form at the end of the Early Miocene, whilst the Kazhdumi reached the oil window over most of the Dezful Embayment as the result of the deposition of the thick flysh-type Agha Jari Formation, in the Late Miocene to Pliocene. Secondary migration of hydrocarbons, essentially vertical, was guided by areas of drainage which were similar to those evidenced by present-day seismic. Moreover, oil/gas migration was enhanced by the fracturing of limestone and marls which resulted from the Zagros folding. Such fracturing is easily observed in outcropping breached anticlines.

The Kazhdumi origin of the oil of the Dezful Embayment main fields such as Agha Jari, Ahwaz, Bibi Hakimeh, Gachsaran, Mansuri, Marun, and Rag-e Safid was verified from stable isotope oil-to-oil and oil-to-source rock correlation and biomarker signatures as described in a former paper (Bordenave and Burwood 1990).

Introduction

Source rocks are not distributed randomly, but are preferentially located in specific geological settings, such as grabens, passive margins, intracratonic marine or lacustrine depressions, and deltas, where environmental conditions were favorable for the accumulation and preservation of organic matter. The example of the Zagros Foothills described in this chapter (Figs. 1 and 2) is one of the most efficient "oil generating systems", covering a 40000 km^2 intrashelf silled depression which has generated and emplaced 65 billion barrels of recoverable oil, i.e., 7.3% of the world oil reserves, from the Albian Kazhdumi Formation.

This chapter discusses three aspects of the Kazhdumi oil-generating system:

[1] Total, Cedex 47, 92069 Paris la Defense, France

[2] Fina Exploration and Production, Zone Industrielle C, 7181, Seneffe (Feluy), Belgium

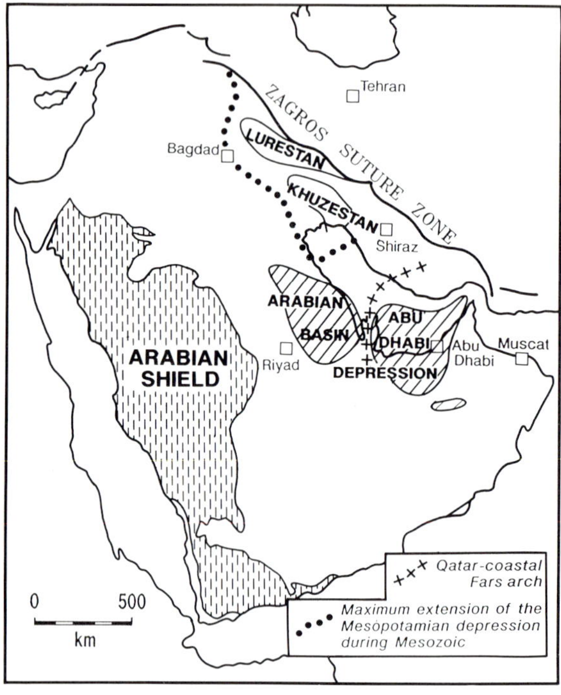

Fig. 1. Middle East sketch map

1. The areal extent, thickness and organic richness of the Kazhdumi source rocks in terms of environment of deposition, which resulted from the coincidence of climatic, oceanographic, and local geological factors.
2. The post-Albian geological history, i.e., progressive burial which led to the genesis, migration, and entrapment of hydrocarbons, originated from the Kazhdumi source rocks, including the relative timing of structure growth and oil generation, the source rock and reservoir relationship, and caprock occurrence.
3. Oil-to-source rock and oil-to-oil correlations using stable isotopes and biomarkers which confirm the Kazhdumi origin of the oil emplaced in the main Iranian oil fields.

Most of the geological and geochemical data used in this chapter were obtained from 1967 to 1978 through field studies in the Fars Province, Lurestan, and Khuzestan Mountain Front (Fig. 2). About 1500 samples were analyzed, covering all the Cambrian to Upper Miocene formations exposed at outcrops and 3000 core and cutting samples were studied from 35 wells. Extrapolation to determine the extent of organic-rich layers was achieved using radioactive and electric logs and by visual examination of thin sections. A database was assembled which allowed the regional distribution of source rocks to be established. Basin modeling procedures using seismically defined synclinal source rock drainage areas, and Pliocene orogenic overburdens

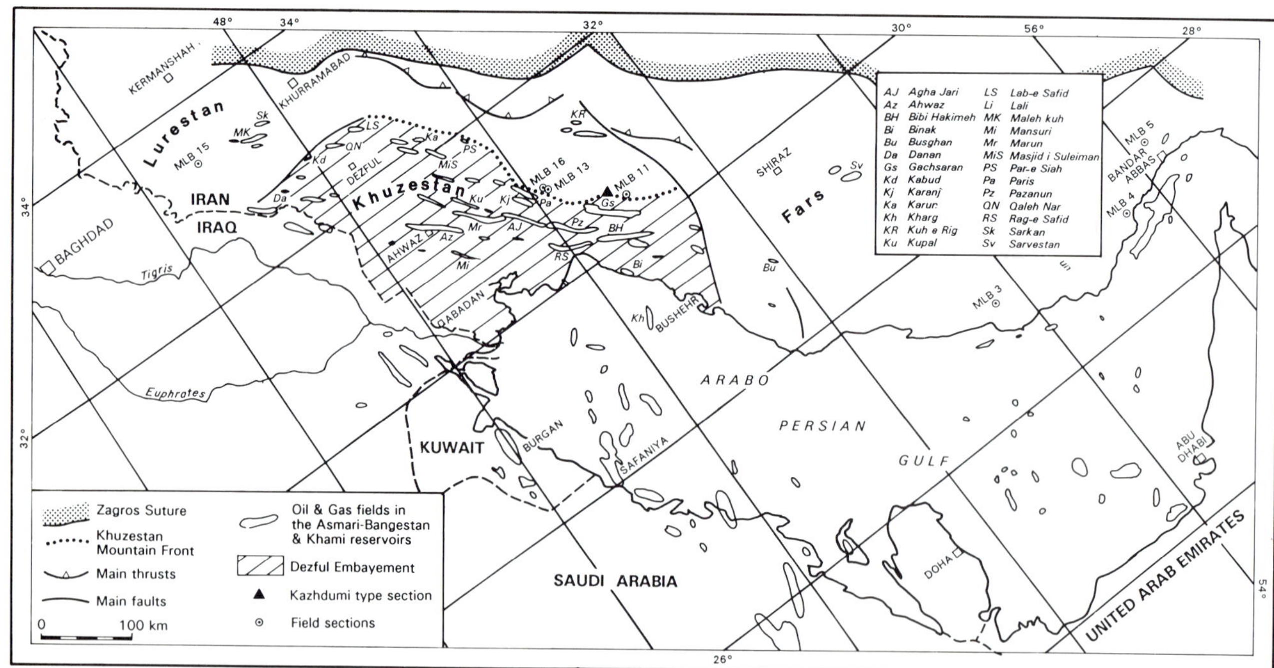

Fig. 2. Location map of the Zagros Foothills, Dezful Embayment, and northeastern part of the Arabo-Persian Gulf

were used to generate a predictive charge model as early as 1971. Subsequently, maturity mapping and timing of generation for candidate source rock units were calibrated against observed vitrinite reflectance and T_{max} data. Stable isotope analyses (^{13}C, ^{34}S) allowed an initial segregation into generic oil families and their correlation with progenitor source rocks. Biomarker analyses were used to corroborate these assignments (Bordenave and Burwood 1990).

Material and Methods

More than 4500 field specimens, cores, and closely spaced ditch cuttings were screened for source potential using LECO total organic carbon (TOC wt.%) analysis. Determinations were performed on decarbonated residues after quantitative calcimetry (10% HCl ambient temperature, 50% HCl at 50 °C). Some of the more attractive rocks (TOC >1%) were then subjected to pyrochromatography (Bordenave et al. 1970; Giraud 1970) and Rock-Eval pyrolyses. Kerogen pyrolyses were performed on pre-extracted rock powders (100 mesh, CH_2Cl_2: MeOH) using a semimicro adaptation of the Fischer assay (oil shale) procedure. These data were subsequently supplemented by a kerogen concentrate pyrolysis and combustion procedure as described by Burwood et al. (1988).

Oil analyses were performed using conventional column chromatographic (SiO_2/Al_2O_3) and capillary gas chromatographic procedures. All isotopic data ($\delta^{13}C$, $\delta^{34}S$) are reported against PDB (NBS = −29.8 ppt) and the Canyon Diablo standards, respectively. GC-MS analyses were conducted using high resolution and metastable ion (SMIM) techniques using whole oils (Bordenave and Burwood 1990).

Middle East Geological Setting Overview

The study area covers the Iranian part of the Zagros Foothills and the Dezful Embayment, including the Lurestan, Khuzestan and Fars provinces and the northeastern half of the Arabo-Persian Gulf (Figs. 1 and 2). This area is part of the old Arabo-Nubian Platform that extends over most of the Middle East from Turkey to Oman. This platform was bordered to the northeast by the Southern Tethys Ocean since its opening in the Late Permian-Early Triassic (Stöcklin 1968) until the Platform collided with the Central Iran Block at the cnd of thc Early Miocene.

Essentially stable, the Platform has persisted since the Early Cambrian. Its northeastern edge, however, was extensively folded and thrusted as the consequence of the Zagros collision, especially during the Late Miocene-Pliocene paroxysmal orogenic phase. Regional structural features persisted from the beginning of the Middle Jurassic until the end of the Early Miocene. A wide, low relief arch extended over Qatar and Coastal Fars (Fig. 1). It was bordered to the E-SE by a semipermanent depression located over onshore and offshore areas of Abu Dhabi. Another extensive depression centered on Lurestan, temporarily extended over the Mesopotamian basin, the northern part of the Gulf and the northeastern limit of Saudi Arabia (Murris 1980). In addition, some NW-SE- elongated highs persisted along the northeastern margins of the continental shelf bordering the South Tethys Ocean (Setudehnia 1978).

The platform was covered by a wide, generally shallow intracratonic sea, the shore of which fluctuated widely according to sea level changes and to low-amplitude differential subsidence. Sedimentation remained dominantly calcareous, consisting of aerobic high energy limestones with frequent breaks in sedimentation on highs and low energy argillaceous limestones in the depressions (see Murris 1980 for a review).

Calcareous sedimentation was temporarily interrupted either by evaporitic episodes, or by sudden influxes of clastics (Fig. 3). Evaporites developed as the result of arid climate during the Late Proterozoic, with the deposition of the thick Hormuz Salt, in the Permo-Triassic, end-Jurassic and the end-Early Miocene. Large amounts of clastics coming from the erosion of the Arabian Shield under a humid climate invaded the shallow marine habitat, during short periods in the Rhaetian, mid-late Barremian, early to middle Albian, Cenomanian, and Oligocene-Early Miocene (James and Wynd 1965; Murris 1980).

Excellent source rocks were deposited in intrashelf basins, when the environment became anoxic due to sea level rises, locally increased subsidence, and either a humid climate or the existence of upwelling in the Tethys Ocean. Anoxic conditions extended over large areas during the Middle Jurassic, Oxfordian, Neocomian, Albian, Campanian/Maestrichian, and Middle Eocene to Early Oligocene. Among these events, two are of particular importance to the oil industry as they are

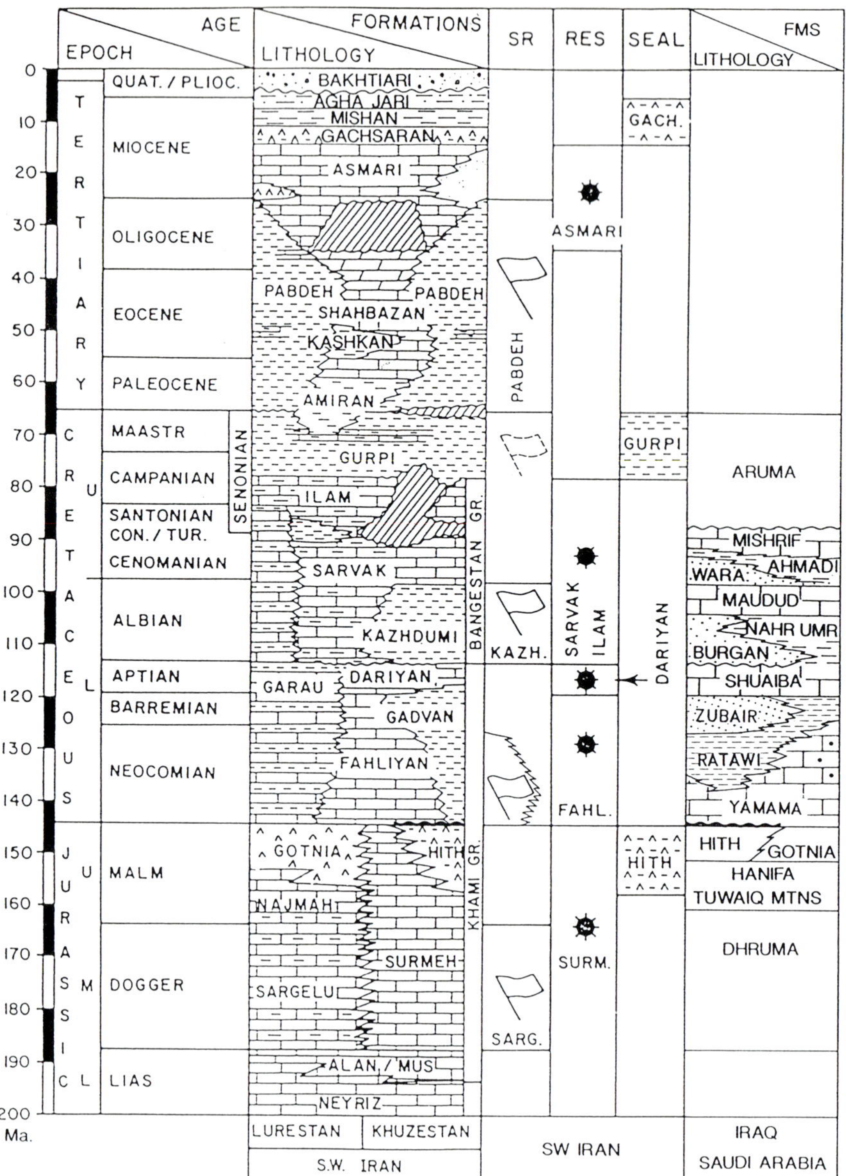

Fig. 3. Schematic stratigraphy and source-reservoir-seal relationship for the Dezful Embayment and neighboring areas. Main source rocks (*SR*), reservoirs (*Res*), and *seals* are indicated

responsible for the accumulation of about half of the world's oil reserves: the Oxfordian anoxia resulted in the deposition of the main source rocks of Saudi Arabia, the Hanifa Formation (Ayres et al. 1982; Lehner et al. 1984; Droste 1990), and of Abu Dhabi, the Diyab Formation (Hassan and Azer 1985); later, the Albian event provided the major source rock of Iran, i.e., the Kazhdumi Formation (Bordenave and Nili 1973; Ala et al. 1980) and the Natih E source rock of Oman (Grantham et al. 1987). The Shilaif source rocks of Abu Dhabi and the Natih A/B of Oman belong to a slightly younger episode dated as Cenomanian.

Sedimentation During the Early Cretaceous

The latest Jurassic was marked by a widespread regression and arid climate which led to the accumulation of evaporites over most of the Arabo-Nubian Platform, with the exception of the NE part of the Zagros Foothills, where open marine conditions persisted (Ricou 1974; Setudehnia 1978). The nature of the evaporites reflected the continuation of the paleoenvironmental differences observed since the Middle Jurassic between platform and depression facies. Shallow marine upper tidal and sabkha-style sediments of the Hith Formation, often associated with high-energy grainstones, covered most of the platform, while deep water laminated primary anhydrite and halite (Gotnia Fm.), together with subordinate dark gray shales were deposited in the Mesopotamian-Lurestan depression (James and Wynd 1965; Murris 1980).

The Jurassic-Cretaceous boundary corresponded to a drastic change of environment. The arid Jurassic conditions were replaced by a warm and humid climate as the northward drifting Afro-Arabian plate reached equatorial latitude (Roth and Bowdler 1981). The alternating carbonate and evaporitic episodes observed during the Jurassic were replaced by a succession of widespread thin carbonates interspersed with clastic influx. The rapid alternation of lithology reflected sudden uplifts of the Arabian Shield (Koop and Stoneley 1982) and a persistent humid climate. The common plant remains and the occurrence of coals in the clastics confirm the humid climate (Lehner et al. 1984).

As intracratonic seawater depths were rarely deeper than 100 m, except in the Lurestan depression, clastics formed thin deltas, or rapidly prograding sand wedges. In the more distal areas, silts and clays admixed with lower energy carbonates, and carbonate sedimentation retreated eastward. Subsequently, restoration of carbonate sedimentation over the Arabo-Nubian Platform at the end of a clastic episode was as sudden as its disappearance. Three major sand wedges were emplaced during the middle-late Barremian (Zubair Sands), mid Albian (Nahr Umr-Burgan Sands), which is described in this chapter, and the early Cenomanian (Wara Sands and Ahmadi Formation).

The early Aptian sea level rise caused deposition of a widespread, shallow water limestone known as the Shuaiba Formation in Saudi Arabia, Iraq, Kuwait, and Abu Dhabi, and the Dariyan Formation in Iran (James and Wynd 1965). In the Lurestan depression, however, deeper water conditions persisted throughout the Neocomian-Aptian interval with the deep Gotnia Anhydrite being overlain by low-energy, radiolarian-rich marls and argillaceous limestones of the Garau Formation. These contain excellent source rocks in their lower part. In Abu Dhabi, a local intrashelf depression resulted in the deposition of the Shuaiba Formation 'Bab Member' source rocks. The rudistid reefs which flanked the depression provided an excellent and prolific reservoir facies (Koop and Stoneley 1982).

The basinwide Aptian transgression was abruptly terminated by a major pre-Albian regression which resulted into the higher areas of the Arabo-Nubian Platform becoming emergent. In Saudi Arabia, the rudistid limestones of the Shuaiba were subjected to weathering and karstification (Lehner et al. 1984) as in the Abadan area. In Coastal Fars, micropaleontological zonation of the Dariyan Formation indicated a considerable loss of limestone section when compared to Khuzestan. There, sedimentation between the Dariyan Limestone and the overlying Albian Kazhdumi marls seems to be continuous (James and Wynd 1965), although the top Dariyan Formation surface is often strongly corroded, with irregular ferruginous deposits, sometimes sandy and glauconitic and rich in reworked fossils. The major pre-Albian regression did not apparently affect sedimentation in the Lurestan area where basinal Garau facies persisted. Here, the Dariyan time-equivalent is recognized as being slightly more calcareous. Sedimentation between the Aptian and Albian also appears to have been continuous in central Khuzestan; however, micropaleontological data are insufficient to confirm continuity.

The short, but major, pre-Albian regression was followed by a low-amplitude transgressive event, marked by a sudden influx of clastics during the early to middle Albian. The clastic facies distribution in the second episode (Nahr Umr-Burgan Sand) mirrored that of the Barremian Zubair event, i.e., an alluvial plain covered all of western Saudi Arabia and Iraq, reaching the western shore of the present Gulf, and the Euphrates River (Powers 1968). The more distal part of the basin, between the Burgan-Safaniya delta located on the western half of the Gulf, northwest of Qatar, and the shallow water carbonate shelf bordering the open sea was the locus for deposition of the Kazhdumi Formation (Fig. 4).

The Kazhdumi Formation

Definition and Facies

The Kazhdumi Formation is defined in the Khuzestan Mountain Front (Fig. 2) as a low-lying argillaceous interval between the top of the Aptian Dariyan Limestone and the base of the cliff formed by the Sarvak Formation of Cenomanian-lower Turonian age (Fig. 3). The type section is located in

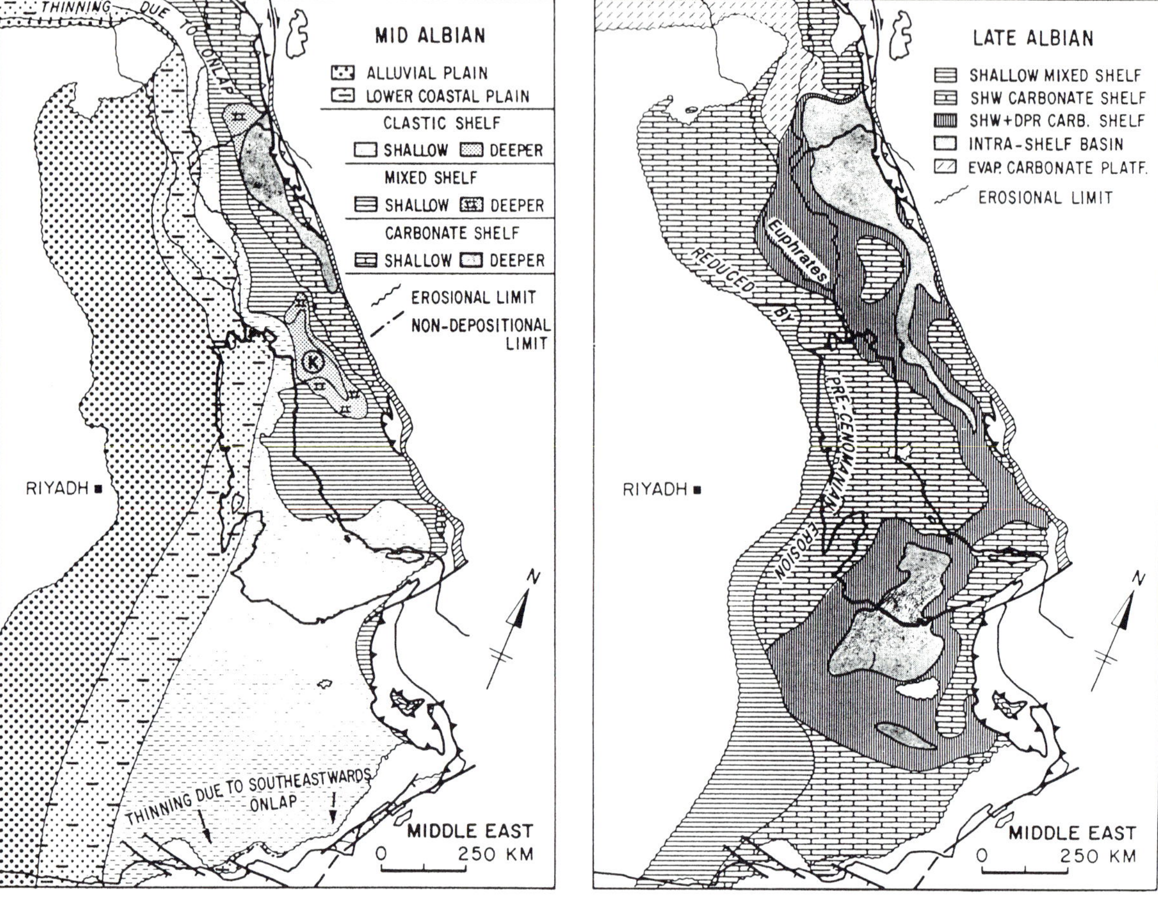

Fig. 4. The Paleogeography of the Middle East in the mid and late Albian (After Murris 1980). (*K*) indicates the Khuzestan subbasin centered on the Dezful Embayment

the Tang-e Gurguda, 10 km north of Gachsaran (Kent et al. 1951) and comprises 210 m of marls, often bituminous, with subordinate black limestone. Glauconite is common in the lower half of the formation, whilst the lowest 30 m contain several reddish, oxidized ferruginous zones which may indicate a diastem (James and Wynd 1965).

At the type locality, the microfauna is predominantly planktonic, but includes conical *Orbitolina* in the basal beds. A rich fauna of *Hedbergella*, *Ticinella*, *Biglobigerinella*, and *Planomalina* spp., together with radiolaria and spicules is found in the lower 80 m. The upper 150 m contains *Globigerina washitensis* Carsey. The Kazhdumi Formation also contains a rich ammonite fauna, including *Knemiceras uhligi* Choffat, *Knemiceras syriacum* Stoliczka, *Spathiceras*, *Oxytropidoceras*, *Stoliczkaia*, and *Parahoplites* spp. (James and Wynd 1965).

The Kazhdumi Formation is considered to be of Albian age. In the deeper part of the basin, where no break in sedimentation is recorded between the Dariyan Limestone and the Kazhdumi shales, deposition may have commenced in the late Aptian. The upper part of the formation, which passes gradationally to the Sarvak Limestone, may belong to the early Cenomanian. However, the Albian-Cenomanian boundary cannot be defined with accuracy (Setudehnia 1978). As there is a possible interference of facies in the definition of the faunal assemblage zones of the Zagros Foothills, a revision of micropaleontological data is needed to definitively establish the limits of the Kazhdumi within a global sequence stratigraphy.

For geochemical purposes, the Kazhdumi Formation was studied at three surface sections of the Khuzestan Mountain Front: section MLB 11, 7 km to the SE of the type section, in the Kuh-e

Mish, MLB 13 and 16, being located about 100 km to the NW, at Kuh-e Bangestan (Fig. 2). One field section, MLB 15, located in Lurestan, covered the deeper water Upper Garau, Kazhdumi time-equivalent facies, whilst a further six surface sections were studied in the Fars Province (ARN 2, 3, 4, and MLB 3, 4, 5).

Kuh-e Mish Section (MLB 11)

The Dariyan Formation, deposited in a shallow oxygenated environment, consists of rather fine-grained limestones in massive and nodular beds interbedded with gray marls. Microfossils include conical forms of *Orbitolina*, *Choffatella* spp., dasyclad algae, bryozoa, and a few associated planktonic fauna of *Globigerina* and tintinnids. Macrofossils includes ostreids, pectinids, gastropods, corals, echinids and a few ammonites (Parahoplites) (Fig. 5).

Toward the top of the Dariyan cliff, a thin bed, indicated as R in Fig. 5, is made of reworked, rounded limestone pebbles. It may correspond either to a short emergent episode, or to a change of sedimentation with submarine erosion, in a shallow, oxic environment. The reworked bed is overlain by nodular limestones similar to those of the upper Dariyan, but more argillaceous, i.e., including progressively thicker gray marl interbeds. The environment remains oxic and shallow with a benthic fauna comprising *Hemicyclammina*, *Cyclammina*, *Lenticullina* spp., arenacids, dasyclad algae, echinoderms, gastropods, pectinids, and ammonites. *Orbitolina* are still present.

Higher in the section, argillaceous limestones and marl become dark and laminated, with the former including some crinoids. The microfauna is mixed with co-existing benthic and pelagic forms

In the upper part of the Kazhdumi, greenish-gray fine-grained and bioturbated earthy marls are very rich in well preserved macrofossils, including *Exogyra*, *Terebratula*, *Rhynchonella*, and corals. The microfauna consist of arenacids and a few planktonic forms including *G. Washitensis* Carsey and *Flabella minima* sp. This fauna shows a return to rather shallow environment. Later, the sedimentation again became calcareous with deposition of the thick Sarvak Formation (250 m) of Cenomanian age, which showed at this location a predominance of planktonic forms (a mixture of *Rotalipora* and *Hedbergella* known as the 'Oligostegina facies').

All the oxic shallow facies are devoid of organic matter (TOC $\leq$ 0.3%). Out of a Kazhdumi total thickness of 230 m, only 40 m of slightly deeper, low-energy dysoxic facies, with a dominantly planktonic fauna contained 1 to 4.5% of organic carbon (Fig. 6).

Kuh-e Bangestan Sections (MLB 13 and 16)

At MLB 13, the top of the *Orbitolina*-rich Dariyan Limestone is marked by a hardground with an irregular, rusty concretion zone. The overlying Kazhdumi Formation is 160 m thick. The lower 50 m consists of intercalations of greenish gray, fine-grained marls, bioturbated gray marls and nodular argillaceous limestones. The fauna is benthic and includes pelecypods, echinids, *Choffatella*, *Orbitolina*, *Epistomina and Cyclamina* spp. Above, 80 m of paper-like, thinly laminated dark-gray marls intercalated with dark-gray, fine-grained argillaceous limestones contain a pelagic fauna including *Globigerinelloides breggiensis*, *Biglobigerina barri*, plus *Hedbergella*, *Schakoina*, and *Saccocoma* spp. *Globigerina washitensis* and radiolaria were observed toward the top of the section. The uppermost beds, below the Sarvak cliff, consist of greenish gray marls, alternating with light gray argillaceous limestones. They contain ammonites and a pelagic microfauna. At MLB 13, about half of the Kazhdumi Formation (80 m) was deposited under pelagic anoxic conditions with TOCs in the vicinity of 3%.

At MLB 16, 6 km north of MLB 13, the pelagic, anoxic facies begin few centimeters only above the hardground and the organic layers are much thicker (120 m), with maximum TOC values of 5.5%. (The Kazhdumi is immature at MLB 13 and MLB 16 with vitrinite reflectance Ro = 0.4%).

The three mountain front sections show a hardground at the top of the Dariyan Limestone, due either to an emergence, or to an abrupt change in sedimentation conditions with strong currents, leading to the deposition of thin argillaceous layers in an oxic environment. The progressive deepening of the water column, demonstrated by the replacement of benthic by pelagic species, corresponds to the establishment of anoxia and the deposition of low-energy, organic-rich marls and argillaceous limestones. The upper part of the Kazhdumi contains a mixture of benthic and pelagic fauna. Although being still dominantly argillaceous, it corresponds to the return of oxic depositional conditions. Oxic

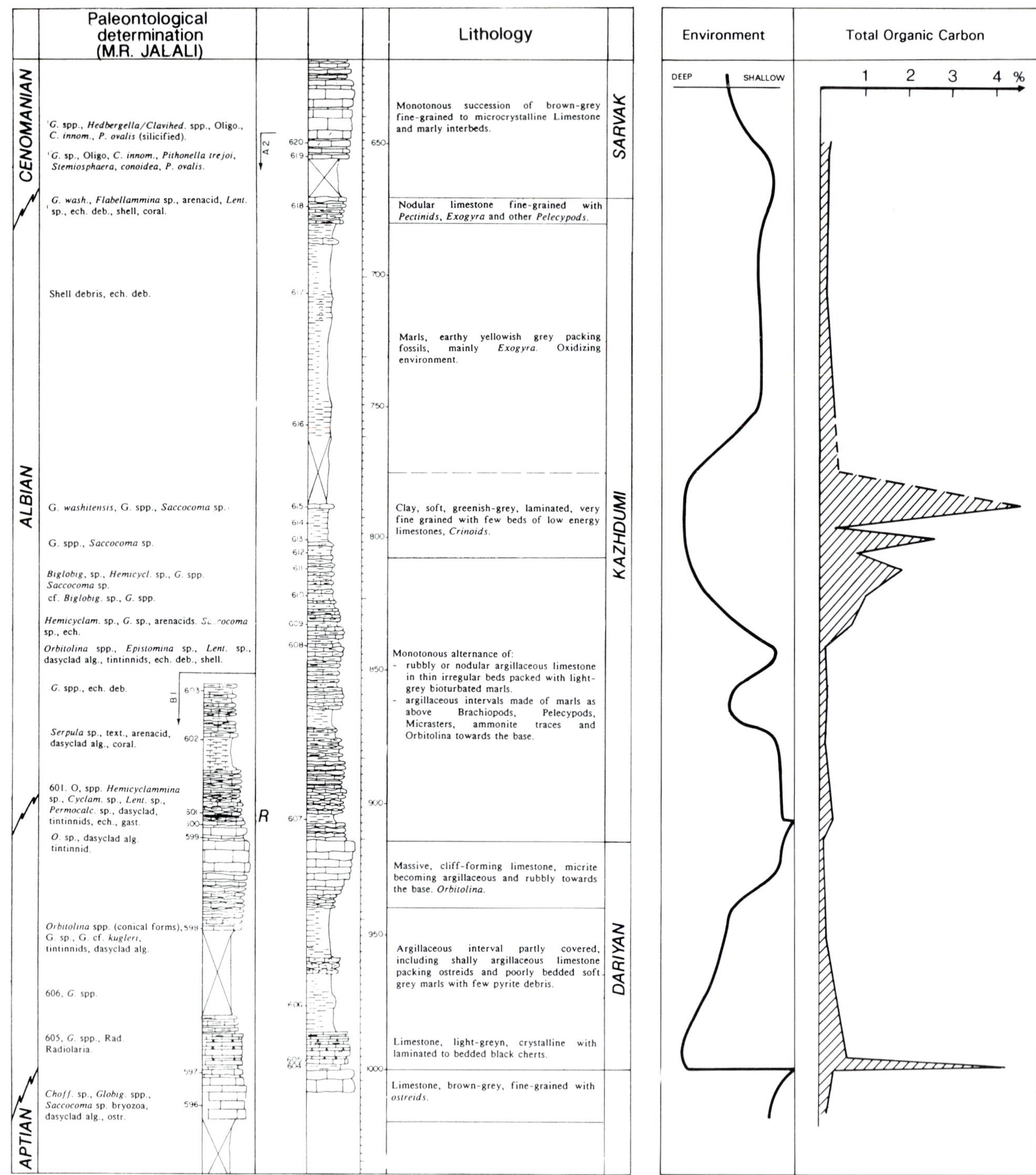

Fig. 5. Lithological description and stratigraphy of the Kazhdumi Formation at Kuh-e Mish, close to the type section

facies continued during the Cenomanian with the deposition of the Sarvak Formation.

A comparison between the mountain front outcrops shows that the Kuh-e Mish section contains sediments deposited in water slightly shallower than that of Kuh-e Bangestan. Organic-rich layers range from 40, 80, and 120 m thick at MLB 11, MLB 13 and MLB 16 respectively (Fig. 6).

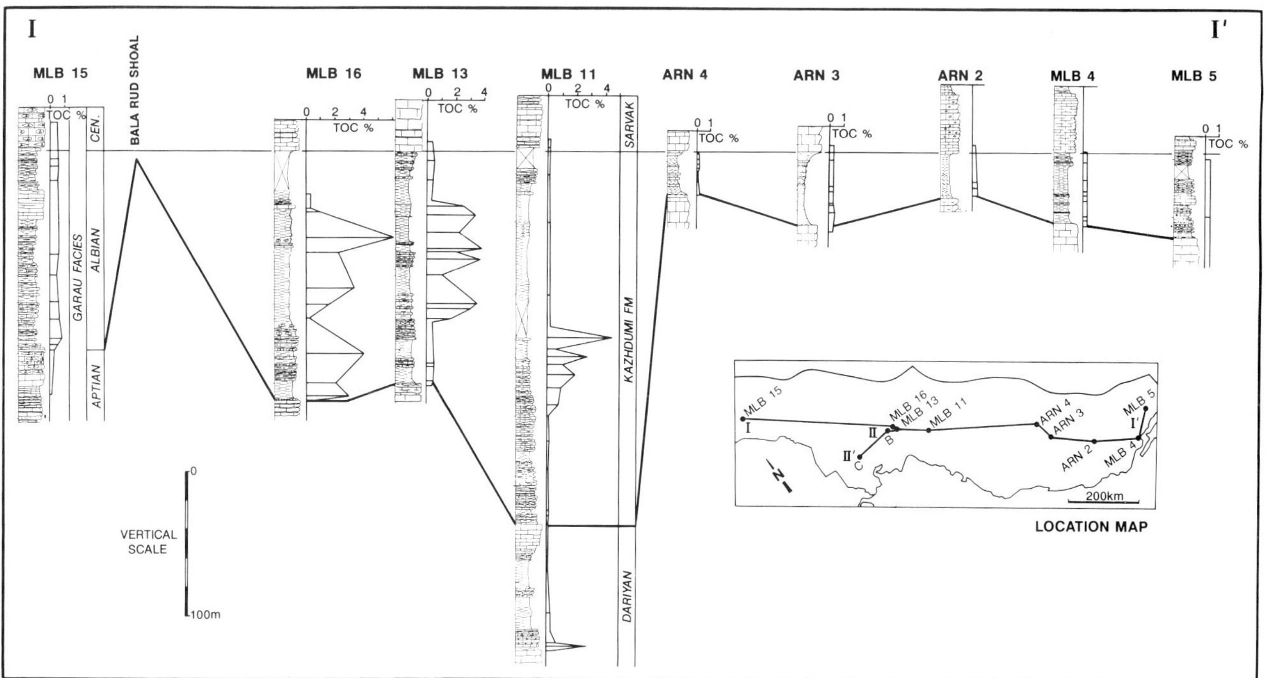

Fig. 6. Cross section of the Kazhdumi from Lurestan to Fars, indicating the evolution of the organic content from northwest to southeast

Kabir Kuh Section (MLB 15)

The Kazhdumi Formation time equivalent was studied at one location in Lurestan, in the Tang-e Garau of the Kabir Kuh (MLB 15) (Fig. 2). Above thick highly organic Garau facies of Neocomian and low-energy cherty limestone of the Dariyan Formation, the Kazhdumi equivalent is represented by 140 m of alternating slightly silty gray marls and fine grained argillaceous limestones. The limestones become sandy and sometimes glauconitic up-section. No fauna was found, except *Biglobigerinella barri* and *Globigerinelloides algeriana* toward the base of the formation and "*Oligostegina*" at the top. The argillaceous interval corresponding to the Kazhdumi Formation is overlain by a thick sequence of cherty limestone, containing *Globigerina washitensis* of the Sarvak Formation.

Although the depositional environment was slightly reducing, as shown by the occurrence of small pyritic concretions, the organic content of the Kazhdumi equivalent is lean (TOC $< 0.8\%$) (Fig. 6).

Fars Province Sections

In the six surface sections studied in Fars, the Aptian shallow water Dariyan Limestone was emergent and eroded during the pre-Albian regression and the basal Kazhdumi may belong to the late Albian. This late marine transgression is generally accompanied by a regime of strong currents with submarine erosion, irregular ferruginous surface, thin ferruginous beds, sometimes sandy and glauconitic, and rich in reworked fossils including a large number of *Orbitolina*, gastropods, pelecypods, and echinids.

In Fars, the Kazhdumi is thin, from a few meters to a maximum of 100 m. Greenish gray marl deposited in oxic, but low-energy environment comprises subordinate light gray argillaceous limestones and irregular reworked beds. The very shallow water microfauna includes *Orbitolina*, *Hemicyclamina*, *Trocholina*, *Textularia* spp. plus algae. Brackish influence was indicated by the occurrence of ostracods. The Kazhdumi marls are overlain by shallow water carbonates of the Sarvak Formation. This formation is sometimes very thin,

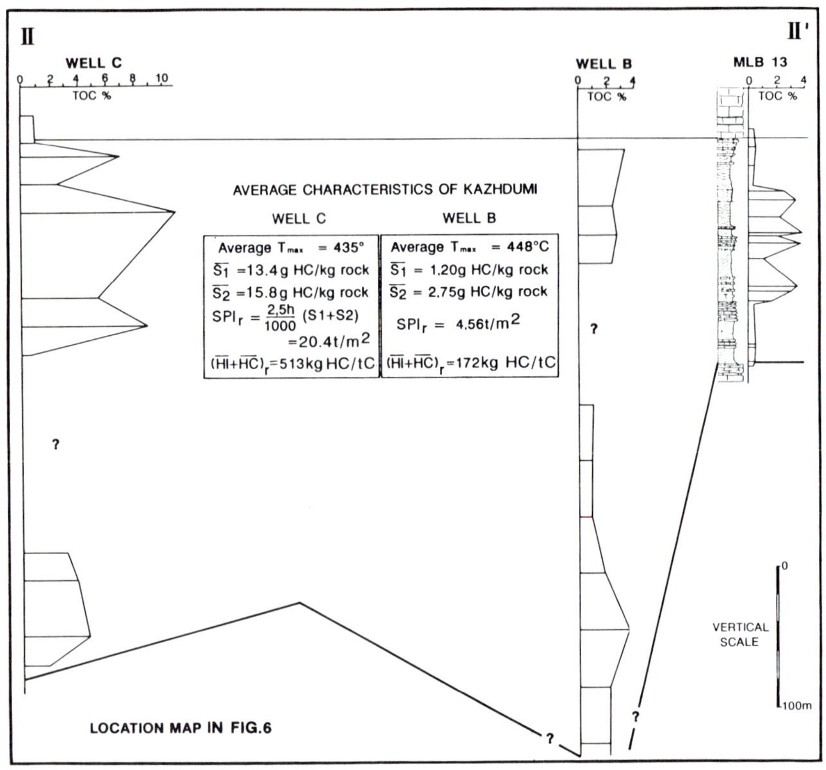

Fig. 7. Cross section of the Kazhdumi from the Khuzestan mountain front to the southwest Dezful Embayment, indicating the evolution of the organic content from northeast to southwest. Note that *well B* corresponds to already mature source rocks ($T_{max} = 448\,°C$) and that initial TOC values were significantly higher

showing both breaks in sedimentation and periods of erosion.

The Kazhdumi marls/argillaceous limestones deposited on the Fars Platform, under very shallow, oxic conditions contain little organic matter (TOC $\leq 0.3\%$) (Fig. 6).

Facies Evolution in the Dezful Embayment

The Kazhdumi marls thicken from the mountain front field sections toward the southwest, reaching more than 300 m in wells drilled in the central part of the Dezful Embayment (Fig. 7). Above the Dariyan shallow water limestones, characterized by *Orbitolina*, *Choffatella*, and *Hensonella* spp., deeper water low-energy dark-gray laminated marls contain a pelagic fauna, including *Globigerina* and radiolaria. The upper zone is marked by the occurrence of *Globigerina washitensis*. The transition between the Kazhdumi marls and the Sarvak is progressive in well II, with a slow decrease of the clay content as the Sarvak Limestone is developed there as an "*Oligostegina*" low-energy facies (Fig. 8).

Gamma ray log interpretation and examination of thin sections were used to estimate the extent of organic-rich intervals in unanalyzed well sections. Uranium in solution, originated by washing of granitic rocks from the Arabian Shield, was transported in rivers in hexavalent soluble form. Hexavalent uranium precipitated as the tetravalent insoluble form when they reached the reducing conditions which prevailed in the Dezful Embayment, resulting in a good correlation between carbon and uranium contents (Bordenave 1993).

Paleogeography and Distribution of the Kazhdumi Formation

An isopach map of the Kazhdumi Formation, (Fig. 9a) was constructed from surface observations, well control, and published data (James and Wynd 1965; Setudehnia, 1978; Koop and Stoneley 1982). Thicknesses do not include the shallow water "Conical *Orbitolina* Limestones", deposited either toward the base and/or in the uppermost part of the Albian interval, along the Bala Rud Flexure, in the Masjid-i Suleiman (MIS) and Kuh-e Rig areas (Fig. 2). There, high energy limestones were deposited during Aptian, Albian, and Cenomanian, without a marked argillaceous break (Danan, Kabud,

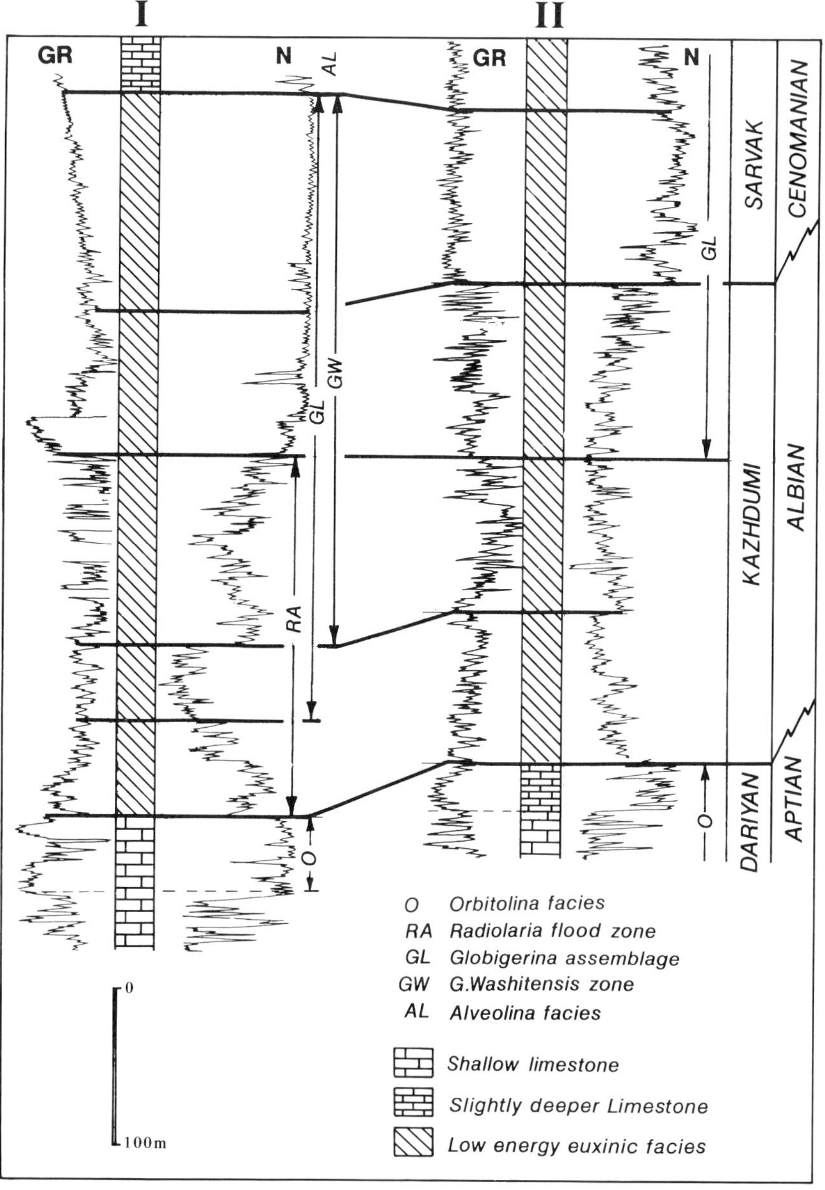

Fig. 8. Gamma ray logs of the Kazhdumi Formation in two wells from the central part of the Dezful Embayment (these wells are located between well B and well C)

Lab-e Safid) areas) (Fig. 2). This predominance of calcareous shallow sedimentation is indicated in Fig. 9a. In the MIS 306 well, for instance, only the argillaceous interval of Albian age was taken into account (60 m), whilst Albian carbonates are much thicker (Setudehnia 1978).

The maximum thickness of the Kazhdumi marls ranges from 300 to 400 m in the central part of the Dezful Embayment in a depression coincident with the main fields area, e.g., Marun, Agha Jari, Paris, Pazanan, Gachsaran, and Bibi Hakimeh (Fig. 2). This area of maximum thickness is bordered to the northeast by the present-day mountain front. The depression extends toward Mansuri and Binak. Elongated NNE-SSW highs, most likely related to old basement faults and deep-seated Hormuz Salt ridges are conspicuous. Such ridges continued to grow during the Upper Cretaceous, forming large traps prior to the Zagros folding, now accommodating fields in Saudi Arabia and the northern part of the Gulf.

To the north, the area of Kazhdumi marl deposition is bordered by the Bala Rud flexure which separates the Dezful depression from the pre-existent Lurestan depression in which the Garau facies continued to accumulate during Albian. Access to

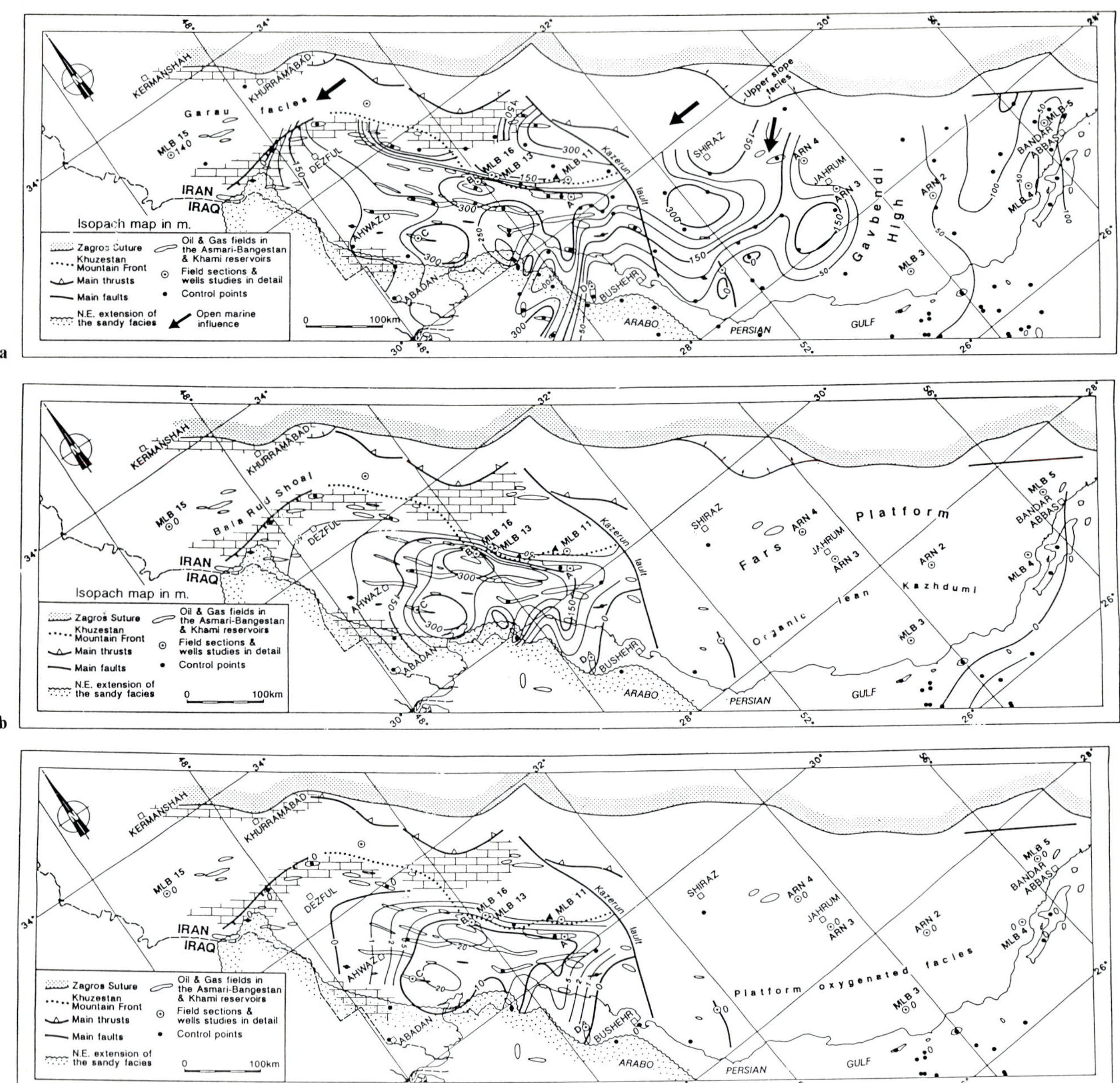

Fig. 9a–c. The Kazhdumi Formation. **a** Isopach map. **b** Isopach of bituminous facies. **c** Cumulative carbon content (h TOC). (*h* is the thickness in metres of layers having TOC values higher than 0.5%; *TOC* is a decimal number)

the open sea of the Lurestan depression is seen southeast of Khurramabad where the local facies corresponded to a deeper marine environment.

Toward the southwest, the Kazhdumi depression was limited by the Safaniya-Burgan delta which extended over the northern part of the Gulf. The sandstones extend toward the southwestern end of the Dezful Embayment, where they reach 60 m in thickness. They pinch out to the southwest of the Mansuri-Binak line, where only argillaceous material accumulated as a prodelta facies.

The Fars platform limits the Kazhdumi depression to the east. Here, generally thin marls deposited in shallow water, oxic conditions, were often interrupted by temporary emergent episodes. The Gavbendi High (Fig. 9a) is well marked and a few salt-related highs show a complete lack of Kazhdumi deposition. A more subsident zone was located west of Shiraz, whilst access to the open sea was likely both to the north and east of Shiraz, where a slope facies was described (Ricou 1974; Murris 1980).

Organic Composition of the Kazhdumi Source Unit

The organic content of the Kazhdumi is given in two cross sections, one almost parallel to the Zagros suture going from Lurestan to Fars (Fig. 6), the other from the Khuzestan Mountain Front to the southwestern side of the Khuzestan depression (Fig. 7).

TOC values measured in field sections are considered to be representative as only fresh samples were collected. A core drill campaign in 1968-1969, which led to the drilling of 26 wells, 10 to 30 m deep, showed the TOC values of surface samples and their equivalent at depth to be similar. Moreover, surface samples are immature with the exception of the MLB 15 section in Lurestan, where samples are marginally mature.

An estimated thickness of the "bituminous" anoxic facies is presented in Fig. 9b. As already mentioned, the Kazhdumi is organic-lean throughout Fars and on the Bala Rud shoal. In Lurestan, the upper part of the Garau Formation, Kazhdumi time-equivalent is organic-lean, although the sediments were deposited in a low-energy environment. Conversely, in the central part of the Dezful Embayment, almost all the marls are bituminous, and the source rock thickness may exceed 300 m. Sometimes, as shown in Fig. 8, the basinal anoxic facies continues in the early Cenomanian with the deposition of the "*Oligostegina* facies" of the Sarvak Formation. However, this Sarvak basinal facies generally shows only a moderate organic content. Over the southeastern part of the Gachsaran field thick low-energy Kazhdumi marls contain relatively little organic matter in the 1-2% range.

TOC values measured in the mountain front sections and in wells A (Gachsaran), B (Paris) and C (Mansuri) are given in Table 1. The cumulative organic carbon content (h TOC/100), h being the cumulative thickness of layers having TOC values in excess of 0.5%, is given in Fig. 9c.

In well C, the Kazhdumi source rock has already reached the top of the oil window ($T_{max} = 435\,^{\circ}C$, $R_0 = 0.71\%$) and a large amount of expelled oil accumulated in the Asmari and Bangestan reservoirs. S_1 varies from 1.18 to 28.8 kg HC/t rock (average value 13.4), S_2 varies from 0.34 to 38.7 kg HC/t rock (average value 15.8). Using these values, the residual Source Rock Potential Index (SPI_r) can be calculated (Demaison and Huizinga 1991):

$$SPI_r = \frac{\overline{2.5(S_1 + S_2)_r h}}{1000} = 20.4^+ \ t/m^2.$$

Unfortunately, only part of the Kazhdumi sections was sampled. The average sum of the hydrogen index ($HI = S_2/TOC$) and of the hydrocarbon index ($HC = S_1/TOC$) is quite high, 513 kg HC/t C, classifying the Kazhdumi as a type II source rock (Fig. 7). In well B, located very close to the mountain front, the source rock characteristics seem rather poor at first glance: the average value of $S_1 + S_2$ is only 4.0 kg HC/t rock, the average HC + HI is of 172 kg HC/t C, and the SPI_r of 4.56. However, these source rocks have already attained the maximum oil generation zone ($T_{max} = 448\,^{\circ}C$, $R_0 = 0.84\%$) and expelled oil, as witnessed by the large Paris field oil accumulation. Therefore the initial SPI was much higher than the residual one.

It is estimated from available TOC, S_1, and S_2 values and from the isopach map of the bituminous

Table 1. TOC values in Mountain Front sections, well A (Gachsaran), well B (Paris), and well E (Mansuri)

Location	Thickness of organic layers: h (m)	TOC% Minimum	TOC% Average	TOC% Maximum	h TOC/100 (if TOC > 0.5%)
MLB 11	50	1	2	4.5	1.0
MLB 13	80	1	2.6	4	2.1
MLB 16	120	1	3	3.5	3.6
Well A	300	1	1.2	2	3.6
Well B	240 +	1	2.2	4	5.3 +
Well C	280 +	1.6	5.1	11	14.3 +

Table 2. Examples of average source rocks potential indices for individual source rocks from various basins. (After Demaison and Huizinga 1991)

Basin (country)	Source rock sequence	Kerogen Type	Average SPI[a] (t/m²)
Junggar (China)	Upper Permian	I	65
Lower Congo (Cabinda)	Lower Cretaceous	I	46
Santa Barbara Channel (USA)	Miocene	II	39
San Joaquin (USA)	Miocene	II	38
Central Sumatra (Indonesia)	Eocene-Oligocene	I	34
Eastern Venezuela FTB	"Middle" to Upper Cretac.	II	27
Kazhdumi Fm. (Iran)	Albian	II	25
Offshore Santa Maria (USA)	Miocene	II	21
Mid. Madgalena (Colombia)	"Middle" to Upper Cretac.	II	16
North Sea (UK)	Upper Jurassic	II	15
Central Arabia (Saudi Arabia)	Upper Jurassic	II	14
Niger Delta (Nigeria)	Tertiary	III	14
Gulf of Suez (Egypt)	Upper Cretaceous-Eocene	II	14
San Joaquin (USA)	Eocene-Oligocene	II to II-III	14
Maturin (Venezuela)	"Middle" to Upper Cretac.	II	12
Maracaibo (Venezuela)	"Middle" to Upper Cretac.	II	10
West Siberia (USSR)	Upper Jurassic	II	8
Cuyo (Argentina)	Triassic	I	8
Paris (France)	Lower Jurassic	II	7
Barrow-Dampier (Australia)	Middle to Upper Jurassic	II-III to III	6
E. Browse-W. Bonaparte (id)	Middle to Upper Jurassic	II-III to III	6
Illinois (USA)	Lower Carboniferous	II	6
Oriente (Ecuador)	"Middle" to Upper Cretac.	II	6
Northwest Arabian (Syria)	Triassic and Upper Cretac.	II	5
Plato (Columbia)	Oligocene-Miocene	II-III to III	5
Northwest Arabian (Turkey)	Upper Silurian-L. Devonian	II	4
Celtic Sea (Ireland)	Lower Jurassic	II to II-III	4
Malvinas (Argentina)	Lower Cretaceous	II to II-III	3
Williston (USA)	Lower Cretac.-L. Carbonif.	II	3
Senegal (Senegal)	Paleocene-Eocene	II-III	2
Cantabrian (Spain)	Lower Jurassic	II	2
Ogaden (Ethiopia)	Upper Jurassic	II to II-III	2
Magallanes-Austral (Chile)	Lower Cretaceous	II-III	1
Metan (Argentina)	Upper Cretaceous	II	1
Parana (Brazil)	Lower Permian	II	1
Palawan (Philippines)	Eocene	II	1
Pelagian (Tunisia)	"Middle" Cretaceous	II	1
Tres Cruces (Argentina)	Upper Cretaceous	II	< 1
Pelagian (Tunisia)	Eocene	II	< 1

[a]SPI numbers based on limited database.

Kazhdumi that in the deeper part of the Dezful depression the initial SPI may have reached 25 to 30, which would classify it amongst the higher values observed in the world (Table 2).

The Source Rock Depositional Habitat

The geological setting of the Kazhdumi source rock system is a 40 000 km^2 intracratonic depression, limited to the north and to the south by the Bala Rud carbonate shoal and the wide Fars platform/Qatar Arch respectively. To the west-southwest, the depression was limited by the Safaniya-Burgan prograding delta. Access to the South Tethys Ocean, was possible only through one or two channels, north or east of Shiraz, where open marine slope facies were described (Fig. 10a). Water circulation was hampered by a submarine high, located almost on the present mountain front, as observed at Kuh-e Mish, upon which shallow water facies continued

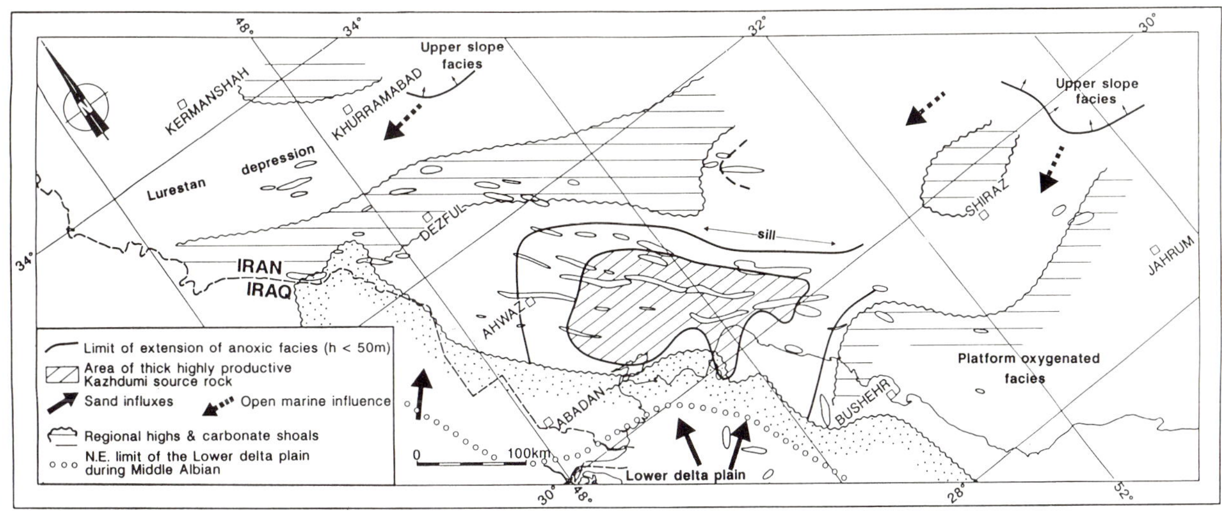

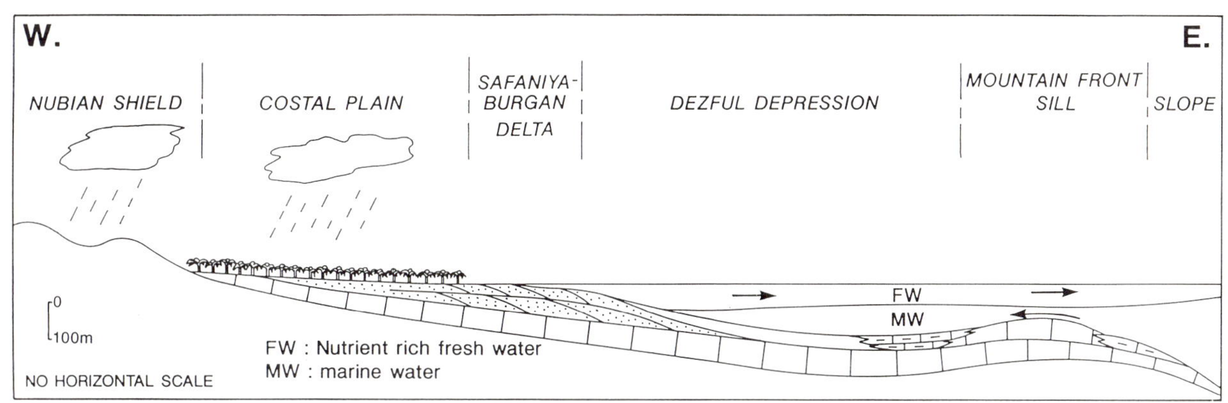

Fig. 10a, b. The Kazhdumi Petroleum system. **a** Paleogeography of the Dezful Depression. **b** The positive water balance mechanism

to accumulate until maximum Albian sea level rise.

As already discussed, the area was under an equatorial humid climate during Lower-Middle Cretaceous and a large amount of fresh water flowed into the Dezful depression by a system of rivers. These fluviatile waters, outwash from the wide Arabo-Nubian landmass, were nutrient-rich (nitrates and phosphates). Ideal conditions were realized for high phytoplankton productivity in the nutrient-rich, low-density, oxygenated, fresh surface water.

The worldwide Albian sea level rise allowed the deeper part of the Dezful Depression to be below the euphotic zone, although the water depth was probably not much greater than 100 m. Conversely, areas which remained slightly shallower, either located on regional highs or related to salt swells, accommodated a shallow water benthic fauna of the *Orbitolina* and arenacid type.

It is assumed (Fig. 10b) that a counter-current fed heavier oceanic waters into the basin, creating a density stratification between the fresh water input from the Nubian Shield and salty oceanic water, resulting in a positive water balance as described by Demaison and Moore (1980). Density stratification and the fall of organic matter in the oxygen-deprived zone is a classical mechanism for the creation of anoxic conditions. Sulfate-reducing bacteria were extremely active as demonstrated by the sulfur incorporated into the Kazhdumi source rock and the sulfur-rich crudes derived from it (Fig. 11). The hydrogen sulfide produced by the bacteria poisoned the bottom waters, which explains the disappearance of benthic fauna and the occurrence of only planktonic forms (*Globigerina* and radiolaria).

Whilst the delta prograded from the northwest, fine-grained clastics, silts, and clays accumulated in the depression in an overall low-energy

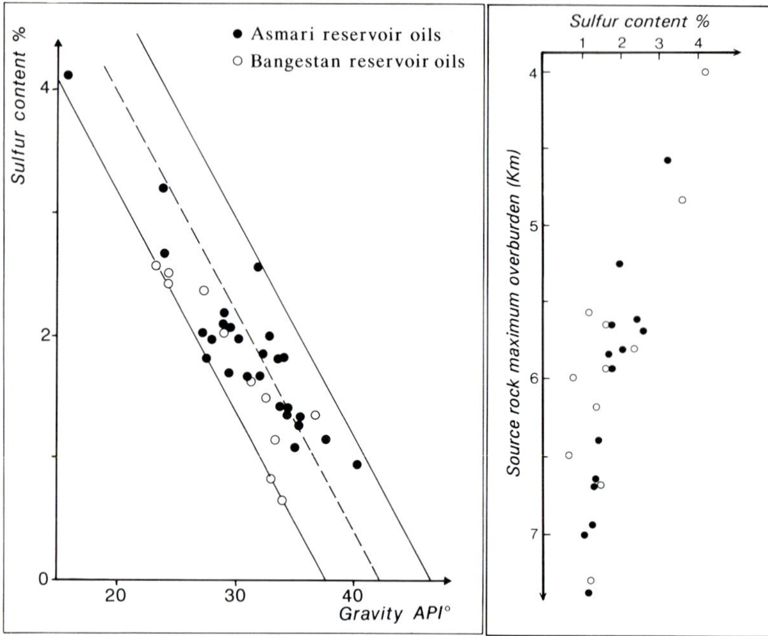

Fig. 11. Correlation of sulfur content/API gravity of oils derived from the Kazhdumi source rocks.

The oil sulfur content is higher than 3% for immature oils. It decreases with the increase of maximum depths to which the Kazhdumi source rock was buried prior to the paroxysmal phase of the Zagros orogeny. These maximum depths were estimated from seismic for each field area of drainage. The present geothermal gradient of the Dezful Embayment is low, about 2.5 °C/100 m (Clarke 1975) and it may have been even lower in the past due to the rapid subsidence of the area during the Pliocene

environment, enhancing the incorporation of organic matter into these sediments.

Although the deep water facies persisted in the Lurestan depression, the organic content of sediments remained low, possibly because the conditions were not adequate for the water stratification to be established, e.g., absence of sill?

Evolution of the Kazhdumi Source Rocks and Oil Generation

The Post-Kazhdumi Events

The evolution of the Kazhdumi source rock depended upon subsequent stratigraphic and tectonic events, which influenced its maturation, petroleum generation, migration, and entrapment. Post-Kazhdumi source rocks are also briefly mentioned to complete the matter of oil-to-source rock correlations.

Local dysoxic conditions occured in Khuzestan and northern Fars during Campanian to Maestrichtian with the deposition of marginal source rocks of the Gurpi Formation. A last anoxic episode led to the deposition of the excellent potential source rock of the Pabdeh Formation, which developed throughout northwest Lurestan, central Khuzestan, and northern Fars during the Middle to Late Eocene and locally during the Early Oligocene in Lurestan (Bordenave and Burwood 1990).

At the end of the Asmari deposition (end Oligocene to Early Miocene) the thickness of the post-Kazhdumi rocks varied between 1500 to 2100 m in the main fields area of the Dezful Embayment. This variation in thickness was related to the existence of a series of NNE-SSW-trending elongated highs and depressions which already existed during the deposition of the Kazhdumi Formation (Fig. 9a). Assuming that the geothermal gradient was not very different from the present, i.e., 2 to 2.8 °C/100 m (Clarke 1975), the temperatures at the end of Asmari deposition were too low for the Kazhdumi Formation to have attained the onset of the oil window.

Although indications of compression at the N-E edge of the Arabo-Nubian Plate were observed at the northeastern limit of Lurestan and Fars in the Late Cretaceous, the collision with the Central Iran block happened from Early Miocene onwards. One consequence of the progressive closure of the South Tethys was the development of the shallow water evaporites of the Gachsaran Formation. These consist of more than 1500 m of anhydrite and salt in an elongated depression which extended from the Lurestan to the Dezful Embayment and the Interior Fars (Shiraz area). Shallow water depositional environments were confirmed by a brackish fauna of ostracods, rotalids and miliolids. By the end of Gachsaran deposition, the overburden of the

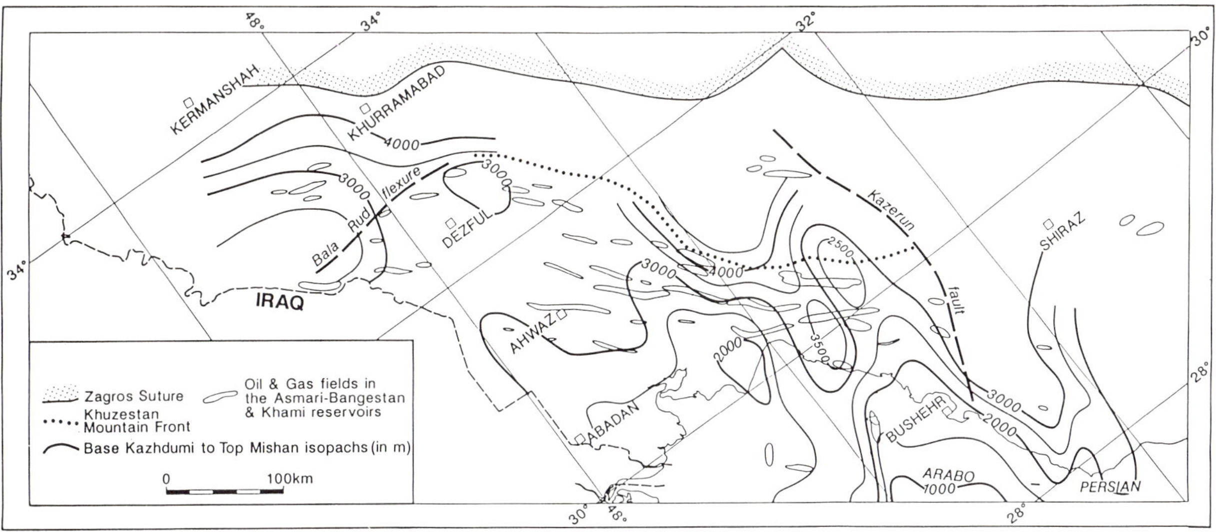

Fig. 12. Isopach map of the base Kazhdumi-top Mishan interval

Kazhdumi Formation varied between 2500 to 3200 m. It was still insufficient to achieve the onset of the oil window, especially at a time when high subsidence rates could have temporarily lowered the geothermal gradients.

The Zagros folding began during the deposition of the Gachsaran Formation. A slight growth of northwest-southeast-trending structures in the Dezful Embayment is observed from the Gachsaran Formation isopachs over the main fields, where a large number of wells have provided close control.

Subsequent deposition of the Mishan Formation as gray marls and shallow water argillaceous limestones, including reefal development in southeastern-most part of the Fars, increased overburden. In the Dezful Embayment and Northern Fars, Mishan thicknesses vary from 300 to 1000 m. By the end of Mishan times (Mid-Miocene), Kazhdumi overburden in the Dezful Embayment had increased to about 2600 to 3500 m. Onset of the oil window was most likely attained in the deepest areas (Fig. 12).

The Mishan was overlain by the Agha Jari Formation, a very thick section of brown calcareous sandstones and gypsum-veined reddish silty marls, deposited in a brackish environment over all Lurestan and Khuzestan, but in a more marine setting in Fars. The sands are partly made of chert fragments, which resulted from the erosion of the uplifted suture zone. The maximum thickness measured was about 4000 m at a site south of Lali. A subsiding trough, oriented northwest-southeast, was centered on the Dezful-Gachsaran area. At the Agha Jari type locality, the thickness of the formation is over 3000 m. The Agha Jari Formation is dated as Late Miocene to Pliocene from its brackish to marine fauna (James and Wynd 1965). Isopach mapping shows that the northwest-southeast Zagros structures were slowly growing during the deposition of the Agha Jari Formation. Overburden on the Kazhdumi source rock increased extremely rapidly during the Agha Jari times, resulting, almost everywhere, in maturity levels sufficient for oil to be generated in the Dezful Embayment. At the end of Agha Jari deposition, the Kazhdumi overburden ranged from to 5200 m at Mansuri to over 6500 m at Pazanan, where the gas window was reached.

The paroxysmal phase of the Zagros orogeny was marked by the unconformable deposition of the Late Pliocene to recent Bakhtiari conglomerates, which correspond to the infill of synclinal areas by coarse material eroded from the highest part of the growing structures. The conglomerates comprise round pebbles and cobbles of Oligocene, Eocene, and Cretaceous limestones and cherts. Thicknesses are extremely variable, as expected for this type of synorogenic deposit, and range from zero to 2500 m in some synclinorial areas. Most of the structures which now accommodate the largest fields have experienced erosion, exposing the Agha Jari and even the Gachsaran Formations in some areas. Agha Jari Formation thicknesses are

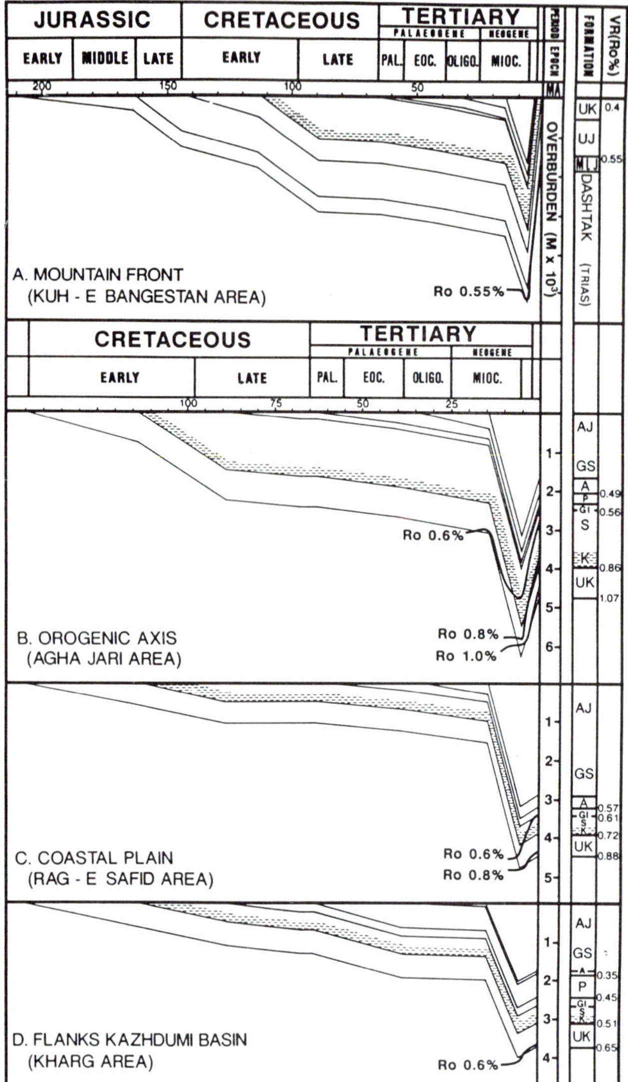

Fig. 13 A–D. Burial history profiles for Dezful Embayment traverse A-A′ showing the maturation history of the Kazhdumi Fm. Location of A-A′ is given in Fig. 14a. Formation abbreviations: *AJ* Agha Jari; *GS* Gachsaran; *A* Asmari; *P* Pabdeh; *GI* Gurpi-Ilam; *S* Sarvak; *K* Kazhdumi; *UK* Upper Khami; *UJ* Upper Jurassic. (Bordenave and Burwood, 1990)

therefore best measured on the flanks of neighboring synclines, or when drilling smaller, low relief structures detected by seismic in the large synclinorial areas between the main fields trends.

Maximum overburden on the Kazhdumi for most of the main structures was reached by the end of the Pliocene, certainly before the very recent Bakhtiari synchronous erosion and redeposition. Conversely, the deepest part of the syncline continued to subside until the present time.

As the result of the early growth of the Zagros structures, from the end of Early Miocene onwards, the actual areas of drainage for the main Dezful Embayment structures, observed by seismic, already existed at the onset of Kazhdumi oil generation at the end of Miocene. Moreover, the Zagros folding caused the fracturing of limestone as well as marl and facilitated the expulsion of petroleum out of source rocks and its vertical migration to Bangestan and Asmari reservoirs.

The Present Kazhdumi Maturity

The vitrinite reflectance available in numerous wells and calibrated subsidence history modeling (Fig. 13), allowed maturity maps to be established for the Kazhdumi (Fig. 14a) and the Pabdeh (Fig. 14b) source rocks (Bordenave and Burwood 1990). However, available data were obtained on the tops of anticlines, and maturity levels could be significantly higher in synclinal deeps. This is valid for the

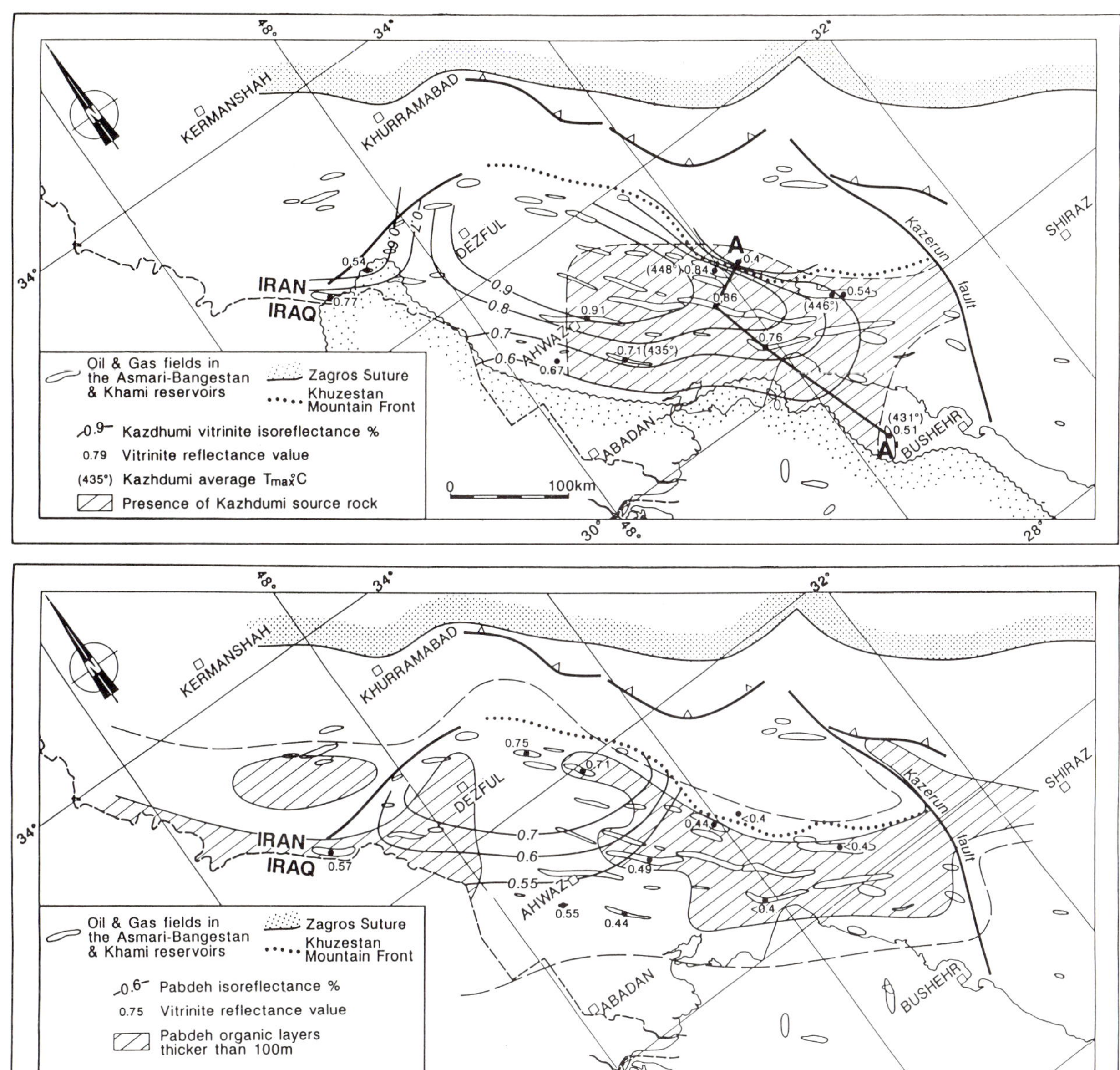

Fig. 14a, b. Maturity of source rocks as shown by vitrinite reflectance profiles. (After Bordenave and Burwood 1990). **a** Kazhdumi Formation. **b** Pabdeh Formation

Gachsaran field where Kazhdumi vitrinite reflectances hardly reaches 0.6%. This corresponds to calculated overburdens of 2650 and 3200 m at the end of Gachsaran and Mishan respectively. A thin Agha Jari Formation exists on the top of the structure. Conversely, nearly 3000 m of Agha Jari was measured in a well drilled on a deep structure immediatly west of Gachsaran. Here, the Kazhdumi had already attained the oil window.

Figure 14a shows that the Kazhdumi was mature enough for optimum oil generation in the Dezful Embayment. On the contrary (Fig. 14b), the Pabdeh has not attained onset oil generation in the main fields area (Ro varies from 0.4 to 0.5% for wells located on the structural highs), even if marginal maturity had been reached in the centers of some synclinal deeps. It can therefore be concluded that the Pabdeh source did not significantly contribute to the charge of the main fields. Conversely,

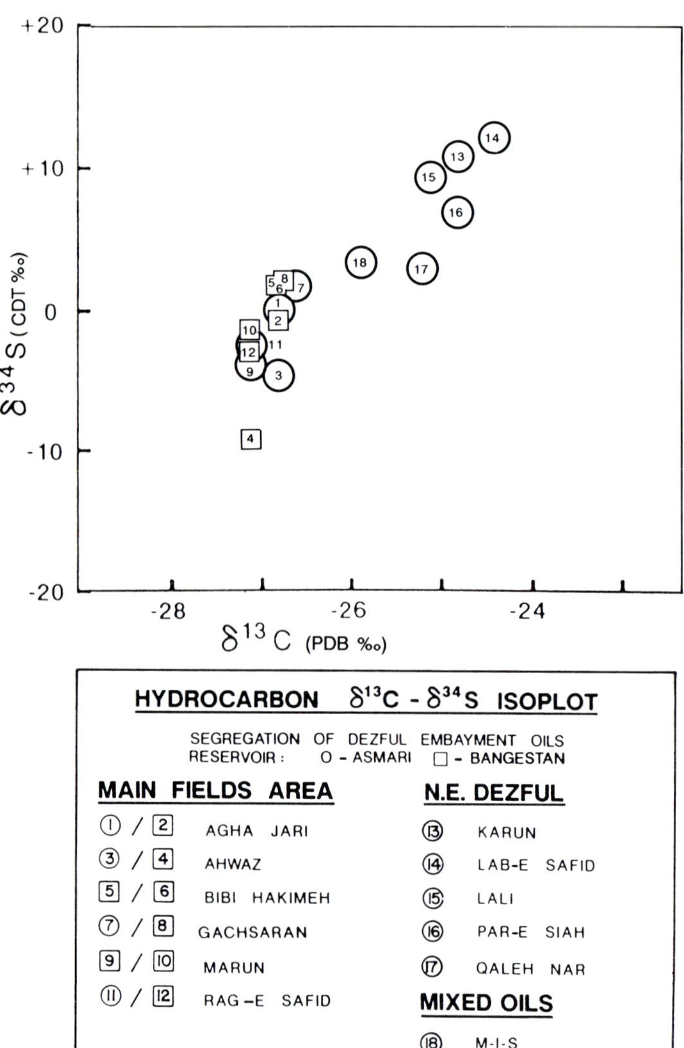

Fig. 15. Carbon and sulfur isotope cross plot of Dezful Embayment oils. (After Bordenave and Burwood 1990)

a maturity sufficient to reach the maximum oil generation zone was attained in the northeastern part of the Dezful Embayment, namely at Masjid-i Suleiman (MIS), Karun, Par-e Siah, Lab-e Safid, Qaleh Nar, and probably at Delhuran.

In the northeast Dezful, where oil is likely to have been generated from the Pabdeh Formation, the Kazhdumi is present as an organic-lean limestone facies. The Kazhdumi may contain some marginal source rocks only to the south of the MIS drainage area (Fig. 9b).

The geological/geochemical scenario for the charge of the Zagros anticlines from known source rocks was confirmed by the results of a generation model and later by correlations between oils and oil and source rocks, using stable isotopes (carbon and sulfur) and biomarkers (Bordenave and Burwood 1990).

Oil-to-Oil and Oil-to-Source Rock Correlations

Results obtained can be summarized as follows:

1. A carbon/sulfur isotope ratio plot (Fig. 15) of the oils trapped either in the Asmari or Bangestan reservoirs clearly shows two main families, one including the main fields, namely Agha Jari, Ahwaz, Bibi Hakimeh, Gachsaran, Marun, and Rag-e Safid, and the other including the fields located in the northeast Dezful (Karun, Lab-e Safid, Lali, Qaleh Nar, and Par-e Siah). An intermediate composition is found for the MIS crude oil. This segregation was confirmed by carbon isotopic compositions of the aromatic versus saturated fractions.

2. The similarity of the carbon isotopic composition of the oils in the main fields with kerogen pyrolysates derived from Kazhdumi source rocks

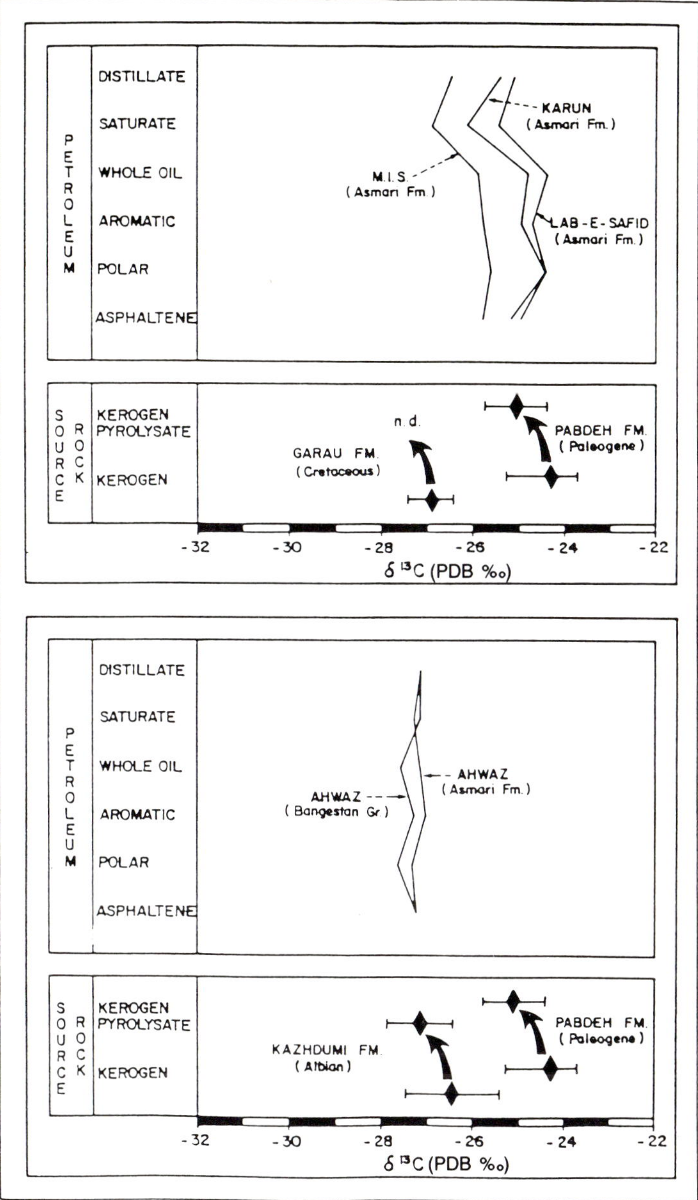

Fig. 16. Crude oil and kerogen pyrolysate carbon isotopic correlation profile showing the similarity between both the main fields (Ahwaz) oils with the Kazhdumi Fm. and the NE Dezful Fields with the Pabdeh Fm. (After Bordenave and Burwood 1990)

provides a positive correlation, whilst the isotopic composition of the NE Dezful oils is close to those of the Pabdeh source kerogens (Fig. 16).

3. The presence of significant amounts 18α(H)-oleanane in the Pabdeh-sourced northeast Dezful oils is characteristic of a terrigenous angiosperm plant contribution, which implied an Upper Cretaceous or a younger source age (Ekweozor et al. 1979). The angiosperm flora appeared in the Cretaceous (Peters and Moldowan 1993) but developed rapidly from the Early Tertiary onwards. 18α(H)-oleanane is not present in the Kazhdumi-sourced main fields oils (Fig. 17). The preference of norhopane compared to hopane in the main field oils, and a high C_{35} hopane in the NE Dezful oils also segregates these oils.

4. Sterane composition provided αα20S:R ethylcholestane data indicating that the oils are fully mature. Sterane contents also exhibit a greater relative abundance of C_{28} compounds in the NE Dezful oils, which may indicate a younger source origin compared to the main field oils (Grantham and Wakefield 1988). The same pattern observed in Iraq (Connan, 1988) and in the Aquitaine basin

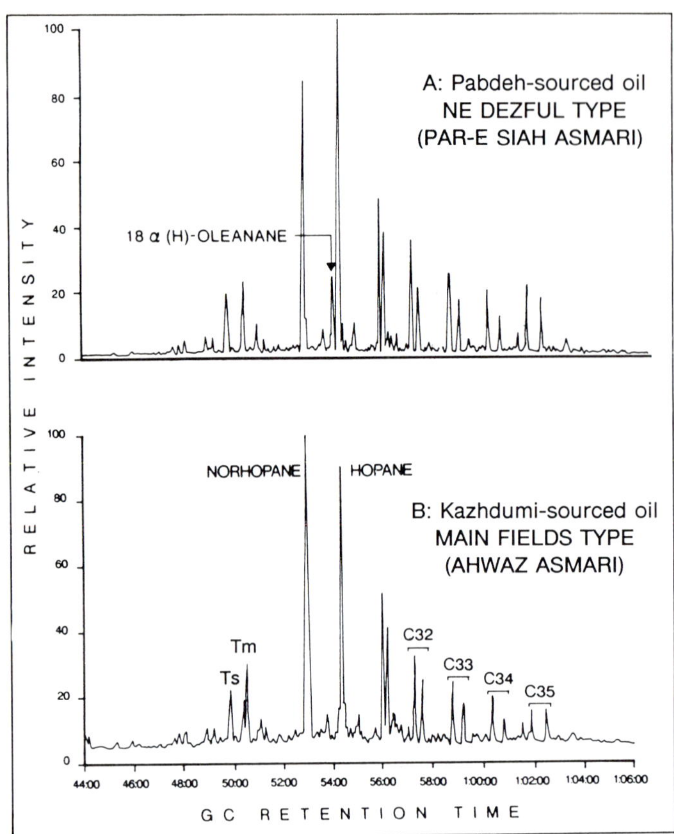

Fig. 17. Terpane m/z 191 fragmentograms for typical main fields (*Ahwaz-Asmari*) and NE Dezful (*Par-e Siah-Asmari*) oils. (After Bordenave and Burwood 1990)

(Connan and Lacrampe-Coulomme 1993) may be characteristic of evaporitic contributions (Fig. 18), which were known to be present during the Pabdeh deposition.

Conclusions

Three aspects of the Kazhdumi source rock geochemistry were discussed in this chapter:

1. The environment of deposition of one of the richest potential source rock.
2. The post-Kazhdumi geological history (Fig. 19) which allowed the productivity of this source rock to be extremely high, leading to the accumulation of about 7.3% of world reserves.
3. Oil-to-oil and oil-to-source rock correlation which allowed the confirmation of the geological/geochemical model derived from the source rock distribution and their maximum overburden toward the end of Pliocene (Fig. 20).

The Dezful Embayment was a preferential geological setting where anoxic conditions developed at least five times, during the Middle Jurassic (Sargelu Fm.), Neocomian (Garau Fm.), Albian (Kazhdumi Fm.), Campanian (Gurpi Fm.), and Mid-late Eocene (Pabdeh Fm.).

During the Albian, the Dezful Embayment corresponded to an intrashelf depression bordered laterally by a shallow shoal to the north, the regional Qatar-Fars arch to the south and separated from the South Tethys Ocean by a submarine sill to the northeast. To the southwest, it was limited by the prograding Safaniya-Burgan delta. The humid equatorial climate caused rivers to carry large amounts of nutrients, which were washed out from the wide Nubian landmass. Nutrients encouraged phytoplankton blooms and extremely high biological activity in the surface waters. The influx of freshwater over a deeper retro-current of oceanic water caused density water stratification. A worldwide sea level rise resulted in the bottom of the depression being significantly deeper than the euphotic zone and assisted in the establishment of euxinic conditions.

Therefore, the key words defining the Kazhdumi anoxia can be summarized as: intracratonic

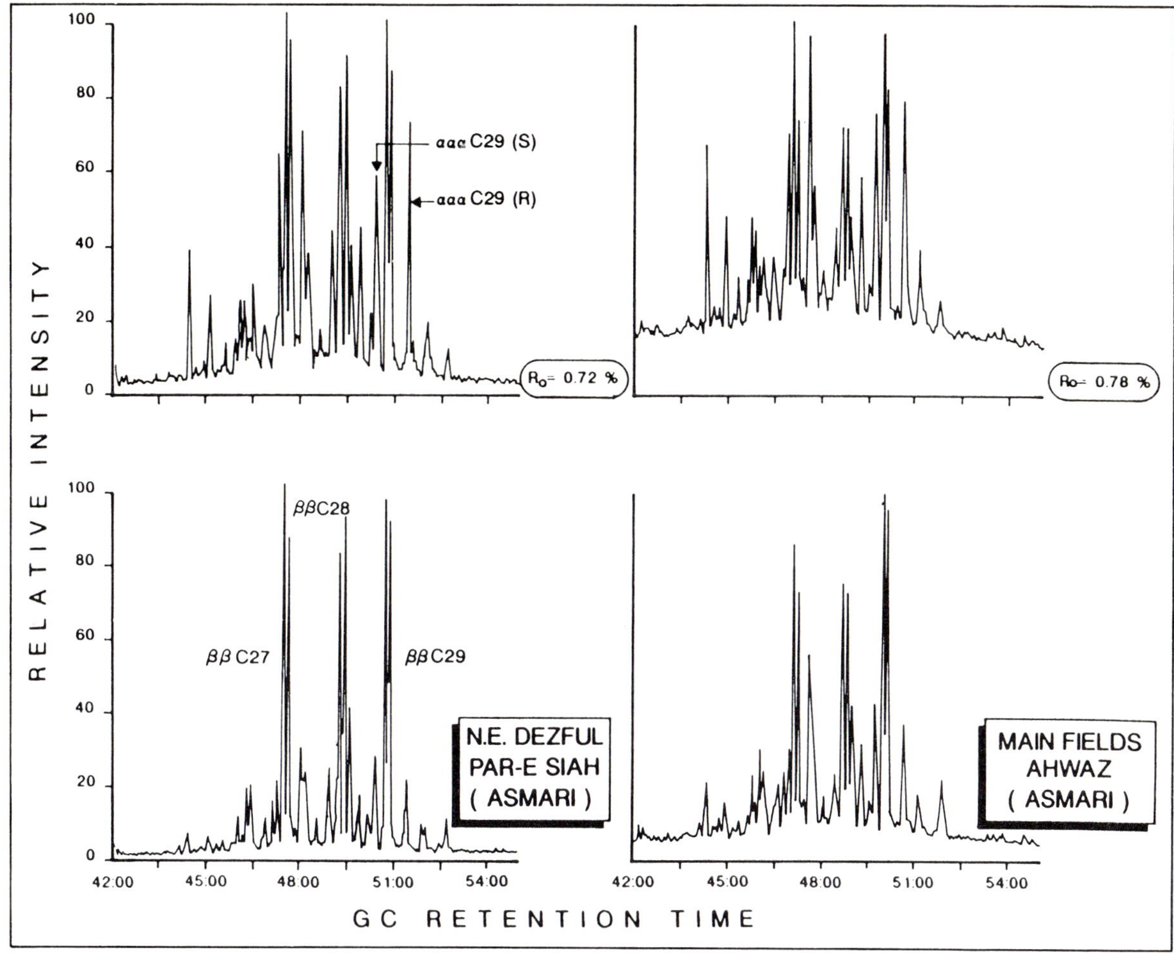

Fig. 18. Sterane m/z 217 and m/z 218 for typical main fields (*Ahwaz-Asmari*) and NE Dezful (*Par-e Siah-Asmari*) oils. (After Bordenave and Burwood 1990)

silled depression, humid climate, nutrients, sea level rise, and water stratification.

Some of the post-Albian geological factors important for the optimization of the Dezful Embayment 'oil generation system' were recognized early in the exploration of the Zagros Foothills, including the existence of excellent calcareous reservoirs (Asmari and Bangestan), large whale-back structures, efficient cap rock provided by the Gachsaran Formation (Dunnington 1967), and the proximity of the Kazhdumi source rock with the Bangestan reservoir. Other factors emerged later, including the early growth of the Zagros structures since the end of the Early Miocene. This preceded the onset of the oil window, providing large traps before the active generation of petroleum. Moreover the early definition of the areas of drainage allowed a semi-quantitative modeling of oil generation. Another favorable factor was the coincidence of oil generation with the late paroxysmal phase of the Zagros folding. This resulted in fracturing, observed during field work in marls and argillaceous limestones of the Pabdeh-Gurpi interval as well as in reservoirs (Fig. 19). This fracturing facilitated the vertical migration processes.

Much of the information used in this chapter was collected before 1978. It is therefore urgent to revise the stratigraphy of the Zagros Foothills within the framework of global sequence stratigraphy, and to re-examine the source rocks penetrated by the numerous wells available. Moreover, more contemporary, kinetic-based generation modeling procedures need to be applied. The Dezful Embayment is not only a prolific petroleum province, but one of the best geological laboratories to calibrate geochemical models in relation to hydrocarbon generation.

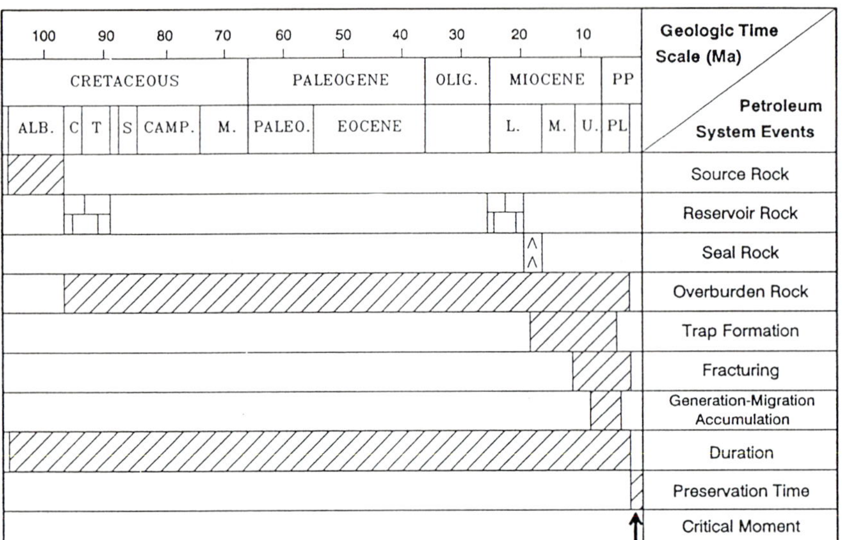

Fig. 19. Kazhdumi petroleum generation timing chart. (After Magoon and Dow 1991)

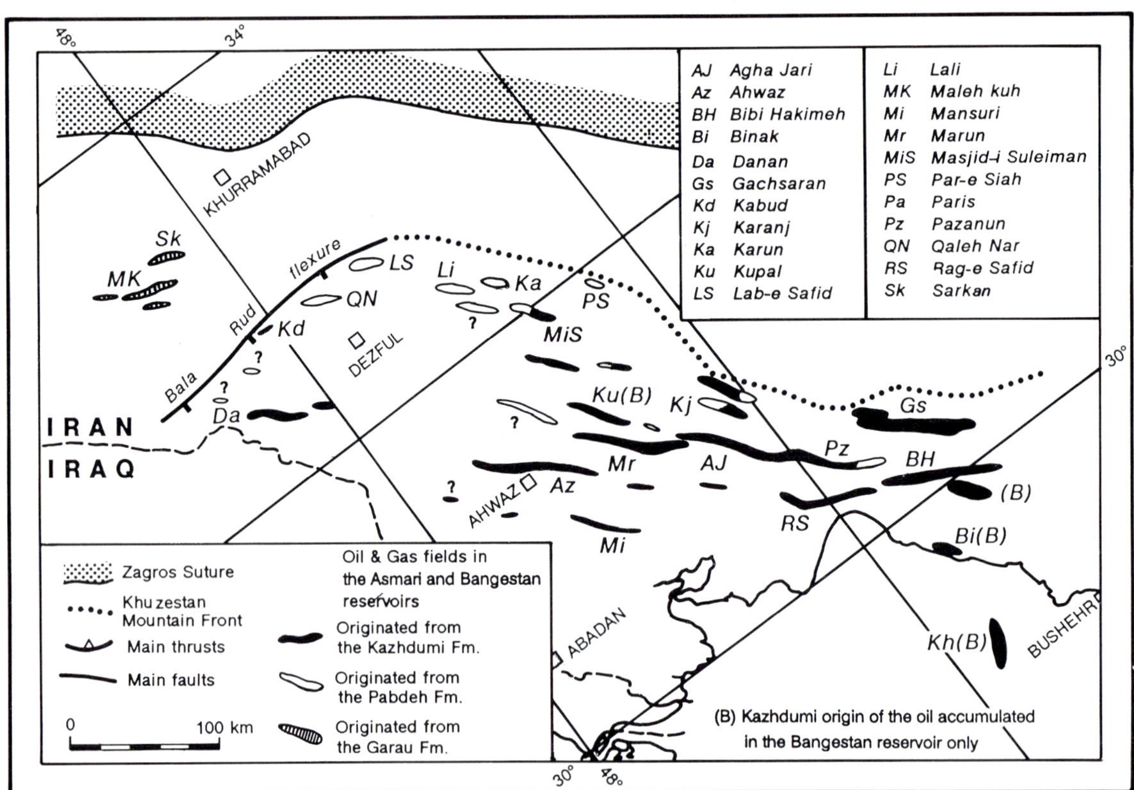

Fig. 20. Lurestan and Dezful Embayment parentage between source rocks and oil accumulations in Asmari and Bangestan reservoirs

Acknowledgments. The authors wish to thank all their colleagues who participated in the geochemical project between 1968 and 1978. Amongst the geologists, we would like to acknowledge A.R. Nili, who participated in a very efficient and dedicated manner between 1967 and 1974, for both the field work and for the review of the data, F. Sahabi, who participated in the Lurestan geochemical appraisal, and M.C. Fozoonmayeh for field work in North Fars. We are indebted to M. Jalali and H. Zavieh, who provided, under the direction of A. Gollestaneh, useful age dating and facies indications from micropaleontological study. R. Stoneley and the regional team led by W. Koop are thanked for their encouragement over the latter stages of the study. We are grateful to R.R.F. Kinghorn, K. Magara and K.E. Peters for their review and their comments which have helped to improve the manuscript.

References

Ala MA, Kinghorn RRF, Rahman M (1980) Organic geochemistry and source rock characteristics of the Zagros petroleum province, Southwest Iran. J Pet Geol 3: 16–89

Ayres MG, Bilal M, Jones RW, Slentz LW, Tartir M, Wilson AO (1982) Hydrocarbon habitat in main producing areas, Saudi Arabia. Am Assoc Petrol Geol Bull 66: 1–19

Bordenave ML (1993) Screening techniques for source rock evaluation—sampling and validity of results of analyses. In Bordenave ML (ed) Applied petroleum geochemistry, Technip, Paris, 225–233

Bordenave ML, Burwood R (1990) Source rock distribution and maturation in the Zagros belt; provenance of the Asmari and Bangestan reservoir oil accumulations. Org Geochem 16: 369–387

Bordenave ML, Nili AR (1973) Geochemical project, review and appraisal of the Khuzestan province, report 1194. Geological and Exploration Division. Iranian Oil Operating Companies, cited by Khosravi Said, 1987

Bordenave ML, Combaz A, Giraud A (1970) Influence de l'origine des matières organiques et de leur degré d'évolution sur les produits de pyrolyse du kérogène. In: Hobson GD, Speers GC (ed) Advances in organic geochemistry 1966 Pergamon, Oxford, p. 389–405

Burwood R, Drozd RJ, Halpern HI Sedivy RA (1988) Carbon isotopic variations of kerogen pyrolysates. Org Geochem 12: 195–205

Clarke RH (1975) Petroleum formation and accumulation in Abu Dhabi. 9th Arab Pet. Cong. Proc. Dubai. B-3, 20–29

Connan J (1988) Quelques secrets des bitumes archéologiques de Mésopotanie révélés par les analyses de géochimie organique pétroliéres. Bull Centre Rech Explor-Prod Elf Aquitaine, 12: 759–787

Connan J, Lacrampe-Coulomme G (1993) The origin of the Lacq Supérieur heavy oil accumulation and the giant Lacq Inférieur gas field (Aquitaine, SW France). In: Bordenave ML (ed) Applied Petroleum Geochemistry. Technip, Paris, p. 465–488

Demaison GJ, Huizinga BJ (1991) Genetic classification of petroleum systems. Am Assoc Petrol Geol Bull 75: 1626–1643

Demaison GJ, Moore GT (1980) Anoxic environments and oil source bed genesis, Am Assoc Petrol Geol Bull 64: 1179–1209

Droste H (1990) Depositional cycles and source rock development in an epeiric intra-platform basin. The Hanifa Formation of the Arabian peninsula. Sed Geol 69: 281–296

Dunnington HV (1967) Stratigraphical distribution of oil fields in the Iran-Iraq-Arabia Basin. J Inst Petrol 53: 129–161

Ekweozor CM, Okogun JI, Ekong DEU, Maxwell JR (1979) Preliminary organic geochemical studies of samples from the Niger Delta (Nigeria). 1: Analyses of crude oils for triterpanes. Chem Geol 27: 11–28

Giraud A (1970) Application of pyrolysis and gas chromatography to geochemical characterisation of kerogen and sedimentary rocks. Am Assoc Petrol Geol Bull 54: 439–455

Grantham PJ, Wakefield LL (1988) Variation in the sterane carbon number distribution of marine source rock derived crude oils through geological time. Org Geochem 12: 61–73

Grantham PJ, Lijmbach GWM, Posthuma J, Hughes Clark MW, Willink RJ (1987) Origin of crude oils in Oman. J Petrol Geol 11: 61–80

Harris PM, Frost SH, Seiglie GA Schneidermann N (1984) Regional unconformities and depositional cycles. Cretaceous of the Arabian Peninsula In: Schee JS(ed) Interregional unconformities and Hydrocarbon Accumulatlon: Am Assoc of Petr Geol, Tulsa, Memoir 36: 67–80

Hassan TH, Azer S (1985) The occurrence and origin of oil in Offshore Abu Dhabi. S.P.E. Middle East Technical Conference and Exhibition, Bahrain. pp 143–160

James GA, Wynd JG (1965) Stratigraphic nomenclature of Iranian Oil Consortium Agreement Area. Am Assoc Petrol Geol Bull 49: 2182–2245

Kent PE, Slinger FCP, Thomas AN (1951) Stratigraphical exploration surveys in south-west Persia: Proc. 3d World Petroleum Cong, The Hague, sect 1, 141–161

Khosravi Said A (1987) Geochemical concepts on origin, migration and entrapment of oil in Southwest Iran. In:Kumar RK et al. (eds) Petroleum geochemistry and exploration in the Afro-Asian Region, Balkema, Rotterdam, p. 531–539

Koop WJ, Stoneley R (1982) Subsidence history of the Middle East Zagros basin. Phil. Trans. R. Soc. Lond. A 305: 149–168

Lehner P, Meijer B, Kooper H (1984) Mesozoic source rocks of the Arabian Peninsula. OAPEC seminar on habitat of petroleum in Arab countries, pp. 73–118

Magoon LB, Dow WG (1991) The petroleum system—from source to trap (Abs), Am. Assoc. Pet Geol Annual convention program, p. 162

Murris RJ (1980) Middle East. Stratigraphic evolution and oil habitat. Am Assoc Petrol Geol Bull 64: 587–618

Peters KE, Moldowan JM (1992) The biomarker guide. Prentice Hall, Englewood Cliffs, 363 p

Powers RW (1968) Lexique stratigraphique international, III, 106, 1, Arabie Saoudite

Ricou LE (1974) L'étude géologique de la Région de Neyriz (Zagros Iranian) et l'évolution structurale des Zagrides. Ph.D. Thesis, Paris-Orsay, 321 p

Roth PH, Bowdler JL (1981) Middle Cretaceous calcareous Nannoplankton, biogeography and oceanography of the Atlantic Ocean, Soc Econ Paleont and Mineralog, Tulsa, Spec Publ 32: 517–546

Setudehnia A (1978) The Mesozoic sequence in South-West Iran and adjacent areas. J Petrol Geol 1: 3–42

Stöcklin J (1968) Structural history and tectonic of Iran, Am Assoc Petrol Geol Bull 52: 1229–1258

Organic-Rich Chalks and Calcareous Mudstones of the Upper Cretaceous Austin Chalk and Eagleford Formation, South-Central Texas, USA

G.J. Grabowski, Jr.

Abstract

The Austin Chalk and Eagleford Formation are widespread Upper Cretaceous deposits of the Gulf Coast region of south-central Texas. The upper part of the Austin is formed by prograding highstand sequences of light-colored bioturbated chalks. The lower part of the Austin and the Eagleford are composed of organic-rich chalks and calcareous mudstones that form the transgressive systems tracts of three sequences. These organic-rich rocks are laminated or sparsely burrowed and contain few benthic fossils, suggesting deposition in a basin with oxygen-deficient sediments.

The organic-rich rocks of the lower Austin are basinally restricted deposits that reach 200 ft (61 m) in thickness. They contain an average of 3.7% TOC composed of a mixed type II/III kerogen (HIo = 481 mg/g), dominated by amorphous organic matter but with 5–40% vitrinite, inertinite, and spore-pollen. Pristane/phytane ratios of 0.6–1.3, minor amounts of oleanane, a predominance of moretane and hopane over normoretane and norhopane, and predominance of hopanes over steranes indicate deposition of mixed marine and terrestrially derived organic matter under dioxic conditions.

The Eagleford is generally thinner (40–150 ft or 12–46 m) but more widespread than the lower Austin. TOC values average 3.7% in shallow cores and 4.5% in one deep core. The organic matter in the shallow cores is a mixed type II/III (HIo = 414 mg/g). However, predominance of normoretane and norhopane over moretane and hopane, increased amounts of tricyclic terpanes, absence of oleanane, and predominance of steranes over hopanes indicate deposition of mainly marine-derived organic matter under anoxic conditions.

Generation of liquid hydrocarbons has occurred below 6000–7000 ft (1830–2130 m) present-day burial for both formations. Rocks at depths below 9000–10 000 ft (2740–3050 m) are now overmature for generation of liquid hydrocarbons. Calculated specific oil yields are 0.3 –0.4 bbls/acre-ft (0.3 bbls/km^2/m) for the lower Austin and the shallow Eagleford cores and 1.2 bbls/acre-ft (1.0 bbls/km^2/m) for the deep Eagleford core. Where thick lower Austin is present, it probably has generated and expelled hydrocarbons. However, the more widespread Eagleford has most likely generated the majority of the oil found in the Austin Chalk.

Oil has migrated into more permeable intervals of the upper Austin and can be discriminated by composition from in situ EOM. Pristane/phytane ratios and stable-carbon isotopes do not differentiate between the lower Austin and the Eagleford as the source rocks for these oils.

Introduction

Chalks are fine-grained carbonate rocks (> 50% calcite) composed primarily of the skeletal remains of calcareous nannoplankton, specifically coccoliths. Most chalks are light in color and contain little organic matter. However, several occurrences of dark-colored, organic-rich chalks and calcareous mudstones (25–50% calcite) are known (Table 1). This chapter summarizes the distribution and composition of the organic-rich chalks and calcareous mudstones of the Austin Chalk and Eagleford Formation of south-central Texas.

The Austin Chalk and Eagleford Formation span an interval of some 12 million years duration from Cenomanian to Santonian (Fig. 1). This was a time of high sea level worldwide (Haq et al. 1987). A large epeiric sea existed on the western North American craton (Williams and Stelck 1973), with chalks forming in the central and southern parts of

P.O. Box 4778, Exxon Exploration Company, Houston, Texas 77210-4778, USA

Table 1. List of some organic-rich chalks described in the literature

Formation	Age	Location	%TOC	OMT
Alcanar Fm. (1)	Middle Miocene	Offshore eastern Spain	0.5–2.5	III
Mauwaqqar-Ghareb Fm. (2)	Maastrichtian	Levant Coast, Jordan, and Israel	5–40	I
La Luna Fm. (3)	Santonian to Turonian	Colombia and Venezuela	3.4–4.9	II
Niobrara Fm. (4)	Santonian to Caniacian	Western United States	0.4–14	II
Tarfaya oil shales (5)	Santonian to Cenomanian	Coastal Morocco	1–14	II
Toolebuc Fm. (6)	Upper Albian	Queensland, Australia	1–25	I

(1) Demaison and Bourgeois 1984; (2) Abed and Amireh 1983; (3) Zumberge 1984; (4) Rice 1984; (5) Leine 1986; (6) Pevear and Grabowski 1985

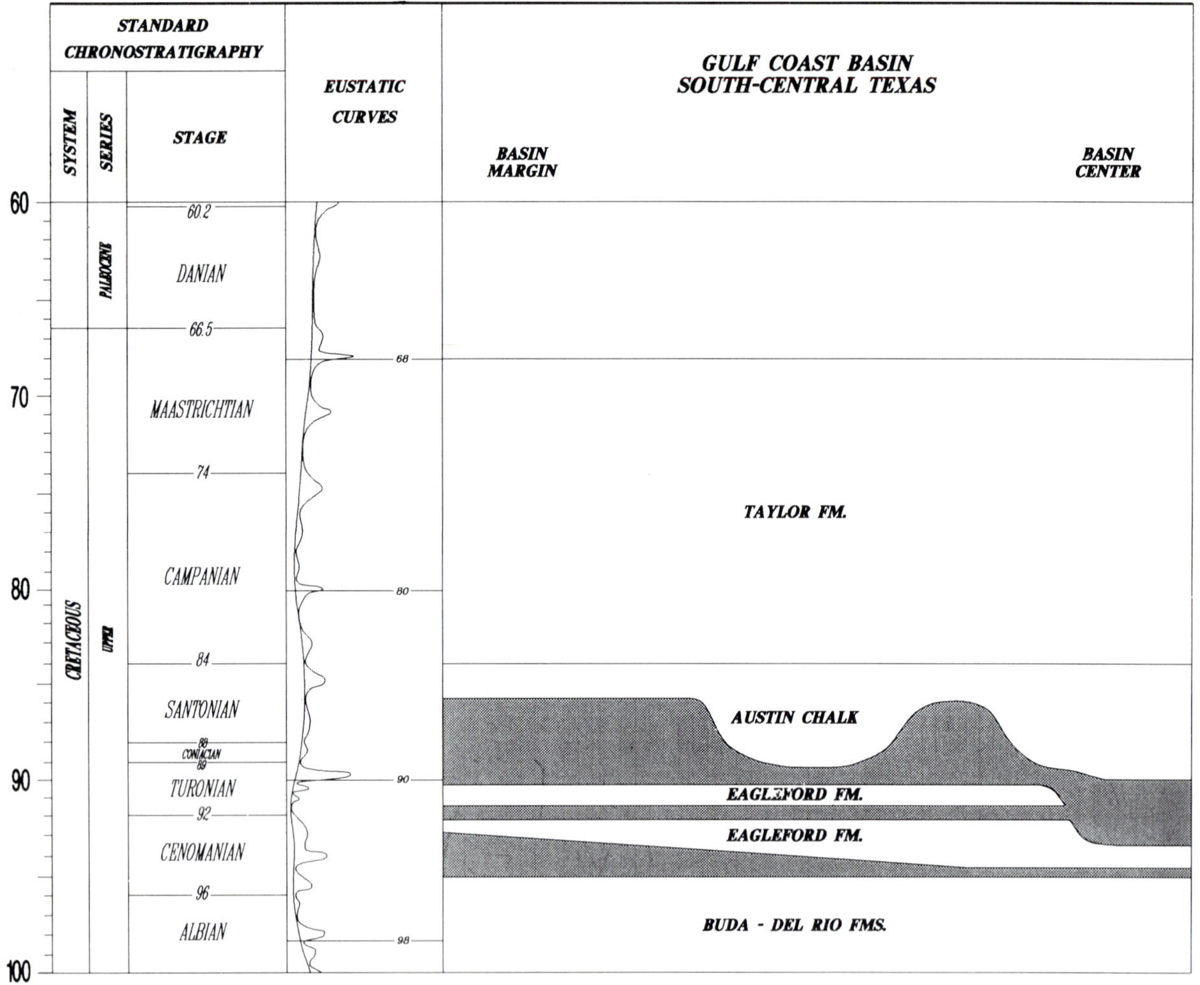

Fig. 1. Chronostratigraphic diagram of the Upper Cretaceous of the Gulf Coast basin of Texas (Barrier 1980). Ages and eustatic curve from Haq et al. (1987)

this seaway (Kauffman 1979). The Austin Chalk and Eagleford Formation formed in the southern end of this epeiric sea, on a gently sloping ramp that deepened into the Gulf of Mexico.

The Austin Chalk is a reservoir for petroleum in south-central Texas. The reservoir potential of the Austin is discussed in Scholle (1977a,b), Scholle and Cloud (1977), Dravis (1979), Czerniakowski et al.

(1984), and Corbett et al. (1987). Aspects of the organic geochemistry of the Austin Chalk are presented in Grabowski (1981a,b, 1984) and Hunt and McNichol (1984). The geochemistry of oils produced from the Austin Chalk is discussed in Thompson (1991), Thompson et al. (1990), and Comet et al. (1993).

Stratigraphy and Depositional Environments

Austin Chalk

The Austin Chalk ranges in thickness from 100 to 650 ft (30 to 200 m) in outcrop and thickens basinward to over 1200 ft (365 m) in the subsurface of south-central Texas (Dravis 1979). Figure 2 shows the distribution of the Austin Chalk in outcrop and a generalized isopach of the formation in the subsurface, based on well data.

The Austin Chalk of south-central Texas can be divided into (1) a widespread upper part, consisting mainly of light-colored, shelfal chalks, and (2) a basinally restricted lower part, that, where present, consists of darker colored chalks. The stratigraphy and sedimentology of these two intervals are described below. Further basinward, an older part of the Austin Chalk is recognizable but will not be described in this chapter.

Upper Part of the Austin Chalk

The Austin Chalk is thinnest in the shallow updip and outcrop trend, where it is less than 6000 ft (1830 m) deep (Fig. 3). In these areas, the Austin is mainly late Santonian in age (Barrier 1980). The Austin unconformably overlies the Eagleford Formation and in turn is overlain by the Taylor Formation (Figs. 4, 5).

The Austin Chalk thickens in a basinward direction, to the southwest of San Antonio and east of Austin (Fig. 2). This thickening occurs by addition of older strata which onlap the underlying

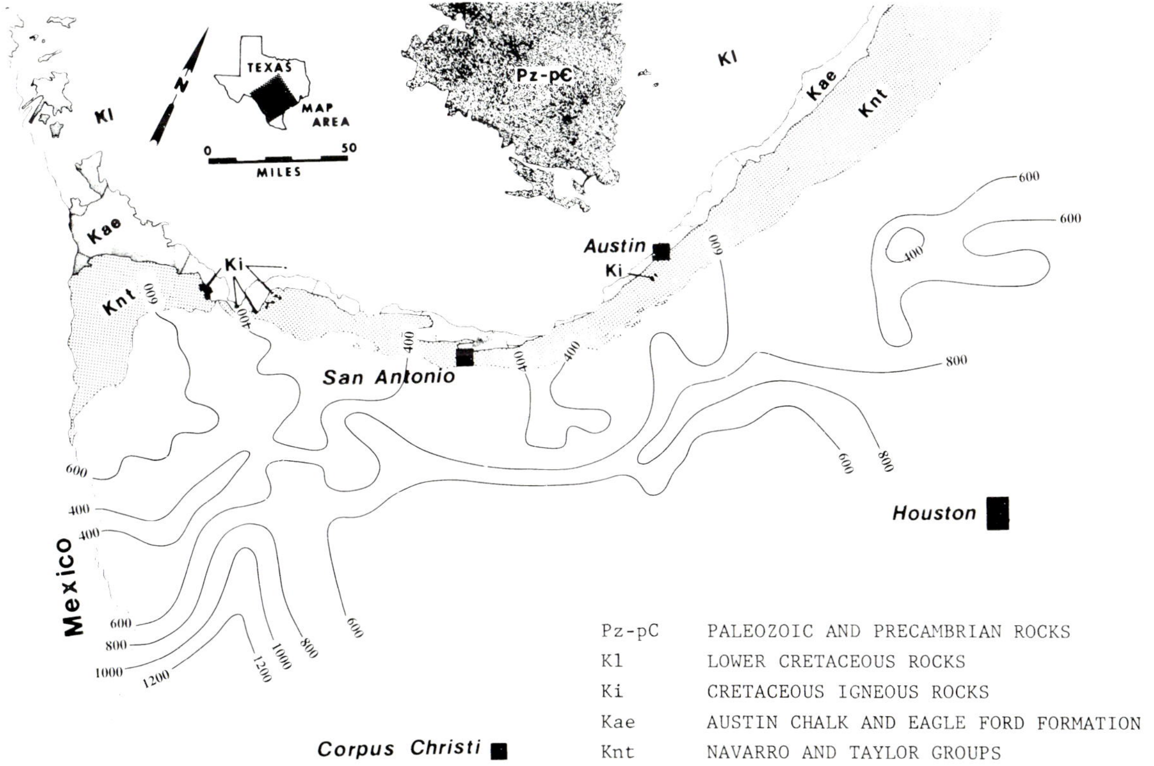

Fig. 2. Map of south-central Texas, showing the outcrop distribution of the Austin Chalk and generalized thickness of the formation in the subsurface, based on well data

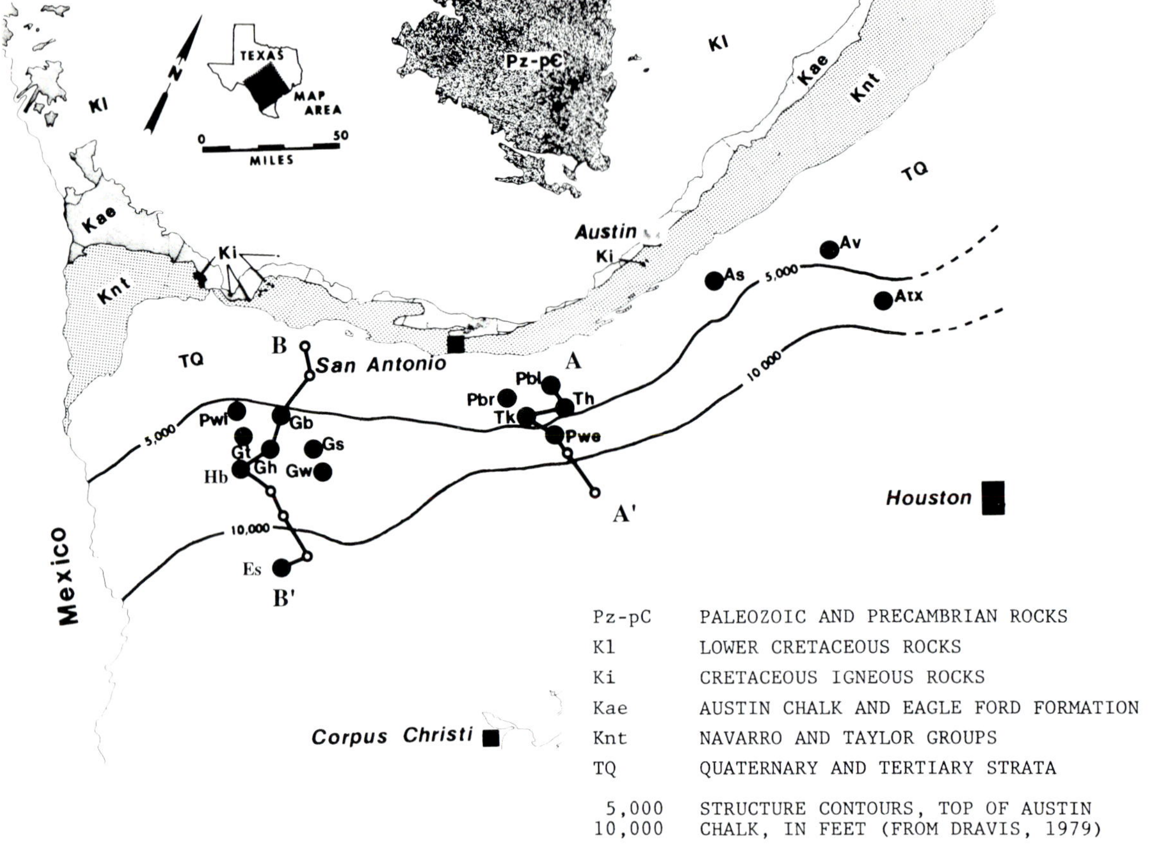

Fig. 3. Map showing the cores studied and the lines of sections in Figs. 4 and 5

Eagleford Formation (Figs. 4, 5). This lower part of the Austin is Santonian to Coniacian in age. Further basinward, a thick wedge of sediment overlies the Eagleford Formation, as shown in Fig. 6.

The sedimentology of the Austin Chalk has been described by Dravis (1979). The upper part of the Austin is seen in updip cores and in outcrops. An example of the chalks of the upper part of the Austin is shown in Fig 7.

Chalks of the upper Austin are typically light colored, thoroughly bioturbated, and contain many coarse benthic skeletal grains (Fig. 8). These chalks are composed principally of calcite (an average 79% acid-soluble material), with lesser amounts of clay minerals, pyrite, glauconite, and quartz, and very little organic matter. Laminae and thin beds of calcareous shale and zones of wispy microstylolites formed by pressure solution are present throughout the upper Austin but form a small percentage of the formation. These argillaceous layers contain an average of 57% acid-soluble material (calcite).

Dravis (1979) interpreted that these light-colored, skeletal chalks formed in a shelfal environment, probably in water only tens of meters deep. Seawater salinity was normal, as indicated by abundant marine fossils including echinoderms. The bottom was oxygenated and supported a diverse benthic epifauna and a burrowing infauna. Deposition of the chalk matrix was mainly by settling of coccoliths and pellets from the water column.

These light-colored, shelfal chalks are widespread across the updip and outcrop trends of the Austin and extend downdip above the older portions of the Austin Chalk. Marker beds within the upper part of the Austin are readily correlated on electric logs in the subsurface (Figs. 4, 5). They show subtle basinward clinoform geometries that record progradation of these shelfal chalks into the Gulf Coast basin. This upper part of the Austin Chalk is interpreted to be a highstand deposit consisting of several prograding sequences.

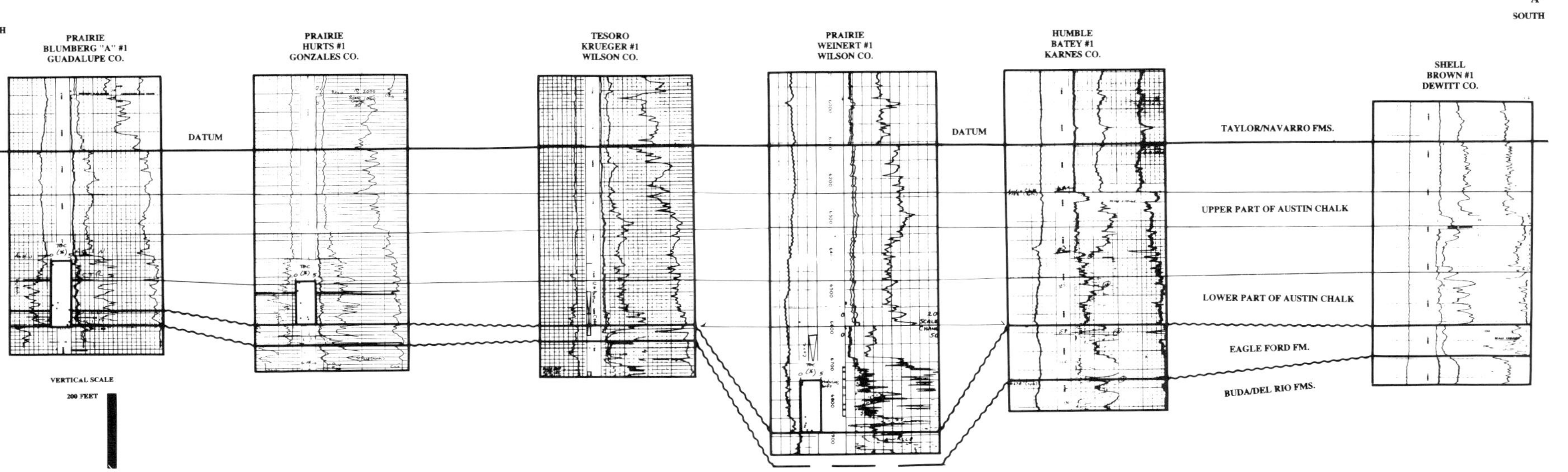

Fig. 4. Well-log cross section, datumed on top of the Austin Chalk, showing the onlapping relationship of the lower portion of the Austin onto the Eagleford Formation. Note that the marker beds correlated in the upper part of the Austin are approximately horizontal regionally. *Box* within wells is measured TOC (%) from cores. Location of cross section given in Fig. 3

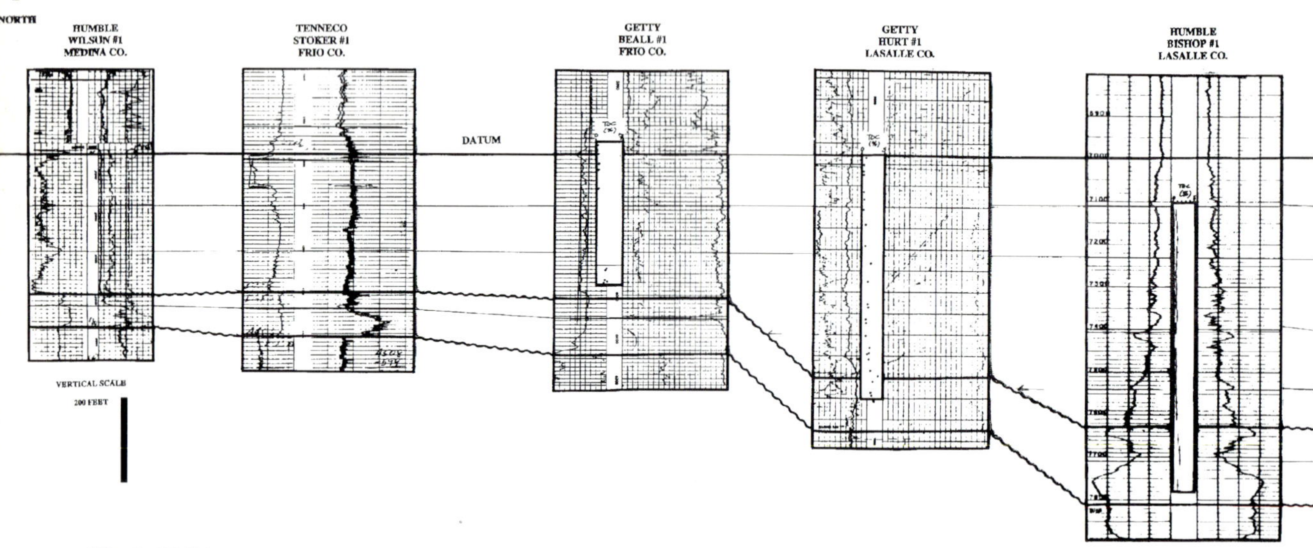

Fig. 5. Well-log cross section, as in Fig. 4

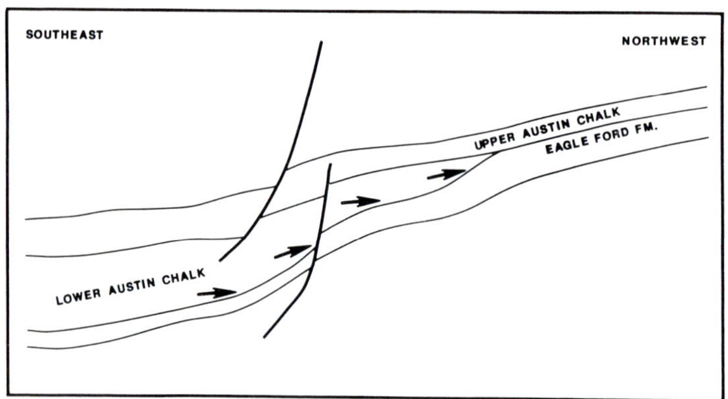

Fig. 6. Line tracing of seismic section from the Gulf Coast basin of Texas, illustrating the thick wedge of Austin Chalk above the Eagleford Formation. *Arrows* indicate onlap recognized on seismic line

Lower Part of the Austin Chalk

The lower part of the Austin Chalk is present only in localized parts of the deeper subsurface of south-central Texas. In these areas, the lower Austin may reach 300 ft (91 m) in thickness. It consists mainly of darker colored, generally laminated chalks (Fig. 9), with sparse interbeds of lighter colored, burrowed chalks and thin beds of brecciated chalk. Descriptions of two cores of the lower Austin are shown in Figs. 10 and 11.

These chalks contain 65–91% acid-soluble material that is mainly calcite. The darker color of these chalks is due to pyrite, organic matter, and clay minerals. The siliciclastic minerals and organic matter are heterogeneously distributed in the lower Austin, although there is a general tendency for the lower 100 ft (33 m) to contain less calcite.

The lamination in these rocks is caused by alternations of light-colored laminae of silt- and sand-sized planktic foraminifera and calcispheres, cemented by calcite, and dark-colored, finer-grained laminae composed mainly of coccoliths and clay. A few laminae of pyrite are present in these rocks.

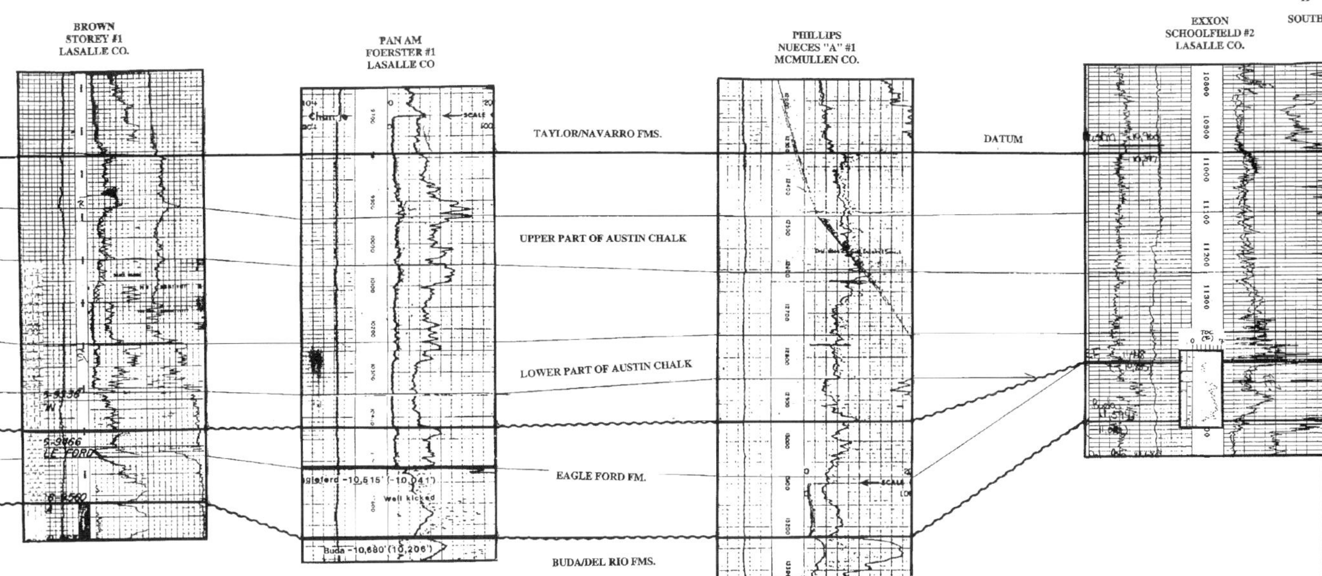

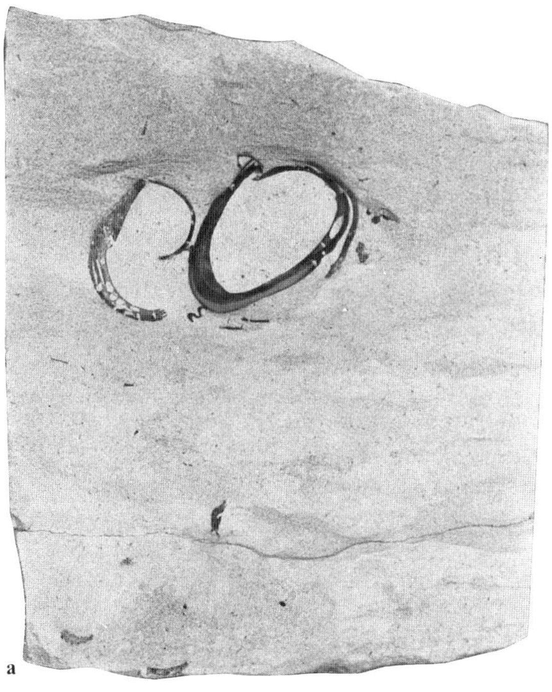

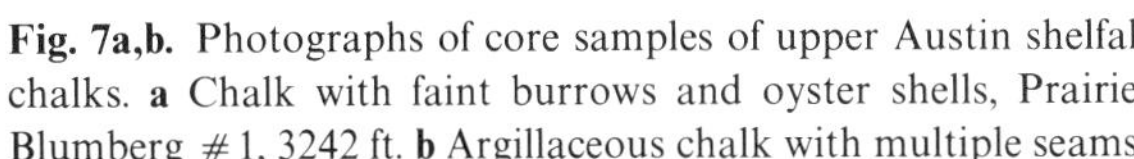
Fig. 7a,b. Photographs of core samples of upper Austin shelfal chalks. **a** Chalk with faint burrows and oyster shells, Prairie Blumberg # 1, 3242 ft. **b** Argillaceous chalk with multiple seams of microstylolites and nodular fabric of chalk containing solitary burrows, Prairie Blumberg # 1, 3272.5 ft.

Wispy microstylolites and rare thin beds of shale occur sporadically in these chalks. These clayey zones contain about 57% acid-soluble material.

Burrows are rare in these chalks, generally consisting of discrete *Chondrites* or *Zoophycus*. At least partially preserved lamination is present in burrowed rocks. Skeletal fragments of benthic

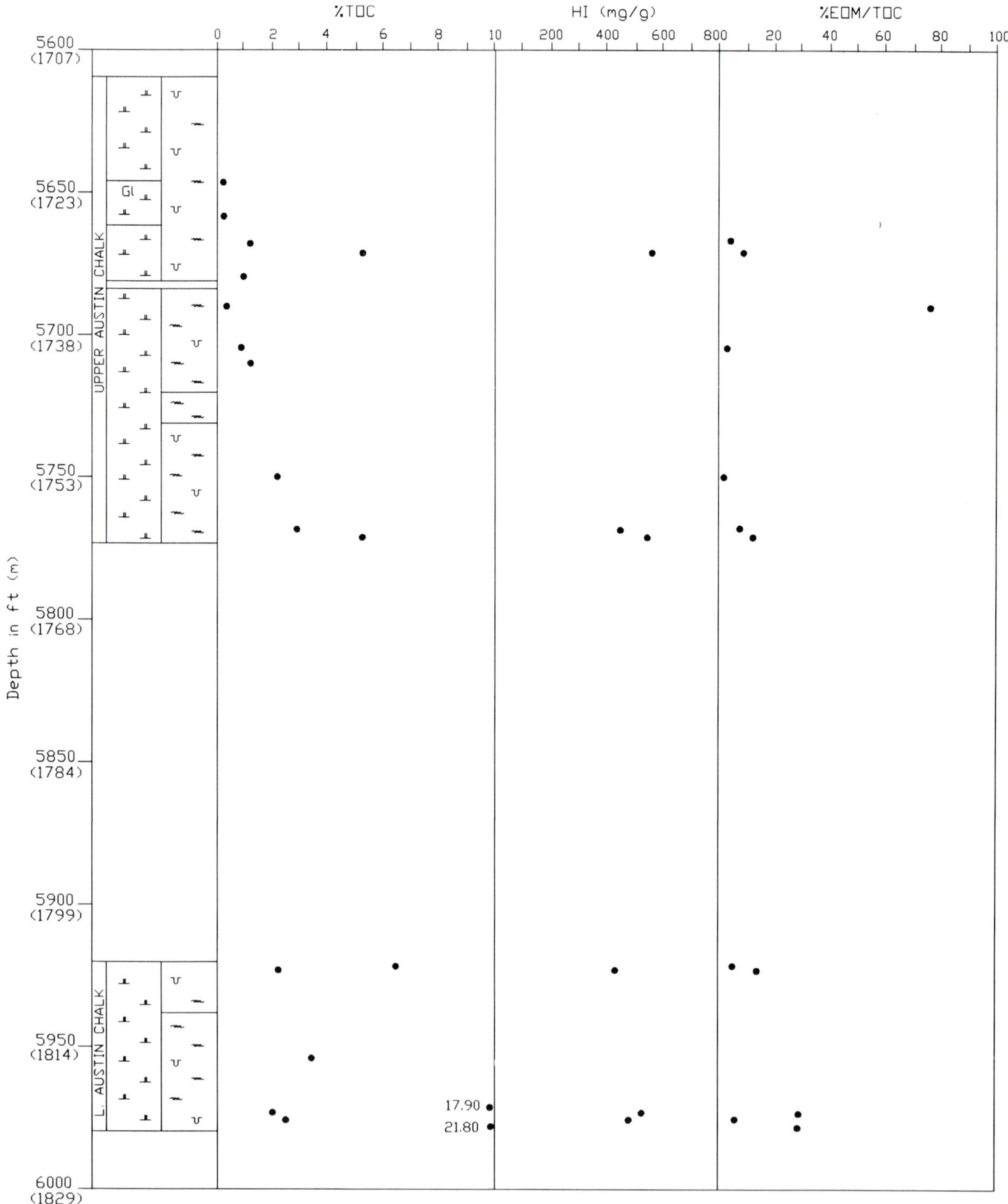

Fig. 8. Description of core from the Getty Beall #1 well, including bioturbated light-colored chalks of the upper part of the Austin Chalk

Fig. 9a-c. Photographs of core samples of lower Austin basinal chalks. **a** Laminated chalk with lenticular laminae, Prairie Weinert #1, 6889.6 ft. **b** Laminated chalk, Getty Hurt #1, 7204.2 ft. **c** Laminated chalk with distinctive pyritic laminae, Amoco Texas A&M #1, 9101 ft.

fauna are absent or extremely rare in the lower Austin.

Thin beds of breccia consisting of intraclasts of lighter colored chalk occur in some downdip wells. Dravis (1979) interpreted these beds as shallow-water sediments that were transported downslope during deposition.

Dravis (1979) interpreted that the darker colored, laminated chalks formed in a basinal environment, below wave base in perhaps 100 m of water. Oxygen deficiency in the bottom sediments is indicated by the paucity of infaunal skeletal remains, by the rarity of burrows and prevalence of lamination, and by the presence of sporadic pyrite laminae. However, the bottom probably was not totally anoxic.

These dark-colored basinal chalks are basinally restricted and onlap the underlying Eagleford Formation (Figs. 4, 5). Time-equivalent chalks are absent from the updip shelfal regions, which at the time of deposition were either subaerially or subaqueously exposed. This lower part of the Austin is the transgressive deposit upon the unconformity surface at the top of the Eagleford Formation. However, the top of the Eagleford is not deeply incised but has sagged, and the lower Austin fills these sag basins.

Eagleford Formation

The Eagleford Formation forms a thin sheet, 40–300 ft (12 91 m) thick, across most of south-central Texas. The Eagleford thins mainly by onlap upon the underlying strata, with generally lesser amounts of erosion beneath the Austin. It is thickest where the overlying lower Austin is thickest (Figs. 4, 5), indicating that these areas were sagging during Eagleford deposition as well as later.

The Eagleford can be divided into upper and lower parts at a distinctive break on electric logs. The upper part is considered Turonian in age, whereas the lower part is predominantly Cenomanian (Fig. 1). Both parts are present across most of south-central Texas.

The Eagleford Formation has been studied in several cores from south-central Texas, one of which is described in Fig. 12. The Eagleford consists mainly of dark-colored, laminated calcareous mudstones and chalks with sparse thin beds of *Inoceramus* fragments. The mudstones and chalks are coccolith-rich and contain some foraminifera, similar to the lower part of the Austin Chalk, but the carbonate content generally is lower in the Eagleford (40–76% calcite by XRD). The lowermost laminae of each sequence of the Eagleford contain more siliciclastic material, less calcite, and less organic matter. No burrows were observed in the cores.

The Eagleford Formation onlaps the underlying strata at a low angle (Figs. 4, 5). This geometry and the basinal nature of the rocks indicate that the Eagleford formed as the transgressive deposits of two sequences, the highstand portions of which have not been preserved.

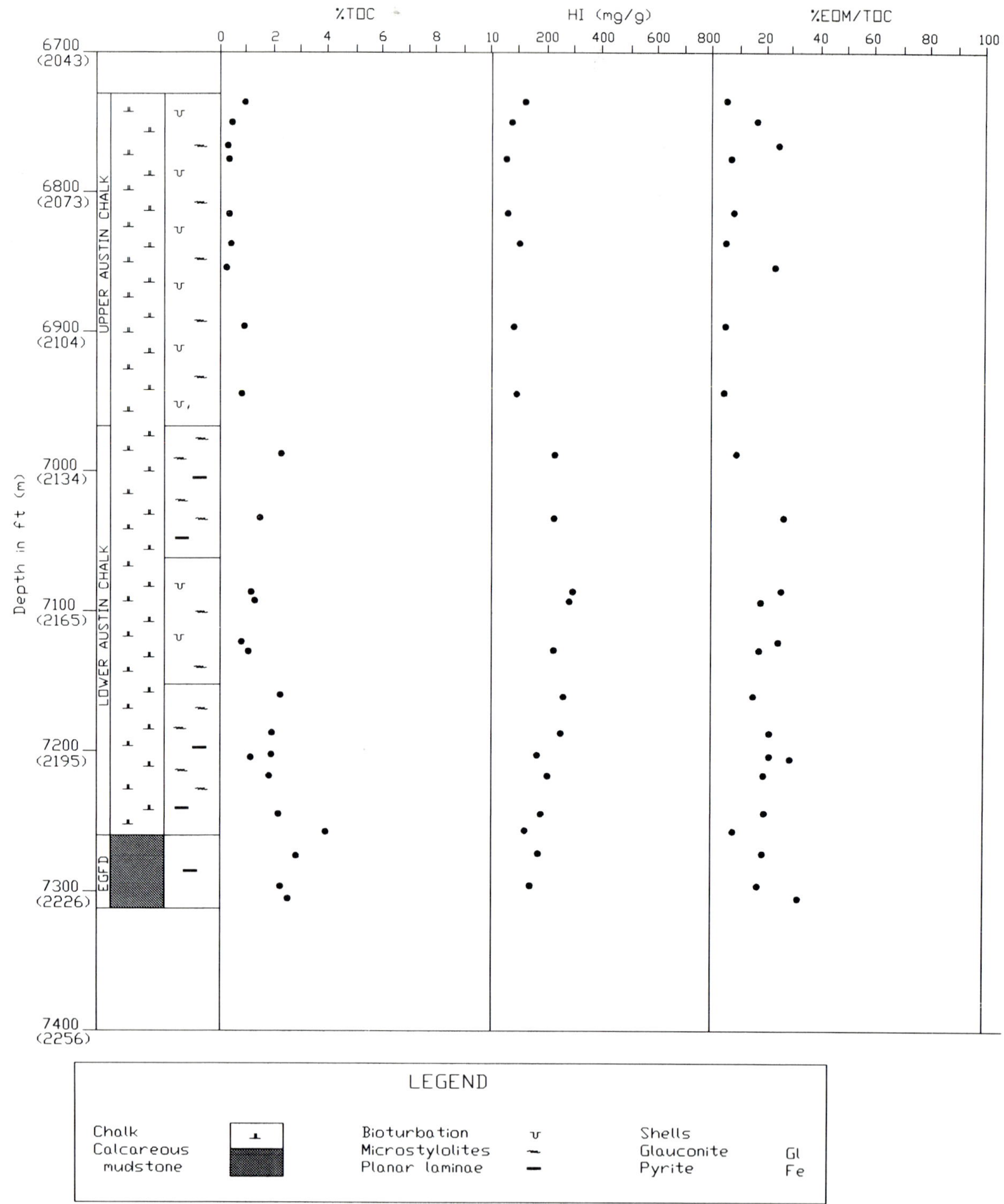

Fig. 10. Description of core from the Getty Hurt #1 well, which includes all of the upper and lower parts of the Austin Chalk and a portion of the upper part of the Eagleford Formation

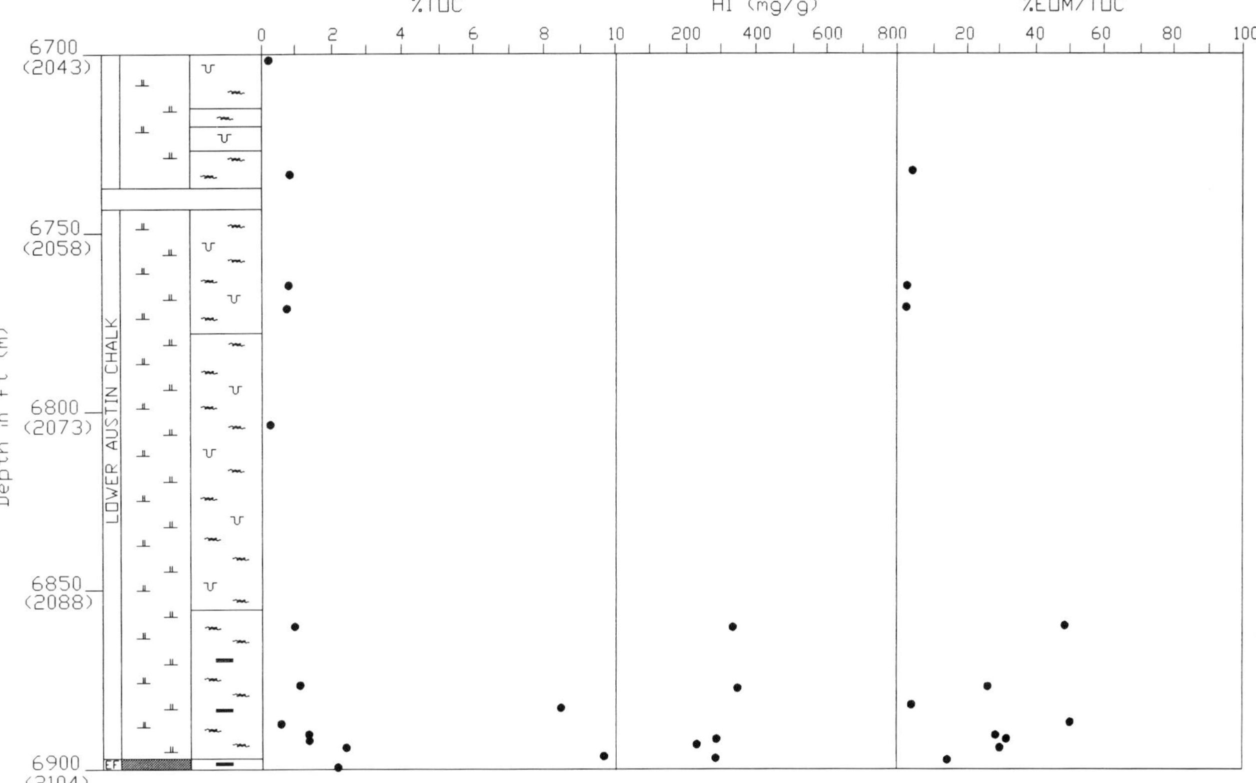

Fig. 11. Description of core from the Prairie Weinert #1 well, which is composed of microstylolitic and sparsely burrowed chalks underlain by laminated and microstylolitic chalks, both of which form the lower part of the Austin Chalk

Characteristics of the Organic-Rich Rocks

Upper Part of the Austin Chalk

Only minor amounts of organic matter are present in the upper part of the Austin Chalk, which is dominated by light-colored, bioturbated chalks. Less than 0.15% TOC is present in the matrix of these chalks. The distribution of TOC in two cores of the upper Austin is shown in Figs. 8 and 10.

The organic matter is concentrated in microstylolites and stylolites that are common throughout the Austin Chalk in the subsurface. These pressure-solution features contain an average of 1.7% TOC. However, thick seams of microstylolites are rare, and stylolites and microstylolites comprise less than 5% of the total thickness of the formation and have little or no influence on the organic content of the upper part of the Austin as a whole.

Visual inspection of kerogen separates from light-colored chalks of the upper Austin shows that the kerogen has the following composition: 40–70% amorphous organic matter, 15–30% vitrinite, 10–25% inertinite, 5% pollen and spores.

Only selected samples of light-colored chalks from the upper part of the Austin Chalk contain sufficient TOC for Rock-Eval pyrolysis (Table 2). Shallow, immature samples range from a mixed type-II/III to type III kerogen (Fig. 13). Hydrogen indices (HI) from these rocks range from 19 to 546 mg hydrocarbons/g TOC, but the average is 272 mg/g (Espitalié et al. 1977).

The proportion of TOC that is extractable (EOM) varies from less than 1 to 84% of TOC (Table 3). The highest proportions of EOM/TOC occur in light-colored chalks with low TOC contents (Fig. 14), whereas samples with high TOC contents have low percentages of EOM/TOC. The bulk of the EOM in these oil-stained, light-colored chalks have migrated into the rock and not been produced form the minor amount of in situ kerogen.

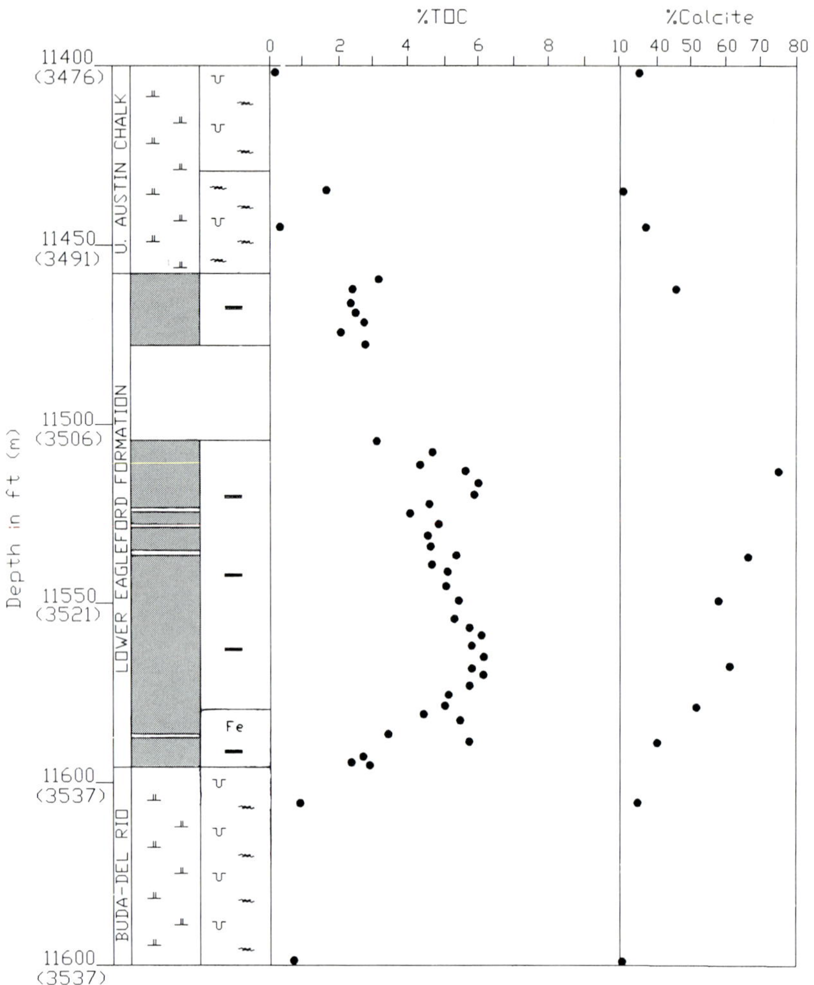

Fig. 12. Description of core from the Exxon Schoolfield #2 well, which covers the lower part of the Eagleford Formation

These migrated oils change the composition of the EOM systematically. The EOM from the oil-stained chalks is generally enriched in hydrocarbons, especially saturated hydrocarbons (Table 3), relative to the chalks with low values of EOM/TOC, reflecting the greater proportion of hydrocarbons in the migrated oils.

In a similar fashion, the composition of the saturated-hydrocarbon fraction of the EOM reflects the addition of migrated oil in these chalks. Pristane/phytane ratios are between 1.3 and 3.1 for chalks with low values of EOM/TOC (Fig. 15), reflecting the mixed organic-matter type that is indigenous to these rocks (Didyk et al. 1978). In contrast, pristane/phytane ratios are between 0.7 and 0.9 in the oil-stained chalks, more like the composition of oils produced from the Austin Chalk.

Lower Part of the Austin Chalk

The TOC content of the darker-colored chalks of the lower part of the Austin Chalk range from 0.9 to 21.8%, with an average of 3.7%. The organic matter is disseminated in the matrix of these chalks. TOC content generally decreases upwards from the base of the Austin Chalk, as illustrated in Figs. 10 and 11.

Clayey laminae and thin beds occur at intervals within the dark-colored chalks of the lower part of the Austin. These clayey horizons are mainly swarms of microstylolites (Dravis 1979) formed by pressure solution of the fine-grained chalk. The TOC in these microstylolite seams is similar to the surrounding chalks (average of 1.9%).

Visual inspection indicates that the kerogen in the dark-colored chalks of the lower part of the

Table 2. TOC and Rock-Eval pyrolysis data for samples of the upper part of the Austin Chalk. (Data marked by * for Getty Hurt well courtesy Simon Petroleum Technology Ltd.)

Core	Depth	TOC	S_1	S_2	S_3	T_{max}	FI	HI	OI	TI	GP
P. Blumberg	3274	1.63	0.08	0.95	0.28	399	5	58	17	0.08	1.03
P. Blumberg	3290	1.33	0.12	4.19	0.93	416	9	315	70	0.03	4.31
P. Blumberg	3292	5.54	0.89	19.00	0.31	416	16	343	6	0.04	19.89
P. Brechtel	3350	1.36	0.08	0.26	0.26	409	6	19	19	0.24	0.34
T. Hurts	4560	3.03	11.40	1.24	0.29	413	376	41	10	0.90	12.64
T. Hurts	4566	3.17	0.30	5.87	0.29	411	9	185	9	0.05	6.17
G. Beall	5672	5.26	1.30	28.70	0.61	427	25	546	11	0.04	30.00
G. Beall	5768	2.76	0.63	12.20	0.53	427	23	442	19	0.05	12.83
G. Beall	5770	5.07	1.49	27.30	0.56	425	29	538	11	0.05	28.79
G. Hurt	6735*	0.77	0.12	0.81	0.48	441	16	105	62	0.13	0.93
G. Hurt	6751*	0.30	0.05	0.19	0.45	439	17	63	150	0.21	0.24
G. Hurt	6775*	0.26	0.03	0.13	0.52	436	12	50	200	0.19	0.16
G. Hurt	6815*	0.31	0.05	0.17	0.70	437	16	55	226	0.23	0.22
G. Hurt	6833*	0.42	0.05	0.41	0.43	439	12	98	102	0.11	0.46
G. Hurt	6893*	0.80	0.09	0.53	1.13	435	11	66	141	0.15	0.62
G. Hurt	6945*	0.74	0.08	0.58	0.83	438	11	78	112	0.12	0.66
G. Hurt	6987	2.14	0.58	4.81	7.15	440	27	225	334	0.11	5.39
G. Hurt	7032*	1.42	1.99	3.20	0.55	439	140	225	39	0.38	5.19
G. Talbutt	7225	3.37	0.93	3.71	0.22	410	28	110	7	0.20	4.64
G. Talbutt	7230	4.30	2.21	7.42	0.41	422	51	173	10	0.93	9.63
G. Talbutt	7233	2.35	1.25	6.18	0.38	438	53	263	16	0.17	7.43
G. Talbutt	7253	1.54	0.71	2.74	0.44	435	46	178	29	0.21	3.45
G. Talbutt	7266	2.17	1.26	6.54	0.37	435	58	301	17	0.16	7.80

TOC is in %; S_1, S_2, S_3 are in mg hydrocarbons/g rock; T_{max} is °C; FI (S_1*100/TOC), HI (S_2*100/TOC) and OI (S_3*100/TOC) are in mg hydrocarbons/g TOC; TI is $S_1/(S_1 + S_2)$; GP ($S_1 + S_2$) is in mg hydrocarbons/g rock.

Austin Chalk has the following composition: 60–85% amorphous organic matter, 5–20% vitrinite, 10–15% inertinite, 0–5% pollen and spores.

The kerogen in the laminated chalks is classified as a mixed type-II/III as determined by Rock-Eval pyrolysis (Fig. 16). HI values range from 432–517 mg hydrocarbons/g TOC (Table 4) for samples of laminated chalk from the lower part of the Austin above 6000 ft (1830 m). The average value of 481 mg/g is considered to be characteristic of the immature kerogen. HI values decrease to 65 mg/g at 9000 ft (2740 m) as shown in Fig. 11 due to hydrocarbon generation from the kerogen. Corresponding H/C atomic ratios are 1.2–1.4 for the shallow samples and 0.6–1.0 for the deepest samples.

The proportion of EOM/TOC ranges from 3 to 55% (average 20.1%) in the lower Austin (Table 4). EOM/TOC generally varies inversely with TOC (Fig. 17), although the relationship is not as dramatic as for the upper Austin. This variation occurs in all cores of the lower Austin that were studied. The EOM is enriched in hydrocarbons, and in particular saturated hydrocarbons, in samples with higher values of EOM/TOC. The "additional" EOM in these rocks results from storage or migration of oil in the more permeable rocks.

Gas chromatograms and mass chromatograms of samples of the lower Austin from below 6000 ft (1830 m) are shown in Figs. 18 and 19. Pristane/phytane and pristane/nC_{17} ratios range from 0.6 to 2.8 and 0.4 to 1.7 respectively (Table 3). These values are fairly constant within samples from a given core, suggesting a facies control (Didyk et al. 1978).

Pentacyclic terpanes are more abundant than steranes in the lower Austin in the Getty Hurt well (Fig. 19). Norhopane is half as abundant as C_{30} hopane in these rocks. Minor amounts of oleanane are present, and C_{29} steranes predominate over C_{27} and C_{28} steranes. The data (Table 5) suggest deposition of mixed marine and terrestrially derived kerogen under low-oxygen, but not totally anoxic conditions (Grantham 1986; Huang and Meinschein 1979; Moldowan et al. 1985). Several biomarker ratios indicate that these samples are within the oil-generative window (Mackenzie 1984; Peters and Moldowan 1993; Seifert 1977;. Waples and Machihara 1991).

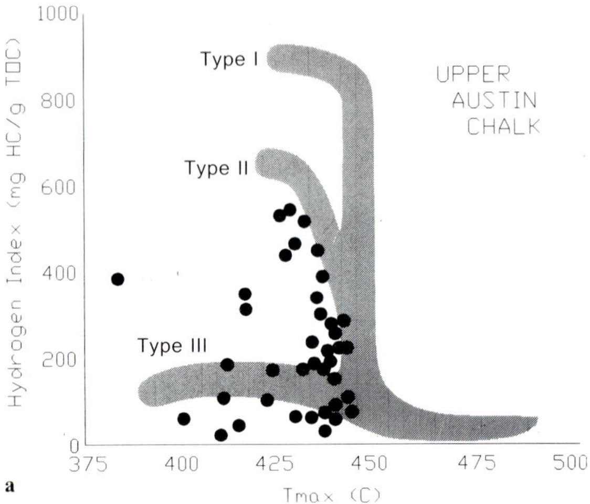

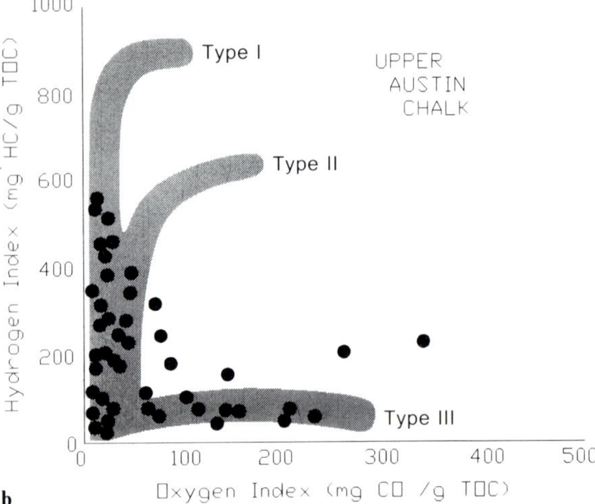

Fig. 13. Plots of HI versus OI and HI versus T_{max} from Rock-Eval pyrolysis of samples of the upper part of the Austin Chalk

The values of stable-carbon isotopes of the kerogen and EOM fractions for the lower Austin are shown in Table 6. The isotope values are all light, as noted by Thompson (1991) and Comet et al. (1993) for Austin Chalk oils.

Eagleford Formation

The dark-colored, laminated chalks and calcareous mudstones of the underlying Eagleford Formation contain 0.6 to 6.1% TOC (average 3.7%). TOC typical decreases upwards within each sequence of the Eagleford, as shown in Fig. 12.

Visual inspection indicates that the kerogen in the laminated chalks and mudstones of the Eagleford Formation has the following composition: 45–85% amorphous organic matter, 5–20% vitrinite, 10–30% inertinite, 0-5% pollen and spores.

Selected results of Rock-Eval pyrolysis are plotted in Fig. 20. HI values (Table 7) are 89–562 mg/g (average 414 mg/g) in shallow cores (2000–3000 ft, 610–915 m) and decrease to 150–280 mg/g (average 215 mg/g) at 7000–8000 ft (2130–2440 m) due to generation of hydrocarbons from the kerogen. HI values of all samples from the Exxon Schoolfield core contain less than 10 mg/g , indicating that they have no remaining potential to generate hydrocarbons.

Pristane/phytane and pristane/nC_{17} ratios are low (0.7–1.0) in the few samples of Eagleford studied. Mass chromatograms of two samples of Eagleford are shown in Fig. 21. Steranes predominate over terpanes, and the biomarkers norhopane and normoretane are prominent on the 191 m/z trace. Minor amounts of C_{24} tetracyclic terpanes are present, whereas oleanane is absent, and there are nearly equal amounts of C_{28} and C_{29} steranes. These data (Table 5) suggest deposition of marine organic matter in calcareous sediments under anoxic conditions. Several biomarker ratios indicate that these rocks are within the oil-generative window.

Interpretation of Organic-Rich Rocks

Depositional Environment

Organic-rich chalks and calcareous mudstones are present in the lower part of the Austin Chalk and the Eagleford Formation. In both formations, the dark-colored, laminated carbonates are onlapping transgressive deposits. However, their geometry and composition are slightly different.

The Eagleford consists of two sequences of laminated, organic-rich calcareous mudstones and chalks that are widespread and onlap the underlying strata at a low angle. The setting at this time was a gently sloping ramp. The highstand portion of each sequence has been removed by erosion, especially in areas that were not subsiding rapidly. No lowstand deposits were formed on this shelf.

TOC and HI values generally decrease upwards within each sequence of the Eagleford. This is most likely due to poorer preservation of organic matter, related to more oxygenated conditions during deposition. The general decrease in siliciclastic

Table 3. TOC and extract data for samples of the Austin Chalk and Eagleford Formation. (Data marked by * for Getty Hurt well courtesy Simon Petroleum Technology Ltd.)

Core	Depth	Unit	TOC	EOM/TOC	SAT	AROM	NSO	ASPH	Pr/Ph	Pr/C_{17}
P. Blumberg	3274	UAC	1.63	1.7	9	10	80	1		
P. Blumberg	3290	UAC	1.33	8.6	6	15	71	8	0.8	0.7
P. Brechtel	3350	UAC	1.36	2.9	10	24	28	38	1.4	0.7
T. Hurts	4520	UAC	1.22	1.2	10	16	58	16	2.1	1.0
G. Beall	5672	UAC	5.26	9.5	23	26	37	14	2.1	1.5
G. Beall	5691	UAC	0.36	75.0	37	36	25	2	0.8	1.4
G. Beall	5750	UAC	2.14	1.3	17	30	38	15	2.3	1.1
G. Beall	5768	UAC	2.76	7.9	30	21	46	3	2.6	1.5
G. Beall	5770	UAC	5.07	11.7	15	26	51	8	2.6	1.6
G. Beall	5944	LAC	6.47	4.1	25	25	35	15	2.1	1.7
G. Beall	5946	LAC	2.22	13.4	16	24	49	11	2.8	1.6
G. Beall	5974	LAC	2.57	28.1	26	18	49	7	2.7	1.6
P. Weinert	6860	LAC	1.00	52.8	6	26	37	31	0.7	0.5
P. Weinert	6878	LAC	1.09	26.3	18	21	50	11	0.6	0.5
P. Weinert	6883	LAC	8.65	4.1	9	24	3	34	0.7	0.4
P. Weinert	6888	LAC	0.65	54.6	29	20	24	28	0.6	0.4
P. Weinert	6891	LAC	1.49	27.4	28	20	30	22	0.7	0.5
P. Weinert	6898	LAC	2.65	28.8	19	24	44	13	0.7	0.5
P. Weinert	6900	LAC	9.72	13.1	17	20	57	6	0.7	0.6
G. Hurt	6767	UAC	0.17	24.7	61	19	16	4	1.1	0.9
G. Hurt	6858	UAC	0.12	23.3	51	12	33	4	1.7	0.9
G. Hurt*	6953	UAC							2.4	
G. Hurt	6987	UAC	2.14	9.8	19	26	23	32	1.7	1.0
G. Hurt*	7022	LAC							0.8	
G. Hurt	7086	LAC	1.37	26.9	49	21	29	1	1.5	1.0
G. Hurt	7204	LAC	1.30	28.9	46	23	27	4	1.1	0.8
G. Hurt	7258	LAC	3.95	7.6	37	32	30	1	1.2	0.9
G. Hurt*	7276	LAC							1.3	
G. Hurt	7304	LAC	2.54	30.1	54	21	23	2	1.0	0.6
G. Talbutt	7225	UAC	3.37	4.1	44	24	29	3	1.7	0.9
G. Talbutt	7230	UAC	4.30	2.2	36	29	32	3	1.6	0.7
G. Talbutt	7233	UAC	2.35	17.0	48	25	21	6	1.8	1.1
G. Talbutt	7253	UAC	1.54	16.2	40	26	25	9	1.1	0.9
G. Talbutt	7255	UAC	0.97	66.0	62	24	14	0	1.1	0.9
G. Talbutt	7266	UAC	2.17	23.6	42	18	38	2		
G. Wilson	8062	EGFD	2.82	29.3	45	23	31	1	1.7	

UAC is upper Austin Chalk, LAC is lower Austin Chalk, EGFD is Eagleford Formation. TOC and EOM/TOC are in %; SAT, AROM, NSO, and ASPH are normalized %. Pr/Ph is pristane/phytane ratio, Pr/C_{17} is pristane/n-C_{17} alkane.

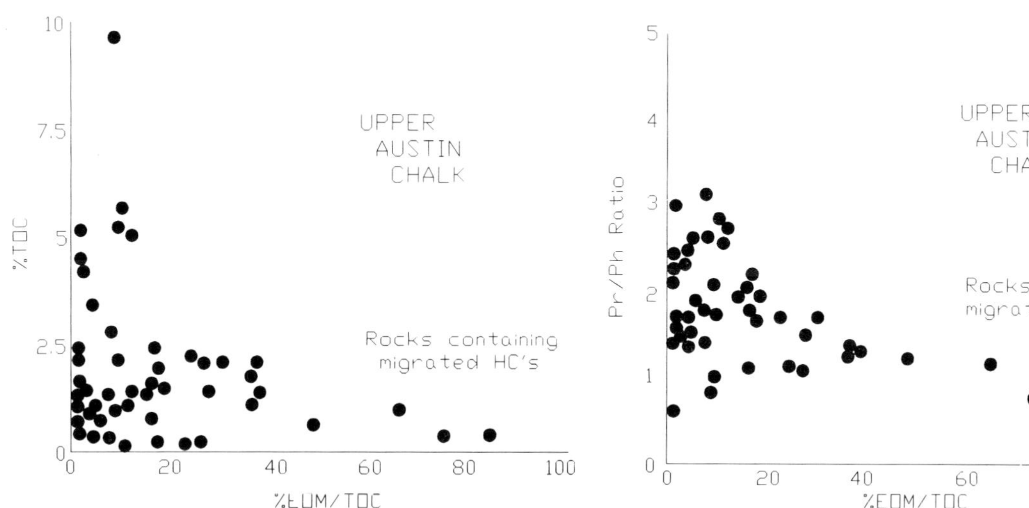

Fig. 14. Plot of TOC versus %EOM/TOC for samples of the upper part of the Austin Chalk

Fig. 15. Plot of pristane/phytane ratio versus %EOM/TOC for samples of the upper part of the Austin Chalk

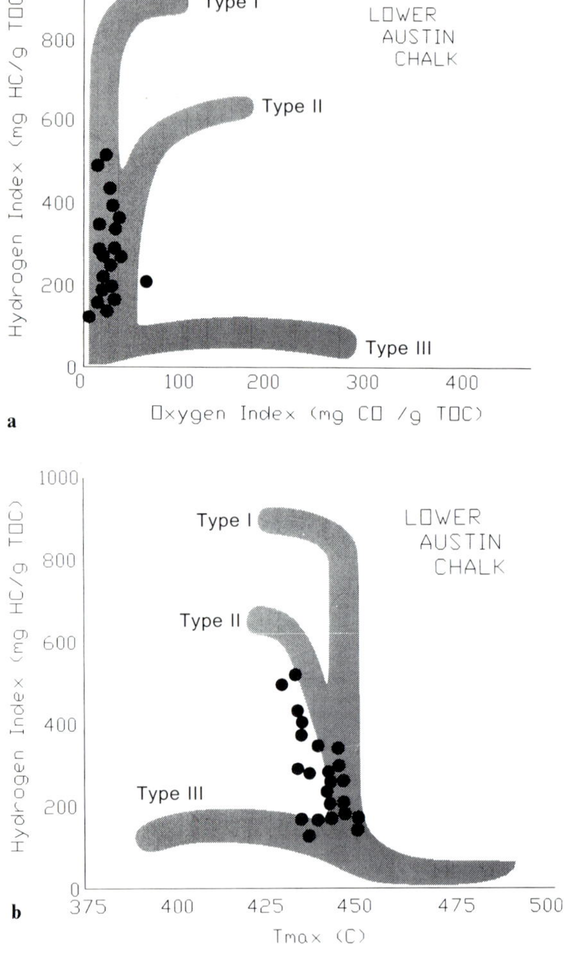

Fig. 16. Plots of HI versus OI and HI versus T_{max} from Rock-Eval pyrolysis of samples of the lower part of the Austin Chalk

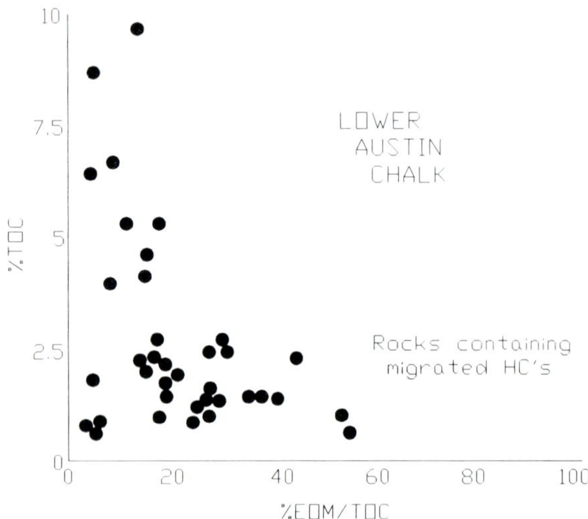

Fig. 17. Plot of TOC versus %EOM/TOC for samples of the lower part of the Austin Chalk

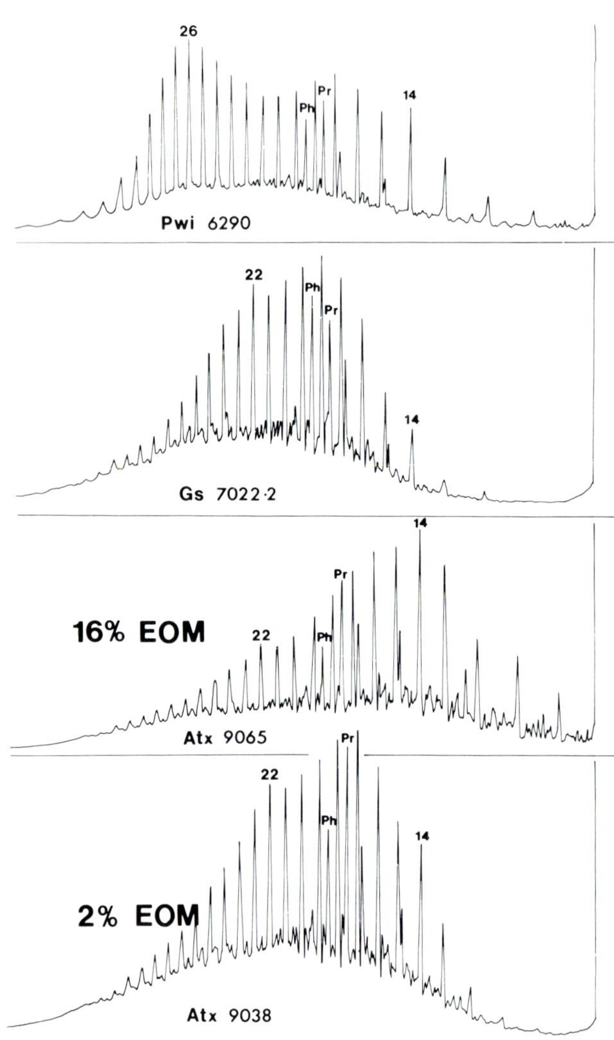

Fig. 18. Gas chromatograms of extracted saturated-hydrocarbon fraction of samples of the lower part of the Austin Chalk. Pr Pristane: Ph Phytane

content upwards suggests that dilution by sediments is not the cause of the change in TOC and HI.

The organic matter in the Eagleford is mainly marine in origin, with lesser amounts of terrestrially derived material. Biomarkers confirm the interpretation of deposition of predominantly marine organic matter in reducing, carbonate sediments.

The lower part of the Austin Chalk, in contrast, shows pronounced onlap onto the underlying Eagleford. The lower Austin is a thicker transgressive deposit than the Eagleford and is restricted to a few basins that formed by subsidence of the Eagleford and older sediments. A thick wedge of sediments present in the deeper part of the Gulf Coast basin is the lowstand deposit of the sequence.

Table 4. TOC and Rock-Eval pyrolysis data for samples of the lower part of the Austin Chalk. (Data marked by * for Getty Hurt well courtesy of Simon Petroleum Technology Ltd.) Units as in Table 2

Core	Depth	TOC	S_1	S_2	S_3	T_{max}	FI	HI	OI	TI	GP
G. Beall	5946	2.22	0.62	9.60	0.69	432	28	432	31	0.06	10.22
G. Beall	5973	2.01	3.74	10.40	0.51	431	186	517	25	0.26	14.14
G. Beall	5974	2.57	1.63	12.70	0.41	428	63	494	16	0.10	14.33
P. Weinert	6860	1.00	0.45	3.39	0.31	443	45	339	31	0.02	3.84
P. Weinert	6878	1.09	1.01	3.75	0.21	438	93	344	19	0.21	4.76
P. Weinert	6891	1.49	0.95	4.17	0.24	436	64	280	16	0.19	5.12
P. Weinert	6892	1.49	1.03	3.43	0.37	440	69	230	25	0.23	4.46
P. Weinert	6898	2.65	1.73	7.58	0.41	442	65	286	15	0.19	9.31
G. Hurt	7086	1.37	1.36	3.90	0.49	432	99	285	36	0.26	5.26
G. Hurt	7091*	1.47	1.04	4.05	0.53	441	71	276	36	0.20	5.09
G. Hurt	7129*	0.99	0.53	2.11	0.69	441	54	213	70	0.20	2.64
G. Hurt	7160	2.16	0.77	5.77	0.66	441	36	267	31	0.12	6.54
G. Hurt	7188*	1.93	0.98	4.93	0.51	444	51	255	26	0.07	5.91
G. Hurt	7203	1.84	1.21	3.00	0.41	437	66	163	22	0.27	4.21
G. Hurt	7217*	1.72	1.21	3.44	0.42	443	70	200	24	0.26	4.65
G. Hurt	7245*	2.13	1.93	3.81	0.54	445	91	179	25	0.34	5.74
G. Hurt	7258	3.95	6.25	4.73	0.25	435	158	120	6	0.57	10.98
G. Hurt	7271*	2.72	1.57	4.50	0.51	448	58	165	19	0.26	6.07
G. Hurt	7298*	2.32	1.16	3.39	0.45	448	50	146	19	0.25	4.55
G. Samuels	7119	1.49	2.28	5.95	0.48	433	153	399	32	0.28	8.23
G. Samuels	7130	1.15	0.79	4.23	0.41	433	69	368	36	0.11	5.02

Table 5. Selected data and ratios from gas chromatography-mass spectrometry of samples of the lower part of the Austin Chalk and the Eagleford Formation

a Core	Depth	Unit	Steranes	$\%C_{27}$	$\%C_{28}$	$\%C_{29}$	
G. Hurt	6953	UAC	39	33	28	39	
G. Hurt	7022	LAC	25	20	35	46	
G. Hurt	7276	EGFD	51	24	37	40	
G. Wilson	8065	EGFD	51	26	36	38	
b Core	**Depth**	**Unit**	**Tricyclics**	**Pentacyclics**	**Ol/H**	**Mo/H**	**C_{24}tetra/tri**
G. Hurt	6953	UAC	8	50	5.2	18.3	1.4
G. Hurt	7022	LAC	17	54	2.6	14.5	0.3
G. Hurt	7276	EGFD	18	30	0	0.2	0.3
G. Wilson	8065	EGFD	21	28	0	7.0	0.5
c Core	**Depth**	**Unit**	**$\%C_{31}$ 22S**	**$\%C_{32}$ 22S**	**$\%C_{29}$ 20S**	**%Ts**	
G. Hurt	6953	UAC	52	54	52	58	
G. Hurt	7022	LAC	59	55	42	52	
G. Hurt	7276	EGFD	57	57	50	83	
G. Wilson	8065	EGFD	61	55	45	72	

(a) Percent total steranes and relative percents of C_{27}, C_{28}, and C_{29} steranes. (b) Percent total tricyclic and pentacyclic terpanes, oleanane/C_{30}-hopane and moretane/C_{30}-hopane ratios, and ratio of C_{24} tetracyclic/tricyclic terpanes. (c) Percent sinistral isomer of C_{31} and C_{32} triterpanes and C_{29} sterane, and percent of Ts to Ts + Tm.

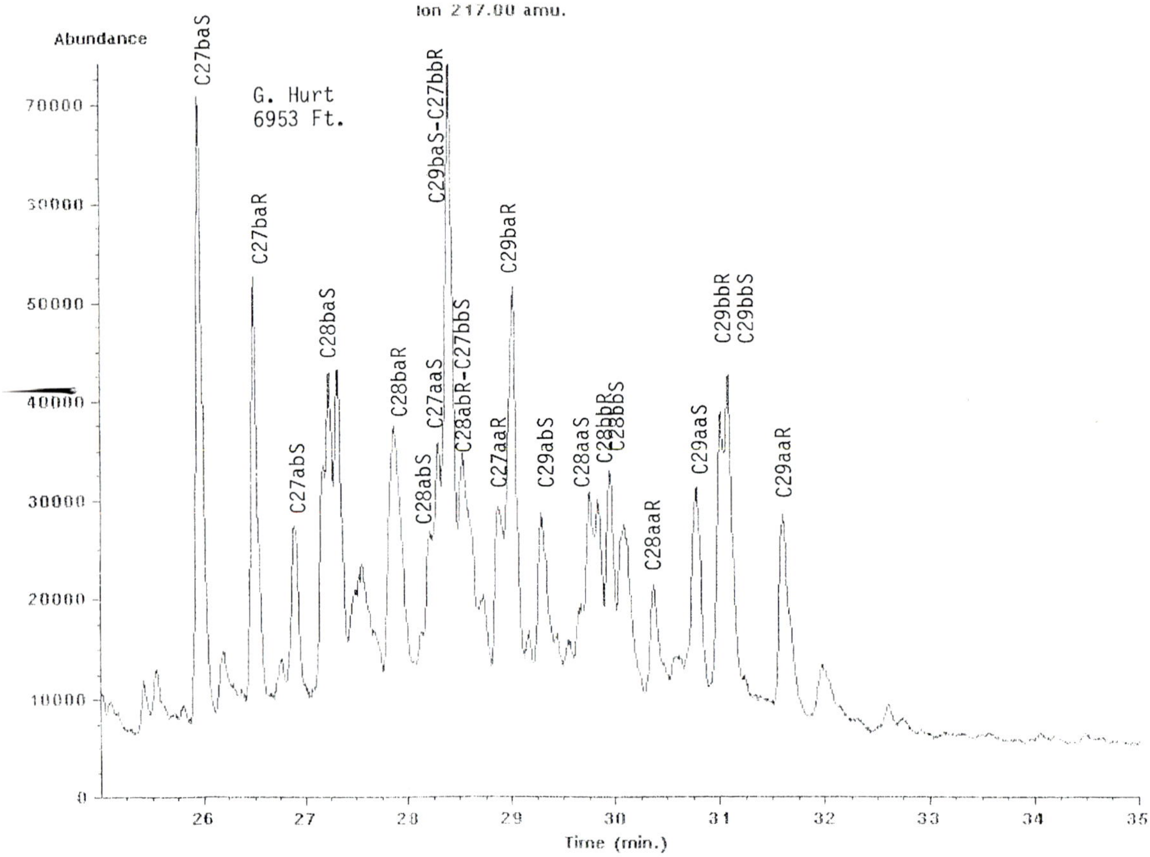

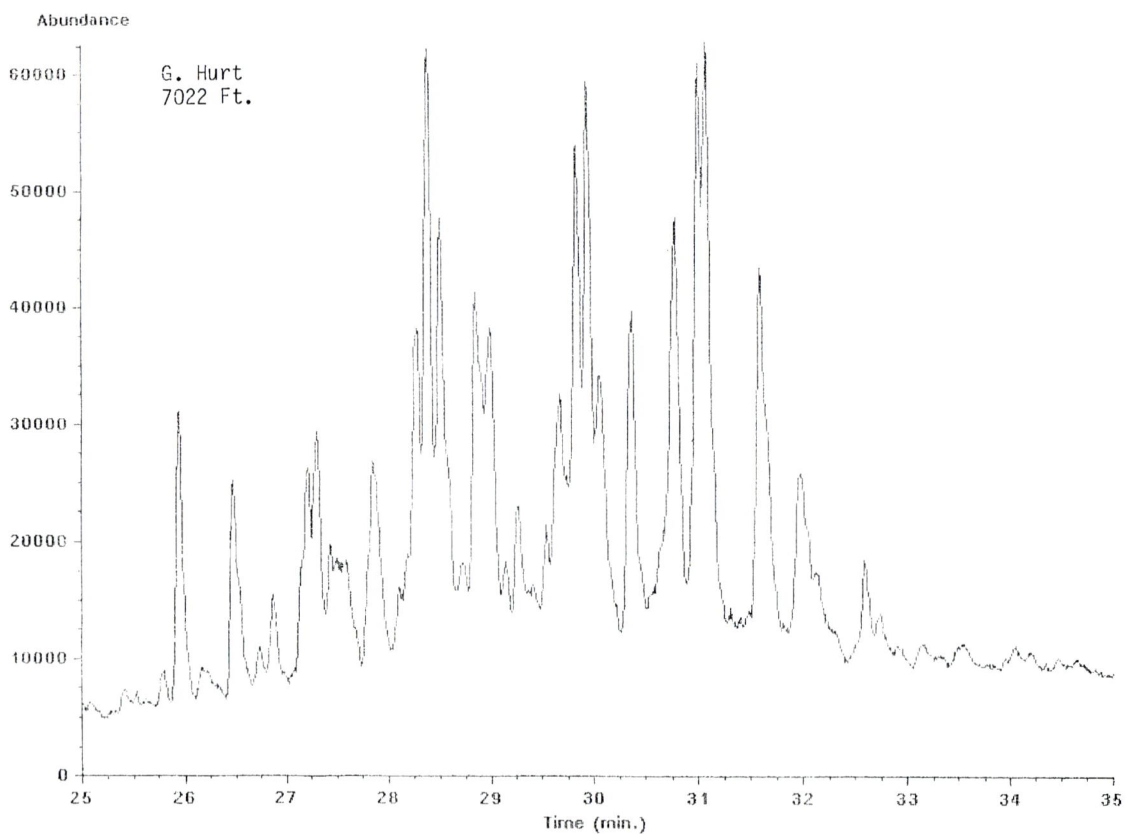

Fig. 19. Mass chromatograms (191 and 217 m/e) of extracted saturated-hydrocarbon fraction of a samples of the Austin Chalk

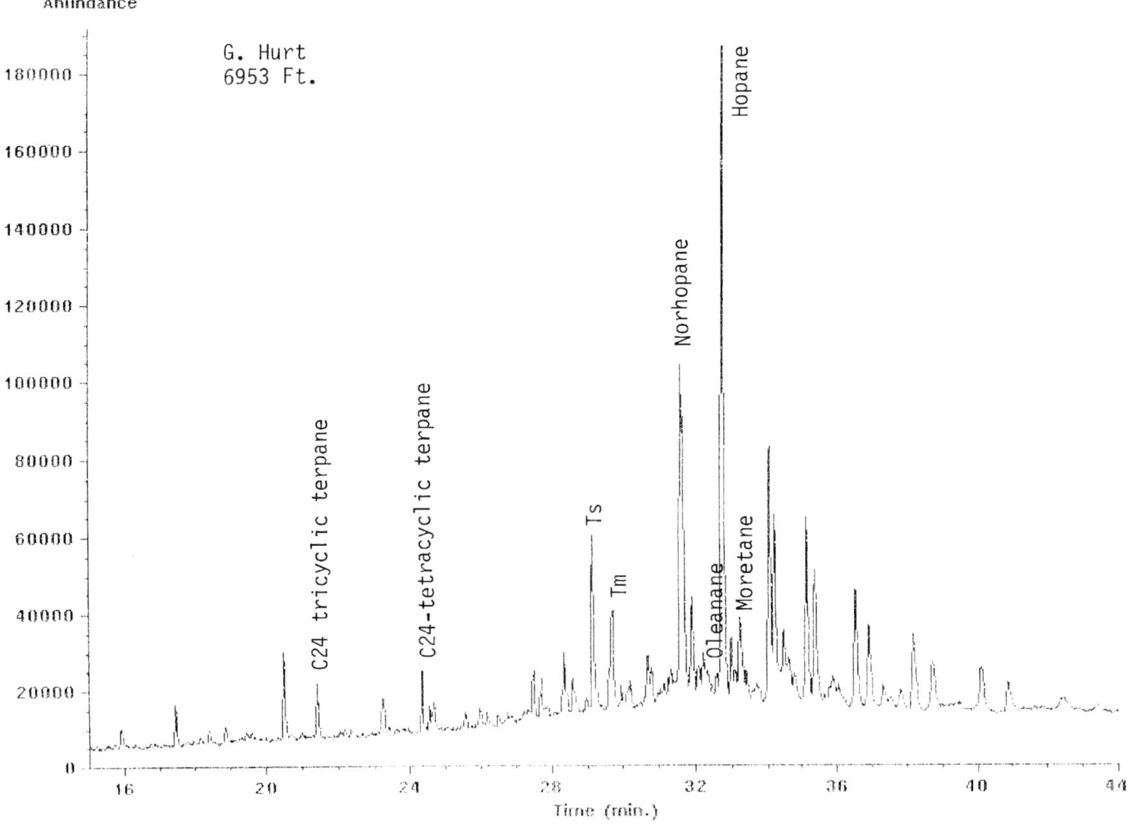
Ion 191.00 amu.
Abundance
G. Hurt
6953 Ft.
180000
160000
140000
120000
100000
80000
60000
40000
20000
0
C24 tricyclic terpane
C24-tetracyclic terpane
Ts
Tm
Norhopane
Oleanane
Hopane
Moretane
16
20
24
28
32
36
40
44
Time (min.)

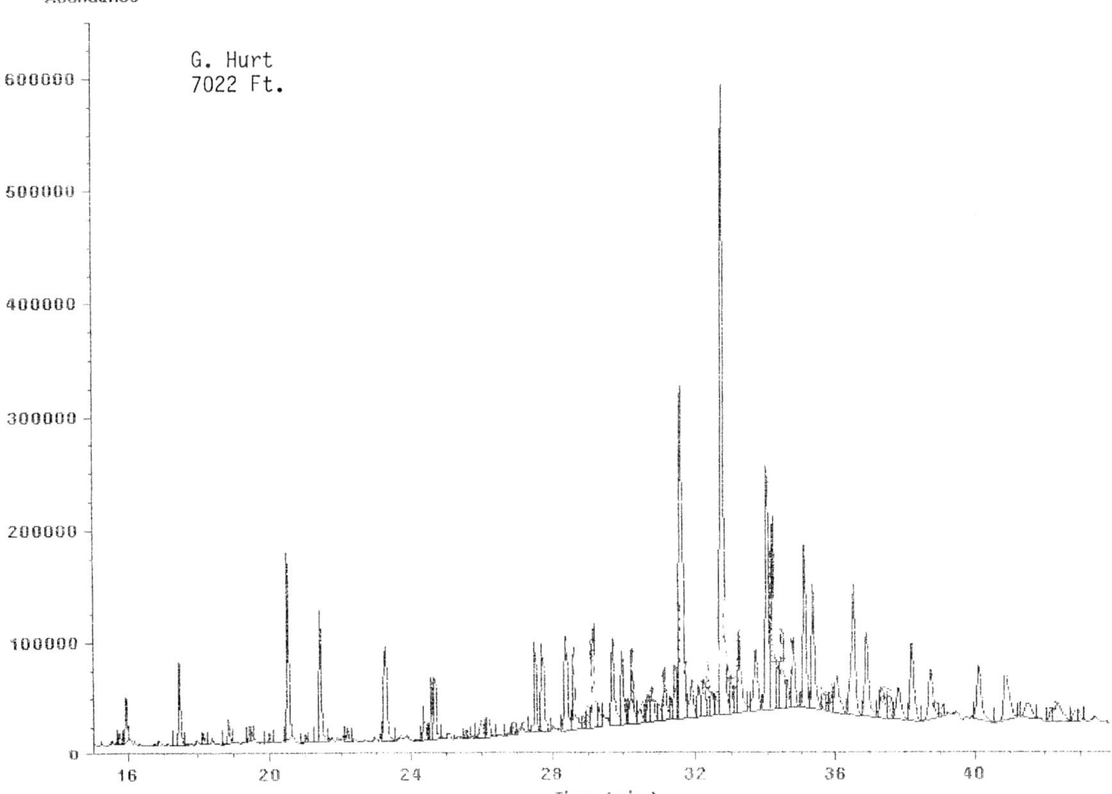
Ion 191.00 amu.
Abundance
G. Hurt
7022 Ft.
600000
500000
400000
300000
200000
100000
0
16
20
24
28
32
36
40
Time (min.)

Table 6. Stable-carbon isotope values of kerogen and fractions of EOM from samples of the lower part of the Austin Chalk, all values reported in per mil relative to PDB

Core	Depth	Kerogen	SAT	AROM	NSO	ASPH
G. Beall	5946	– 27.2	– 29.6			
G. Beall	5973	– 27.1	– 27.8	– 27.3	– 27.0	– 26.9
G. Beall	5961		– 28.0	– 27.2	– 27.0	
G. Hurt	7086	– 26.3	– 27.6	– 26.5	– 26.2	– 26.2
G. Samuels	7115	– 26.6	– 28.1			
G. Samuels	7119	– 26.6	– 28.1			
G. Samuels	7130	– 26.7	– 28.4			
P. Weinert	6878	– 27.2	– 27.3	– 27.2	– 26.9	– 27.8
P. Weinert	6891.3	– 27.1	– 27.4	– 27.0	– 26.9	– 27.7

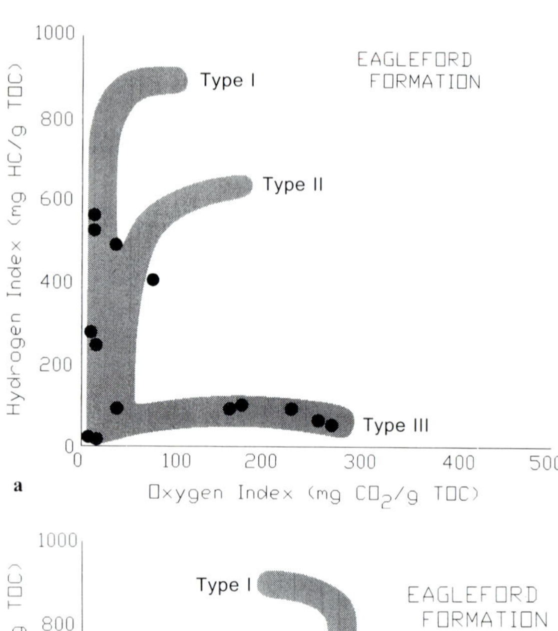

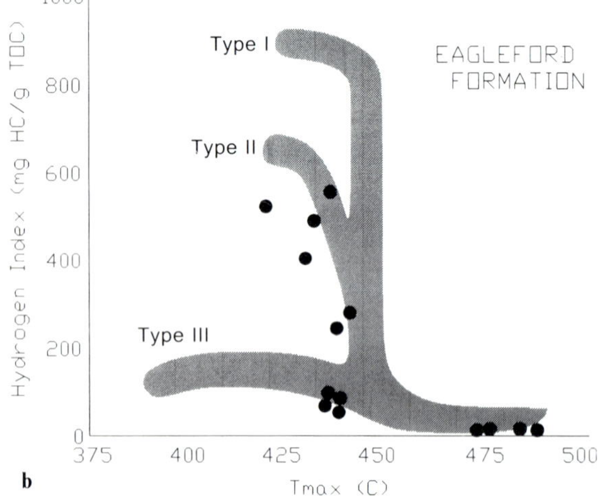

Fig. 20. Plot of HI versus OI and HI versus T_{max} from Rock-Eval pyrolysis of samples of the Eagleford Formation

The portions of south-central Texas where the lower Austin is missing were subaerially or subaqeously exposed during deposition of the lower Austin in adjacent basins. The lower part of the Austin Chalk and the time-equivalent unconformity surface are overlain by prograding highstand deposits of the upper part of the Austin Chalk.

The chalks of the lower Austin are laminated or sparsely burrowed, and TOC and HI values are generally the same as or lower than those of the underlying Eagleford. TOC and HI show an overall increase upwards in the lower Austin, most likely due to increasing oxygenation of the sediments through time.

The organic matter in the lower part of the Austin Chalk is a mixture of marine and terrestrially derived material. Biomarkers confirm the interpretation that deposition of marine organic matter with some terrestrially derived material occurred in oxygen-deficient but not anoxic sediments.

The lower part of the Austin Chalk and the Eagleford Formation are interpreted to have formed under similar conditions. However, deposition of mineral matter and preservation of organic matter were better in the Eagleford. The cause of anoxic bottom conditions in south-central Texas has not been determined.

Maturation and Hydrocarbon Generation

At less than 6000 ft (1830 m) burial, both the lower Austin and the Eagleford are immature for generation of hydrocarbons. Below this depth, hydrocarbons have been generated from the organic matter in both rocks, as indicated by these parameters: decrease in HI, decrease in H/C atomic ratio, increase in %EOM/TOC, increase in $S_1/S_1 + S_2$, increase in % saturated hydrocarbons in the EOM, increase in *n*-alkanes in the saturated

Table 7. TOC and Rock-Eval pyrolysis data for samples of the Eagleford Formation. Data on E. Schoolfield are averages of 46 samples. (Data on H. Bandera samples courtesy Simon Petroleum Technology Ltd.) Units as in Table 2

Core	Depth	TOC	S_1	S_2	S_3	T_{max}	FI	HI	OI	TI	GP
H. Bandera	1670	1.08	0.82	4.40	0.79	428	76	407	85	0.16	5.22
H. Bandera	1810	2.82	0.53	13.74	1.03	431	19	487	291	0.04	14.27
H. Bandera	2010	5.97	2.62	33.55	0.75	435	44	562	449	0.07	36.17
P. Blumberg	3336	4.20	2.25	22.10	0.56	419	54	526	13	0.09	24.35
G. Wilson	8062	2.82	2.34	6.99	0.41	437	83	248	15	0.25	9.33
G. Wilson	8136	5.67	4.60	15.70	0.48	439	81	277	8	0.23	20.30
E. Schoolfield	11538	4.51	0.09	0.32	0.36	451	2	7	9	0.20	0.41

Table 8. Input and output of calculations of hydrocarbon generation from the lower part of the Austin Chalk and the Eagleford Formation from south-central Texas

Unit	TOCo	HIo	S_2o	Oil	Rock	Sp. yield	B/af	B/km^2	Ft	m	1000B/a	1000B/k^2
UAC	1.0	272	2.72	0.83	2.4	0.005	60	50				
LAC	3.7	481	17.6	0.83	2.4	0.04	340	280	143	44	115	28
EGFD-1	3.7	414	15.1	0.83	2.4	0.05	400	320	73	22	25	6
EGFD-2	9.0	600	53.6	0.83	2.4	0.15	1200	980	290	88	172	42

TOCo is original TOC (%), HIo is original HI (mg HC/g TOC), S_2o is original S_2 (mg HC/g rock). Oil is oil density (g/cc), Rock is rock density (g/cc). Sp. yield is specific yield (cc/cc rock), B/af is specific yield in barrels/acre.ft, B/km^2 is specific yield in barrels/km^{2}m; ft and m are average thicknesses of units in feet and meters; 1000B/a and 1000B/k^2 are yields for the average thickness of units, as thousands of barrels/acre and thousands of barrels/km^2 of surface area.

fraction, increase in %C_{29} 20S, %C_{31} 22S and %C_{32} 22S. Between 9000 and 11 500 ft (2740 and 3500 m), the kerogen has yielded all of the hydrocarbons possible, and hydrocarbon generation has ceased.

Calculation of the yield of hydrocarbons that is possible from the lower part of the Austin Chalk and the Eagleford Formation is given in Table 8. Using average values of TOC and HI from immature cores, the specific oil yields calculated are about the same for each formation (340–400 barrels/acre-ft or 280-320 barrels/km^2/m).

However, the biomarker data suggest that the deeper Eagleford may contain more TOC and a higher proportion of marine organic matter. An original HI of 600 was assumed for this "deep" facies of the Eagleford, and present-day TOC was multiplied by 2.0 to arrive at the assumed original TOC content. The specific oil yield calculated for this rock is 1200 barrels/acre-ft (1000 barrels/km^2/m). Although these oil yields are significantly greater for this "deep" facies of the Eagleford compared with the lower part of the Austin Chalk, the validity of these calculations and the distribution of the "deep" facies of the Eagleford is questionable.

Calculations of oil yield from volumes of rock are even more difficult to make, due to uncertainties in the distribution and thickness of these rocks. The results of these calculations using average values of thickness are given in Table 8.

So far, the ability of these rocks to expel hydrocarbons has not been considered. Both formations contain significant amounts of EOM and have probably expelled hydrocarbons, although parts of the Austin Chalk clearly contain oil that has been stored in place or has migrated into the rock.

In conclusion, both the lower part of the Austin Chalk and the Eagleford Formation most likely are capable of generating hydrocarbons, and hydrocarbons most likely are expelled from both source rocks. The specific oil yields and oil yields per area of rock that are calculated fall within an order of magnitude for these rocks. It is probable that where both the lower part

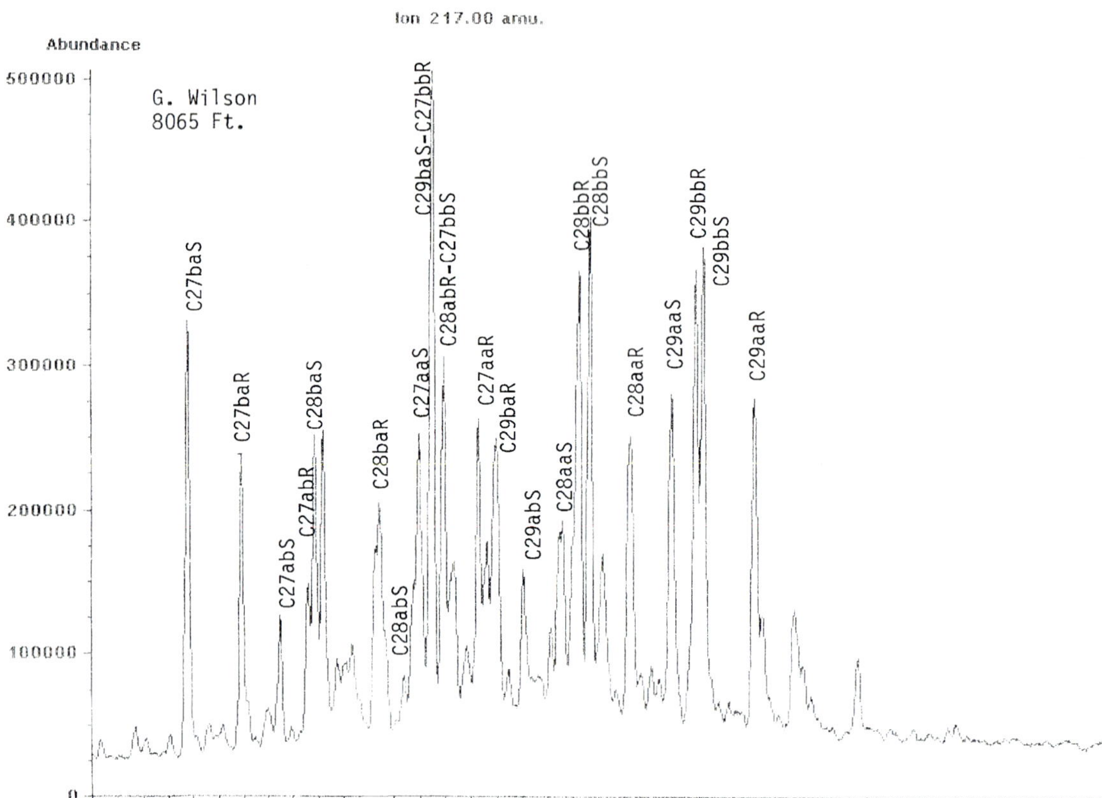

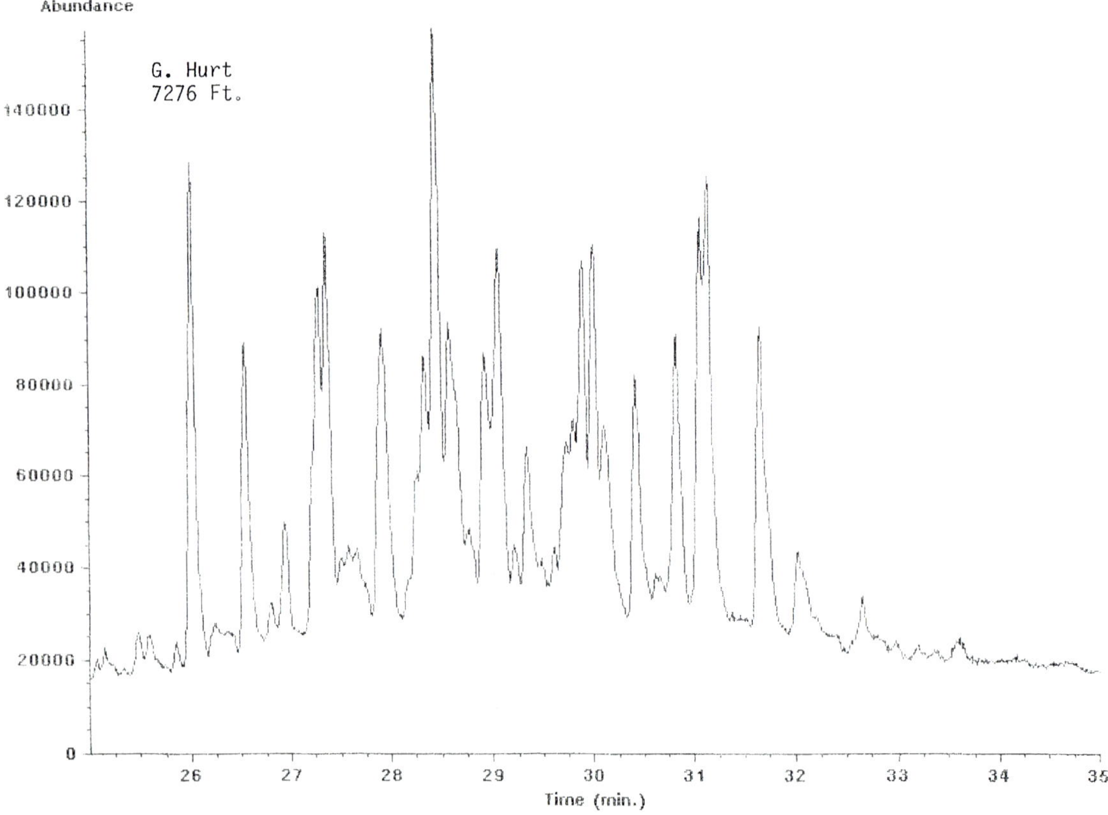

Fig. 21. Mass chromatograms (191 and 217 m/e) of extracted saturated-hydrocarbon fraction of samples of the Eagleford Formation

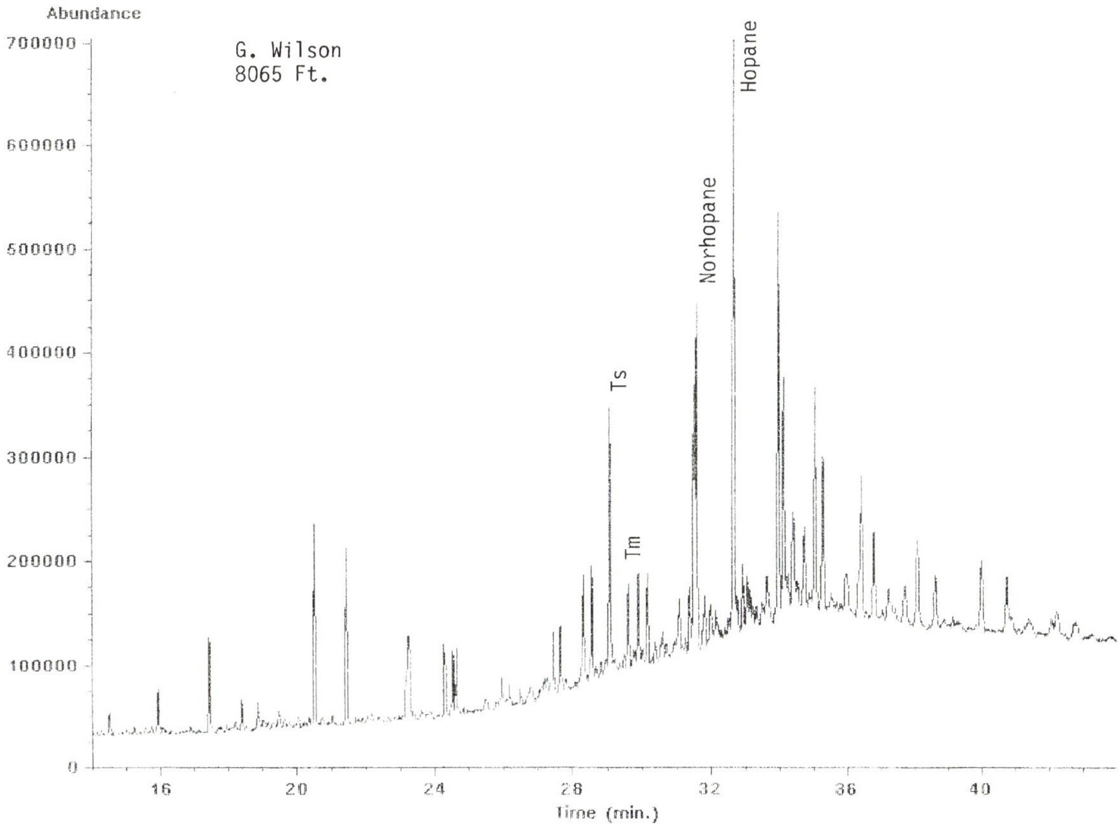

Ion 191.00 amu.

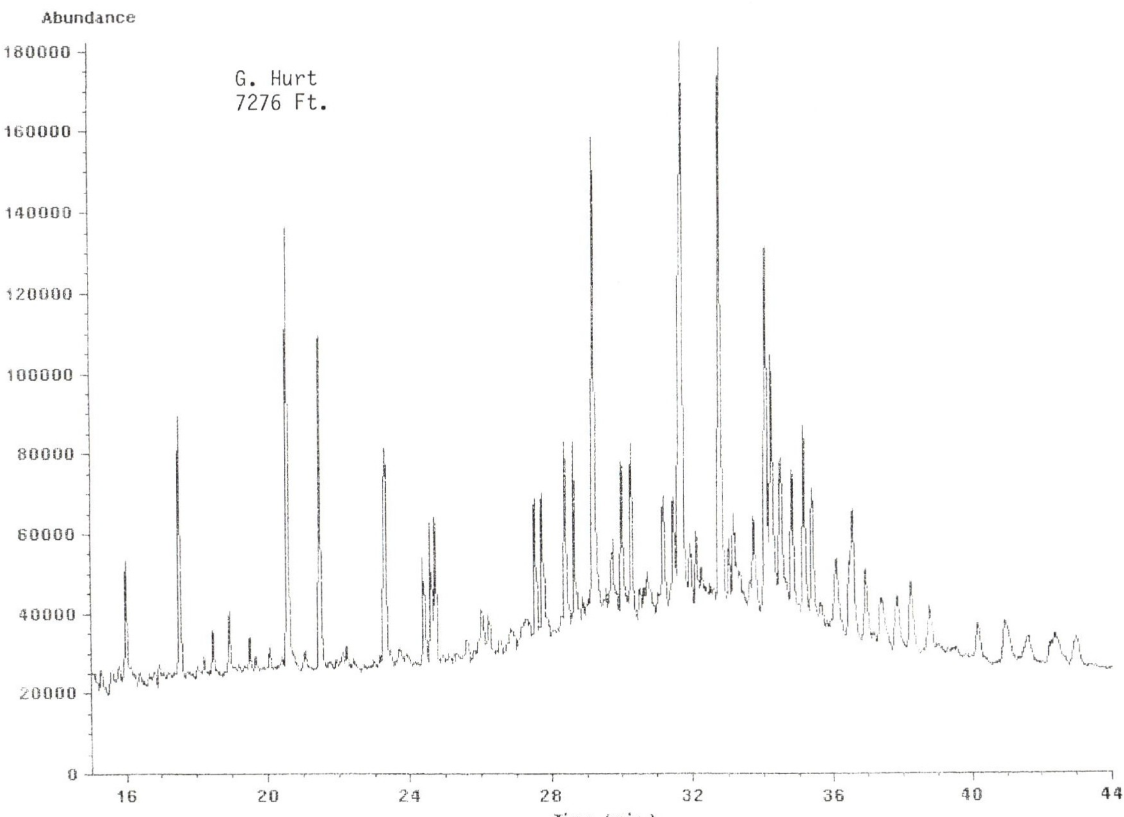

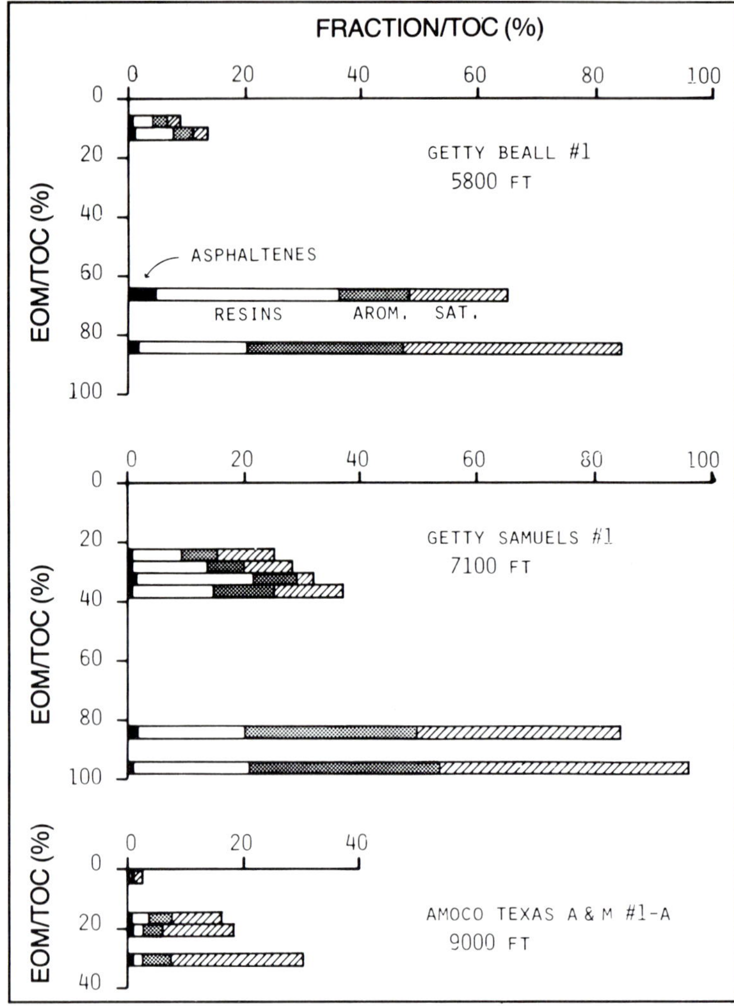

Fig. 22. Bar plot of composition of extractable organic matter for several samples of Austin Chalk from three wells, versus the content of extractable organic matter relative to TOC. Note that with increasing content of EOM/TOC, both the proportion and absolute amounts of resins and hydrocarbons increases, while the absolute content of asphaltenes remains fairly constant

of the Austin Chalk and the Eagleford Formation are present and mature for hydrocarbon generation, they have both generated and expelled hydrocarbons.

Hydrocarbon Migration

The composition of the EOM in the Austin Chalk is highly variable due to migration of oil into permeable rocks. These migrated oils contain a higher proportion of saturated hydrocarbons and fewer nonhydrocarbons relative to in situ EOM (Fig. 22).

Samples of crude oils from the Austin Chalk of south-central Texas were not analyzed in this study. Data was provided by Amoco Production Company on several crude oils.

Four groups of crude oils can be recognized in Austin Chalk reservoirs (Fig. 23). Groups A and B include most of the crude oils of the trend. These oils occur over a wide range of depths. Group C crude oils are light oils or condensates. Group D crude oils are shallow, biodegraded oils.

Figure 24 plots some geochemical parameters of crude oils from groups A and B. Group A crude oils have low pristane/phytane ratios and low saturate/aromatic hydrocarbon ratios, similar to the "autochthonous" EOM from the lower part of the

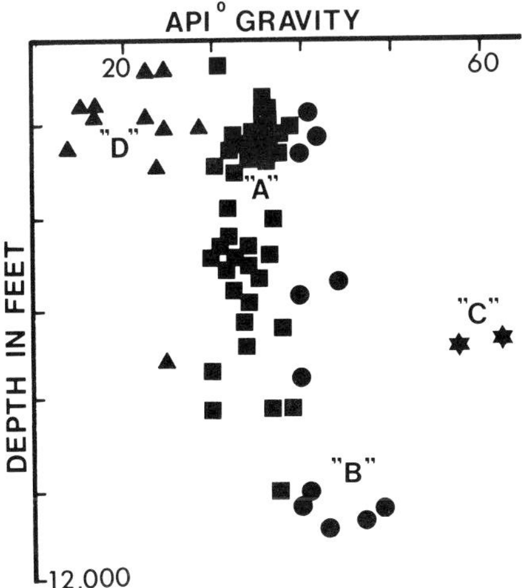

Fig. 23. Plot of API gravity of crude oils versus depth of reservoir for Austin Chalk oils (Scholle 1977b), showing how the four groups generally plot

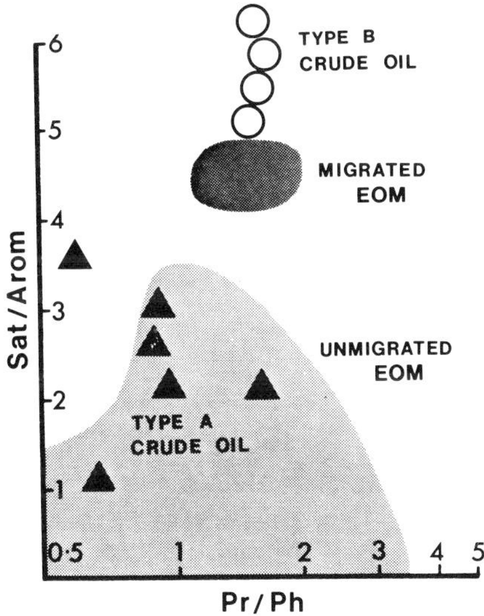

Fig. 24. Saturate/aromatic hydrocarbon ratio versus pristane/phytane ratio for crude oils of the Austin Chalk. Group A is solid triangles, group B is open circles. Shaded regions show the range of compositions of unmigrated bitumen (samples with low EOM/TOC) and migrated bitumen (samples with high EOM/TOC)

Austin Chalk and the Eagleford Formation. Group B crude oils are different and most likely were generated from other source rocks.

Summary

Organic-rich chalks and calcareous mudstones are present in both the lower part of the Austin Chalk and in the Eagleford Formation. The organic-rich sediments of the Eagleford were deposited as widespread transgressive deposits of two sequences on a ramp into a basin with anoxic sediments. Organic-rich chalks were deposited in restricted basins with an oxygen-deficient bottom sediments during the transgression of the Austin Chalk across the top of the Eagleford Formation.

These chalks and calcareous mudstones contain an average of 3.7% TOC composed predominantly of marine organic matter mixed with lesser amounts of terrestrially derived material. The organic matter is thermally mature and has generated hydrocarbons at depths greater than 6000 ft (1830 m). Calculated specific oil yields are within an order of magnitude for these rocks, suggesting that, where present and mature, the lower Austin and Eagleford should both be considered source rocks. However, the Eagleford is more widespread and therefore is the volumetrically more important source rock for most of south-central Texas.

Acknowledgments. The original research on the Austin Chalk was for the author's doctoral dissertation at Rice University. I thank D.R. Baker, the examining committee, and several outside readers for their guidance and support of that project. Cores were provided by the Texas Bureau of Economic Geology, Getty Oil Company, and Amoco Production Company. GCMS data were provided by Texaco. Simon Petroleum Technology Ltd. kindly allowed use of data on samples from two cores from the 1983 Robertson Research (US) Inc. report on the Cretaceous of the Gulf Coast.

Funding for the original research was provided by the US Department of Energy Grant DE-AS05-80ER10764, US Geological Survey Grant 14-08-0001-G0649, Geological Society of America Grant 2454-79, Gulf Coast Association of Geological Societies, Amoco Production Company, Chevron Oil Company, and the Gulf Oil Foundation. Exxon Exploration Company approved release of this publication.

References

Abed AM, Amireh BS (1983) Petrography and geochemistry of some Jordanian oil shales from North Jordan. J Petrol Geol 5: 261–274

Barrier J (1980) A revision of the stratigraphic distribution of some Cretaceous coccoliths in Texas. J Paleontol 54: 289–308

Comet PA, Rafalska JK, Brooks JM (1993) Sterane and triterpane patterns as diagnostic tools in mapping of oils, condensates and source rocks of the Gulf of Mexico region. Org Geochem 20: 1265–1296

Corbett K, Friedman M, Spang J (1987) Fracture development and mechanical stratigraphy of Austin Chalk, Texas. Am Assoc Petrol Geol Bull 71: 17–28

Czerniakowski LA, Lohmann KC, Wilson JL (1984) Closed-system marine burial diagenesis: isotopic data from the Austin Chalk and its components. Sedimentology 31: 863–877

Demaison G, Bourgeous FT (1984) Environment of deposition of Middle Miocene (Alcanar) carbonate source beds, Casablanca Field, Tarragona Basin, offshore Spain. In: Palacas GJ (ed) Petroleum geochemistry and source-rock potential of carbonate rocks. Am Assoc Petrol Geol, Tulsa, Stud Geol 18: 151–161

Didyk BM, Simoneit BRT, Brassell SC, Eglinton G (1978) Organic geochemical indicators of palaeoenvironmental conditions of sedimentation. Nature 272: 216–222

Dravis JJ (1979) Sedimentology and diagenesis of the Upper Cretaceous Austin Chalk Formation, south Texas and northern Mexico. PhD diss, Rice Univ, Houston, 513 pp

Espitalié J, Laporte JL, Madec M, Marquis F, Leplat P, Paulet J, Boutefeu A (1977) Méthode rapide de characterization des roches meres, de leur potentiel petrolier et de leur degré d'evolution. Rev Inst Français du Pétrole 32: 23–42

Grabowski GJ, Jr (1981a) Origin, distribution and alteration of organic matter and generation and migration of hydrocarbons in Austin Chalk, Upper Cretaceous, Southeastern Texas. PhD Diss, Rice Univ, Houston, 262 pp

Grabowski GJ, Jr (1981b) Source-rock potential of the Austin Chalk, Upper Cretaceous, southeastern Texas. Gulf Coast Assoc Geol Soc Trans 31st Annu Meet, pp 105–113

Grabowski GJ, Jr (1984) Generation and migration of hydrocarbons in Upper Cretaceous Austin Chalk, south-central Texas. In: Palacas JG (ed) Petroleum geochemistry and source- rock potential of carbonate rocks. Am Assoc Petrol Geol, Tulsa, Stud Geol 18: 97–115

Grantham PJ (1986) Sterane isomerization and moretane/hopane ratios in crude oils derived from Tertiary source rocks. Org Geochem 9: 293–304

Haq BU , Hardenbol J, Vail PR (1987). Chronology of fluctuating sea levels since the Triassic. Science 235: 1156–1167

Huang WY, and Meinschein WG (1979) Sterols as ecological indicators. Geochim Cosmochim Acta 43: 739–745

Hunt JM, McNichol AP (1984) The Cretaceous Austin Chalk of south Texas – a petroleum source rock. In: Palacas J G (ed) Petroleum geochemistry and source-rock potential of carbonate rocks. Am Assoc Petrol Geol, Tulsa, Stud Geol 18: 117–135

Kauffman EG (1979) Cretaceous marine cycles of the western Interior. Mountain Geol 6: 227–245

Leine L (1986) Geology of the Tarfaya oil shale deposit, Morocco. Geol Mijnbouw 65: 57– 74

Mackenzie AS (1984) Application of biolocial markers in petroleum geochemistry. In: Brooks J, Welte DH (eds) Advances in petroleum geochemistry. Academic Press, London, pp 115–214

Moldowan JM, Seifert WK, Gallegos (1985) Relationship between petroleum composition and depositional environment of petroleum source rocks. Am Assoc Petrol Geol Bull 69: 1255–1268

Peters KE, Moldowan JM (1993) The biomarker guide – interpreting molecular fossils in petroleum and ancient sediments. Prentice Hall, Englewood Cliffs, 363 pp

Pevear DR, Grabowski GJ, Jr (1985) Geology and geochemistry of the Toolebuc Formation, an organic-rich chalk from the Lower Cretaceous of Queensland, Australia. In: Crevello PD, Harris PM (eds) Deep-water carbonates: buildups, turbidites, debris flows and chalks – a core workshop, Soc Econ Paleontol Mineral, Tulsa, Core Workshop 6: 303–341

Rice DD (1984) Occurrence of indigenous biogenic gas in organic-rich, immature chalks of Late Cretaceous age, eastern Denver basin. In: Palacas JG (ed) Petroleum geochemistry and source rock potential of carbonate rocks. Am Assoc Petrol Geol, Tulsa, Stud Geol 18: 135–150

Scholle PA (1977a) Chalk diagenesis and its relation to petroleum exploration: oil from chalks, a modern miracle? Am Assoc Petrol Geol Bull 61: 982–1009

Scholle PA (1977b) Current oil and gas production from North American Upper Cretaceous chalks. US Geol Surv Circ 767: 51 pp

Scholle PA, Cloud K (1977) Diagenetic patterns of the Austin Group and their control of petroleum potential. In: Bebout D G, Loucks R G (eds) Cretaceous carbonates of Texas and Mexico, applications to subsurface exploration. Univ Texas Bur Econ Geol, Austin, Rep Invest 89: 257–259

Seifert WK (1977) Source rock/oil correlations by C27-30 biological marker hydrocarbons. In: Campos R, Goni J (eds) Advances in organic geochemistry 1974. ENADIMSA, Madrid, pp. 21–44

Seifert WK, Moldowan JM (1978) Applications of steranes, terpanes and monoaromatics to the maturation, migration and source of crude oils. Geochim Cosmochim Acta 42: 77–95

Thompson KFM (1991) Petroleum classification based on the ratio of sulfur to nitrogen: application in the East Texas basin. Gulf Coast Assoc Geol Soc Trans 41: p 602

Thompson KFM, Kennicutt MC, II, Brooks JM (1990) Classification of offshore Gulf of Mexico oils and condensates. Am Assoc Petrol Geol Bull 74: 187–198

Waples DW, Machihara T (1991) Biomarkers for geologists – a practical guide to the application of steranes and triterpanes in petroleum geology. Am Assoc Petrol Geol, Tulsa, Methods Explor Series 9: 91 pp

Williams GD, Stelck CR (1973) Speculations on the Cretaceous paleogeography of North America. Geol Assoc Can Spec Pap 13: 1–20

Zumberge JE (1984) Source rocks of the La Luna Formation (Upper Cretaceous) in the Middle Magdalena Valley, Colombia. In: Palacas JG (ed) Petroleum geochemistry and source rock potential of carbonate rocks. Am Assoc Petrol Geol, Tulsa, Stud Geol 18: 127– 133

Petroleum Geochemical Characterisation of the Lower Congo Coastal Basin Bucomazi Formation

R. Burwood, S.M. De Witte, B. Mycke, and J. Paulet[1]

Abstract

An integrated biostratigraphic and geochemical study of the Lower Congo Coastal basin Bucomazi Fm. has been performed on the type section penetrated in well CABGOC 86-1.

Contributing in large part to a reserves estimate of 4.5 billion BO (recoverable) in the coastal states (northern Angola, Cabinda, Congo and Zaire), such source quality rocks were deposited under lacustrine conditions accompanying the rift phase opening of the South Atlantic margin. Reaching thicknesses up to 1.8 km of Pre-Salt[2] sediments, localised graben/half graben initiated subsidence provided the focus for accumulation of organic-rich sequences during the latter Early Cretaceous.

Paleo-environmental fauna/flora (ostracod spp.) from 86-1 suggested a limited late Barremian-Aptian age range and an initially freshwater, evolving into a saline, depositional regime. With an anticipated diversity in depositional style reflecting salinity, water depth and paleo-climatic control, the Bucomazi Fm. reduces to a generic sedimentological term.

Lower Congo oils reflect compositional diversity with previous geochemical characterisation of the Bucomazi Fm. amply confirming a high level of vertical and lateral heterogeneity in these sediments. Embracing both source potential, organofacies development and kerogen kinetic considerations, such factors can exert important constraints on hydrocarbon generation and the realisable prospectivity of a basin.

Investigation of these factors has been pursued as a regional benchmark study to systematically evaluate the source characteristics of the "type section" sequence.

High density Rock-Eval based analyses (3-m frequency) revealed considerable source inhomogeneity, this being recognised in terms of variable organic richness, kerogen type and kinetic parameters. Similarly, zones of enhanced source richness, delineated by conspicuous carbon isotope excursions to ^{13}C-enriched kerogens, reflected periods of intense biomass turnover deriving from stimulated carbon dioxide budget and nutrient influx.

Stratigraphic delineation of the section, incorporating these data and supporting bitumen biomarker signatures, allowed segregation of the section into two major sequences comprising four subordinate sub-members. The lower, Middle Bucomazi "organic-rich" zone corresponding to "basin fill" sedimentation, exhibited a fine structure of cyclical depositional events with the accumulation of kerogens of higher activation energy and variable carbon isotopic signature. By contrast, the "sheet drape" Upper Bucomazi sediments, although lacking in persistent organic richness, were isotopically more homogeneous and revealed kerogens with lower, generatively more labile, activation energies.

As illustrated by a simple subsidence model, this dichotomy in activation energy has important implications in understanding the sequential maturation and subsequent hydrocarbon expulsion from these heterogeneous source subunits. In this way, a mechanism is available for the mixing of different charges, the aggregate oils thus formed accounting for the inherent diversity of the emplaced petroleums, all ostensibly from a common source. Application of these concepts has led to a more detailed understanding of the Ponta Vermelha Trough hydrocarbon habitat and the character of the coastal Zaire oils.

[1]Fina Exploration and Production, Fina Research, Zone Industrielle C, 7181 Seneffe (Feluy), Belgium

[2]Pre-Salt stratigraphic terminology used in this chapter denotes Lower Cretaceous sediments older than the Loeme Fm. and collectively embracing the Lucula, "Bucomazi", Toca and Chela Fms of Neocomian through Aptian age.

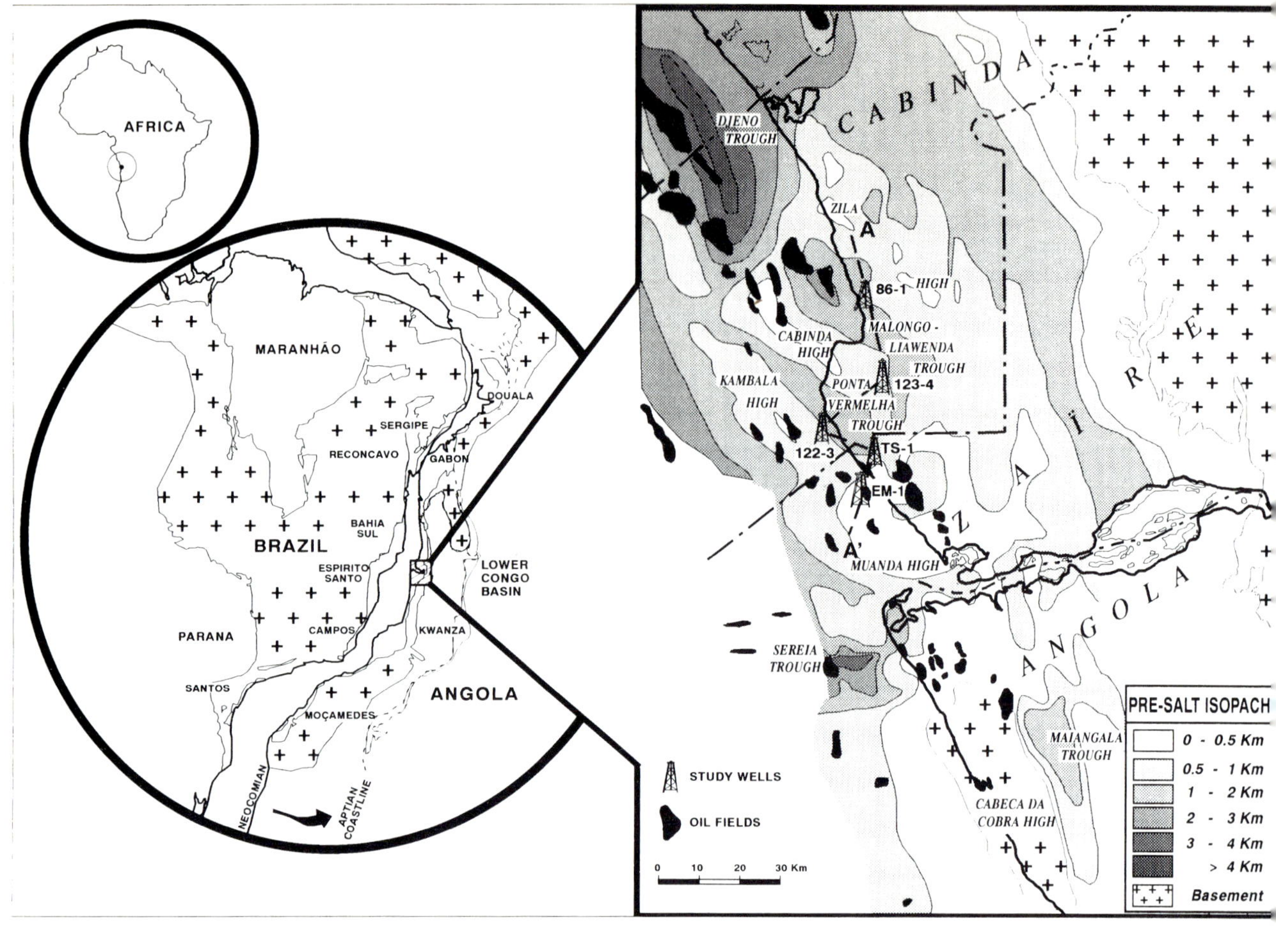

Fig. 1. Location map showing Lower Congo Coastal basin study area in terms of pre-separation south Atlantic margin, respective positions of the Pre-Salt paleo-depositional centres for the Malongo/Liawenda and Ponta Vermelha Troughs and wells; being: *86-1* Bucomazi Fm. type section; *123-4; 122-3* depocentre reconstruction; *TS-1* Tshiende-1; *EM-1* East Mibale-1 as discussed in text

Introduction

Early Cretaceous rift-phase sediments of the West Africa-Brazil Atlantic margin have economic importance in hosting both reservoir facies and, most notably, highly prolific source rocks for hydrocarbon generation. In the Lower Congo Coastal basin, being the hinterland to the discharge of the Zaïre River (Fig. 1), such source potential is recognised in the bituminous shales of the Bucomazi Fm. Regarded to be of lacustrine depositional origin (Brice et al. 1982; McHargue 1990), these sediments accumulated during the Neocomian-Barremian Synrift I/II opening of the southern Atlantic.

Reserve estimates of ca 4.5 billion BO (recoverable) for the coastal states (northern Angola, Cabinda, Congo and Zaire) confer major petroleum province status on this region. With production from drift phase clastic, carbonate platform and pre-salt reservoirs, a diversity in oil type and character indicates a complex hydrocarbon habitat charged by multiple sourcing.

Elsewhere around the margin time equivalents include the Marnes Noires (Congo), Melania (Gabon) plus the Brazilian analogues Guaratiba (Santos basin), Lagoa Feia (Campos basin), Mariricu (Espirito Santo basin), and Candeias (Recôncavo basin) formations, respectively.

Named after the coastal Cabinda location of the type section well (Bucomazi-1; now systematically 86-1), Bucomazi is used in the Lower Congo area to denote the variable lithostratigraphic unit between the regional unconformity at the Chela clastics/Toca carbonates level and the massive Lucula sands (Fig. 2). Although hosting considerable shale

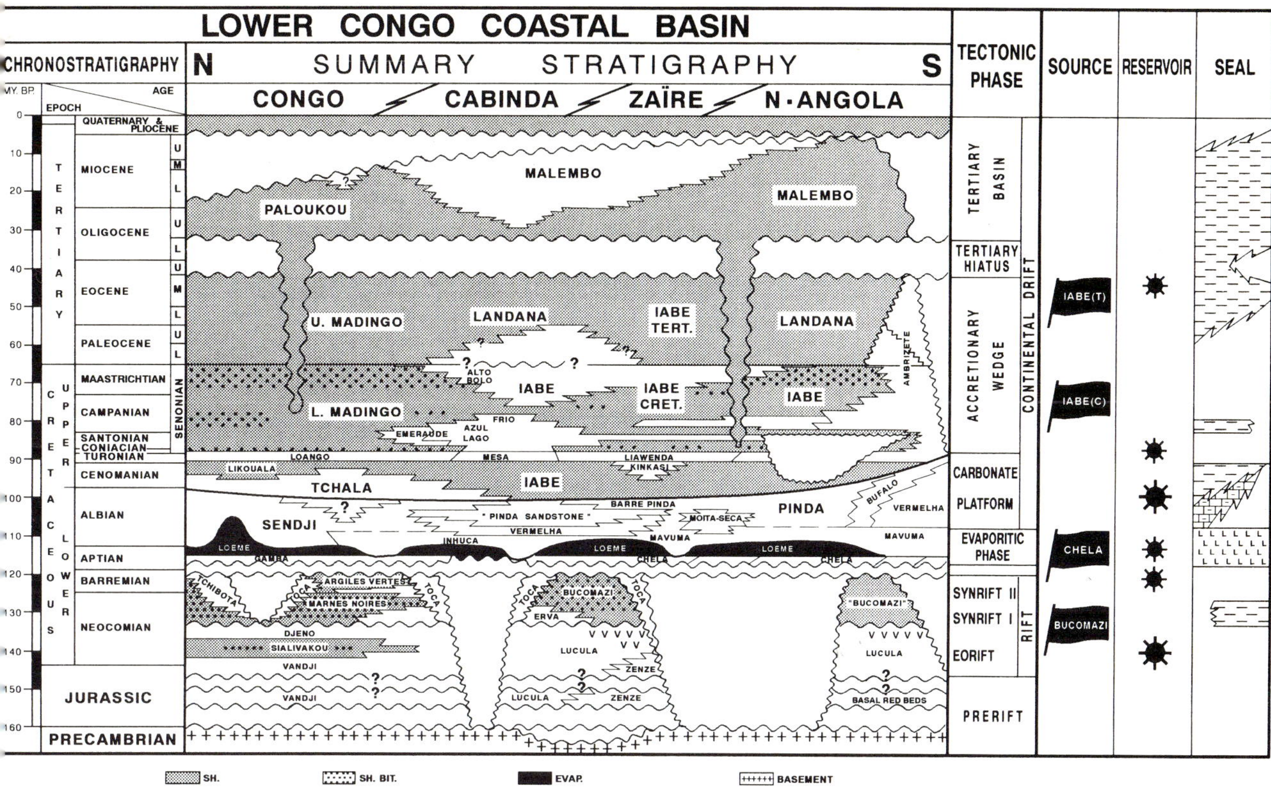

Fig. 2. Schematic stratigraphy for the Lower Congo Coastal basin summarising local source-reservoir-seal relationships

thickness, up to 1800 m being reported by McHargue (1990), lithology can fluctuate rapidly between fine-texture, bituminous-calcareous clastics to silts and massive sands elsewhere. This variability undoubtedly reflects tectonic and paleoenvironmental/climatic control on lake-fill sedimentation processes with source quality shales not necessarily being ubiquitous to the whole area. "Bucomazi" is similarly used as a litho-stratigraphic term for fine-texture, pre-salt clastics found elsewhere along the Angolan margin to the south. Although possibly time equivalents, use of this term in this context is conjectural as development of the *de stricto* Bucomazi "organic-rich" facies appears to be restricted to Lower Congo depositional centres, only. In addition to the pre-salt Lucula Sands, other receptive reservoir facies are found in the Chela and Toca Fms plus post-salt carbonate platform and drift phase deep marine clastic sediments (Fig. 2). Significant petroleum accumulations of Bucomazi Fm. provenance are distributed amongst fields located in northwest Angola (e.g. N'Zombo, Quinguila, Quinfuquena etc.); on-/offshore Zaire (e.g. Liawenda, Kinkasi, Tshiende, Mibale etc.); offshore Cabinda (North, South, West Malongo, Kambala etc.) and on-/offshore southern Congo (e.g. Kunji, Mengo and Emeraude). Recoverable reserves range from marginally economic (e.g. Liawenda with 0.5 MMBO) to supergiant field proportions (e.g. Malongo North with 500 MMBO). Crude oil type and quality is highly variable ranging, for instance, from low to high wax compositions (Burwood et al. 1990). Shallow, in-reservoir biodegraded, heavy oils (e.g. Emeraude) versus condensate-like (e.g. N'Siamfumu) to gas accumulations (e.g. Livuite) are observed elsewhere.

Noting the economic importance of the Bucomazi Fm., these sediments warrant detailed investigation thereby more fully defining the generative aspect of their operative hydrocarbon habitat. This stems from their complex source characteristics as controlled by the variable paleoenvironment of lacustrine/post-lacustrine deposition.

Recent investigations (Burwood et al. 1990, 1992) have revealed considerable source potential and organofacies variation within a diverse selection of Bucomazi sections (Fig. 3). In addition to demonstrating both environmental changes and

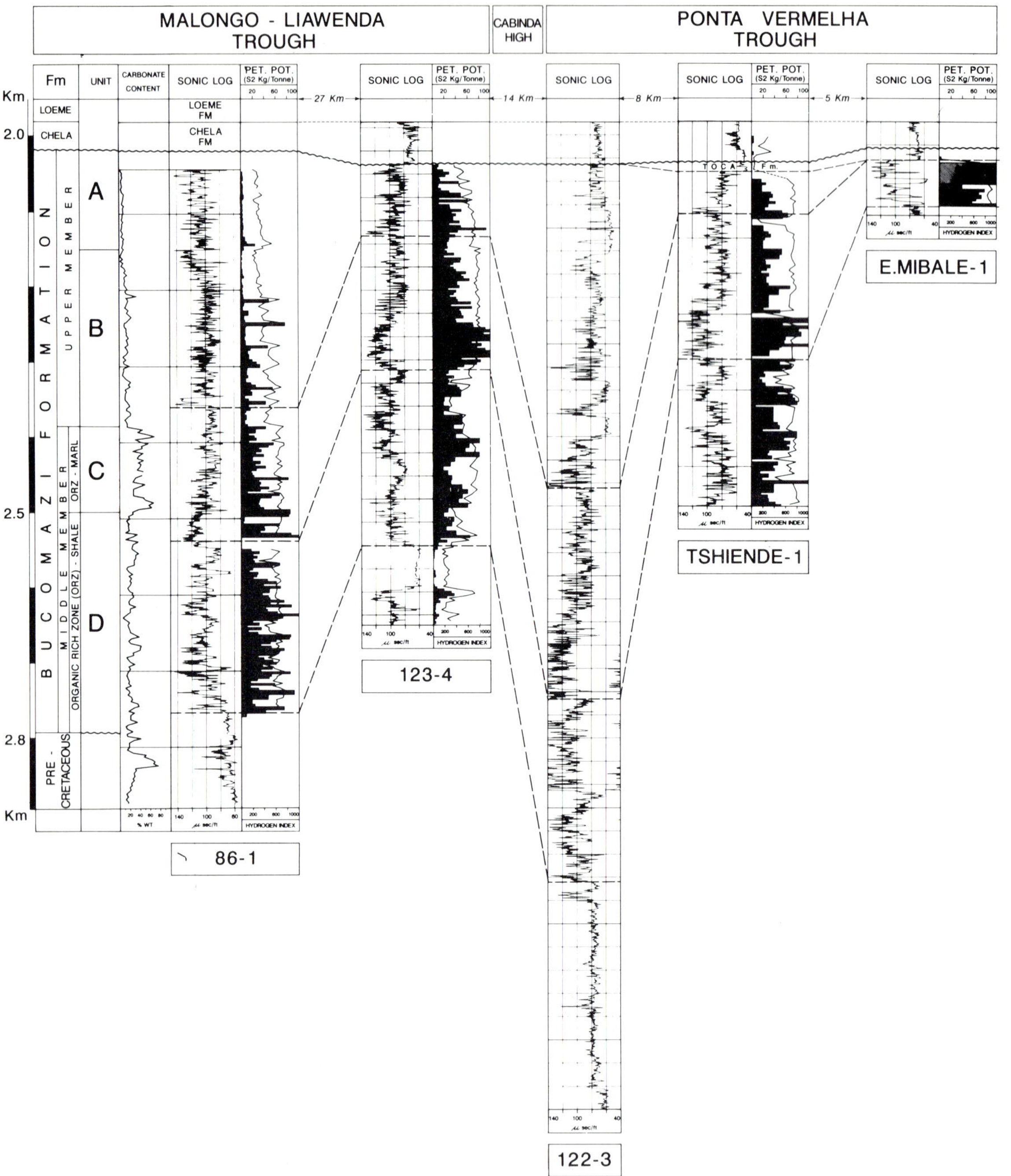

Fig. 3. Development of the Bucomazi Fm. in CABGOC 86-1 and sonic log/organic richness (S_2) correlation across the Malongo-Liawenda-Ponta Vermelha Troughs. Note cross-basin continuity of section broadly equivalent to the micritic Unit C (ORZ-marl) in terms of organic richness and sonic character

cyclicity in sedimentation patterns, these data have practical importance in highlighting the variability in kerogen abundance, type and attendant transformation kinetic parameters. The latter are crucial in the understanding of the process of sequential source maturation and hydrocarbon expulsion during basinal subsidence. With discontinuous generation of hydrocarbons from differing organofacies, the concept of an aggregate oil charge takes on real significance. Such information has proved useful in unravelling the internal diversity of the northwest Angola (Burwood et al. 1990) and analogous coastal Zaire oils (unpubl. data).

With the objective of presenting a systematic Bucomazi source evaluation data set, supplemented by *de novo* biostratigraphic control, the current study focuses upon the type section in well CABGOC 86-1. Although this section penetrates only the middle and upper members of the formation, it provides a useful regional benchmark amply illustrating source heterogeneity and attendant organofacies variation.

Regional Setting

Location of the Lower Congo Coastal basin with the distribution of the prominent pre-salt depositional areas and major oil fields is illustrated in Fig. 1. Figure 2 provides a summary stratigraphy and tectonic history for the study area which has been described in detail by Brice et al. (1982) and McHargue (1990).

During the Early Cretaceous separation of South America and central-southern Africa, development of the proto-Atlantic margin evolved through a series of rift (Eorift, Synrift I and II) and continental drift events. Initially, a range of localised graben/half-graben stimulated troughs (subbasins) developed along the rifted margins. These fault-bounded sub-basins were formerly isolated and acted as the paleo-depositional centres for active lacustrine sedimentation (Synrift I, Neocomian through Barremian times). In the coastal Lower Congo area, these centres can be identified as the northwest-southeast-trending Djeno, Malongo/Liawenda, Ponta Vermelha and Sereia Troughs, each of which constituted a separate lacustrine system (Fig. 1). Tectonism and rate of subsidence would have controlled the influx of coarse detrital fill versus fine-texture clastic sediments. Climatic mediated paleoenvironmental conditions would similarly complement this evolution. Thus, aridity/humidity, nutrient input, salinity/pH, water depth and the attendant levels of anoxia would have dictated biomass turnover, preservation and the deposition of organic rich sediments.

From later Barremian times onwards (Synrift II) subsidence rates declined with the lakes coalescing, essentially into an extended, sometimes ephemeral, inland sea. More oxygenated waters limited the formation of organic-rich sediments and encouraged the deposition of calcareous mudstones and carbonates, particularly in shallow water-shore edge locations.

McHargue (1990) considered that rifting gave way to the drift event in early Aptian times in concert with a major regional uplift and erosion. During resumed subsidence, deposition of the Chela Fm. was accompanied by increasingly persistent early marine incursions. These events preceded the Loeme Salt-forming desiccation event which eventually gave way to permanent marine conditions and carbonate-dominated sedimentation accompanying the drift phase, passive margin opening of the southern Atlantic Ocean.

Stratigraphy of the Bucomazi Type Section in CABGOC 86-1

The Bucomazi section as developed in well 86-1 penetrates 750 m of dominantly medium/fine-texture clastics, highly carbonaceous over the lower section but also sometimes calcareous and frequently interspersed with coarser siltstones. The lower section is traditionally termed the organic-rich zone or ORZ.

Figure 3 illustrates the 86-1 pre-salt section in context of a Lower Congo regional traverse across the Malongo-Liawenda and Ponta Vermelha Troughs. It should be noted that all wells are more or less peripheral to their deepest paleo-depositional centres and, excluding possibly 122-3, only a partial pre-salt section is penetrated. In the case of 86-1, McHargue (1990) records that the type section contains the upper and middle members of the Bucomazi, only.

Various stratigraphic information assembled in the present study for the type section is summarised in Fig. 4. Here, note that a contemporary biostratigraphic approach (Weston 1992) identified the Cretaceous section in 86-1 as being exclusively of Late Barremian age only.

In terms of ostracod biostratigraphic interpretation all but the very uppermost sediments corresponded to the AS9 zone (Krommelbein and Weber 1971; Grosdidier and Bignoumba 1984). Palynomorph zonation after Doyle et al. (1977, 1982) similarly identified a range corresponding to the transition from the CV to CVI microfloras. In terms of the Harland et al. (1989) chronostratigraphy, this is thought to correspond to 750 m of

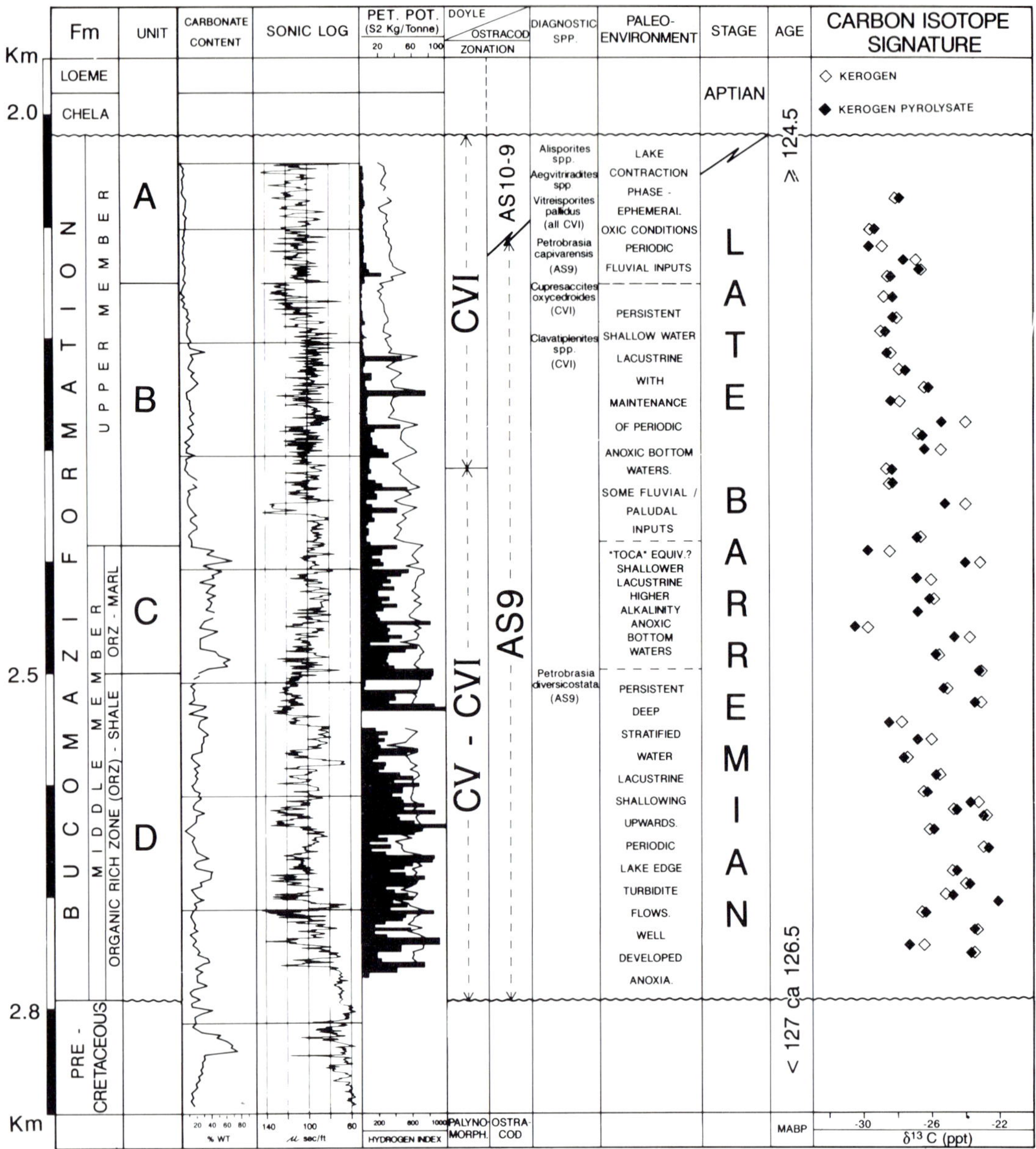

Fig. 4. CABGOC 86-1 Bucomazi Fm. type section : composite of stratigraphic, paleo-depositional, lithological and kerogen carbon isotopic information

compacted sediments not older than 127 Ma or younger than 124.5 Ma, respectively. Translation of these data into sedimentation rate yields a phenomenal accumulation of 0.6 to 0.85 km/Ma of uncompacted sediment.

Such depositional rates are not atypical for rifted lacustrine environments, comparable values being recorded for contemporary/Holocene sedimentation in the East African rift (Carbonnel et al. 1983; Tiercelin and Vincens 1987). However, noting the richness particularly of the ORZ (organic-rich zone) such rapid sedimentation rates imply an enormous turnover in biomass with accumulation being assisted by a very effective preservation process.

Paleoenvironmental interpretation of the section, with its predominantly argillaceous lithologies, amorphous kerogens and non-marine ostracod content, suggested the whole was deposited in a non-marine lacustrine environment (cf. Fig. 4).

Changes in these factors and the micropaleontological assemblages implied that the lake hinterland to 86-1 was deeper and more distal, and of a more humid climatic setting, during deposition of the lower part of the section. Subsequently, this became shallower and more marginal with ephemeral episodes showing evidence of desiccation and probable subaerial exposure during deposition of the post-ORZ section.

It is considered that climate and lake water depth were the primary control on the character of the sediments penetrated. There is also evidence that rocks from the upper part of the well section were dominated by "sheet drape" sediments, with those below 2400 m showing evidence for a major period of "basin fill" sedimentation as defined by Bouroullec et al. (1991). These conclusions are expanded below in conjunction with a detailed description and discussion of the various source intervals recognised.

Electric log, primarily sonic and density, descriptions of Bucomazi Fm. sections gave highly diagnostic traces showing a strong positive correlation between the acme of organic content and low velocity/density intervals. In conjunction with lithological description and calcimetry, this provided a first pass at segregation into paleoenvironmental significant depositional sections (Fig. 4).

The lower "Middle Member" broadly equated to a consistently organic rich zone and is trivially known as the ORZ. A deeper, stratified water depositional regime is anticipated. Note towards the top of this unit mineral carbonate contents increase, the sediments becoming calcareous shales and/or argillaceous limestones (marls). Onset of the "Upper Member" was marked by a sudden decline in carbonate character and is believed to be a shallower-water, periodically emergent/ephemeral, subsident regime.

The top of the Bucomazi Fm. is marked by onset of the Chela depositional regime, this following a short-duration but widespread regional unconformity, the sediments being characterised by a marked change in E-log character.

Material and Methods

This chapter is based upon a large, recently assembled data set using contemporary geochemical procedures for CABGOC 86-1 materials sampled at a high frequency of 3 m or less.

Similar data for the Chela Fm. were assembled on returns from wells CABGOC 123-4, East Mibale-1 and Tshiende-1. Abundant (i.e. primary) hydrocarbon source potential for the Chela and Bucomazi Fms was ascribed to those sediments with $S_2 \geq 5$ kg/t. Petroleum potential evaluated by Rock-Eval procedures is expressed in metric ($\times 10^6\,m^3$ (oil equiv.)/km^3 (rock) or traditional (bbl/acre foot) units as illustrated in Table 1. Kerogen activation energy distributions were determined on pre-extracted sediments using a Rock-Eval 5 and OPTKIN 1 software.

Stable isotope measurements ($\delta\ {}^2H$, ${}^{13}C$) were made on whole-oils, bitumens, fractions, kerogens and kerogen pyrolysates, as appropriate. Results are reported vs PDB and SMOW standards (NBS 22 = -29.8 ppt and -119 ppt, respectively).

Biomarker analyses were performed on oils and bitumens using a VG Autospec high resolution GC-MS system employing SIR data acquisition mode.

Detailed lithological, palynological, palynofacies and ostracod analyses were carried out on ditch cuttings and conventional core specimens over the interval 2035-2760 m. Kerogen residues (HCl, HF procedure) were examined by transmitted light microscopy.

Hydrocarbon Characteristics of Lower Congo Basin Oils

An insight into the source characteristics of the Bucomazi Fm. is naturally manifest in petroleums derived from this rock unit as developed in the generatively mature catchments of the Lower Congo Coastal basin.

The hydrocarbon system charged by this source (estimate 4.5 BBO recoverable reserves to-date) shows a preponderance of Cretaceous carbonate platform, followed by pre-salt, reservoired occurrences (Fig. 2). This is in marked contrast to the equivalent Brazilian Espirito Santo and Campos basin habitats (ca 100 MMBO and 6 BBO recoverable, respectively), with the latter being dominated by a late drift phase, Tertiary deep water clastic reservoir style.

An investigation of Lower Congo petroleum composition is instructive in that, contrary to wide quality variations in terms of API gravity and waxiness (Fig. 5), oils of pre-salt provenance show

Table 1. Summary source characteristics: Pre-Salt Bucomazi well (86-1) type section

Candidate source unit	Thick-ness (m)	Primary source net/gross	Mean TOC (% wt)	Mean hydrogen index GOPR	Kerogen type	Kinetic parameter E_a/A (kcal/mol)	Petroleum potential			Souce signature	
							$S_1 + S_2$ kg/t	m^3(oe)/km^3 (rock) ($\times 10^6$)	bbl (oe)/ af (rock)	Pyrolysate δ13C (ppt)	Bitumen biomarker
Chela Fm.	ca 10 < 50	≤ 10%	(1.5)	400 0.16	II	52 2.862E + 13	≤ 6	≤ 18	≤ 78	– 28.4 to – 25.4	25,30-BNH prominent
Bucomazi Fm. Upper member zone A	117	16%	1.3	(570) (0.17)	II	52 5.238 + 13	5 (12)	9 (35)	70 (272)	– 30.0 to – 27.0	Gammacerane prominent
Upper member zone B	226	61%	2.9	(500 +) (0.13)	II to II (sup.)	52 3.887E + 13	13 (20)	37 (60)	287 (466)	– 29.0 to – 25.0	Gammacerane Ethyl-cholestane prominent
Middle member (ORZ) zone C	13	100%	4.2	700 + 0.13	II (sup.)	54 1.035E + 14	36	102	791	– 30.0 to – 24.0	30-NH 28, 30-BNH 25,28,30-TNH prominent Low Ts/Tm ratio
Middle member (ORZ) zone D	294	100%	7.9	700 + 0.13	I to II (sup.)	56 4.852E + 14	57	165	1280	– 29.0 to – 22.0	High cholestane abundance

Values in parenthesis represent mean for those sediments with primary source potential ($S_2 > 5$ kg/t), only.

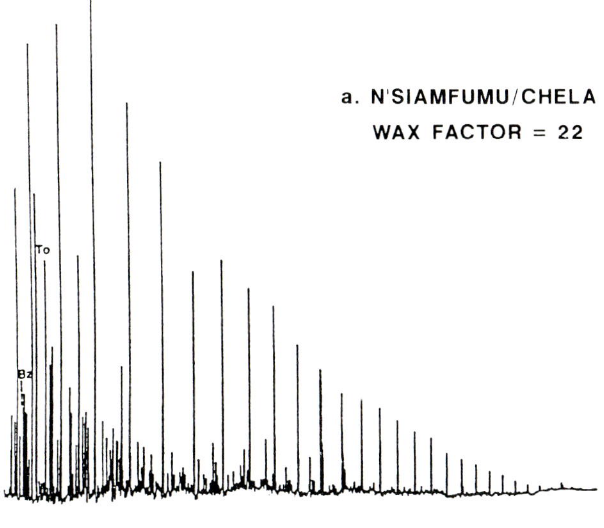

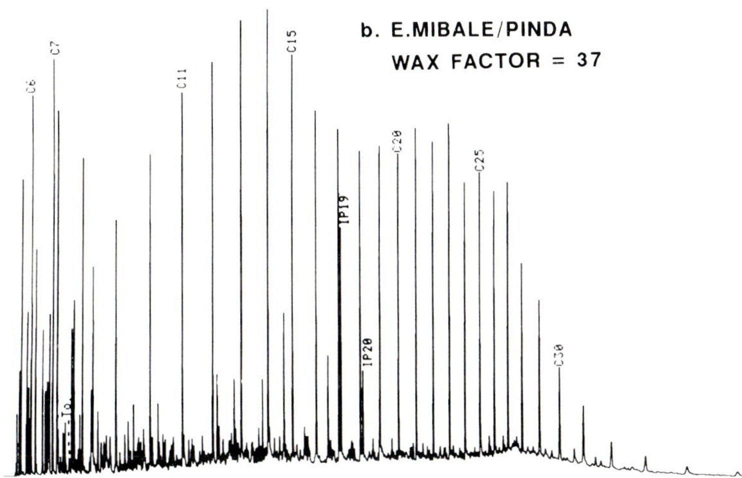

Fig. 5a,b. Lower Congo Coastal basin petroleums : pre-salt sourced **a** waxy versus **b** non-waxy Zaire crude oil chromatograms. Wax factor calculated on n-C_{25} + /n-C_{18} + normalised n-alkane content basis. n-Alkanes and acyclics identified by carbon numbers; *Bz*, benzene; To, toluene

many common features suggesting a generic, if but diffuse, co-identity. Such diversity has been recognised by Mello et al. (1988a,b) for oils of the Brazilian basins and has led Burwood et al. (1990, 1992) to relate this for the Lower Congo analogues to the large vertical and lateral organofacies variations within the Bucomazi Fm. Such a situation supports and accounts for the prevalence of aggregate oils sourced by a repertoire of end-member organofacies types. Organofacies variation in itself is undoubtedly an expression of the variable paleoenvironmental conditions prevailing during source deposition (Burwood et al. 1990, 1992).

Examination of a miscellaneous group of Lower Congo oils (50) in terms of stable isotopic composition (δ^2H, δ^{13}C) revealed two differentiable populations (Fig. 6). Whereas both groups showed an extended continuum in ^{13}C range (ca 5 ppt), segregation into ^{2}H-enriched (major) and -depleted (minor) populations was evident. With Bucomazi Fm. bitumen extracts also consistently showing ^{2}H-enriched values ($-95 +/- 10$ ppt), and ^{2}H-depleted equivalents from post-salt candidate source units, the pre-salt Bucomazi sourcing of a majority of these oils was firmly established.

A similar level of segregation between oils of pre- and post-salt provenance was also evident in the ^{2}H versus canonical variable cross-plot (Sofer 1984) illustrated in Fig. 6.

Collectively, these two graphic displays established not only a clear-cut segregation between pre- and post-salt sourcing but additionally implied a considerable internal organofacies variation within the contributory Bucomazi Fm.

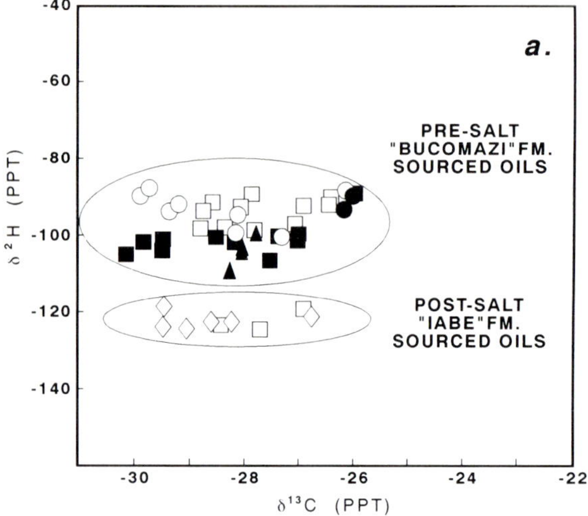

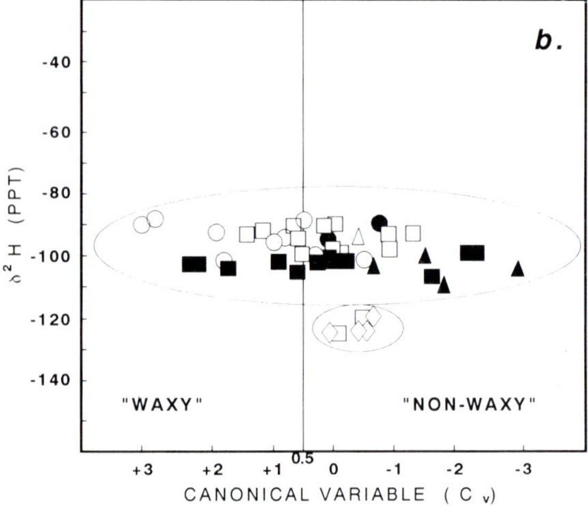

Fig. 6a,b. Lower Congo Coastal basin petroleums: differentiation of a study group of Angolan, Cabinda, Congo and Zaire (*solid symbols*) as to pre-salt Bucomazi, or post-salt, source provenance in terms of **a** deuterium (^{2}H) isotopic signature and **b** canonical variable (Sofer 1984) comparisons. Reservoir identity denoted by ○ (Pre-Salt); □ (Early Drift Carbonate Platform); △ (Late Drift Deep Marine Clastic) and ◇ (Neogene Marine Clastic)

by virtue of the extended ^{13}C and canonical variable ranges.

Examination of the biomarker compositions for the Bucomazi-derived oils similarly showed considerable variations in diagnostic signatures. Typical triterpane and sterane fragmentograms, and a range of data for a selection of parameters, are illustrated in Figs 7–10. Gammacerane, probably derived from tetrahymanol, is related to the abundance of *Tetrahymena sp*., a primitive organism that is particulary abundant in hypersaline waters (ten Haven et al. 1989; Venkatesan 1989). The abundance of gammacerane in some of the oils may therefore be related to their provenance from source rocks deposited under hypersaline conditions (ten Haven, 1989); (Fig. 8). The range covered by the Zaire oils, for instance, suggests highly disparate sourcing with markedly different end-member organofacies, further supporting the concept of mixed, aggregate oils in place.

Alternatively, desmethylhopanes (e.g. 28,30-bisnorhopane: 28,30-BNH and 25,28,30-trisnorhopane: 25,28,30-TNH) are associated with depositional situations supporting high levels of anoxia (Katz and Elrod 1983; Curiale et al. 1985). The former bisnorhopane has been found in abundance in Jurassic marine anoxic sediments (Farrimond et al. 1990), a feature that can also probably be related to non-marine (i.e. lacustrine) anoxic sediments (see later section, Middle Member ORZ-Unit C).

The interplay of anoxic versus hypersaline source inputs is best illustrated by the hopane normalised gammacerane vs 25,28,30-TNH differential (Fig. 8) demonstrating that there are oils of more or less anoxic source. The existence and mixing of charges from disparate organofacies within the "Bucomazi" is again supported.

Amongst other diagnostic biomarker component ratios, Zumberge (1987) has demonstrated an empirical marine/non-marine relationship in terms of the C_{25} and C_{26} tricyclic terpane abundances. When applied to the Lower Congo study suite of "Bucomazi"-derived oils (Fig. 9), a majority showed a non-marine, lacustrine source affinity. A group of the Angolan oils, however, were at variance with this trend in showing hints of a "marine" input, this feature requiring explanation in terms of source depositional considerations.

Other such complicating factors contributing to the provenance of Lower Congo oils are also manifest in the unusual abundance of 25,30-bisnorhopane (25,30-BHN) in many of these petroleums (Fig. 10). Although 25-norhopanes are traditionally associated with in-reservoir oil biodegradation (Volkman et al. 1983; Noble et al. 1985), on this occasion the presence of this component in the oils is believed to be source-related (Burwood et al. 1990; Blanc and Connan 1992). At this stage it is sufficient to record that 25,30-BNH appears to be intimately associated with source development in the immediate post-Bucomazi Chela Fm.

Overall, Lower Congo pre-salt "Bucomazi" sourced oils were consistently ^{2}H enriched (Fig. 6) supporting a generic co-identity, this being modified by the observed diversity in ^{13}C and biomarker

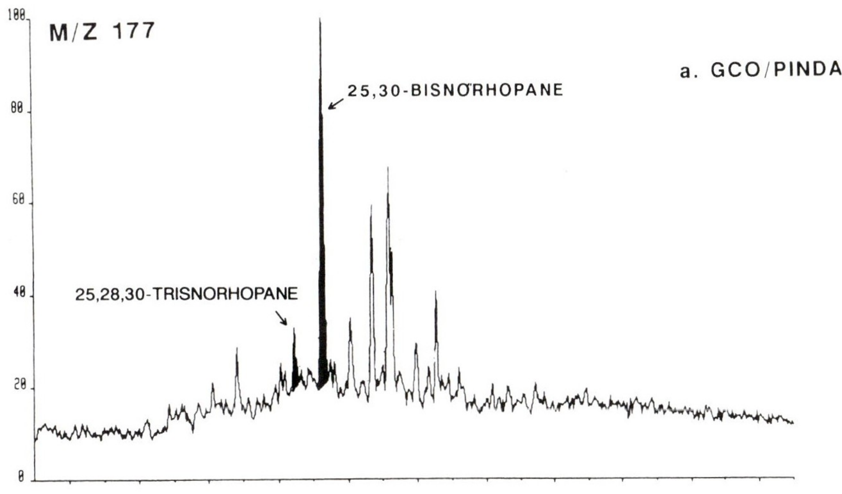

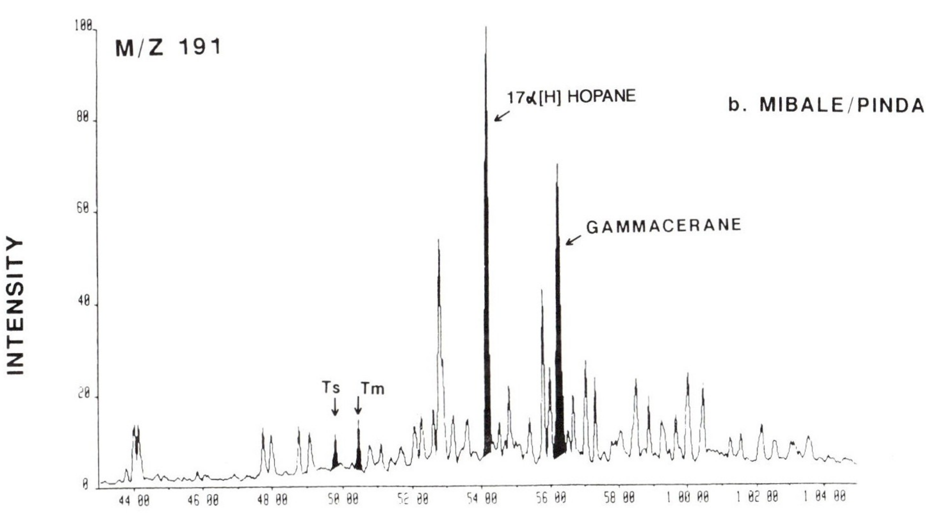

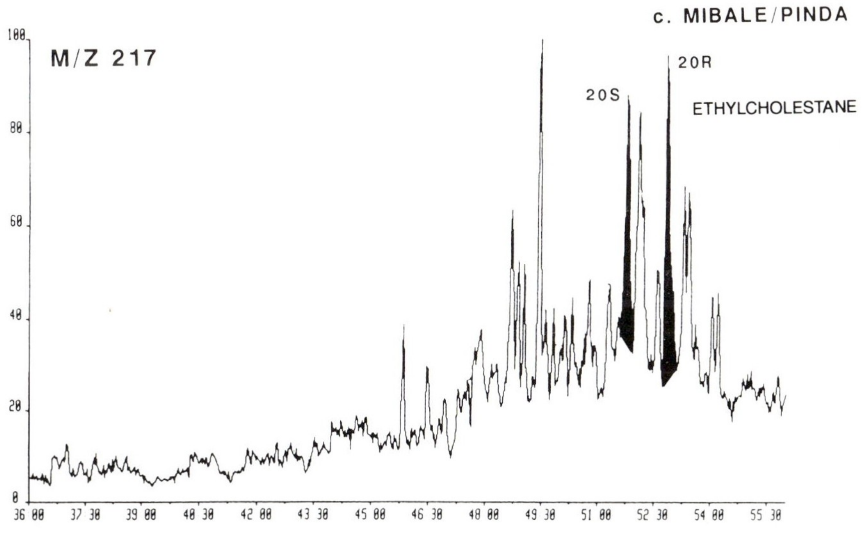

Fig. 7a-c. Lower Congo Coastal basin petroleums: biomarker composition as revealed by typical m/z **a** 177 (25-norhopanes); **b** 191 (triterpanes) and **c** 217 (steranes) fragmentograms for the captioned oils

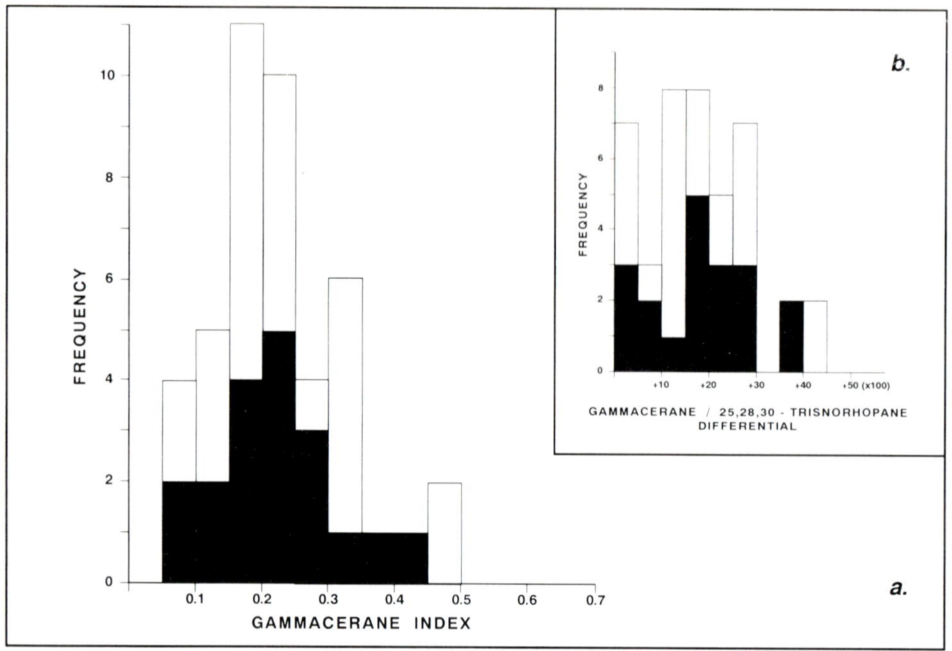

Fig. 8a,b. Lower Congo Coastal basin petroleums: segregation of pre-salt Bucomazi Fm.-derived oils in terms of **a** gammacerane index (gammacerane/gammacerane + hopane), and **b** gammacerane – 25,28,30-trisnorhopane differential (for definition of this parameter see Fig. 15). Zaire coastal oils highlighted as the *solid population*

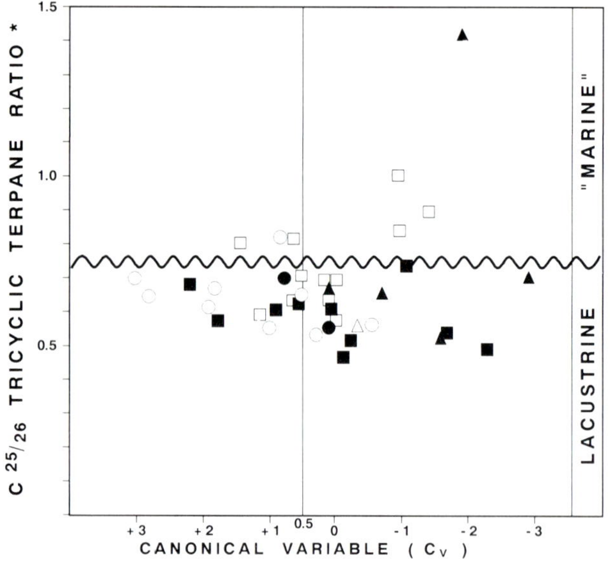

Fig. 9. As in Fig. 8, on basis of canonical variable versus C25/26 tricyclic terpane ratio. (After Zumberge 1987)

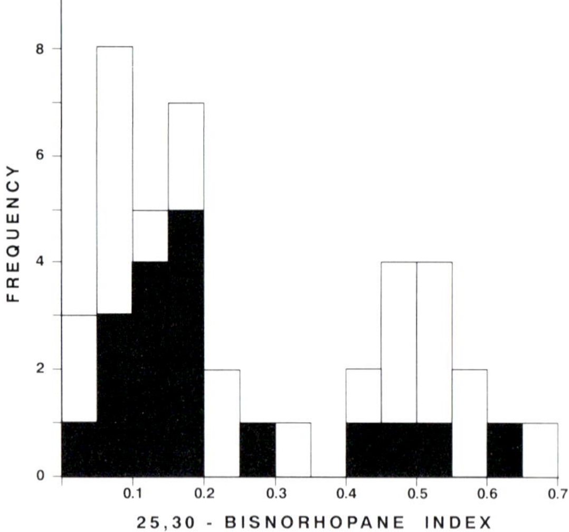

Fig. 10. As in Fig. 8, on basis of 25,30-bisnorhopane (25,30-BNH/25,30-BNH + norhopane) ratio. Although absent in the pre-salt reservoired example (denoted ● in Fig. 9), note that a majority of the Zaire oils contain this marker, sometimes in high abundance

signature. This diversity was suggestive of oils of aggregate composition showing wide ranging ^{13}C signatures these reflecting, amongst many factors, variable source hypersalinity, anoxicity, reworking and even marine influences.

Such diversity is thought to derive from disparate sedimentary units showing a considerable paleoenvironmental control over source organofacies development and variability. On maturation, the generation and expulsion of hydrocarbons of variable, aggregate compositions from such a source melange is to be anticipated.

Characteristics of the CABGOC 86-1 Bucomazi Fm. Type Section

Paleoenvironmental typing of the section penetrated in well 86-1 permits differentiation of these sediments into four units broadly consistent with the trivial terminology illustrated in Fig. 4. Working up-section, the Middle Member ORZ is sub-divisible into a lower highly carbonaceous, argillaceous siltstone lithology (Unit D). This is superceded by an interval of more evidently laminated and calcareous (micritic) character (Unit C). Each of these units showed characteristic, cyclical low velocity sonic (LVU) responses, these often being coincident with the acme in organic richness.

Progression into the Upper Member is marked by an abrupt decline in micrite and consistently high organic content. Again, of dominantly siltstone lithology, the lower part of this unit exhibited intermittent and localised intervals of organic richness, only (Unit B). After one last productive episode, much of the uppermost unit showed relatively low organic contents ($S_2 \leq 5$ kg/t) consistent with evidence for ephemeral, sub-oxic conditions (Unit A).

This zonation provides a convenient basis for the detailed description of the source characteristics of the type section as summarised in Table 1. Each of the units is described below and considered in turn, working up-section, thereby following basin development.

Thermal Maturity Status of the CABGOC 86-1 Bucomazi Section

Before describing each Bucomazi Unit in detail, some general thermal maturity trends are provided. Assessment of the section, employing vitrinite reflectance measurements, were often most successfully achieved for those intervals showing the higher sonic velocity and lower hydrogen indices (cf. Fig. 4). Data at top Unit D (2500 m) gave values at 0.4% Ro, increasing to ca 0.8% for the basal interval (2760 m).

Bitumen-based biomarker maturity revealed comparable values. Drimane isomerisation, for instance, indicated the maturity interval to start from a vitrinite reflectance equivalent of 0.48% in Unit A and to increase to 0.7% at depth. This value is reached at about 2600 m equivalent to Unit D and marks the maximum maturity measurable by drimane isomerisation (Noble et al. 1987). The ethylcholestane isomerisation is grossly paralleling this trend and the vitrinite reflectance equivalent increases from 0.4 in Unit A to 0.7 in Unit D.

More informal data such as production indices ($S_1/S_1 + S_2$) and T_{max} values similarly attested to pre-generative levels of maturity.

These data are important in that they demonstrate a majority of the Bucomazi section to be immature for hydrocarbon generation in the conventional sense (i.e. VRE $\leq 0.6\%$ Ro). Similarly, the less labile type I kerogens encountered in the basal, higher maturity, Unit D are adjudged to be still immature. Of consequence, the 86-1 section provides not only a type section but one in which the full, realisable petroleum potential of Bucomazi-style sediments can be evaluated without complicating maturity considerations.

CABGOC 86-1 Bucomazi Middle Member ORZ-Unit D

These sediments, comprising 294 m of fissile, dark brown argillaceous siltstones grading into silts showing microlamination, represent the "basin fill" lithology of Bouroullec et al. (1991). Persistently carbonaceous, organic carbon contents (TOC) are consistently $> 5\%$ and routinely in the range 10%, peaking at 24 wt%. Petroleum potential measured as the Rock-Eval $S_1 + S_2$ parameter is always in excess of 10 kg/t, giving an exceptional average of 57 kg/t for the unit. Volumetric yield parameters equate to 165 metric petroleum potential units, or 1280 bbl/acre foot, as defined in Table 1. A summary source quality description is illustrated in Fig. 11d.

Sonic log description of the section showed cyclical velocity contrasts, the acme of organic input and preservation being marked by conspicuous low velocity (LVU) sub-units. Here, TOC and S_2 reach values in excess of 15% and 100 kg/t, exceptionally 24% and 191 kg/t being recorded. Hydrogen indices are frequently high (ca 700), suggesting very attractive hydrocarbon convertibility on maturation. The corresponding high velocity (HVU) source sub-units often show a decline (relative) in organic content and hydrogen index leading

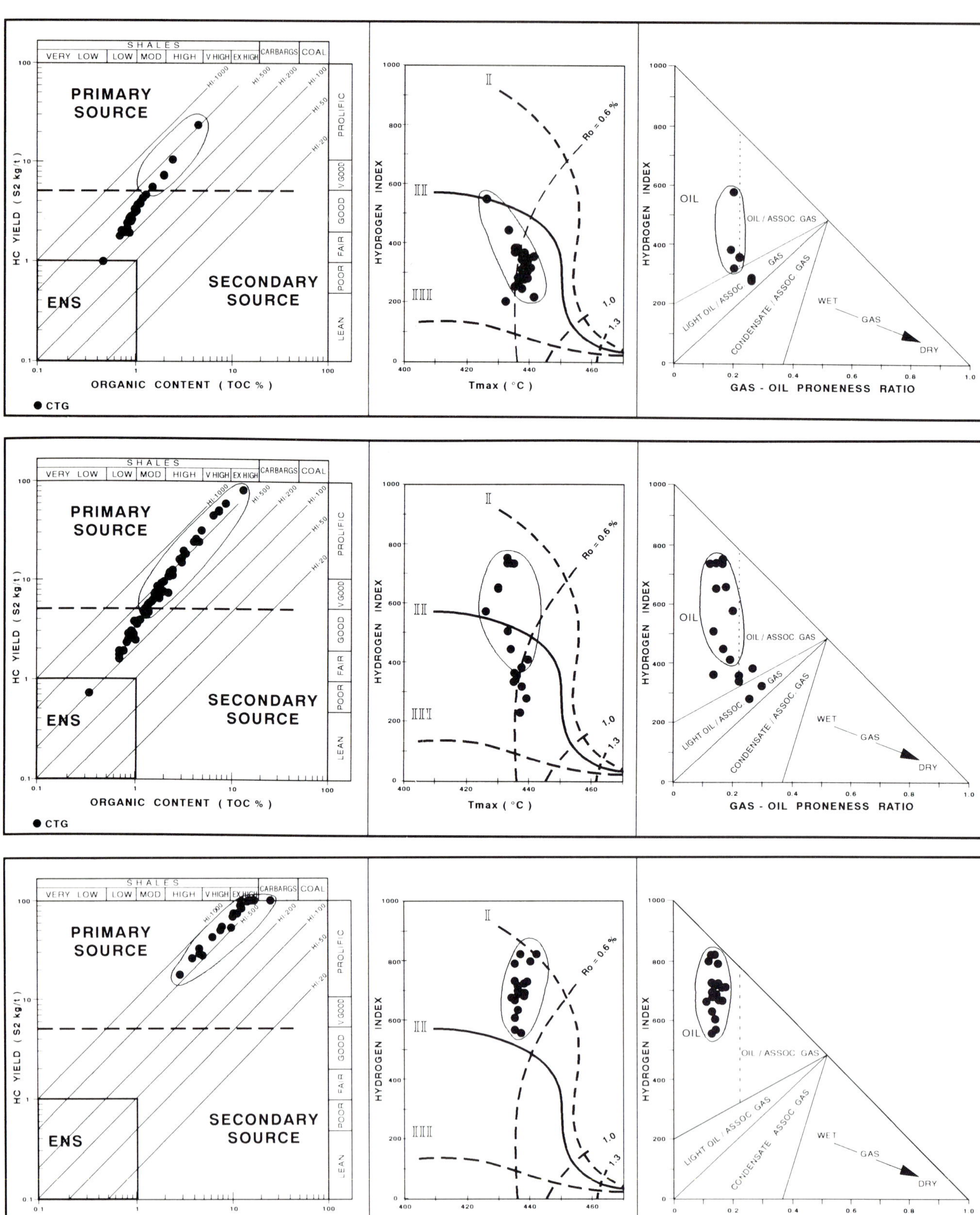

Fig. 11a-d. Summary source rock quality information on basis of Rock-Eval and GeoFina Hydrocarbon Meter measurements **a** Unit A. **b** Unit B. **c** Unit C. **d** Unit D. Primary source ascribed to those sediments with $S_2 \geq 5$ kg/t; *ENS* Effective non-source. Definition of Gas-Oil Proneness Ratio given by PGC C_1–C_5 content normalised to total pyrolysate

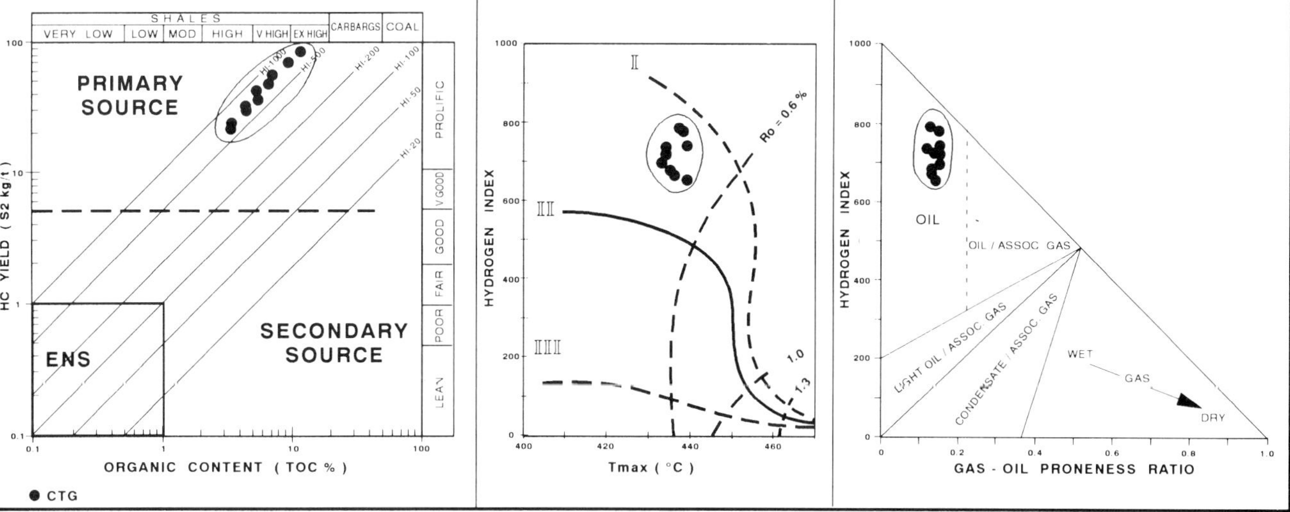

Fig. 11d

to the interpretation of turbidite influx. Here, proximal, organic leaner and terrestrially influenced sediments are thought to have periodically slumped into the more distal, deeper and stratified lake centre environments.

Whatever the cause, it is evident that each HVU is followed by an intense period of biomass turnover and preservation, this of necessity being stimulated and supported by an increased availability of nutrients. It is interesting to note that the relatively "coarse" siltstone texture of the supporting mineral matrix is capable of preserving such high organic contents. This is perceived to be a consequence not only of the lake centre anoxia, but the dramatic input of both biomass and mineral matrix. Although stimulated by very rapid tectonic subsidence, optimal paleoenvironmental conditions, particularly a humid, high precipitation hinterland also play a crucial role.

An organofacies description for these super-rich assemblages suggested a continuum of type II through I kerogens (Fig. 11d) of persistently high hydrogen indices and oil proneness ratio (e.g. see kerogen pyrogram, Fig. 12). As mentioned above, the lower hydrogen indices coincident with the sonic HVU episodes were more typical of "type II" behaviour, in reality the assemblages probably being admixtures of oil prone algal (I) and herbaceous (III) organic matter.

Nevertheless, visual examination of these kerogens revealed the presence of semi-degraded plant/woody tissue in what were otherwise consistently amorphous, sapropelic assemblages (Fig. 11d). With the occasional recognition of black wood/vitrinite clasts, this provided further evidence for the periodic ingress of proximal, shoreline sediments into the deeper lacustrine areas as turbidity flows.

Despite this variation in compositional make-up, kerogen transformation kinetic parameters often revealed less labile, type I behaviour with higher activation energy (Ea) distributions peaking at 56 kcal/mol (Arrhenius factor $\times$ 4.852 E + 14 s^{-1}) (e.g. see Sundaraman et al. 1988).

It is again noteworthy that those HVU assemblages showing more of a "type II" behaviour gave discernibly broader distributions peaking at lower Ea values (54 kcal/mol), but showing comparable Arrhenius factors. The concept of a mixed assemblage, incorporating type III detritus, is again supported (Fig. 13).

Further description of the organofacies diversity of Unit D is demonstrated by the carbon isotopic composition of the component kerogen and kerogen pyrolysates as illustrated in Fig. 4. Such observations have shown considerable utility in revealing variable organofacies in what could otherwise be described as homogeneous sediments (Burwood et al. 1989, 1990; Bailey et al. 1990).

The advantage of these isotopic measurements lies in providing not only a mean description of the whole kerogen ($\delta^{13}C$ kerogen) but also that of the oil prone sub-component of the assemblage ($\delta^{13}C$ pyrolysate). Variation in these values reflects and records fluctuations in detrital biomass input to the sediment.

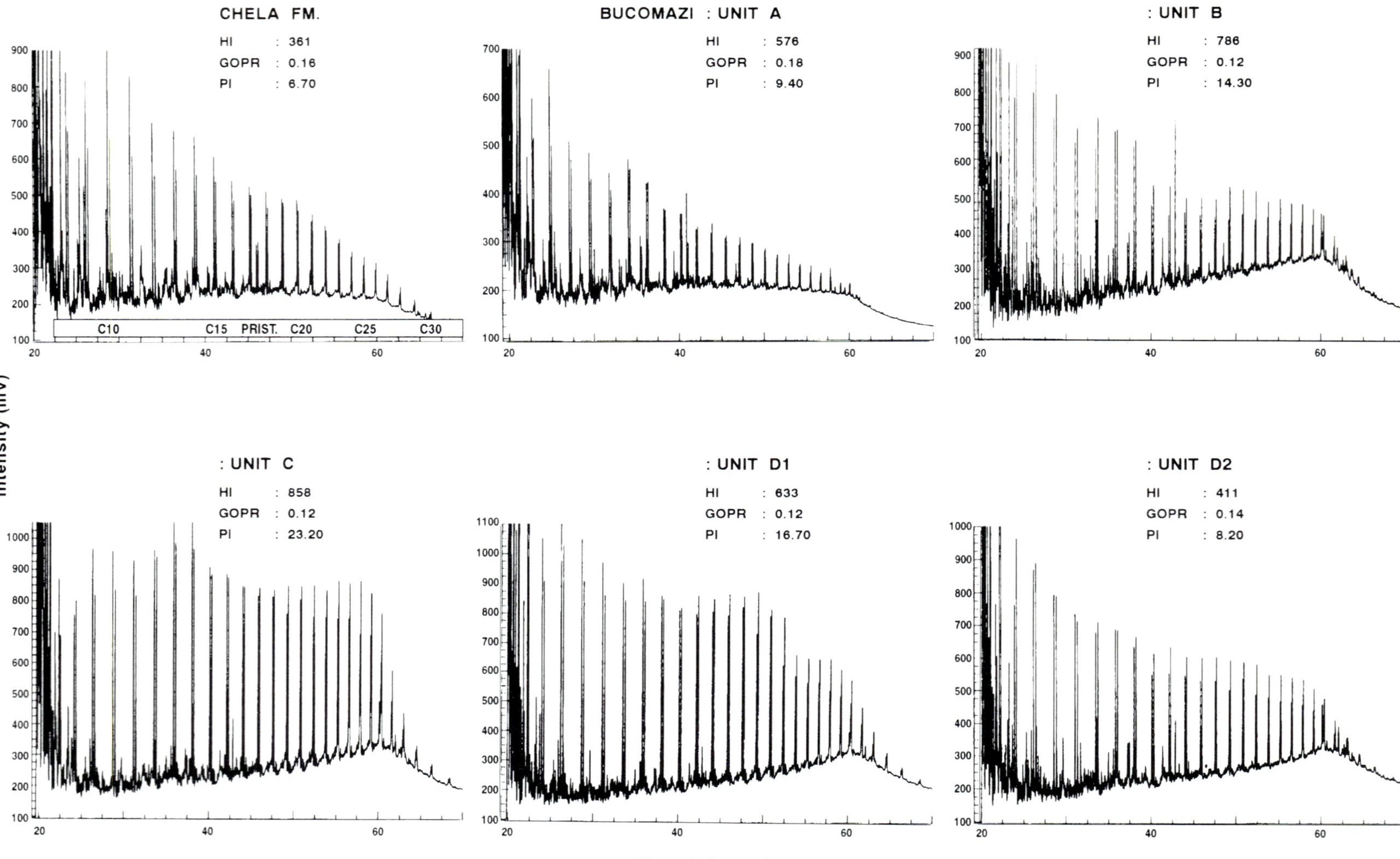

Fig. 12. GeoFina Hydrocarbon Meter kerogen (S_2) pyrograms for typical candidate source units representative of Chela Fm. and Bucomazi Units A, B, C and D. For Unit D, examples of type I (*D1*) versus mixed type I/III (*D2*) assemblages are illustrated. For definition of Paraffin Index (PI) see Larter and Senftle (1985)

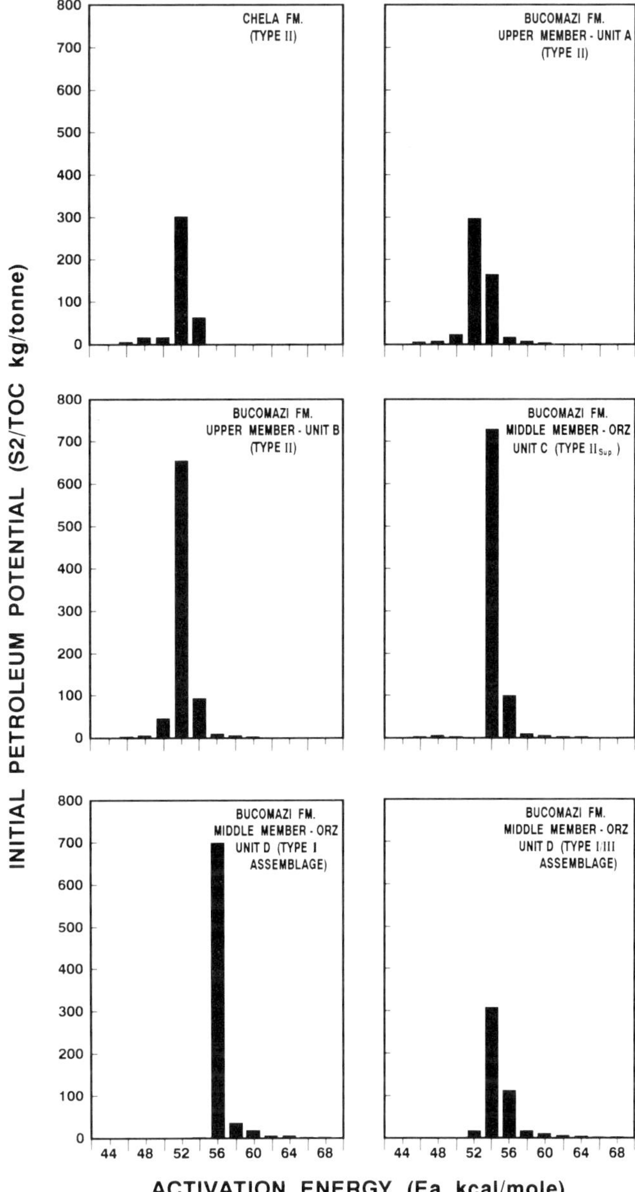

Fig. 13. Rock-Eval 5/Optkin-derived kerogen transformation activation energies (Ea kcal/mol) for Chela and Bucomazi Fm. candidate source units. Note contrast between Chela and Bucomazi Fm. Units A plus B (lower Ea, type II) and Units C and D (higher Ea, type I). An internal diversity within Unit D is also evident with the mixed type I/III assemblage showing a distinctively broader distribution

The organofacies trend over the whole Bucomazi section, showing at least three macro-isotopic excursions, is broadly similar to that described for the well CABGOC 123-4 (Burwood et al. 1992). The excursions correlate with maxima in organic production and preservation and undoubtedly have paleoenvironmental significance in terms of nutrient availability and carbon dioxide budget.

Concerning that part of the trend for Unit D, this embraces an isotopic range of 7 ppt (ca − 29 to − 22 ppt) and contains two of the macro-excursions. On this occasion no strong relationship was observed between other organofacies attributes, such as Ea or HI, and respective carbon isotopic composition other than the organic richest sediments often showed the most enriched ^{13}C values. This is seen as a facet of CO_2 budget control, the lake surface waters at time of maximum biomass turnover being depleted in ^{12}C. Following the observations of Hollander and McKenzie (1991), this can be rationalised in terms of accelerating biomass production and turnover resulting in an increasing disequilibrium between the ^{13}C composition of atmospheric versus lake surface dissolved CO_2. With enhanced nutrient input stimulating ever

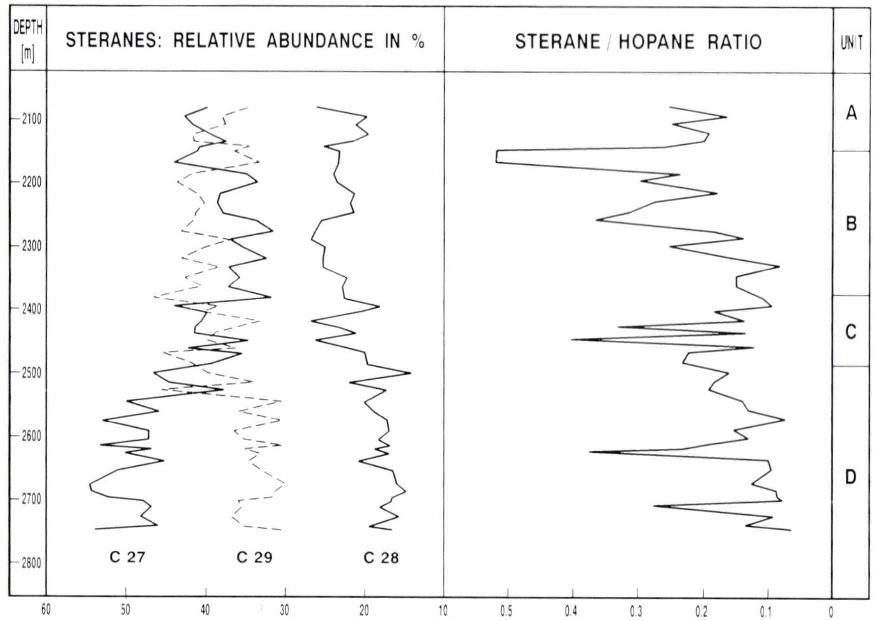

Fig. 14. Well CABGOC 86-1 Bucomazi Fm. bitumen biomarker composition: sterane/hopane ratio and sterane abundance variations with depth/unit. Note abundance of cholestane, implying a dominant algal input, over Unit D

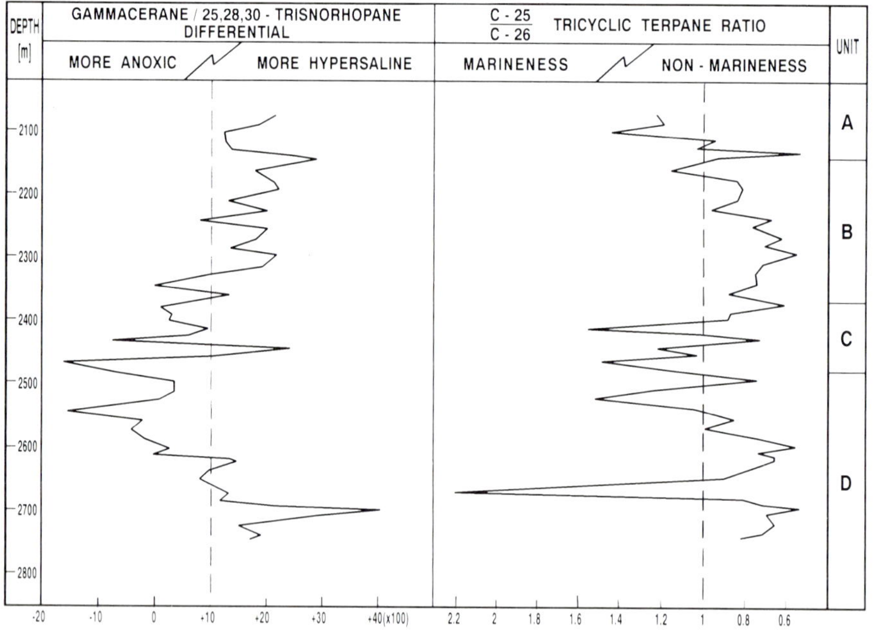

Fig. 15. As in Fig. 14. Variation of C_{25}/C_{26} tricyclic terpane ratio and gammacerane/25,28,30-trisnorhopane differential by depth/unit. Differential defined by subtraction of the hopane normalised gammacerane and 25,28,30-trisnorhopane ratios. Note concurrence of negative differential and tricyclic ratios > 1 over Upper Unit D and C with "more marine-like" implications (cf. Burwood et al. 1992)

more prolific algal blooms, these organisms have the ability to adapt to bicarbonate (HCO_3^-) utilisation as their primary photosynthetic carbon source. With the dissolution of atmospheric CO_2 in mildly alkaline waters accompanied a large fractionation in favour of ^{13}C, the HCO_3^- pool would become increasing enriched in the heavier isotope. In consequence, the photosynthetic products, hence co-sedimented detrital input, would similarly become more depleted in ^{12}C. The nature of the long-term paleoenvironmental event, stimulating greater nutrient availability and algal bloom intensity is presently equivocal but presumably related to humidity and fluviatile input to the lake. Thus during deposition of Unit D, two periods of increasing precipitation, interspersed by a drier interlude, could be invoked.

It is interesting to note that the two HVU episodes, roughly corresponding to the end of these proposed wetter periods, could therefore represent lake highstands. Such conditions would similarly be conducive for the co-occurrence of turbiditic events.

Biomarker characterisation of the associated bitumens for Unit D sediments are summarised in Figs 14 and 15. From earlier discussion, it should be remembered that these hydrocarbon compositions are representative of inherent bitumens and not those deriving from intense thermogenic transformation of precursor kerogens.

In addition to the ubiquitous and dominant hopane series, the relative abundances of the steranes, gammacerane, desmethylhopanes and the tricyclic terpanes were of greatest diagnostic significance. Sterane/hopane ratios were consistently low (≤ 0.3; Fig. 14), as could be anticipated for a lacustrine-derived biota (Mello et al. 1988b).

Whereas methylcholestane contents remained constant over Unit D, a conspicuous differential between cholestane and ethylcholestane, in which the former dominated, was evident (Fig. 14). This was interpreted in terms of Huang and Meinschein (1976, 1979), where cholestane is representative of predominantly algal input, whereas ethylcholestane is derived from higher land plants. This concept was, however, established on the basis of sterols and the use of steranes instead is of limited value and not warranted in all cases, as has been shown by numerous studies (Volkman 1986; Wünsche et al. 1987; Grantham and Wakefield 1988). Cholestane/ethylcholestane ratios therefore need to be supported by other biomarker data. After Zumberge (1987), low C_{25}/C_{26} tricyclic terpane ratios (< 1.0) characterised much of the unit as typically lacustrine (Fig. 15).

From high basal gammacerane contents (peak abundance in well 86-1 is found at ca 2700 m), the gammacerane index steadily decreased towards the top of Unit D. This decrease is accompanied by a steady increase in 28,30-BNH and 25,28,30-TNH. Because of the perceived negative correlation of the gammacerane and 25,28,30-TNH ratios (Fig. 15), up-section progress in the "biomarker differential" (Burwood et al. 1992) classified Unit D as an increasingly anoxic episode. Such behaviour in 25-norhopane enhancement has also been recognised for the acme of marine anoxia (Katz and Elrod 1983; Curiale et al. 1985).

Consolidation of these characteristics with the biostratigraphic based paleoenvironmental interpretation provided a more definitive description of deep, fresh, possibly brackish, water supported anoxic conditions. As witnessed by organic content, such conditions appear to provide the optimum for biomass turnover, co-sedimentation and preservation.

CABGOC 86-1 Bucomazi Middle Member ORZ – Unit C

Concluding the "basin-fill" episode, Unit D is succeeded by 113 m of more calcareous, microlaminated sediments.

Persistently carbonaceous, these sediments were again dominanted by a medium to dark brown argillaceous siltstone lithology but were on this occasion interspersed with white-cream micrite horizons in addition to coarser silts. Organic carbon contents were invariably greater than 2 and typically in the range 4-5%, supporting petroleum potentials peaking from 45 to 92 kg/t (Table 1). The interval mean at 36 kg/t equated to volumetric yields (ultimate) of 102 metric petroleum potential units, being 791 bbl/acre foot. A summary source quality description is illustrated in Fig. 11c.

Sonic log character of the unit showed much fine structure (Fig. 4) but essentially comprised one cycle of increasing velocity up-section. This broadly equated to a relative decline in mean organic content. S_2 yield, however, indicated periodic high-points in richness, these presumably corresponding to episodes of increased nutrient availability and biomass turnover. Although paleoenvironmental evaluation suggested shallower, more alkaline waters conducive to micrite deposition, a high level of anoxia was nevertheless maintained.

Again seen as super-rich assemblages (Fig. 11c), organofacies description suggested superior type II (i.e. HI > 700) kerogens of a highly oil prone nature (e.g. see kerogen pyrogram, Fig. 12). Visual examination of these kerogens revealed a dominant yellow-brown amorphous organic matter with a small, but persistent, occurrence of degraded plant/woody material. Recognisable cuticular tissue was recorded sporadically; however, the palynofacies make-up was noteworthy in its very limited miospore and algal species diversity, the latter implying a restrictive paleoenvironment.

Kerogen transformation kinetic activation energies showed narrow distributions, consistently peaking at a lower 54 kcal/mol ($A = \times 1.035\ E + 14 s^{-1}$) implying that the organofacies present was discernibly different to that in the underlaying Unit D (Fig. 13). Paleoenvironmental control on species adaptation and resulting biomass input is deduced.

Organofacies description was further revealed through the carbon isotopic signatures of kerogens and their respective pyrolysates (Fig. 4). Here the assemblages embrace a 6 ppt range in ^{13}C, but appeared to represent one macro-cycle, only. Isotopically enriched kerogens at the base of the unit progressively become depleted up-section. Assuming a consistency in biomass input, this implied a considerable modification to the dissolved CO_2 budget, hence water chemistry, during sedimentation.

The key to these changes is possibly related to the calcareous nature of the sediments and the way in which micrite precipitation mediated the photosynthetic CO_2 pool. With the observed decline in organic productivity and/or preservation up-section, it is perceived that a sufficient pool of dissolved CO_2, favouring photosynthetic ^{12}C assimilation, was available. Similarly, micrite precipitation, in itself preferentially incorporating the heavier (^{13}C) isotope via the intermediary of HCO_3^- would enhance relative ^{12}C concentrations.

We have here a possible explanation for the perceptible organofacies difference, as manifest in kerogen kinetic character, between Unit C and D assemblages. In the latter case biomass turnover is dominated by a microflora capable of adapting to HCO_3^- assimilation, whereas up-section an alternative flora, photosynthetically needing only the ability to exploit dissolved CO_2, was more dominant. Differences in the nature of the contributory biomass debris, co-sedimented as the kerogen precursors, would be the eventual control on organofacies.

Biomarker characterisation of Unit C bitumens is summarised in Figs. 14 and 15. Sterane/hopane ratios, remained typically low, with the C_{27} and C_{29} steranes showing comparable abundances (Fig. 14). On this occasion the bitumens were notable for their conspicuous desmethylhopane (25,28,30-THN; 28,30-BNH) and norhopane contents. The gammacerane-25,28,30-TNH differential again typified maintenance of a high level of anoxia (Fig. 15). Persistently abundant 28,30-BNH, sometimes thought to have anoxia-related connotations, differentiated these sediments from Unit D. This segregation was further amplified by C_{25}/C_{26} tricyclic terpane ratios >1, these giving a "marine-like" character to the unit (Fig. 15). All these factors were perceived to be a further endorsement of a modified organofacies.

Similarly, a steady increase in gammacerane over this unit could be interpreted either as enhanced salinity or as more favourable conditions for the growth of *Tetrahymena*-type organisms. Abundant norhopane is indicative of a carbonate depositional environment that is, with respect to bitumen isotopic compositions, clearly distinguishable from the sequences above and below.

The onset of Unit C is marked by the most ^{13}C enriched bitumen ($\delta^{13}C$ at -25 ppt), being an isotopic excursion from -29.0 ppt in the sample immediately below (uppermost Unit D). This isotopic excursion is mirrored in the sterane (C_{27}/C_{29} ratio) composition. The top of Unit C is delineated by the most ^{13}C-depleted bitumen observed over the entire section (-30.0 ppt) accompanied by sterane compositions now favouring the ethylcholestanes.

However, over Unit C, sterane composition remains fairly constant, as do a number of biomarker parameters including the tricyclic terpane ratio mentioned above. The elevated abundances of 28,30-BNH and 25,28,30-TNH confirm high levels of anoxia within this section. Additional support for the uniqueness of Unit C comes from the Ts/Tm ratio that is significantly lower when compared with deep Unit D and upper units of the profile, this feature being associated with changed pH and Eh conditions.

Paleoenvironmentally, these data are thought to be representative of a shallower water, alkaline depositional regime, but one still sufficiently persistent to maintain a high level of anoxia. Probably equating to a lacustrine lowstand, conditions were possibly less humid with the overall reduction in lake volume resulting in a higher pH water chemistry.

CABGOC 86-1 Bucomazi Upper Member – Unit B

"Basin-fill" sedimentation, equating to the ORZ proper, ended abruptly with cessation of micrite formation and a switch over to a depositional regime that was only intermittently organic-rich (Fig. 4). This change to "sheet drape" sedimentation (Bouroullec et al. 1991) is believed to correspond to the tectonically less active Synrift II phase of basinal subsidence. In many sections, onset of this event is recognised in terms of a very conspicuous sonic log low velocity multiplet.

Although Unit B was deposited under lacustrine conditions, water depths were shallower, brackish, and supported a less persuasive anoxia. Paludal (fern spores) inputs were recognised, the

Lower Congo area possibly by now comprising an extensive inland sea formed by coalescence of the individual proto-lakes.

Unit B sediments comprise 226 m of mixed claystone, silty claystone through to silt lithologies. Highly organic-rich intervals are located mainly in very fissile, red-brown to dark brown claystones with "earthy" characteristics. Whereas organic richness hovered around $+/-1\%$, these richer intervals showed $7 \leq 13\%$ TOC contents, equating to a mean petroleum potential of 13 kg/t. A summary source quality description is illustrated in Fig. 11b. On this occasion segregation into primary versus secondary source units gave a 61% net/gross for the former.

Mean petroleum potential for the primary unit improved to 20 kg/t equating to volumetric petroleum potentials of 37 metric units or 287 bbl/acre foot (Table 1). Such potential is not insignificant, and indicated that sediments other than the ORZ need to be considered seriously as candidate source units. In the case of the 86-1 section, these data reduce to an effective primary source thickness of 141 m.

Organofacies description of these richer assemblages suggested a continuum of type II kerogens (Fig. 11b) showing middling hydrogen indices (400) and attractive oil proneness ratios (e.g. see kerogen pyrogram, Fig. 12). The less organic-rich sediments equated to type II (inferior) kerogens with hydrogen indices between 200 and 400. Visual examination of the richer intervals again revealed kerogens dominated by yellow-brown amorphous assemblages but with an overall increase in miospore and recognisable algal (*Botryococcus spp.*) components. Subordinate indications of degraded herbaceous material were again noted. Kerogen transformation kinetic data provided rather variable activation energies of broad distribution peaking at 52 kcal/mol and suggestive of mixed assemblages. These lower E_a assemblages ($A = \times 3.887$ $E + 13\,s^{-1}$) would be more labile than those type II (superior)/type I analogues routinely observed in Units C/D (Fig. 13).

Completing organofacies description, carbon isotope data for the kerogens and their respective pyrolysates was again instructive (Fig. 4). Covering a 5-ppt range, data revealed a further macro-isotopic cycle, before reverting to consistently ^{13}C-depleted assemblages over the upper part of the unit. Once more, considerations affecting the available CO_2 budget need to be considered in explanation of this trend.

Here, it may be insufficient to invoke a biomass turnover effect on the measured isotopic composition of the CO_2 pool, because both ^{13}C depleted and enriched assemblages are observed for the organic-rich intervals. Rather, it is possibly a case of increasingly complex water chemistry.

With evident signs of salinity/hypersalinity considerations exerting environmental control on those microfloras that were best able to adapt, lacustrine conditions conceivably changed significantly during the onset of the Synrift II phase. After an initial humid period, mirrored by the excursion in ^{13}C enrichment, arid conditions with all important consequences as to water depth and lake contraction set in.

Sediments of Unit B are characterised by ethylcholestane predominance, indicative of a more terrigenous influence in the organic matter (Fig. 14). However, the sterane/hopane ratio showed a net constant increase with respect to Units D and C, nevertheless still to be considered as being non-marine. Although 28,30-BNH is still very abundant, 25,28-30-TNH is found only in trace quantities, indicating a major facies change and probably a modification in the persistence of anoxia in the water column (Fig. 15). 30-Norhopane steadily decreased up-section, reflecting the diminishing abundance of carbonates and the associated biomass. Gammacerane is found at elevated concentrations.

Paleoenvironmental interpretation for this interval gave floras sometimes indicative of paludal associations, but with ostracod assemblages always suggestive of relatively shallow lacustrine conditions. Periods of greater or lesser humidity controlled the ephemeral character of the lake and in the former case allowed development of a stratified water column with maintenance of anoxic bottom conditions. Such conditions undoubtedly facilitated the periodic acmes in biomass preservation and organic richness.

CABGOC 86-1 Bucomazi Upper Member – Unit A

The Bucomazi cycle concludes with the "sheet drape" deposition of Unit A, these sediments showing strong indications of the ephemeral, periodically sub-oxic, lake contraction events (Fig. 4). Lithologically mixed, the 117-m unit comprised a spectrum of sub-fissile to fissile claystone/shale, variously modified by silt, intervals.

Organic contents were modest (ca 1% TOC), except for a basal interval where values peaked at 5.3%. This equated to a primary source with a net/gross ratio of 16% for the unit. Accumulation of these basal sediments undoubtedly represented the last anoxic episode, of sufficient duration, for effective source deposition. Recognised as olive-grey, laminated claystones, a mean petroleum potential of 12 kg/t equated to volumetric yield (ultimate) values of 35 metric units or 272 bbl/acre foot. This corresponded to a thin (ca 20 m), but not insignificant source unit.

Organofacies description of these sediments identified type II/II (inferior) kerogens, the richer basal interval showing higher hydrogen indices (ca 500) having oil prone characteristics (Figs 11a and 13). The less organic-rich sediments had lower hydrogen indices (≤ 400) of a more mixed oil/gas-prone nature. Kerogen residues, often of poor to moderate recovery only, consisted almost entirely of colourless to pale yellow amorphous organic matter.

Both pyritic inclusions and herbaceous/woody indications were absent or rare and suggested sub-oxic conditions and an arid hinterland. The basal organic-rich interval resembled Unit B with thicker, more abundant yellow-brown amorphous residues conspicuously pitted with pyritic inclusions.

Kerogen kinetic data gave broader activation energy distributions peaking at 52 kcal/mol and typical of type II assemblages (Fig. 13). These observations confirmed the trend that the more labile, and hence generatively active, kerogen assemblages are located in the upper "sheet drape" sedimentary episode of Bucomazi Fm. deposition.

Completing organofacies description, kerogen and pyrolysate carbon isotope data showed a small isotopic excursion to more enriched assemblages over the basal organic-rich interval. Above, depleted ^{13}C values were consistent and not unlike those for the upper part of Unit B-suggesting a consistency in detrital input to these less organic-rich sediments (Fig. 4).

Biomarker signatures for Unit A bitumens revealed the highest sterane/hopane ratios observed, strongly suggestive of marine influences at the boundary of Units B and A. These then decreased to values characteristic of the borderline between marine and non-marine environments (Fig. 14). This is supported by elevated tricyclic terpane (C_{25}/C_{26}) ratios and markedly different sterane compositions, favouring cholestane and generally comparable to those of Unit C (Figs. 14 and 15). Gammacerane ratios are very high at the base of the unit, consistent with the marine-like indications recorded by the corresponding sterane/hopane and tricyclic terpane ratios (Figs. 14 and 15). Whereas the 25,28,30-TNH ratio is minimal over the unit, 28,30-BNH reaches its peak concentration and remains abundant.

The low organic carbon concentration would argue for rapid sedimentation or an oxygenated water column throughout this episode. The lesser variability of the bitumen carbon isotopic signature is indicative of stabilised photosynthetic activity and potentially constant carbon flux. Paleoenvironmental interpretation for Unit A suggested a shallow, low-energy lacustrine environment of deposition.

The paucity of palynodebris and the dominance of gymnosperm pollen over fern spores was consistent with an arid climate and a hinterland supporting sparse plant growth. Reddish inter-laminations and mottling within the sediments, with co-occurrence of evaporitic minerals, were suggestive of periodic sub-aerial exposure, oxidation and intermittent desiccation.

Following the last highstand, supporting the temporary anoxia at the base of the unit, ephemeral conditions ensued with a progressive contraction and drying-out of the lake. Consistent with the arid climate and hinterland, microfloras of low diversity, and adapted to the hypersaline conditions, could be anticipated.

Characteristics of Chela Fm. Sediments

Marking the conclusion to "sheet drape" sedimentation, the Bucomazi episode ended with a major regional regression and truncation in the Aptian. Subsequently, deposition of the Chela Fm. saw the initiation of drift phase tectonics as a prelude to permanent oceanic ingression and the Loeme Salt desiccation event.

During Chela times, the Bucomazi Fm. would have been subject to stripping, particularly on the emergent highs, with possible re-distribution of these sediments. Thus for a complete description of the pre-salt source rock habitat, this episode should not be overlooked. Although deposited exclusively as a sand over a basal gravel conglomerate in 86-1, elsewhere in the Ponta Vermelha Trough thinly bedded shales are often observed in

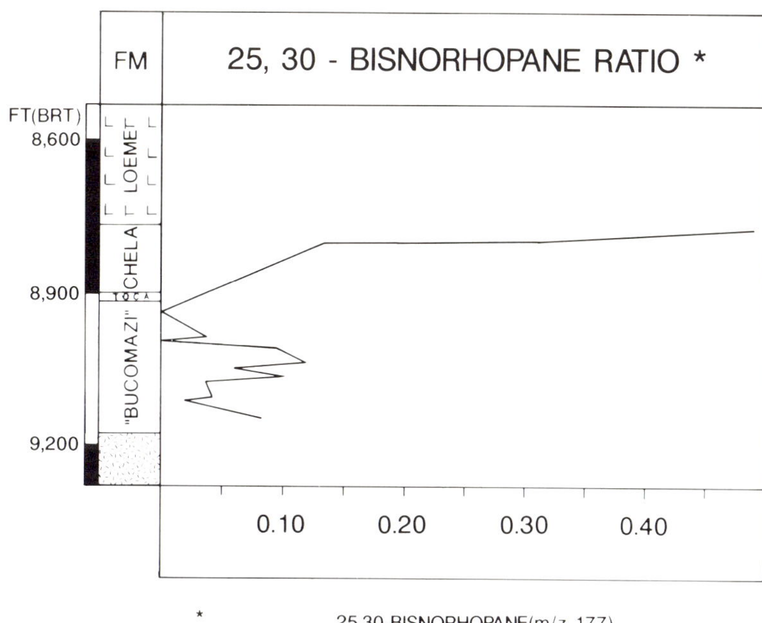

Fig. 16. Bitumen biomarker composition: norhopane normalised content of 25,30-bisnorhopane for the East Mibale-1 pre-salt section. Note the much enhanced abundance of this marker over the Chela Fm., being a characteristic of this candidate source unit

many well sections. Of variable organic richness, these shales constitute thin, primary source units. A composite of data for source quality intervals is summarised in Table 1 and draws on information assembled from wells 123-4 (Burwood et al. 1992), Tshiende-1 and East Mibale-1.

In these organically richer shales, TOC contents ranged from 1 to 2% and equated to primary source status with S_2 values at $5 \leq 10$ kg/t. Corresponding petroleum potentials could thus attain ≤ 18 metric units or 78 bbl/acre foot. With an estimated 10% net/gross effective source thickness, these sediments were essentially insignificant in comparison with Bucomazi ORZ Units C and D. Nevertheless, certain aspects of their composition suggest that they are of importance as a contributory source in the Lower Congo habitat.

Hydrogen indices equated to type II/II (inferior) kerogens with oil proneness confirmed by low GOPR values (cf. pyrogram, Fig. 12; Table 1). Organofacies description identified amorphous residues with carbon isotope values ranging from -25 to -28 ppt. Most significantly, broad low-activation energy distributions, peaking at 52 kcal/mol and typical of more labile type II assemblages, were observed (Fig. 13).

Biomarker characterisation of the Chela-associated bitumens showed higher gammacerane-25,28,30-TNH differentials, suggesting a lesser anoxicity and/or possible hypersaline influence in the deposition of these sediments. However, the bitumens were most noteworthy for their enhanced 25,30-BNH content (Fig. 16). Whereas this biomarker can be detected at low levels in Bucomazi sediments, the greater abundance in the Chela Fm. in order of magnitude was perceived to be a source-related matter. Although 25,30-BNH is frequently found in heavily biodegraded oil residuum (Volkman et al. 1983), it is observed as a member of the homologous series of 25-norhopanes deriving from the corresponding regular hopane series. Its selective occurrence in sediment extracts (e.g. see Noble et al. 1985) is on this occasion thought to denote a facet peculiar to Chela deposition. As previously recorded (Burwood et al. 1990), this marker is often conspicuous in oils of pre-salt provenance that had migrated into post-salt reservoirs.

Blanc and Connan (1992), on the basis of a statistical review of 25,30-BNH occurrence in oils, concur that this marker is not necessarily associated with in-reservoir biodegradation, and can have depositional environment implications.

At this stage we feel that the prevalence of 25,30-BNH in Chela Fm. sediments is possibly related to the reworking and re-deposition of eroded Bucomazi sediments and in situ modification of 30-norhopane, and/or its diagenetic precursor, into 25,30-bisnorhopane.

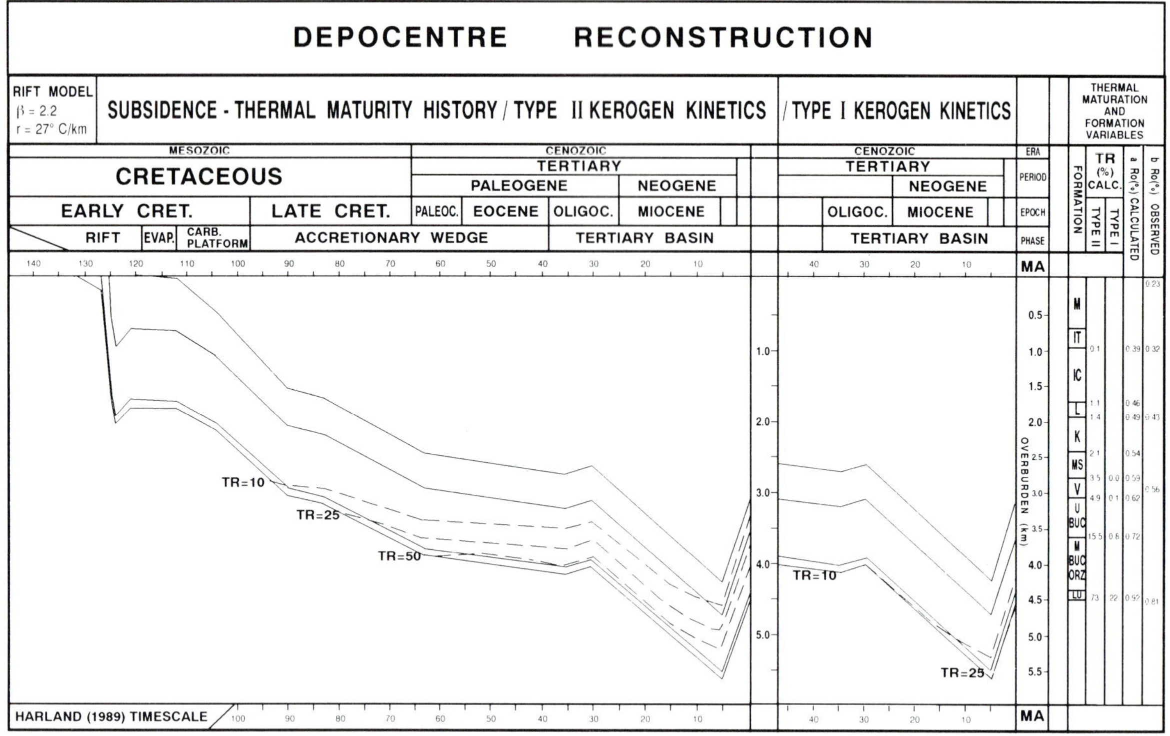

Fig. 17. Subsidence and thermal history model for Bucomazi section at a Ponta Vermelha Trough depocentre location, contrasting behaviour of type I and II kerogen assemblages. Rift model $\beta = 2.2$ over period 132-112 MaBP, present-day geothermal gradient 27°C/km. Maturity isopleths given in terms of transformation ratio (TR%). Present-day vitrinite reflectance equivalent (VRE%) calculated via **a** type III (vitrinite) kinetics, **b** overburden-Ro relationship observed in well Tshiende-1. Guide to overburden sediments (cf. Fig. 2): Malembo (*M*); Iabe Tertiary (*IT*); Iabe Cretaceous (*IC*); Liawenda (*L*); Kinkasi (*K*); Moita Seca (*MS*); Vermelha (*V*); evaporites/Chela (*R*); Bucomazi Upper Member (U. Buc.); Middle Member-ORZ (M. Buc.) and Lucula (*Lu*)

Kerogen Kinetic Criteria: Implications for Hydrocarbon Generation

Systematic screen of the pre-salt Bucomazi-Chela Fm. section in Lower Congo control wells revealed significant variations in kerogen kinetic criteria (cf. Table 1, Fig. 13). Higher activation energies, denoting less labile (type I) kerogens, observed over the extent of the ORZ declined to those typical of type II assemblages for the Upper Member and the Chela Fm.

This contrast in kinetic character can have profound consequences over the circumstances, including timing, of hydrocarbon generation and expulsion, providing the mechanism for the mixing of charges and emplacement of aggregate oils of variable composition.

These processes are illustrated by the simple subsidence and maturation history model illustrated in Fig. 17 in which the generative circumstances of type I and II kerogen assemblages are contrasted. The section modelled represents reconstruction of a depocentre location for the Ponta Vermelha Trough (cf. Fig. 1) and is based on well 122-3 stratigraphy.

Development of a mature pre-salt catchment within this sub-basin is thought to control charging of the coastal on- and offshore Zaire fields such as Mibale, Lukami and Tshiende, being those oils highlighted in Figs 6 to 10.

Although now decayed to modest present-day geothermal gradients (26–30°C/km), a pre-salt syndepositional heat flash (β factor up to 2.2) would have accompanied rift phase tectonism. The subsidence model illustrated embraces late-phase rifting control ($\beta = 2.2$) with the maturation isopleths based on the extent of kerogen conversion (transformation ratio) computations as dictated by measured activation energies for typical type I and II assemblages encountered (see Table 1). The model was additionally controlled by observed overburden-vitrinite reflectance data obtained over the Tshiende-1 post-salt section. Hydrocarbon generation status was defined in terms of transformation ratio equivalence (TRE), being:

Onset generation	10% TRE
Active generation	25%
Optimum/peak generation	50%

For the present, note that the model predicts that the complete pre-salt section is accommodated within the oil window as defined by modelled vitrinite reflectance equivalent (VRE: 0.62–0.92%).

Whereas TRE isopleths similarly placed type II kerogen containing sediments within the oil window, only onset status was achieved for type I analogues over the very basal part of the Middle Member ORZ. With the realisation that the Upper Member was dominated by type II, and the ORZ by type I, kerogens, the most effective generative status was that achieved in the less deeply buried source member. A detailed appreciation of the present model revealed the following for the individual source members,

Upper Member (type II kerogens): onset status achieved ca. 8 MaBP at base of member, with progression towards active generation (TRE-25%) at present.
Basal section of member, inclusive of Zone B sediments, thoroughly early oil mature.
Middle Member-ORZ (type I kerogens): onset status achieved ca. 13 MaBP at base of the 122-3 reconstructed member with progression towards active generation still not attained at present. Basal section of the ORZ equivalent sediments (cf. Fig. 4) above the TRE 10% isopleth and adjudged to be still immature, for effective oil production, at present.

This model thus predicts that under the most favourable conditions, whereas the basal Upper Member of the Bucomazi has attained oil generative status, and became abundantly productive, the ORZ in this context is non-productive. Less deeply buried ORZ, away from the depocentre, would be even more discriminated against as being of effective generative status (e.g. in the up-flank Tshiende- well, where a VRE of less than 0.6% has been recorded).

In reality, these events focus attention on the volumetrically greater catchment of mature Upper Member source which can be anticipated to provide a larger, if not exclusive, input to the expulsion and charging model.

The nature of the hydrocarbons generated and composition of the emplaced aggregate oils will be controlled by their respective progenic source organofacies, timing of generation and subsequent expulsion and mixing processes. Information summarised in Table 1 suggests that wide-ranging ^{13}C compositions and distinctively diagnostic biomarker fingerprints can be anticipated.

Source-Oil Correlation Assignments

Supporting evidence rationalising the proposed charging model is available through comparison of the Ponta Vermelha Trough-derived coastal Zaire oils with their candidate source units.

Assuming development of a similar Bucomazi source sequence in this contiguous sub-basin (cf. unpublished data for well Tshiende-1, see Fig. 3), reference to Table 1 showed the Middle Member ORZ Units C and D to be the dominant potential source units. However, within the Ponta Vermelha catchment these sediments are assessed to have not generally achieved oil productive status and then only marginally so in basal sections of the deepest depocentre. The volume of mature source is thus likely to be severely constrained.

Although of relatively inferior source quality, the more labile kerogen assemblages of the Upper Member are also abundantly oil mature at lesser overburdens and constitute a volumetrically greater productive catchment area.

The study group of Zaire oils were all depleted in ^{2}H (ca -100 ppt SMOW) and embraced a 4 ppt spread in ^{13}C value (Fig. 6), the latter being indicative of inherent source diversity. In terms of biomarker signature, a majority of these oils showed gammacerane indices at ± 0.25 (Fig. 18) suggestive of hypersaline (i.e. lacustrine) source depositional influences. The modest abundance of desmethylhopane indicators (25,28,30-TNH and 28,30-BNH), on the other hand, attested to the absence of highly anoxic conditions as typically associated with the deposition of the ORZ.

The overall inverse relationship and positive differential between gammacerane and 25,28,30-TNH indices in favour of the former (Figs 8 and 15) was thus suggestive of petroleums of hypersaline source depositional provenance and/or of aggregate oils incorporating a significant contribution of such a charge. With the Zaire oils all showing positive differentials, a compelling correlation with

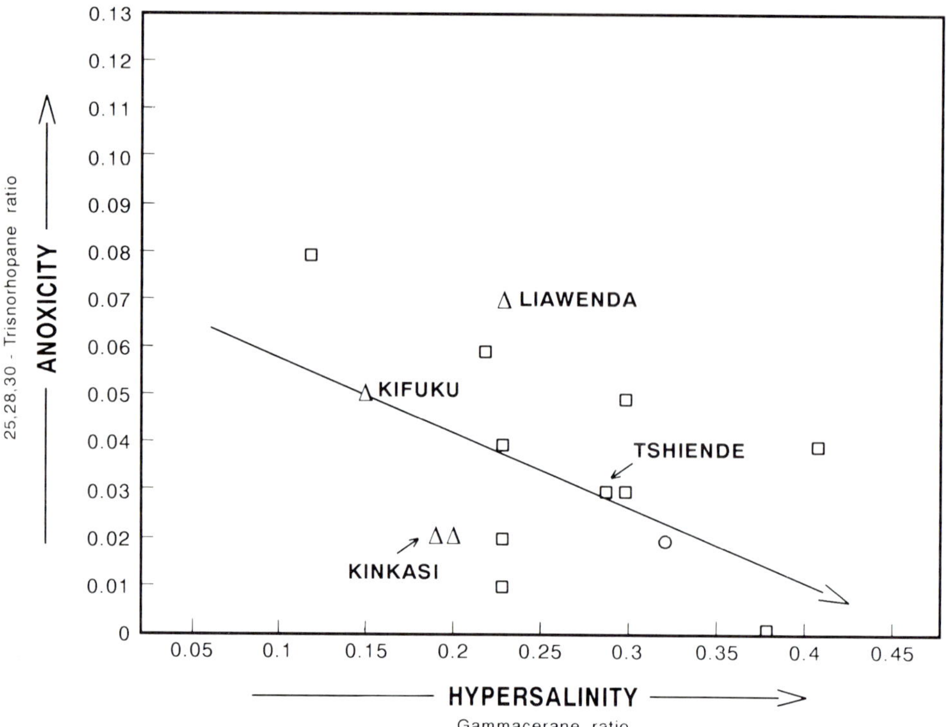

Fig. 18. Zaire study group oils. Variation of hopane normalised 25,28,30-trisnorhopane and gammacerane ratios reflecting the net interplay of hypersaline and anoxicity influences in the source provenance of these petroleums. For key to reservoir identity see Fig. 6

Upper Member (Zone B) Bucomazi sediments was evident in the cross-comparison of Figs. 8 and 15.

Upper Member Bucomazi sediments, particularly those with Zone B-like characteristics, could thus be adjudged to be prime candidates in the provenance of these petroleums. Cumulative ^{13}C isotopic profiles for the oils are illustrated in Fig. 19 and contrasted with the kerogen pyrolysate signatures for both Zone B and A sediments. For the oils, these evidently segregate into three subgroups being isotopically depleted and enriched end-members (ca -30.0 and -26.00 ppt, respectively), and a diffuse intermediate grouping at -28.5 to -27.0 ppt.

The comparison of kerogen pyrolysate and whole-oil ^{13}C values has been found to be a useful source-oil correlation procedure (Burwood et al. 1989) in that it permits comparison of the oil prone component of a candidate source kerogen assemblage with its progenic oil. On this occasion, note that all the petroleums fell comfortably within the Zone B isotopic range, with the petroleum potential weighted mean value (ca -27.0 ppt) broadly correlating with the intermediate group of oils.

This can be interpreted to imply that the ^{13}C intermediate oils are of aggregate composition drawing on contributions, to greater or lesser extent, from the whole range of kerogen assemblages present.

On the other hand, the end-member oil groupings show evidence of greater selectivity in provenance with an origin from either the ^{13}C more depleted and/or enriched assemblages. These constitute the extremities of the Zone B range with ^{13}C depleted kerogens also prominent in the Zone A source unit.

By reference to the stratigraphic distribution of isotopic signatures (cf. Fig. 4) we can thus speculate that the ^{13}C depleted oils derived from uppermost Zone B and/or Zone A facies; intermediate oils draw on variable contributions from the whole of the Upper Member; whilst the ^{13}C enriched oils probably derive from the basal half of Zone B. In this context, the compositional diversity of the Zaire oils can be accounted for by invoking the variable maturation of the type II kerogen assemblages of the Upper Member source unit, only.

In this context it should also be noted that those oils reservoired in post-salt (i.e. carbonate platform

Fig. 19. Carbon isotope based source-oil correlation diagram comparing crude oil ^{13}C profiles for the Zaire study group with candidate source unit kerogen pyrolysate signatures. Note isotopic range for Bucomazi-Unit B accommodates all whole-oil values promoting the *a priori* match between petroleums and source. In the case of the end-member, Tshiende oils note a more selective match with the ^{13}C-depleted organofacies of Unit B and/or A. For the Kinkasi/Lukami oils a match with the ^{13}C-enriched organofacies of Unit B is implied

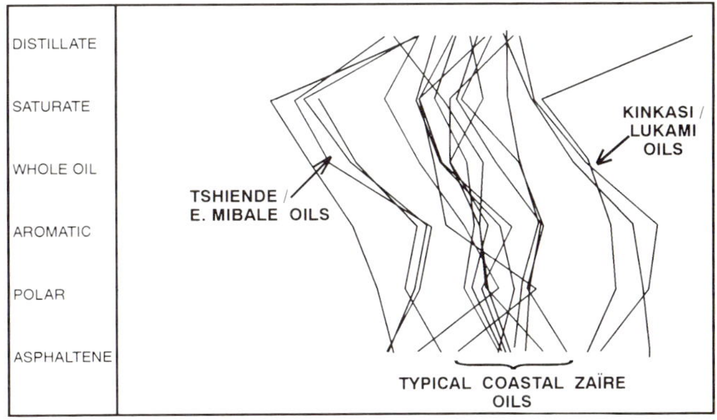

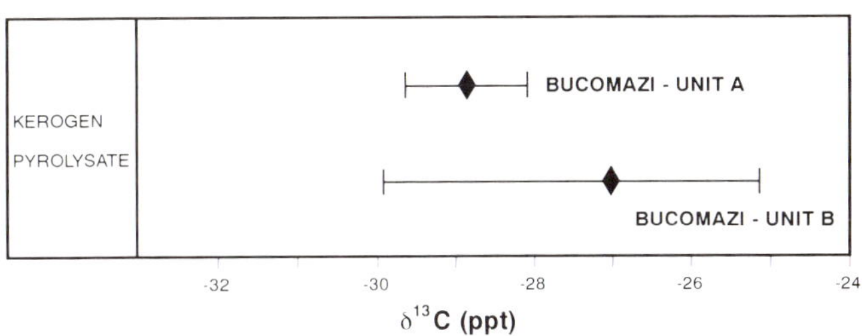

and/or Drift Phase Marine Clastic) reservoirs invariably contained 25,30-BNH (Fig. 10). With its strong association with Chela Fm. candidate source sediments this is seen as implying such a contribution to the migrating charge. These sediments are thus adjudged to be locally oil mature, certainly able to contribute a component to the aggregate petroleum. As observed on this occasion, and previously for a group of analogous Angolan oils (Burwood et al. 1990), this marker is either absent or of insignificant abundance in those pre-salt (i.e. Lucula Fm.) reservoired oils. This is consistent with lateral, intra pre-salt, migration into the stratigraphic older reservoir, thereby precluding contact with the up-section Chela Fm.

It nevertheless follows that any such type II sub-components in the dominantly type I ORZ assemblages would also have become productive and could complicate, or otherwise influence, this interpretation. However, it seems very evident, noting the more modest source potential of the upper Bucomazi and the limited coastal Zaire reserves discovered to date (ca 0.5 billion BO recoverable), that the super-rich ORZ as a whole has not been an effective hydrocarbon generator in this area.

Conclusions

A significant volume of Lower Congo Coastal basin hydrocarbon reserves can be assigned to an Early Cretaceous, Bucomazi Fm. provenance.

These highly prolific, organic-rich sediments accumulated as lacustrine deposits in a series of rift initiated graben depocentres.

For the CABGOC 86-1 type section, the Middle and Upper Members of the Bucomazi Fm. present are represented by Barremian age sediments of limited (ca 3 Ma) timespan. Thickness and organic content (ca 750 m, TOC $\leq 24\%$ wt/wt) attest to exceptional sedimentation rates (0.85 km/Ma), a highly effective environment for organic preservation and an abundant turnover in biomass. Variable organofacies development in these sediments was evidently controlled by paleoenvironmental factors. These included humidity and water depth and the attendant anoxia and aqueous chemistry as to alkalinity and hypersalinity.

The Middle Member, comprising highly carbonaceous clastics of the Organic Rich Zone

(ORZ), corresponded to "basin fill" sediments associated with the earlier, most active tectonism of the Synrift I episode. In the Upper Member, "sheet drape"deposition ensued from the milder subsidence during Synrift II times as a prelude to marine ingression.

ORZ sediments contained assemblages dominated by less labile, higher activation energy type I kerogens. Additionally, intermittent terrestrial organic input was associated with lake highstands. A shallowing of water depth towards the end of the "basin fill" episode resulted in the deposition of more calcareous micrites, these still supporting highly organic rich, superior type II assemblages. Kerogen carbon isotopic signatures attested to dramatic excursions in biomass turnover and adaptation, CO_2 budget and lake chemistry.

"Sheet drape" sediments of the Upper Member, representative of a general contraction to a more ephemeral lacustrine regime, were organically less rich and persistent. Where developed, source quality zones supported more labile, lower activation energy kerogen assemblages in association with indications of hypersalinity.

Ramifications of "basin fill" ORZ, versus "sheet drape" type II, candidate source behaviour has important implications for the kinetic control of hydrocarbon generation. Where overburden or heat flow is limited, as in the Ponta Vermelha Trough, the more labile, but less deeply buried kerogens may realise their hydrocarbon potential in preference to the type I analogues. Yielding a more abundant charge, these sediments would heavily influence the local migration and emplacement model. For effective realisation of ORZ potential, a considerable increment in overburden, as achieved only in the deeper pre-salt troughs, is deduced to be necessary.

In combination with the limited reserves of the coastal Zaire region, source-oil correlation assignments were consistent with charging of this area from upper "sheet drape-style" sediments as opposed to the richer, yet insufficiently matured, ORZ.

Thus, whereas the Bucomazi Fm. can be seen as a super-rich source system in its entirety, full realisation of its potential in practice is very dependent upon kinetic control associated with the subsidence history of these sediments.

Acknowledgments. We thank the management of Fina Exploration and Production for permission to publish this chapter and appreciate the encouragement and assistance given by G. Fortems, G. McLanachan and W.H. Ziegler. C. Gobeau, Ch. Soriani and R. Janssens are thanked for technical assistance and C. Bruggemans together with A. Tondeur for preparing the manuscript. We are grateful to Ministério da Energia, Sonangol and Cabinda Gulf Oil Co. Ltd., through J. Prata, Fina Petróleos de Angola, for provision of materials and to Zaïrep for release of data used in this chapter. We thank J. Espitalié (IFP) for verification of kerogen kinetic data and appreciate the input provided by J. Weston, R. Bate and M. Lester (SSI) over the biostratigraphic dating and paleo-environmental interpretation of the 86-1 well section. The editorial comments provided by J. Stonebraker and R.J. Hwang are gratefully appreciated.

References

Bailey NJ, Burwood R, Harriman G (1990) Application of pyrolysate carbon isotope and biomarker technology to organofacies definition and oil correlation problems in North Sea Basins. In: Durand B, Behar F (eds) Advances in organic geochemistry 1989. Pergamon Press, Oxford. Org Geochem 16: 1157–1172

Blanc Ph, Connan J (1992) Origin and occurrence of 25-norhopanes: a statistical study. Org Geochem 18: 813–828

Bouroullec JL, Rehault JP, Rolet J, Tiercelin JJ, Mondeguer A (1991) Quaternary sedimentary processes and dynamics in the northern part of the Lake Tanganyika Trough, East African Rift System. Evidence of lake eustatism? Bull Cent Rech Explor Prod Elf-Aquitaine 15: 343–368

Brice SE, Cochran MD, Pardo G, Edwards AD (1982) Tectonics and Sedimentation of the South Atlantic Rift Sequence: Cabinda, Angola. In: Watkins JS, Drake CL (eds) Studies in continental margin geology. Am Assoc Petrol Geol, Tulsa, Mem 34: 5–18

Burwood R, Drozd RJ, Halpern HI, Sedivey RA (1988) Carbon isotopic variations of kerogen pyrolysates. Org Geochem 12: 195–205.

Burwood R, Jacobs L, Paulet L (1989) Kerogen pyrolysis-carbon isotope technology : application to source-oil correlation problems. Rev Paleobot Palynol 65: 367–377

Burwood R, Cornet PJ, Jacobs L, Paulet J (1990) Organofacies variation control on hydrocarbon generation: a Lower Congo Coastal basin (Angola) case history. Org Geochem 16: 325–338

Burwood R, Leplat P, Mycke B, Paulet J (1992) Rifted margin source rock deposition: a carbon isotope and biomarker study of a West African Lower Cretaceous "lacustrine" section. Org Geochem 19: 41–52

Carbonnel P, Grosdidier E, Peypouquet JP and Piercelin JJ (1983) Les ostracods, témoins de l'évolution hydrologique d'un lac de rift. Exemple du Lac Bogoria, Rift Gregory, Kenya. Bull Cent Rech Explor Prod Elf Aquitaine 7: 301–313

Curiale JA, Cameron D, and Davis DV (1985) Biological marker distribution and significance in oils and rocks of the Monterey Formation, California. Geochim Cosmochim Acta 49: 271–288

Doyle JA, Biens P, Doerenkamp A, Jardine S (1977) Angiosperm pollen from the pre-Albian Lower Cretaceous of Equatorial Africa. Bull Cent Rech Explor Prod Elf-Aquitaine 1: 451–473

Doyle JA, Jardine S, Doerenkamp A (1982) *Afropollis*, a new genus of angiosperm pollen, with notes on the Cretaceous palynostratigraphy and palaeoenvironments of northern Gondwana. Bull Cent Rech Explor Prod Elf-Aquitaine 6: 39–117.

Farrimond P, Eglington G, Brassell SC, Jenkyns HC (1990) The Cenomanian/Turonian anoxic event in Europe: an organic geochemical study. Mar Petrol Geol 7: 75–89

Grantham PJ, Wakefield LL (1988) Variations in the sterane carbon number distributions of marine source rock derived crude oils through geological time. Org Geochem 12: 61–73

Grosdidier E, Bignoumba J (1984) Exemple de coupures écostratigraphiques basées sur un renouvellement des faunes d'ostracodes lacustres dans le Crétacé inférieur du rift Atlantique au droit du sud Gabon. Géol Médit 11: 87–95

Harland BW, Armstrong RL, Cox AV, Craig LE, Smith AG, Smith DG (1989) A geological time scale. Cambridge Univ Press, 263 pp

Hollander DJ, McKenzie JA (1991) CO_2 control on carbon-isotope fractionation during aqueous photosynthesis: a paleo-pCO_2 barometer. Geology 19: 929–932

Huang WY, Meinschein WG (1976) Sterols as source indicators of organic materials in sediments. Geochim Cosmochim Acta 40: 323–330

Huang WY, Meinschein WG (1979) Sterols as ecological indicators. Geochim Cosmochim Acta 43: 739–745

Krommelbein K, Weber R (1971) Ostracoden des "Nordost-Brasilianischen Wealden". Beih geol Jahrb 115: 93

Katz BJ, Elrod LW (1983) Organic geochemistry of DSPD site 467, Offshore California, middle Miocene to lower Pliocene strata. Geochim Cosmochim Acta 47: 389–396

Larter SR, Senftle JT (1985) Improved kerogen typing for petroleum source rock analysis. Nature 318: 277–280

Mchargue TR (1990) Stratigraphic development of proto-South Atlantic rifting in Cabinda, Angola – a petroliferous lake basin. In: Katz BJ (ed) Lacustrine basin exploration case studies and modern analogs. Am Assoc Petrol Geol, Tulsa, Mem 50: 307–326

Mello MR, Gaglianone PC, Brassell SC, Maxwell JR (1988a) Geochemical and biological marker assessment of depositional environments using Brazilian offshore oils. Mar Petrol Geol 5: 205–223

Mello MR, Telnaes N, Gaglianone PC, Chicarelli MI, Brassell SC, Maxwell JR (1988b) Organic geochemical characterisation of depositional paleoenvironments of source rocks and oils in Brazilian marginal basins. Org Geochem 13: 31–45

Noble R, Alexander R, Kagi RI (1985) The occurrence of bisnorhopane, trisnorhopane and 25-norhopanes as free hydrocarbons in some Australian shales. Org Geochem 8: 171–176

Noble R, Alexander R, Kagi RI (1987) Configurational isomerization in sedimentary bicyclic alkanes. Org Geochem 11: 151–156

Sofer Z (1984) Stable carbon isotope compositions of crude oils : application to source depositional environments and petroleum alteration. Am Assoc Petrol Geol Bull 68: 31–49

Sundararaman P, Teerman SC, Mann RG, Mertani, B (1988) Activation energy distribution: a key parameter in basin modeling and a geochemical technique for studying maturation and organic facies. Proc 17th Indon Pet Assoc Ann Conv, Jakarta, Oct, 1988, 1: 169–185

ten Haven HL, Rohmer M, Rullkötter J and Bisseret P (1989) Tetrahymanol, the most likely precursor of gammacerane, occurs ubiquitously in marine sediments. Geochim Cosmochim Acta 53: 3073–3079

Tiercelin JJ, Vincens A (1987) Le demi-graben de Baringo-Bogoria, Rift Gregory, Kenya. 30.000 ans d'histoire hydrologique et sédimentaire. Bull Cent Rech Explor Prod Elf-Aquitaine 11: 240–540.

Venkatesan MI (1989) Tetrahymanol: its widespread occurrence and geochemical significance. Geochim Cosmochim Acta 53: 3095-3101

Volkman JK (1986) A review of sterol markers for marine and terrigenous organic matter. Org Geochem 9: 83–99

Volkman JK, Alexander R, Kagi RI, Woodhouse GW (1983) Demethylated hopanes in crude oils and their applications in petroleum geochemistry. Geochim Cosmochim Acta 47: 785–794

Weston J. (1992) Biostratigraphic and paleoenvironmental analysis of the interval 6677-9067′ of well CABGOC 86-1, Cabinda. Stratigraphic Serv Int Ltd, Guildford, 18pp (unpub report).

Wünsche L, Gülaçar F.O. and Buchs A. (1987) Several unexpected marine sterols in a freshwater sediment. Org Geochem 11: 215-219

Zumberge J (1987) Prediction of source rock characteristics based on terpane biomarkers in crude oils: a multivariate statistical approach. Geochim Cosmochim Acta 51: 1625–1637

Source Rock Characterization of the Late Cretaceous Brown Limestone of Egypt

V.D. Robison[1]

Abstract

The Brown Limestone of Egypt is a primary source of hydrocarbons in the Gulf of Suez. It was deposited during the Late Cretaceous prior to Miocene rifting and development of the Gulf of Suez. Organic-rich, age-equivalent units to the Brown Limestone can be found across much of the northeastern margin of ancient Gondwana, and occur presently from Libya to Syria. The Brown Limestone was deposited during the initial marine transgression across the stable platform. Near-shore deposits of the Brown Limestone contain abundant phosphorites and glauconites, in addition to the organic-rich shales and marls. Offshore facies, such as found in the Gulf of Suez, while also organic-rich, do not contain concentrated phosphorite or glauconitic horizons.

The association of phosphorites, glauconites, and abundant organic matter indicates a zone of high primary productivity, which along with the development of low oxygen levels, led to preservation of organic-rich, oil-prone shales and marls. Organic enrichment, hydrocarbon generation potential, and the type of organic matter varies in response to the depositional setting. The most hydrogen-enriched facies occur in the Gulf of Suez, while landward facies, in Cretaceous time, vary from strongly oil-prone to nonsource.

Based on molecular and isotopic geochemical data, all of the oils from the Gulf of Suez appear to have been generated from common or very similar sources. After accounting for differences in maturity, extracts from the Brown Limestone appear to be very similar to the examined oils. However, several potential sources have been identified in the Gulf of Suez and the oils cannot be unambiguously correlated back to the Brown Limestone. These sources range in age from Late Cretaceous to Miocene and contribution from the Eocene limestones is probable.

Introduction

The Late Cretaceous Brown Limestone is considered to be a primary source for the oil and gas discovered to-date in the Gulf of Suez (Chowdhary and Taha 1987; Shahin 1988; Nagati 1992; Mostafa et al. 1993). When adequately buried, as in the Gulf, the unit is capable of generating significant quantities of hydrocarbons. Age-equivalent units to the Brown Limestone occur across much of the northeastern margin of Africa (Bein and Amit 1982; Kemper and Zimmerle 1983; Germann et al. 1985; Mikbel and Abed 1985; Notholt 1985; Abed and Al-Agha 1989). These units, including the Brown Limestone, were deposited in isolated, bathymetric lows beneath areas of elevated levels of primary productivity (Bartov and Steinitz 1977; Glenn and Mansour 1979; Glenn 1980; Kolodny 1980; Soudry et al. 1985; Germann et al. 1987; Abed and Al-Agha 1989; Glenn 1990; Lewy 1990). In general, these units are organic-rich and the primary restriction on generation of volumetrically significant quantities of hydrocarbons appears to be adequate levels of thermal maturity. Only in the Gulf of Suez have depths of burial and geothermal gradients been sufficient to generate hydrocarbons in economically significant volumes.

Here geochemical data available from the Gulf of Suez is combined with onshore data to characterize the Brown Limestone's source potential, variability, and depositional setting. Onshore Egypt, the Brown Limestone is equivalent to the Duwi and lower Dakhla Formations. The origin and distribution of these organic-rich shales and marls is investigated in this chapter.

[1]Texaco E&P Technology Department, 3901 Briarpark, Houston, Texas, USA

Stratigraphy

The basal sediments over most of the stable platform in central and southern Egypt are members of the Nubian Group (Fig. 1). Locally, these facies consist of sands and shales that represent lagoonal to fluvio-marine deposits (Said 1990a; Klitzsch and Squyres 1990). The uppermost facies of the Nubian Group are typically green to red shales of the Matulla Formation or the Variegated and Quseir shales. Klitzsch and Squyres (1990) have recently reviewed the Late Cretaceous stratigraphic nomenclature and suggest that the term Nubian be dropped.

Initial deposits of the Late Cretaceous transgression are characterized by organic-rich shales and marls of the Brown Limestone offshore and by phosphorites, organic-rich shales and marls, and bioclastic carbonates of the Duwi Formation onshore (Glenn 1980; Richardson 1982; Soliman et al. 1986, 1989; El-Kammar et al. 1990; Glenn 1990; Said 1990a) (Fig. 1). The Duwi is equivalent to the Phosphate Formation of the Western Desert (Schroter 1986). Both the Duwi and Phosphate Formations are time-transgressive and become younger to the south. The main period of phosphorite deposition in these units in southern Egypt occurred in the late Campanian to early Maastrichtian (Hendriks and Luger 1987; Issawi 1987). The phosphorite horizons in these formations represent littoral shallow water deposits that mark the ancient shore line. The association of organic-rich shales and marls with phosphatic and glauconitic deposits indicates a highly productive shelf area along the northern Gondwana margin (Baturin 1982).

The Duwi Formation phosphorites and organic-rich shales and marls are thought to have been deposited in a relatively restricted, possibly lagoonal setting (Schroter 1986; Ganz et al. 1990). Periodic storm or tidal action concentrated the phosphates into hardgrounds during drops in sea level (Glenn and Arthur 1990; Robison and Engel 1993). Phosphorites are not as pronounced in distal facies of the Duwi Formation, i.e., the Brown Limestone.

Paleogeographic reconstructions suggest that the Gulf of Suez deposits represent offshore facies (Klitzsch and Squyres 1990). Bathymetric lows on the shallow shelf may have acted as sites of accumulation of organic matter (Lewy 1990). Fluctuations in sea level were not sufficient to expose these shelf sediments to winnowing action by tidal or storm influences. In Egypt, the most favorable conditions for phosphorite formation occurred near the end of the Cretaceous along the landward edge of the broad shelf areas that developed on the northern margin of Africa (Baturin 1982).

Conformably overlying the Duwi Formation is the Late Cretaceous to Early Tertiary Dakhla Formation (Fig. 1). The lower Dakhla is similar to the Duwi in its organic character but lacks the development of phosphorite hardpans even though portions of the Dakhla Formation contain 1–5% fluorapatite as determined by semiquantitative X-ray diffraction (Robison 1986). This lack of phosphorite hardpan development is thought to be due

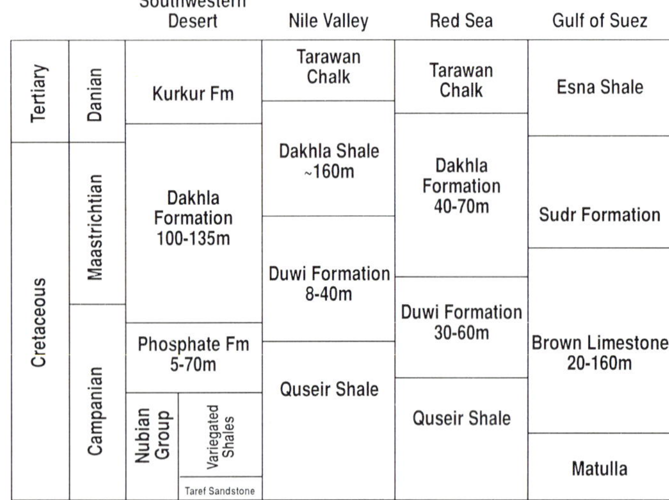

Fig. 1. Stratigraphic nomenclature and thicknesses for Late Cretaceous intervals in Egypt. (Compiled from Chowdhary and Taha 1987; Issawi 1987; Schroter 1986; Soliman et al. 1989; Said 1990a; Robison and Engel 1993)

to the Dakhla Formation being deposited in slightly deeper water than the Duwi Formation (Said 1990b). In the Gulf of Suez, the Brown Limestone rarely contains fluorapatite.

Upper Cretaceous sediments are found in outcrop and the shallow subsurface onshore from localities along the Red Sea coast through central Egypt in the Nile Valley and into the western desert. Similar stratigraphic sequences occur in the Late Cretaceous along the northeastern African margin from Libya to Syria.

Regional Geology

In the Late Cretaceous a broad downwarping of the northeast margin of Africa occurred, resulting in a major transgression that covered most of present-day Egypt (Luger and Schrank 1987). During the time of maximum transgression, the shallow stable shelf that developed across central and southern Egypt was extensive (Klitzsch and Squyres 1990). A series of phosphorites, glauconites, and organic-rich shales and marls were deposited across much of the platform that extended from modern day Libya to Syria.

The association of organic-rich sediments and phosphorites has commonly been used to invoke a zone of high primary productivity. Several authors have postulated the existence of an upwelling zone to explain the proposed levels of productivity (Ganz 1984; Germann et al. 1985). Others have suggested that nutrients supplied by fluvial input from a deeply weathered continental land mass could have supplied the phosphorus necessary to enhance surface water productivity in the Late Cretaceous seaway (Glenn and Arthur 1990). Both of these theories can be supported by paleogeographic and paleoclimatic reconstructions.

During the Late Cretaceous, the African continent was migrating northward and the northeastern margin was at low latitudes (Fig. 2). The continental land mass to the south of the Cretaceous Tethys seaway was in a tropical equatorial position and hence a zone of intense weathering. Danian shorelines of the southern Tethys margin are believed to have been lined by mangrove swamps (Gregor and Hahn 1982). Palynological studies of samples from the Eastern Desert of Egypt suggest that these mangrove swamps may have existed since the Late Cretaceous and been in a region of heavy rainfall (Schrank 1984a,b). Humid climatic conditions are also suggested by the character of the phosphorites (Garrison et al. 1979; Issawi 1987).

Orientation of the coastline also suggests that upwelling may have been stimulated by the prevailing surface winds. Both upwelling and runoff may have acted to supply nutrients to the Late Cretaceous shelf. The supply of nutrients would have sustained high rates of primary productivity. The association of organic-rich sediments with phosphatic horizons not only suggests high levels of organic productivity but also indirectly suggests low levels of oxygen above the benthic boundary (Schlanger and Jenkyns 1976; Waples 1982). Direct evidence of the development of areas of low oxygen is indicated by the highly laminated nature of the shales and marls and the general lack of benthic faunas.

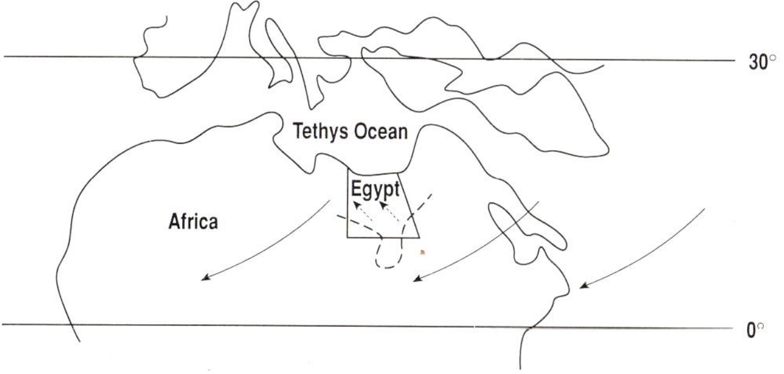

Fig. 2. Paleogeographic reconstruction for the Late Cretaceous, 75 MaB.P. (University of Chicago 1990). Inferred surface winds and upwelling sites derived from the National Center for Atmospheric Research Center's Community Climate Model (CCM0)

Deposition of the organic-rich facies has been suggested to have occurred in broad, isolated, bathymetric lows that existed on the carbonate platform (Schroter 1986; Bock 1987; Ganz et al. 1990). The occurrence of age-equivalent phosphorites and interbedded organic-rich shales in southern Israel and Jordan have also been attributed to deposition in synclinal basins sheltered from the Tethys Ocean (Bartov and Steinitz 1977; Kolodny 1980; Soudry et al. 1985; Abed and Al-Agha 1989; Lewy 1990). In these lows, circulation would have been restricted and the lower portion of the water column would have become isolated and depleted in oxygen due to the influx of organic matter.

Kemper and Zimmerle (1983) describe the occurrence of a similar depositional belt in Syria, northeast Jordan, and northwest Iraq that was deposited during the Late Cretaceous/Early Tertiary. These deposits contain siliceous sediments, phosphorite accumulations, and shales with elevated organic carbon contents. An upwelling system in a warm, subtropical ocean has been invoked to explain the distribution and occurrence of these deposits (Kemper and Zimmerle 1983). Several other age-equivalent deposits of organic-rich shales and associated phosphorites have been described along the southern Tethys margin (Kolodny 1980; Bein and Amit 1982; Mikbel and Abed 1985; Abed and Al-Agha 1989).

Brown Limestone facies in the Gulf of Suez do not exhibit concentrated phosphorites. Similarly, outer shelf deposits of the Ghareb Formation in southern Israel do not contain significant phosphate occurrences (Lewy 1990).

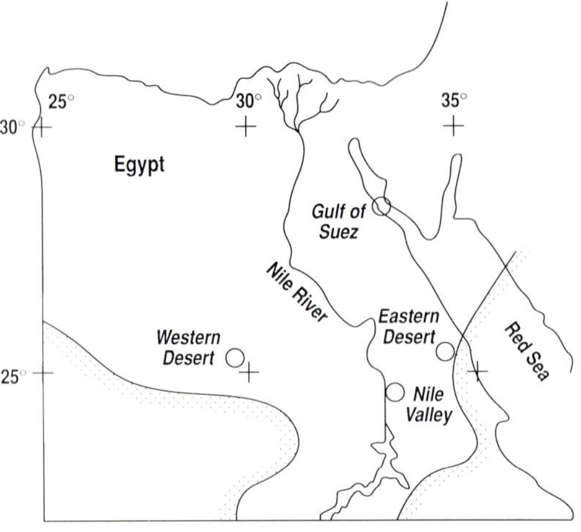

Fig. 3. Locality map for sample collection sites. Onshore localities are described in Robison (1986) and Robison and Engel (1993). Samples from the Gulf of Suez are from East Gharib and Ras Gharib wells

Sample Localities

All samples in this study are of Late Cretaceous age and onshore are from the Duwi and lower to middle Dakhla. No Danian strata were sampled in this study.

Late Cretaceous Brown Limestone samples are included from four different regions of Egypt (Fig. 3). In the Western Desert, samples from the Phosphate and Dakhla Formations were collected in the Abu Tartur mining district (Robison 1986). In the Nile Valley, the Duwi and Dakhla Formations were sampled in the Sibaiya mining district from both outcrops and exposures in the strip mines. Near the towns of Quseir and Safaga on the Red Sea coast, the Duwi and Dakhla Formations were sampled underground in several active phosphate mines (Robison and Engel 1993). Cuttings samples collected at 30-foot intervals from four wells in the Gulf of Suez near Ras Gharib are included as representative of the immature Brown Limestone.

All of the samples reported here are immature with respect to hydrocarbon generation and migration and thus can be used as representative of the original generative potential of the Late Cretaceous interval. Weathering may result in carbon loss in some samples collected in the Nile Valley. These samples were collected at the surface from an active strip mining operation (Robison 1986).

Twelve oil samples recovered from Ras Gharib and East Gharib wells in the Gulf of Suez are included as representative of the produced oils in the region. None of the oils can be attributed directly to the Brown Limestone due to the presence of another very similar source sequence in the Gulf, the Eocene Thebes Limestone.

Source Rock Characterization

The Duwi and Dakhla Formations and their lithostratigraphic equivalents are organically enriched

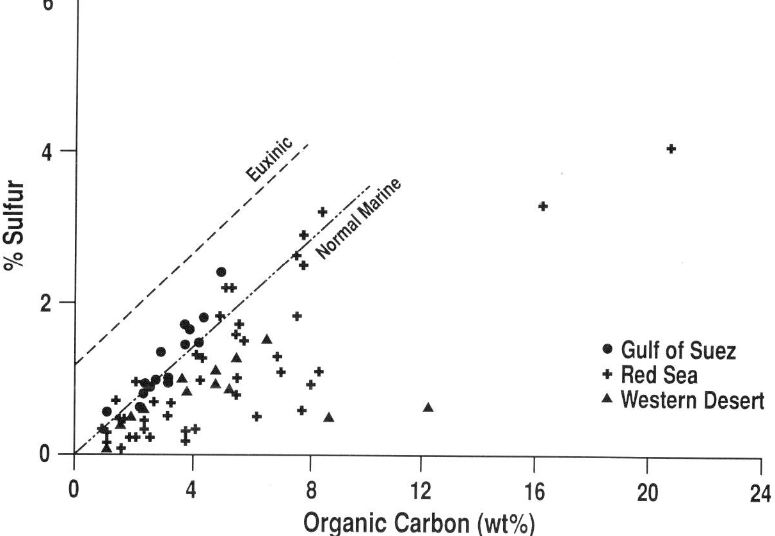

Fig. 4. Organic carbon versus sulfur cross plot for samples from the Western Desert, Red Sea, and Gulf of Suez localities

in the southwestern desert, the Red Sea area, and in the Gulf of Suez. Organic carbon values of over 20% are observed (Fig. 4). Organic carbon versus sulfur contents suggest that the Brown Limestone was deposited under normal marine waters. The laminated, organic-rich nature of the samples suggests that bottom water may have been dysaerobic to anoxic. Pyrite contents, as determined by X-ray diffraction, are in general low, indicating that insufficient iron was available resulting in the formation of sulfur-rich organic matter (Robison 1986).

Available samples of the Duwi and Dakhla Formations from the Nile Valley are organically lean. All of the collected samples contain < 0.3 wt.% organic carbon and are not included on Fig. 4. It is not clear whether this is due to depositional conditions or surface weathering conditions since all of the collected samples from the Nile Valley are from surface exposures (i.e., outcrops and strip mines). Some samples from the Sibaiya region of the Nile Valley contain trace amounts of pyrite, which oxidizes more rapidly than organic matter (Lewan 1980). This suggests that weathering cannot fully explain the paucity of organic matter in these shales. It has been suggested that the Nile Valley region was a bathymetric high during Duwi and Dakhla times (Glenn 1990).

Most of the samples with greater than 1 wt.% organic carbon exhibit excellent hydrocarbon generation potentials as determined by Rock-Eval pyrolysis (Tissot and Welte 1984). When plotted relative to the organic carbon content two distinct populations are apparent (Fig. 5). The lower trend

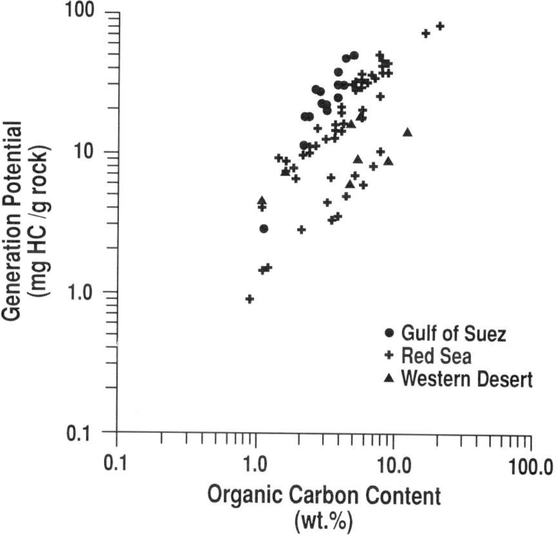

Fig. 5. Organic carbon versus generation potential for samples from the Western Desert, Red Sea, and Gulf of Suez localities

is composed of samples from the Duwi Formation in Red Sea region and the Western Desert. These samples exhibit lower hydrogen indices, typically below 150, and are representative of type III kerogen trend indicating primarily gas-prone organic matter (Fig. 6). Samples from the upper trend are hydrogen-enriched and typically exhibit hydrogen index values greater than 300. Most samples show affinities for the type II reference curve but some samples from the Gulf of Suez fall along the type I reference curve.

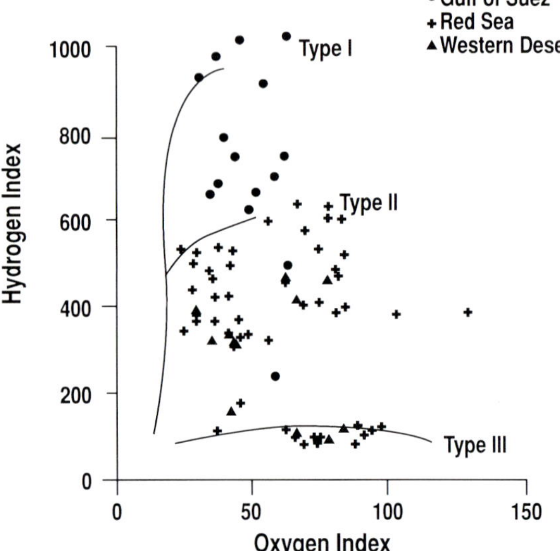

Fig. 6. Modified van Krevelen diagram for samples from the Western Desert, Red Sea, and Gulf of Suez

Visual kerogen analysis indicates that the samples from the lower trend contain significant quantities of detrital organic matter. These samples typically contain between 30 and 85% vitrinite and nonfluorescent, amorphous organic matter. Several also contain some inertinite, typically $\leq 2\%$. Stratigraphically, these samples occur at the base of the Duwi Formation in the Red Sea region and near the top of the Phosphate Formation in the Western Desert. The low hydrogen index values appear to be due to the influx of hydrogen depleted terrestrial organic matter and not to differences in preservation of organic matter.

Samples from the upper trend exhibit significant variation in their hydrogen indices with samples from the Gulf of Suez exhibiting hydrogen indices near 1000. The kerogen is primarily composed of fluorescent amorphous organic matter with only a minor exinitic or occasional vitrinitic component. Exinites and vitrinites were limited to samples with relatively low hydrogen indices (< 450) and are typically an insignificant portion of the total organic matter. This variation in the hydrogen index in the absence of a significant change in organic matter type suggests that preservation of organic matter was the critical factor in determining source quality because the type of organic matter deposited did not change significantly. Preservation of organic matter can be influenced by a number of factors including, primary productivity, bottom water oxygen contents, benthic life forms and the degree of bioturbation, and the rate of burial of the exposed organic matter. The relative importance of each of these is not addressed in this study.

Maturity analyses suggest that all of the samples included in this chapter are immature with respect to hydrocarbon generation and expulsion and therefore represent an undistorted data set for characterization. Vitrinite reflectance measurements on selected samples from the Western Desert and the Red Sea yield values between 0.3 and 0.45%. In the Gulf of Suez, a maturity equivalent to an R_0 of 0.50 to 0.55% was determined. Rock-Eval pyrolysis T_{max} values between 411 and 426 are consistent with these measurements.

Further characterization of the organic matter was accomplished through extraction and characterization of the bitumens. $C_{15}+$ saturated hydrocarbon chromatograms exhibit an immature signature with a bimodal distribution (Fig. 7). Isoprenoid hydrocarbons dominate the chromatograms in the C_{15} to C_{19} molecular weight range while biomarker compounds, primarily steranes and triterpanes, dominate the chromatograms in the C_{26} to C_{31} molecular weight ranges. Normal paraffins are minor components in available samples from all of the onshore localities.

Normal paraffin distributions exhibit two characteristic signatures. Samples from the lower portion of the section in the Red Sea and the upper portion of the section in the Western Desert exhibit a normal paraffin distribution skewed towards the higher homologs, in particular the C_{27}, C_{29}, and C_{31} normal alkanes. This type of signature appears to record the dominant input of terrestrial organic matter in these samples and is consistent with both the pyrolysis and visual kerogen analyses.

In general, the remaining samples exhibit distributions dominated by C_{15}–C_{19} normal paraffins. These samples also commonly exhibit a slight even/odd predominance in this lower molecular weight range. Normal alkane distributions with an even/odd predominance are typical of carbonate source rocks (Palacas 1984). The manifestation of this signature in these samples may be related to the low levels of thermal maturity and may be masked by thermally derived alkanes in the main phase of hydrocarbon generation and migration. However, several oils from the Gulf of Suez exhibit a slight even/odd predominance suggesting that this signature may be recognizable in oils that have experienced less thermal stress and can be used to relate the oils back to their primary source.

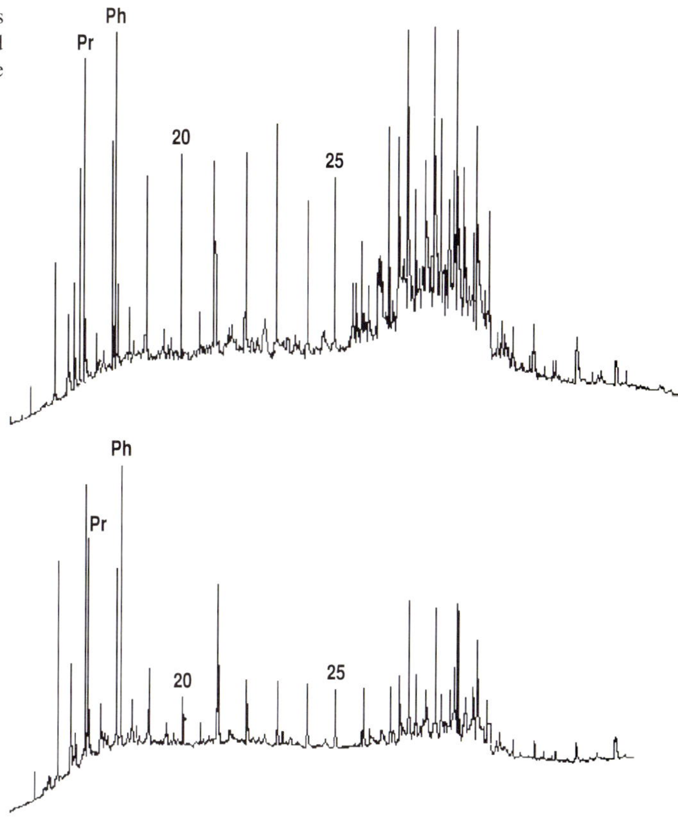

Fig. 7. Typical saturated hydrocarbon chromatograms of the $C_{15}+$ fraction from the Western Desert (*top*) and the Red Sea (*bottom*). Both samples are from oil-prone facies of the Brown Limestone

Most of the samples have pristane/phytane values of < 1 regardless of stratigraphic position. While not always definitive these values suggest low oxygen levels (dysaerobic to anoxic) during deposition (Didyk et al. 1978). This is consistent with the dark, laminated nature of most of the samples, suggesting little bioturbation in the marly horizons.

The triterpane and sterane biomarkers also exhibit an immature distribution. The m/z 191 chromatograms display prominent peaks that correspond to the unsaturated hopenes and moretanes (Fig. 8). The C_{22} carbon site in the extended hopanes reveals primarily an R configuration with little isomerization to the S configuration. C_{35} homohopanes are more abundant than the C_{34} homohopanes, again supporting deposition under low oxygen conditions (Peters and Moldowan 1991). Gammacerane, typically thought to be related to hypersaline depositional environments (Philp et al. 1991), is also very abundant in these immature distributions. Tricyclic terpanes are in low concentrations. Philp et al. (1991) found high concentrations of both tricyclic terpanes and gammacerane in Chinese oils generated from saline lacustrine source facies, and suggested the relative concentrations of these compounds could be related to the salinity of the source depositional environment. However, Moldowan et al. (1985) indicated that gammacerane is ubiquitous in crude oils. The significance of the high concentration of gammacerane in these samples is unclear.

The m/z 217 chromatograms exhibit a predominance of the $5\alpha,14\alpha,17\alpha$ 20R isomers with only minor contributions of the $5\alpha,14\alpha,17\alpha$ 20S configuration. The distribution of C_{27}, C_{28}, and C_{29} $5\alpha,14\alpha,17\alpha$ 20R steranes is very similar for all facies of the Brown Limestone (Fig. 9). One biomarker ratio that does appear to vary with organic matter type as determined by the hydrogen index is the sterane/hopane ratio. A trend of increasing sterane/hopane ratio with increasing hydrogen index is observed (Fig. 10). A similar but less dramatic increase in the C_{27}/C_{29} sterane ratio has been observed with increasing hydrogen indices for samples from the Red Sea area (Robison 1986). The variations in both of these ratios appear to be related to organic-facies and are consistent with both pyrolysis and visual kerogen data.

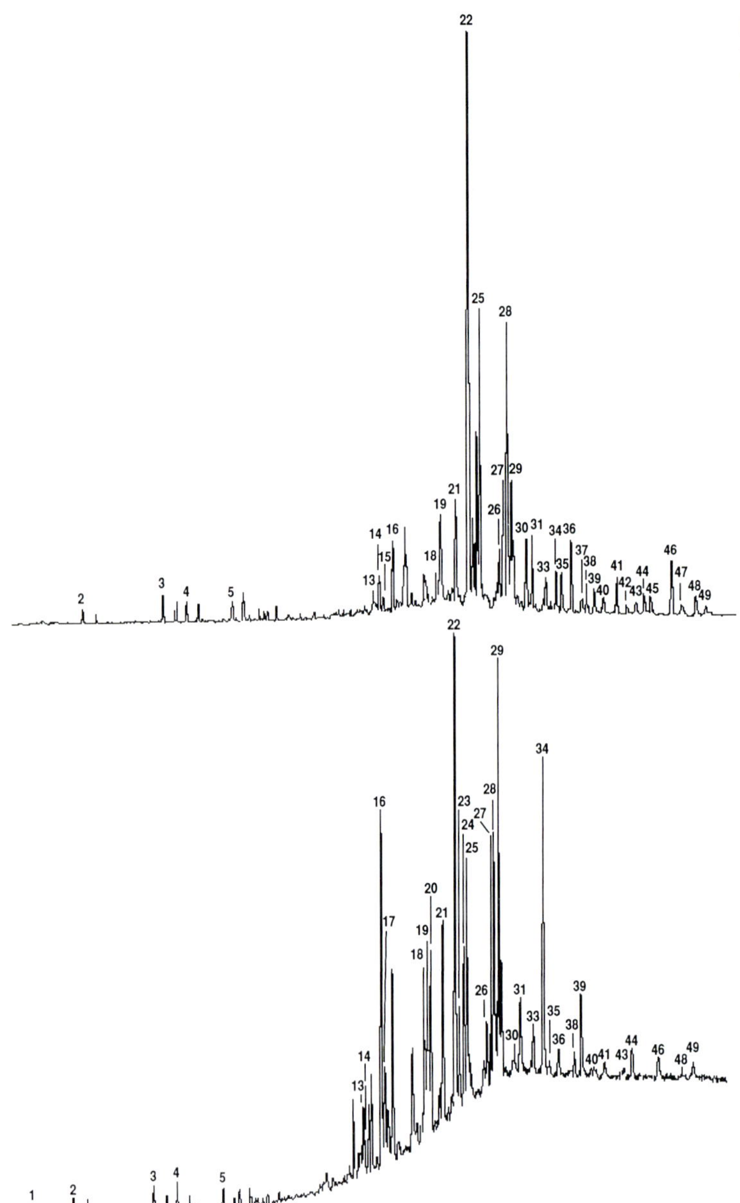

Fig. 8. Typical m/z 191 chromatograms from the Brown Limestone. Samples are same as those in Fig. 7. Peak identifications in Table 2

Presently the Brown Limestone is mature below approximately 10 000 feet (Cooles et al. 1986; Ungerer et al. 1986; Chowdhary and Taha 1987). Geothermal gradients increase gradually from the north to south in the Gulf of Suez ranging up to 55°C/km in the southern half (Shahin 1988) indicating that depths to the top of oil window would be shallower in the south. Mature facies of the Brown Limestone reached optimum depths for generation and expulsion of hydrocarbons near the end of the Miocene (Ungerer et al. 1986; Shahin 1988). This timing was related to the deposition of massive Miocene evaporites and the rapid burial of the Brown Limestone into and in many places through the oil window.

Source Variability

Several organic facies of the Brown Limestone are apparent from the available data. The first is represented by the Gulf of Suez section where hydrogen enrichment is the greatest (Table 1). The average

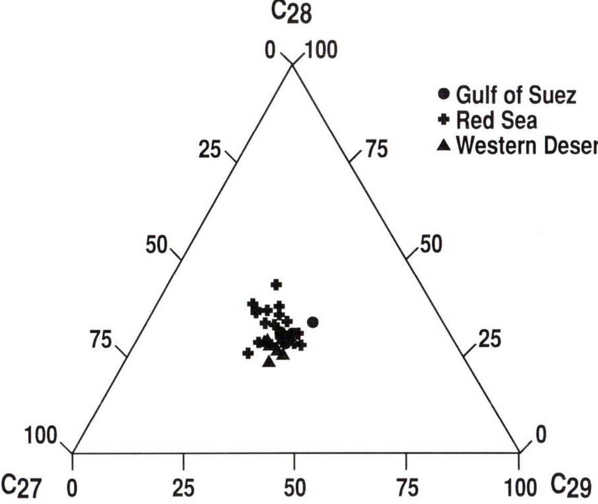

Fig. 9. C_{27}, C_{28}, C_{29} $5\alpha,14\alpha,17\alpha$ (20R) sterane distributions for Brown Limestone

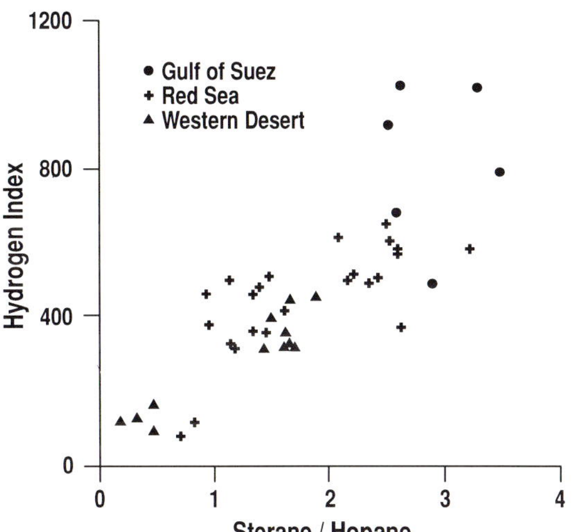

Fig. 10. Hydrogen index versus sterane/hopane ratio for Brown Limestone samples.

Sterane/hopane

$$= \frac{C_{27},C_{28},C_{29} 5\alpha,14\alpha,17\alpha \text{ (20R) steranes}}{C_{30}\ 17\alpha,21\beta + C_{30}\ 17\beta,21\alpha + C_{30}\ 17\beta,21\beta \text{ hopanes}}$$

organic carbon content is 3.2 wt.%. The kerogen appears to be a type I/II based on pyrolysis and visual kerogen data and contains little, if any, detrital organic matter from the landmass to the south. These data are consistent with previously published data on the Brown Limestone from the Gulf of Suez (Chowdhary and Taha 1987; Ungerer et al. 1986).

The second group of samples include much of the section in the Eastern Desert (Robison and Engel 1993) and the lower portion of the sampled section in the Western Desert (Robison 1986). While organic enrichment is slightly higher than that observed in the Gulf of Suez, hydrogen enrichment is lower. The kerogens are mixed type II/III based on pyrolysis data. The slightly elevated organic carbon content may be due to a contribution of terrestrial organic matter even though little recognizable terrestrial organic matter is evident from visual kerogen analysis. However, the 15 to 75% non-fluorescent amorphorous organic matter includes degraded humic organic matter and any unidentifiable material, indicating that there may be a significant unrecognized terrestrial organic matter input. Samples from both of these first two groups are highly laminated, suggesting low oxygen content of the surface sediments during deposition retarding the activity of burrowers (Lee 1992).

The third grouping is represented by samples from the base of the section in Eastern Desert (Robison and Engel 1993) and the top of the sampled section in the Western Desert (Robison 1986). These samples, while commonly organic-rich, are relatively hydrogen-depleted and contain primarily gas-prone organic matter. This gas-prone nature is primarily due to the influx of detrital, land plant sourced organic matter (Robison and Engel 1993). This facies appears to correspond to Glenn's (1990) glauconite facies of the Western Desert.

The last group of samples are represented by the section encountered in the Nile Valley which is organically lean and source potential is low to absent. This portion of the section in the Nile Valley has been interpreted to represent prodelta and delta front deposits (Glenn 1990). While organic-rich facies of the Brown Limestone appear to be restricted to bathymetric lows that existed on the stable platform in the Campanian and Maastrichtian seas, deposition in the Nile Valley may have been influenced by these deltaic processes (Germann et al. 1987; Klitzsch and Squyres 1990).

Alternatively, it has been suggested that sample weathering in the Nile Valley is responsible for the low organic carbon contents (M.D. Lewan, pers. comm.). More severe weathering of the sample suite would presumably be caused by the difference in mining techniques, where in the Nile Valley surface strip mines provided access to the Late Cretaceous section, in the Eastern and Western Deserts underground mines and core samples were obtained (Robison 1986).

Table 1. Organic geochemical attributes of the Brown Limestone in Egypt

Brown Limestone organic facies	Av. TOC	Av. S_2	Av. HI	Representative section
1	3.2	25.2	751	Gulf of Suez
2	5.1	22.4	412	Red Sea – upper portion Western Desert – lower portion
3	4.0	6.7	205	Red Sea – lower portion Western Desert – upper portion
4	< 1	n.d.	n.d.	Nile Valley

TOC Organic carbon in wt%.
S_2 mg hydrocarbons/g rock.
HI Hydrogen Index [(S_2/TOC)*100].
n.d. not determined.

Hydrocarbon Characterization

Oils in the Gulf of Suez range from heavy and asphaltic (≤ 20°) to light, extending to the condensate (≥ 50°) range. As expected several attributes of the oils change with API gravity including sulfur contents and percentage of hydrocarbons (Fig. 11). In general, crudes from the Gulf are sour regardless of gravity with sulfur contents of > 1%. Despite the range in bulk characteristics, all of the examined oils from the Gulf appear to have a common or very similar source based on their molecular and isotopic signatures (Rohrback 1983; Sofer 1984). Twelve oils from the central portion of the Gulf of Suez, examined as part of this study, are very similar to previously described oils and appear to be derived from a similar source or sources. Oils examined in this study are from Ras Gharib and East Gharib wells.

Several characteristics of the oils suggest that they were derived from a marine carbonate or marl source that was deposited under reducing conditions. These include the high sulfur content and the

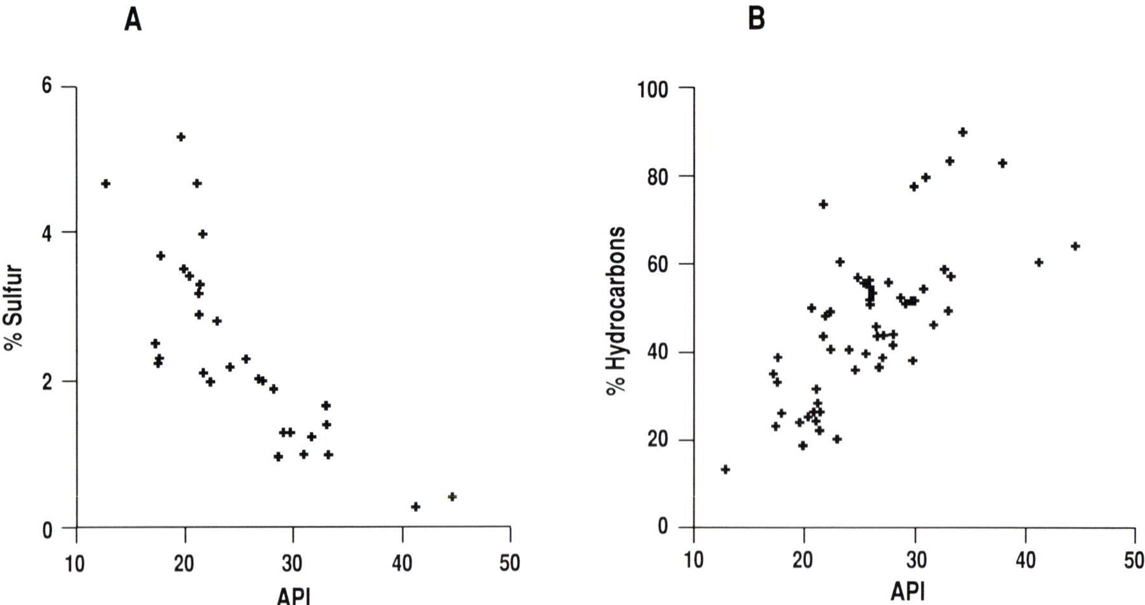

Fig. 11A,B. API gravity versus % sulfur (**A**) and % hydrocarbons (**B**). Includes data from Rohrback (1983)

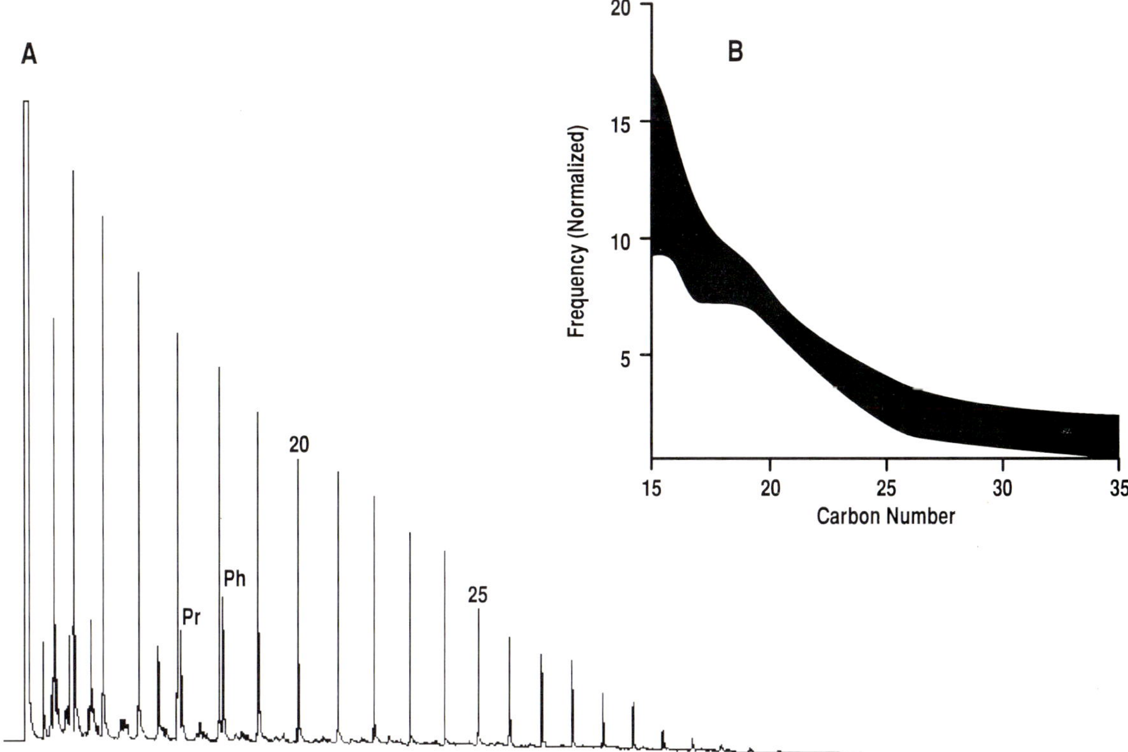

Fig. 12A,B. Saturated hydrocarbon chromatogram for oil from the East Gharib EGB-1 well (drill stem test from 6220 feet, **A**) and normalized distribution of *n*-alkanes (**B**) for oils included in this chapter

$C_{15}+$ saturated hydrocarbon chromatograms which exhibit a typical marine signature (Fig. 12). Several oils also exhibit a slight even/odd normal alkane predominance suggesting a carbonate source (Palacas 1984). Pristane/phytane ratios are < 1 in all oils.

Typical m/z 191 saturated hydrocarbon chromatograms exhibit a low concentration of tricyclic terpanes (Fig. 13). The extended hopanes exhibit a C_{35}/C_{34} homohopane ratio of > 1 suggesting reducing conditions during deposition (Peters and Moldowan 1993). Gammacerane occurs in low but detectable concentrations in all 12 oils.

C_{27}, C_{28}, and C_{29} $5\alpha,14\alpha,17\alpha$ 20R sterane distributions observed in the oils are similar to distributions seen in the Brown Limestone (Fig. 9) with a general dominance of C_{27} steranes over C_{28} and C_{29} steranes (Fig. 14). The m/z 217 chromatograms exhibit low concentrations of diasteranes as would be expected in carbonate sourced oils (Palacas 1984; Peters and Moldowan 1993).

Stable carbon isotopic signatures of the saturated and aromatic hydrocarbons are very similar for nearly all of the examined oils (Fig. 15). Two outliers from the group suggest other families of oils may be present consistent with the interpretations of Mostafa et al. (1993). This clustering of most of the oils suggests little variation in the source facies that generated these hydrocarbons and indirectly suggests the presence of a single or at most two very similar source sequences.

Oil to Source Correlations

No definitive oil to source correlations have been published for the Gulf of Suez. The oils appear to be a single family (Rohrback 1983) with minor differences due to alteration effects and not to differences in the source (Chowdhary and Taha 1987). However, Mostafa et al. (1993) have recently presented results suggesting that there are three distinct oil families in the Gulf. Late Cretaceous, Eocene, and Miocene age sections all appear to contain source quality units in the Gulf of Suez (Barakat 1982 Rohrback 1983; Richardson et al. 1986; Chowdhary and Taha 1987).

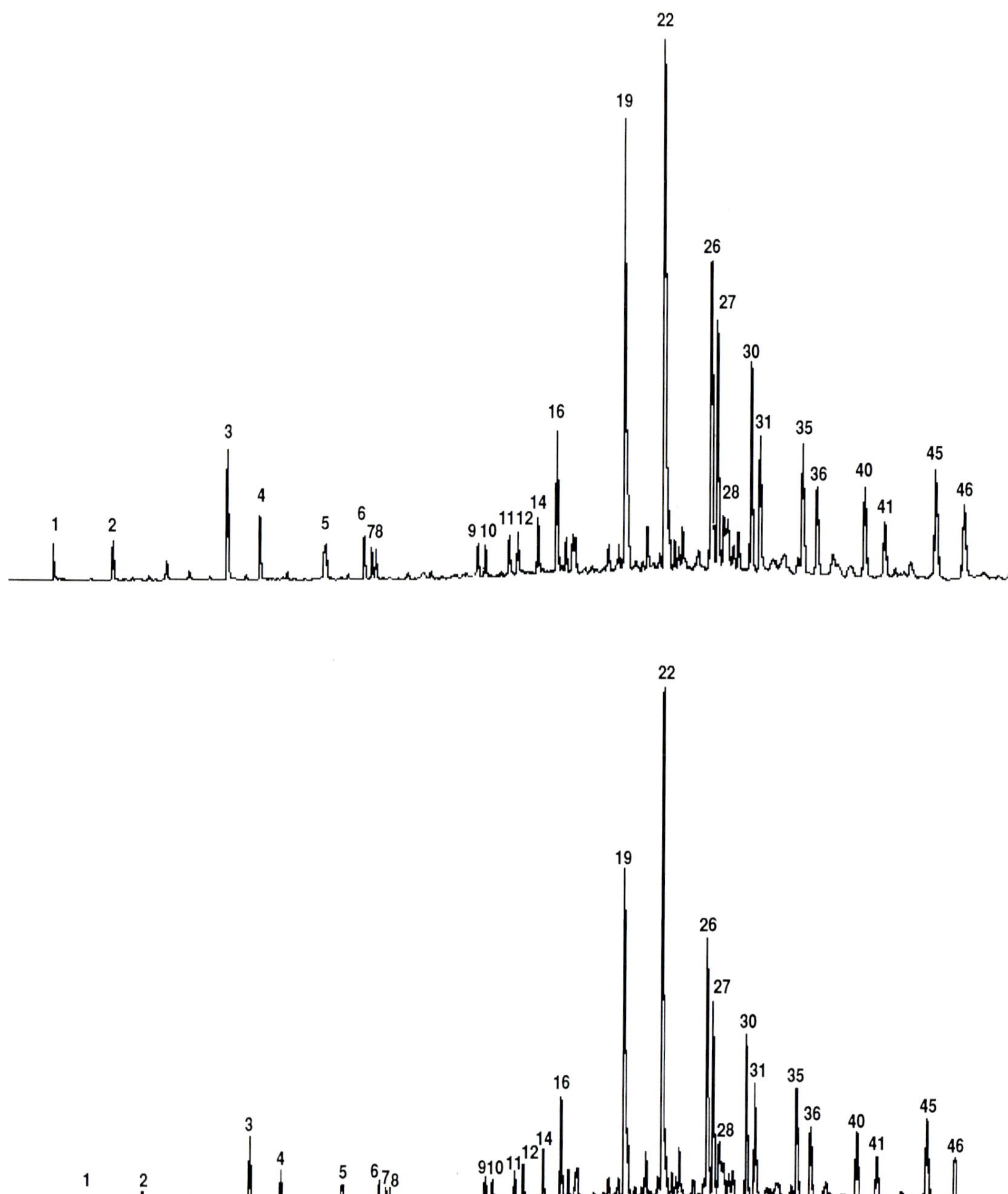

Fig. 13. Typical m/z 191 chromatograms for Gulf of Suez oils. *Top sample* from East Gharib EGB-1 well, 6220 feet. *Bottom sample* from East Gharib EGJ-2 well, 10 555 feet. Peak identifications in Table 2

Klemme and Ulmishek (1991), in a recent survey of source rocks of the world, indicate that the Miocene Rudeis Formation is the primary source of the oil in the Gulf of Suez based on the work of Kholief and Barakat (1986). However, Kholief and Barakat (1986) present no classical organic geochemical data suggesting that their studied samples were either organic-rich or of source quality. While much of the Miocene is composed of organic-poor, evaporitic facies, it does contain discreet intervals that are organic-rich (Richardson et al. 1986). However, across much of the Gulf, the Miocene section is immature with respect to hydrocarbon generation and expulsion (Ungerer et al. 1986; Chowdhary and Taha 1987; Nagati 1992). Mostafa et al. (1993) suggest that the Rudeis has

Table 2. Peak identification for m/z 191 chromatograms

Peak	Compound
1	C_{20} Tricyclic terpane
2	C_{21} Tricyclic terpane
3	C_{23} Tricyclic terpane
4	C_{24} Tricyclic terpane
5	C_{25} Tricyclic terpane
6	C_{24} Tetracyclic terpane
7	C_{26} Tricyclic terpane A
8	C_{26} Tricyclic terpane B
9	C_{28} Tricyclic terpane A
10	C_{28} Tricyclic terpane B
11	C_{29} Tricyclic terpane A
12	C_{29} Tricyclic terpane B
13	C_{27} Trisnorneohop-13(18)-ene
14	C_{27} 18α(H)-Trisnorhopane (Ts)
15	C_{27} Trisnorhop-17(21)-ene
16	C_{27}17α(H)-Trisnorhopane (Tm)
17	C_{27} 17β(H), 21β(H)-Trisnorhopane
18	C_{29} Norhop-17(21)-ene
19	C_{29} 17α(H), 21β(H)-Norhopane
20	C_{30} Hop-17(21)-ene
21	C_{29} 17β(H), 21α(H)-Normoretane
22	C_{30} 17α(H),21β(H)-Hopane
23	C_{30} Neohop-13(18)-ene
24	C_{31} Homohop-17(21)-ene
25	C_{29} 17β(H), 21β(H)-Norhopane + C_{30} 17β(H), 21α(H)-moretane
26	C_{31} 17α(H), 21β(H)-Homohopane (22S)
27	C_{31} 17α(H), 21β(H)-Homohopane (22R)
28	C_{30} Gammacerane
29	C_{30} 17β(H), 21β(H)-Hopane
30	C_{32} 17α(H), 21β(H)-Homohopane (22S)
31	C_{32} 17α(H), 21β(H)-Homohopane (22R)
32	C_{32} 17β(H), 21α(H)-Homomoretane (22S)
33	C_{32} 17β(H), 21α(H)-Homomoretane (22R)
34	C_{31} 17β(H), 21β(H)-Homohopane
35	C_{33} 17α(H), 21β(H)-Homohopane (22S)
36	C_{33} 17α(H), 21β(H)-Homohopane (22R)
37	C_{33} 17β(H), 21α(H)-Homomoretane (22S)
38	C_{33} 17β(H), 21α(H)-Homomoretane (22R)
39	C_{32} 17β(H), 21β(H)-Homohopane
40	C_{34} 17α(H), 21β(H)-Homohopane (22S)
41	C_{34} 17α(H), 21β(H)-Homohopane (22R)
42	C_{34} 17β(H), 21α(H)-Homomoretane (22S)
43	C_{34} 17β(H), 21α(H)-Homomoretane (22R)
44	C_{33} 17β(H), 21β(H)-Homohopane
45	C_{35} 17α(H), 21β(H)-Homohopane (22S)
46	C_{35} 17α(H), 21β(H)-Homohopane (22R)
47	C_{35} 17β(H), 21α(H)-Homomoretane (22S)
48	C_{35} 17β(H), 21α(H)-Homomoretane (22R)
49	C_{34} 17β(H), 21β(H)-Homohopane

contributed to the oils in the south central portion of the Gulf.

Many authors have referred to the Late Cretaceous and Eocene carbonates as the primary source rocks for the region (Shahin and Shebab 1984; Chowdhary and Taha 1987; Elzarka and Mostafa 1988; Shahin 1988; Nagati 1992; Mostafa et al. 1993). The Eocene carbonate is the Thebes Formation which exhibits organic geochemical attributes that are very similar to those of the Brown Limestone throughout the Gulf of Suez. Organic enrichment and hydrocarbon generation potentials are similar. The Late Cretaceous Brown Limestone and Eocene Thebes are typically separated by only a few (usually ≤ 500) feet and little difference in levels of thermal maturity and timing of hydrocarbon generation is seen between the two. While several attributes of the Brown Limestone suggest that it can be correlated to oils in the Gulf of Suez, detailed geochemical data from the Thebes Formation were not available for this study and contribution from the Thebes cannot be discounted based on its apparent levels of organic enrichment and organic matter type (Table 4).

The similarity of the oils examined from the Gulf of Suez suggests the presence of a single source or sources that exhibit similar geochemical signatures. The presence of several discreet source systems contributing hydrocarbons to different traps does not appear to be reasonable (Rohrback 1983). This indirect evidence and the overall similarity of the Brown Limestone's geochemical signature to that of the analyzed oils suggests that it is a major contributor to the oils of the Gulf of Suez. Age-equivalent units have been correlated to heavy hydrocarbon occurrences (tars) in and near the Dead Sea (Spiro et al. 1983; Tannenbaum and Aizenshtat 1984, 1985; Rullkötter et al. 1985). Contribution from the Miocene Rudeis Formation to several oils in the southern part of the Gulf has also been suggested by Mostafa et al. (1993).

Summary

The Brown Limestone is an organic-rich, oil-prone source that has generated economically significant quantities of oil in the Gulf of Suez. While contribution from other sources cannot be ruled out, the Brown Limestone appears to be a major contributor of hydrocarbons in the Gulf.

The hydrocarbon-generating ability of the Brown Limestone varies both stratigraphically and spatially. The facies with the greatest liquid hydrocarbon generation potential occurs in the Gulf of

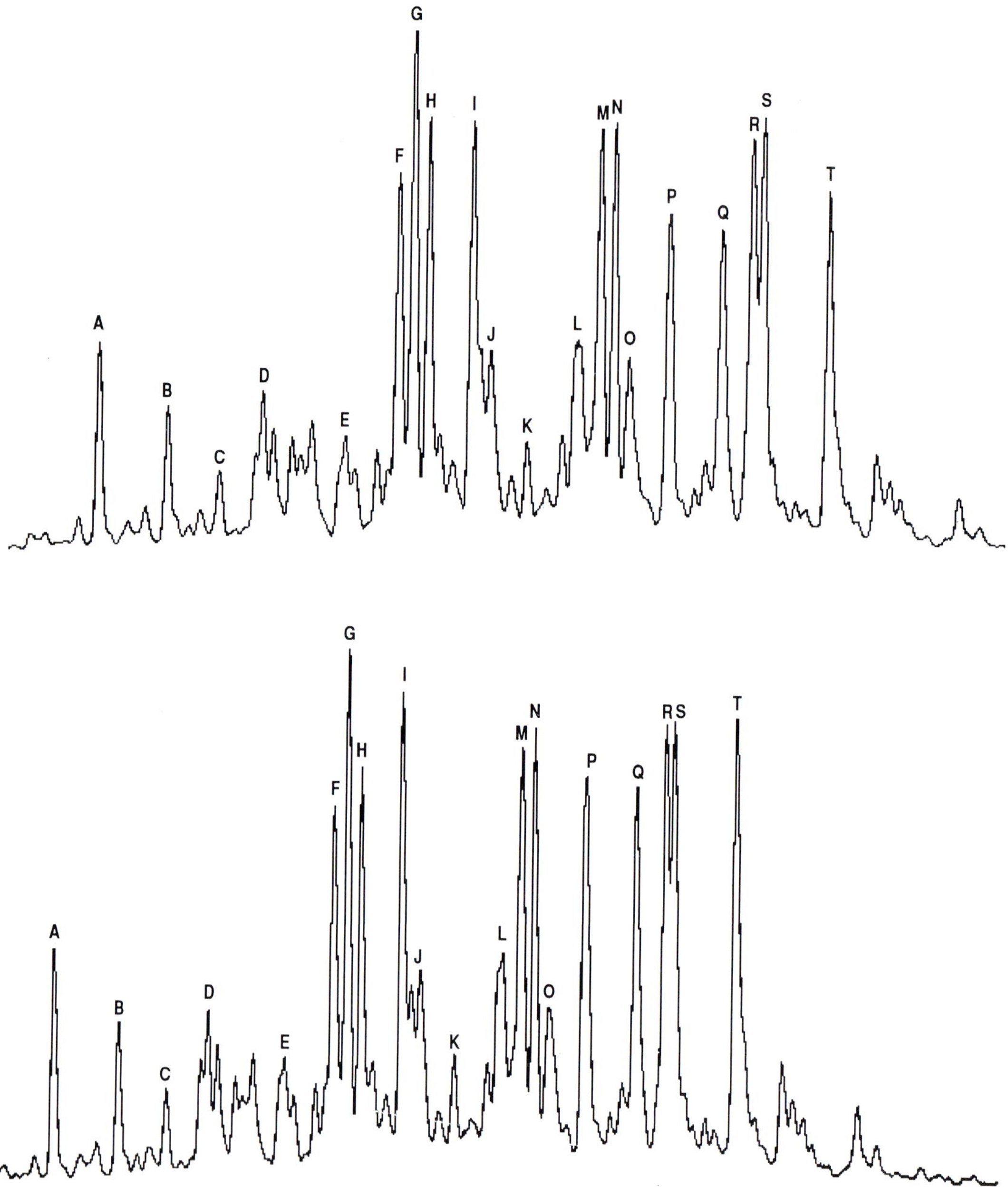

Fig. 14. Typical m/z 217 chromatograms for Gulf of Suez oils. *Top sample* from East Gharib EGB-1 well, 6220 feet. *Bottom sample* from East Gharib EGJ-2 well, 10 555 feet. Peak identifications in Table 3

Suez. Near-shore facies vary from nonsource to strongly oil-prone. In these near-shore regions, where the Brown Limestone contains oil-prone organic matter, it occurs interbedded with nonsource to gas-prone organic matter (Robison and Engel 1993).

The deposition of these organic-rich facies was related to the development of an area of high primary productivity, as evidenced by the association of phosphorites and glauconites with the organic-rich marls. The laminated nature of the sediments suggests the development of dysaerobic to anoxic conditions in response to the supply of organic matter. Preservation of organic matter may also have been aided by the development of semirestricted water bodies in isolated bathymetric

Table 3. Peak identification for m/z 217 chromatograms

Peak	Compound Assignment
A	13β, 17α-Diacholestane (20S)
B	13β, 17α-Diacholestane (20R)
C	13α, 17β-Diacholestane (20S)
D	13α, 17β-Diacholestane (20R) + 24-Methyl-13β, 17α-Diacholestane (20S)
E	24-Methyl-13β,17α-Diacholestane (20R)
F	14α, 17α-Cholestane (20S)
G	14β, 17β-Cholestane (20R) + 24-Ethyl-13β, 17α-diacholestane (20S)
H	14β, 17β-Cholestane (20S) + 24-Methyl-13α, 17β-diacholestane (20R)
I	14α, 17α-Cholestane (20R)
J	24-Ethyl-13β, 17α-Diacholestane (20R)
K	24-Ethyl-13α, 17β-Diacholestane (20S)
L	24-Methyl-14α, 17β-Cholestane (20S)
M	24-Methyl-14, 17β-Cholestane (20R) + 24-Ethyl-13α, 17β-diacholestane (20R)
N	24-Methyl-14β, 17β-Cholestane (20S)
O	24-Propyl-13α, 17β-Diacholestane (20S)
P	24-Methyl-14α, 17α-Cholestane (20R)
Q	24-Ethyl-14α, 17α-Cholestane (20S)
R	24-Ethyl-14β, 17β-Cholestane (20R)
S	24-Ethyl-14β, 17β-Cholestane (20S)
T	24-Ethyl-14α, 17α-Cholestane (20R)

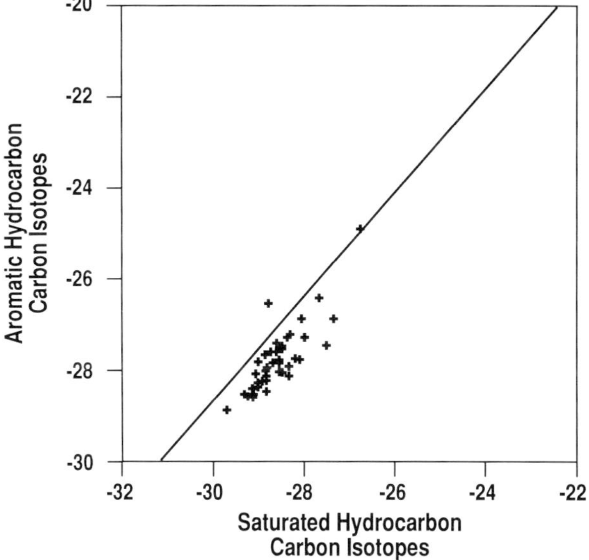

Fig. 15. Saturated versus aromatic hydrocarbon carbon isotopic distributions for Gulf of Suez oils. Includes data from Rohrback (1983)

lows across the platform. The development of this depositional setting led to the widespread deposition of organic-rich rocks across the northeastern margin of Gondwana in the Late Cretaceous.

Table 4. Geochemical data for Eocene Thebes Formation and Late Cretaceous Brown Limestone from the EG-E1 well, Gulf of Suez

Depth (feet)	TOC	$S_1 + S_2$	HI
Thebes Formation			
4880	1.7	8.2	467
4900	2.1	18.7	859
4920	1.6	9.8	581
4940	1.6	6.7	403
4980	1.8	14.6	775
5000	2.2	12.9	558
5020	2.1	20.5	951
5040	2.4	24.9	970
5060	1.6	7.5	449
5080	2.2	19.5	866
5120	1.1	5.0	464
5160	1.1	5.1	445
5180	1.0	4.1	409
5200	1.1	8.5	728
5240	1.3	6.3	482
5260	1.6	10.8	629
5300	3.1	21.3	651
5320	2.6	25.0	885
5360	2.1	11.1	515
5380	1.6	10.9	640
Brown Limestone			
5880	5.0	50.7	987
5900	4.3	30.3	680
5920	4.4	49.7	1055
5960	2.2	18.1	784
5980	2.8	30.2	1010
6000	2.6	31.3	1151
6020	2.2	11.2	502

Where these rocks have been buried and experienced sufficient thermal maturation levels to generate and expel their product, they can be an important contributor to the hydrocarbon resource base.

References

Abed AM, Al-Agha MR (1989) Petrography, geochemistry and origin of NW Jordan phosphorites. Jour Geol Soc Lond 146: 499–506

Bartov Y, Steinitz G (1977) The Judea and Mount Scopus Groups in the Negev and Sinai with trend surface analysis of the thickness data. Isr J Earth Sc 26: 119–148

Baturin GN (1982) Phosphorites on the sea floor; origin, composition and distribution. Elsevier, Amsterdam, 345 pp

Bein A, Amit O (1982) Depositional environments of the Senonian chert, phosphorite and oil shale sequence in Israel as deduced from their organic matter composition. Sedimentology 29: 81–90

Bock WD (1987) Geochemie und Genese der oberkretazischen Phosphorite Ägyptens. Berl Geowiss Abh Reihe A 35: 82–94

Chowdhary LR, Taha S (1987) Geology and habitat of oil in Ras Budran field, Gulf of Suez, Egypt. Am Assoc Petrol Geol Bull 71: 1274–1293

Cooles GP, Mackenzie AS, Quigley TM (1986) Calculation of petroleum masses generated and expelled from source rocks. Org Geochem 10: 235–245

Didyk BM, Simoneit BRT, Brassell SC, Eglinton G (1978) Organic geochemical indicators of paleoenvironmental conditions of sedimentation. Nature 272: 216–222

El-Kammar, AM, Darwish, AM, Philip G and El-Kammar, M.M. (1990). Composition and origin of black shales from Quseir area, Red Sea, Egypt. Journal of the Univ Kuwait 17: 177–189

Elzarka MH, Mostafa AR (1988) Oil prospects of the Gulf of Suez, Egypt a case study. Org Geochem 12: 109–121

Ganz H (1984) Organic-geochemical and palynological studies of a Dakhla shale profile in southeast Egypt. Berl Geowiss Abh Reihe A 50: 363–374

Ganz HH, Luger P, Schrank E, Brooks PW, Fowler MG (1990) Facies evolution of Late Cretaceous black shales from southeast Egypt. In: Huc AY (ed) Deposition of organic facies. Am Assoc Pet Geol, Tulsa, Stud Geol 30: 217–229

Garrison RE, Glenn CR, Snavely PD, Mansour SEA (1979) Sedimentology and origin of Upper Cretaceous phosphorite deposits at Abu Tartur, Western Desert, Egypt. Ann Geol Surv Egypt 9: 261–281

Germann K, Bock WD, Schroter T (1985) Properties and origin of Upper Cretaceous Campanian phosphorites in Egypt. Sci Geol Mem 77: 23–33

Germann K, Bock WD, Ganz H, Schroter T, Troger U (1987) Depositional conditions of Late Cretaceous phosphorites and black-shales in Egypt. Berl Geowiss Abh Reihe A 15: 629–668

Glenn CR (1980) Stratigraphy, petrology and sedimentology of the Duwi formation (Late Cretaceous) eastern Egypt. MS Thesis, Univ California, Santa Cruz, 269 pp

Glenn CR (1990) Depositional sequences of the Duwi, Sibaiya and Phosphate Formations, Egypt: phosphogenesis and glauconitization in a Late Cretaceous epeiric sea. In: Notholt AJG, Jarvis I (eds) Phosphorite research and development. Geol Soc, London, Spec Publ 52: 205–222

Glenn CR, Arthur MA (1990) Anatomy and origin of a Cretaceous phosphorite- greensand giant, Egypt. Sedimentology 37: 123–154

Glenn CR, Mansour SEA (1979) Reconstruction of the depositional and diagenetic history of phosphorites and associated rock of the Duwi Formation (Late Cretaceous) Eastern Desert, Egypt. Ann Geol Surv Egypt 9: 388–407

Gregor HJ, Hahn H (1982) Fossil fructifications from the Cretaceous-Palaeocene boundary of the SW-Egypt (Danian, Bir Abu Mangar). Tertiary Res 4: 121–147

Hendriks F, Luger P (1987) The Rakhiya Formation of the Gebel Qreiya area: evidence of Middle Campanian to Early Maastrichtian synsedimentary tectonism. Berl Geowiss Abh Reihe A 75: 83–96

Issawi B (1987) A review of Egyptian Late Cretaceous phosphate deposits. In: Notholt AJG, Sheldon RP, Davidson DF (eds) Phosphate deposits of the world. Cambridge Univ Press, Cambridge, 2:187–193

Kemper E, Zimmerle W (1983) Facies patterns of a Cretaceous/Tertiary subtropical upwelling system (Great Syrian desert) and an Aptian/Albian boreal upwelling system (NW Germany), In: Thiede J, Suess E (eds) Coastal upwelling – its sediment record, Part B. Sedimentary records of ancient coastal upwelling. Plenum Press, New York, pp 501–533

Kholief MM, Barakat MA (1986) New evidence for a petroleum source rock in a Miocene evaporite sequence, Gulf of Suez, Egypt. J Petrol Geol 9: 217–226

Klemme HD, Ulmishek GF (1991) Effective petroleum source rocks of the world: Stratigraphic distribution and controlling depositional factors. Am Assoc Petrol Geol Bull 75: 1809–1851

Klitzsch EH, Squyres CH (1990) Paleozoic and Mesozoic geological history of northeastern Africa based upon new interpretation of Nubian strata. Am Assoc Petrol Geol Bull 74: 1203–1211

Kolodny Y (1980) Carbon isotopes and depositional environment of a high productivity sedimentary sequence - the case of the Mishash-Ghareb Formations, Israel. Isr J Earth Sci 29: 147–156

Lee C (1992) Controls on organic carbon preservation: the use of stratified water bodies to compare intrinsic rates of decomposition in oxic and anoxic systems. Geochim Cosmochim Acta 56: 3323–3335

Lewan MD (1980) Geochemistry of vanadium and nickel in organic matter of sedimentary rocks. PhD Thesis, Univ Cincinnati, 353pp

Lewy Z (1990) Transgressions, regressions and relative sea level changes on the Cretaceous shelf of Israel and adjacent countries. A critical evaluation of Cretaceous global sea level correlations. Paleoceanography 5: 619–637

Luger P, Schrank E (1987) Mesozoic to Paleogene transgressions in middle and southern Egypt–summary of paleontological evidence. Curr Res Afr Earth Sci, Balkema, Rotterdam, vol. 14: pp 199–202

Mikbel S, Abed AM (1985) Discovery of large phosphate deposits in NW Jordan. Dirasat 12: 125–136

Moldowan JM, Seifert WK, Gallegos EJ (1985) Relationship between petroleum composition and depositional environment of petroleum source rocks. Am Assoc Petrol Geol Bull 69: 1255–1268

Mostafa A, Ganz H, Matheis G (1993) Origin of petroleum in the Gulf of Suez basin, Egypt. In: Øygard K (co-ordinating ed) Poster sessions from the 16th Int Meet Org Geochem. Stavanger 1993, Falch Hurtigtrykk, Oslo, Norway, pp 30–34

Nagati M (1992) Red Sea oil shows attract attention to Miocene salt, post-salt sequence. Oil Gas J 90 (Dec 7): 46, 48–50, 52–53

Notholt AJG (1985) Phosphorite resources in the Mediterranean (Tethyan) phosphogenic province: a progress report. Sci Geol Mem 77: 9–21

Palacas JG (1984) Petroleum geochemistry and source rock potential of carbonate rocks. Am Assoc Petrol Geol, Tulsa, Stud Geol 18: 208

Peters KE, Moldowan JM (1991) Effects of source, thermal maturity, and biodegradation on the distribution and isomerization of homohopanes in petroleum. Org Geochem 17: 47–61

Peters KE, Moldowan JM (1993) The biomarker guide – interpreting molecular fossils in petroleum and ancient sediments. Prentice Hall, Englewood Cliffs, NY, 585pp

Philp RP, Fan P, Lewis CA, Li J, Zhu H, Wang H (1991) Geochemical characteristics of oils from the Chaidamu, Shanganning and Jianghan basins, China. J Southeast Asian Earth Sci 5: 351–358

Richardson M (1982) A depositional model for the Cretaceous Duwi (phosphate) Formation, south of Quseir, Red Sea coast, Egypt. MS Thesis, Univ South Carolina, 395 pp

Richardson M, Arthur MA, Katz BJ (1986) Miocene syn-rift evaporites of the Red Sea; Their deposition and hydrocarbon source potential. Am Assoc Petrol Geol Bull 70: 638–639

Robison VD (1986) Organic geochemical characterization of the Late Cretaceous-Early Tertiary transgressive sequence found in the Duwi and Dakhla Formations, Egypt. PhD Thesis, Univ Oklahoma, 176 pp

Robinson VD, Engel MH (1993) Characterization of the source horizons within the Late Cretaceous transgressive sequence of Egypt. In: Katz BJ, Praft LM (eds) Source rocks in a sequence stratigraphic framework. Am Assoc Pet Geol, Tulsa, Stud Geol 37: 101–117

Rohrback BG (1983) Crude oil geochemistry of the Gulf of Suez. In: Bjoroy M (ed) Advances in organic geochemistry, 1981. Wiley, New York, pp 39–48

Rullkötter J, Spiro B, Nissenbaum A (1985) Biological marker characteristics of oils and asphalts from carbonate source rocks in a rapidly subsiding graben, Dead Sea, Israel. Geochim Cosmochim Acta 49: 1357–1370

Said R (1990a) Part 3: Geology of selected areas; Red Sea coastal plain. In: Said R (ed) The geology of Egypt. Balkema, Rotterdam, 734 pp

Said R (1990b) Part 4: Discussion; Cenozoic. In: Said R. (ed) The geology of Egypt. Balkema, Rotterdam, 734 pp

Schlanger SO, Jenkyns HC (1976) Cretaceous oceanic anoxic events–causes and consequences. Geol Mijnbouw 55: 179–184

Schrank E (1984a) Organic-walled microfossils and sedimentary facies in the Abu Tartur phosphates (Late Cretaceous, Egypt). Berl Geowiss Abh Reihe A 50: 177–187

Schrank E (1984b) Organic-geochemical and palynological studies of a Dakhla Shale profile (Late Cretaceous) in southeast Egypt: Part A: succession of microfloras and depositional environment. Berl Geowiss Abh Reihe A 50: 189–207

Schroter T (1986) Die lithofazille Entwicklung der oberkretazischen Phosphatgesteine Ägyptens – ein Beitrag zur Genese der Tethys-Phosphorite der Ostsahara. Berl Geowiss Abh Reihe A 67: 105

Shahin AN (1988) Oil window in the Gulf of Suez basin, Egypt. Am Assoc Petrol Geol Bull 72: 1024–1025

Sofer Z (1984) Stable carbon isotope compositions of crude oils: Applications to source depositional environments and petroleum alteration. Am Assoc Petrol Geol Bull 68: 31–49

Soliman MA, Habib ME, Ahmed EA (1986) Sedimentologic and tectonic evolution of the Upper Cretaceous-Lower Tertiary succession at Wadi Qena, Egypt. Sediment Geol 46: 111–133

Soliman HA, Ahmed EA, Aref MAM, Rushdy M (1989) Contribution to the stratigraphy and sedimentology of the Upper Cretaceous-Lower Eocene sequences east of Esna, Nile Valley, Egypt. Bull Fac Sci Assiut Univ 18: 41–67

Soudry D, Nathan Y, Roded R (1985) The Ashosh-Haroz facies and their significance for the Mishash palaeogeography and phosphorite accumulation in the northern and central Negev, southern Israel. Isr J Earth Sci 34: 211–220

Spiro B, Welte DH, Rullkötter J, Schaefer RG (1983) Asphalts, oils, and bituminous rocks from the Dead Sea area – a geochemical correlation study. Am Assoc Petrol Geol Bull 67: 1163–1175

Tannenbaum E, Aizenshtat Z (1984) Formation of immature asphalt from organic- rich carbonate rock – II. Correlations of maturation indicators. Org Geochem 6: 503–511

Tannenbaum E, Aizenshtat Z (1985) Formation of immature asphalt from organic- rich rocks – I. Geochemical correlation. Org Geochem 8: 181–192

Tissot BP, Welte DH (1984) Petroleum formation and occurrence. Springer, Berlin Heidelberg New York, 699pp

Ungerer P, Chenet Y, Moretti I, Chiarelli A, Oudin JL (1986) Modelling oil formation and migration in the southern part of the Suez rift, Egypt. Org Geochem 10: 247–260

University of Chicago (1990) Paleogeographic atlas project, Ziegler AM

Waples DW (1982) Phosphate-rich sedimentary rocks: significance for organic facies and petroleum exploration. J Geochem Explor 16: 135–160

Brown Shale Formation: Paleogene Lacustrine Source Rocks of Central Sumatra

P.A. Kelley[1], B. Mertani[2], and H.H. Williams[3]

Abstract

This chapter summarizes several earlier studies on the origin of oil in Central Sumatra basin, Indonesia, by Williams et al. (1985); Katz and Kelley (1987); Robinson (1987); Sundararaman et al. (1988); Katz and Mertani (1989); Longley et al. (1990); and Katz (1991) supplemented with unpublished data.

The approximate ten billion barrels of oil in the Central Sumatra basin of Indonesia have been generated from organic-rich shales of lacustrine origin of the Brown Shale Formation within the Paleogene Pematang Group. During Paleogene time, large freshwater lakes developed within structurally controlled rift troughs.

Geochemical analyses were used to identify the source beds, interpret the depositional environment of the source rocks, correlate oils to source rocks, and determine the maturation history of the source rocks. Systematic variations in organic type reflect depositional environments, and along with organic maturity largely determine the type of hydrocarbons generated.

Asymmetry of the rift troughs is the primary factor governing the predominantly lateral migration of hydrocarbons towards the gentle hinge margin. Minor vertical migration is related to fault and fracture systems that have diverted migration from its preferential lateral mode. Structure, combined with stratigraphy and aerial distribution of sealing shales, controls the entrapment of oil.

Introduction

Most of the established oil reserves in the Central Sumatra basin are contained in Miocene reservoirs, although sourced from lacustrine shales of the Brown Shale Formation in the underlying Paleogene section (Williams et al. 1985). The Brown Shale Formation occurs in Paleogene rift troughs in Central Sumatra basin of which four are identified in Fig. 1: Aman, Balam, Kiri, and Rangau troughs. The Paleogene Brown Shale sediments provide an analogue of lacustrine sediments deposited in a humid tropical climatic setting (Williams et al. 1985).

This chapter summarizes several earlier studies on the origin of oil in Central Sumatra basin, Indonesia by Williams et al. (1985); Katz and Kelley (1987); Robinson (1987); Sundararaman et al. (1988); Katz and Mertani (1989); Longley et al. (1990); and Katz (1991) supplemented with unpublished data.

General Geology

The Central Sumatra basin (Fig. 2) is part of a series of back-arc basins developed linearly along the leading edge of Sundaland, as a result of subduction of the Indian Ocean Plate beneath the Southeast Asian Plate during the Paleogene (de Coster 1974; Mertosono and Nayoan 1974; Cameron 1983; numerous other authors). The basin is considered a large pull-apart basin bounded by major strike-slip faults to the north and south. Within the basin, troughs were formed as a series of asymmetric half graben structures separated by horst blocks. The geometric dimensions of these half grabens are typically about 15 km wide by 40 to 100 km long. These half grabens were filled with non-marine clastics and lacustrine sediments of the Pematang Group which attains maximum thicknesses in excess of 1500 m.

Uplift, folding and faulting occurred during and following the deposition of the Pematang Group sediments. During the Miocene, following Pematang deposition, the region subsided and the sediments became progressively more marine.

[1]Texaco EPTD, Houston, Texas 77215, USA
[2]P. T. CALTEX Pacific Indonesia, Rumbia, Indonesia
[3]National Research Authority, Amman, Jordan

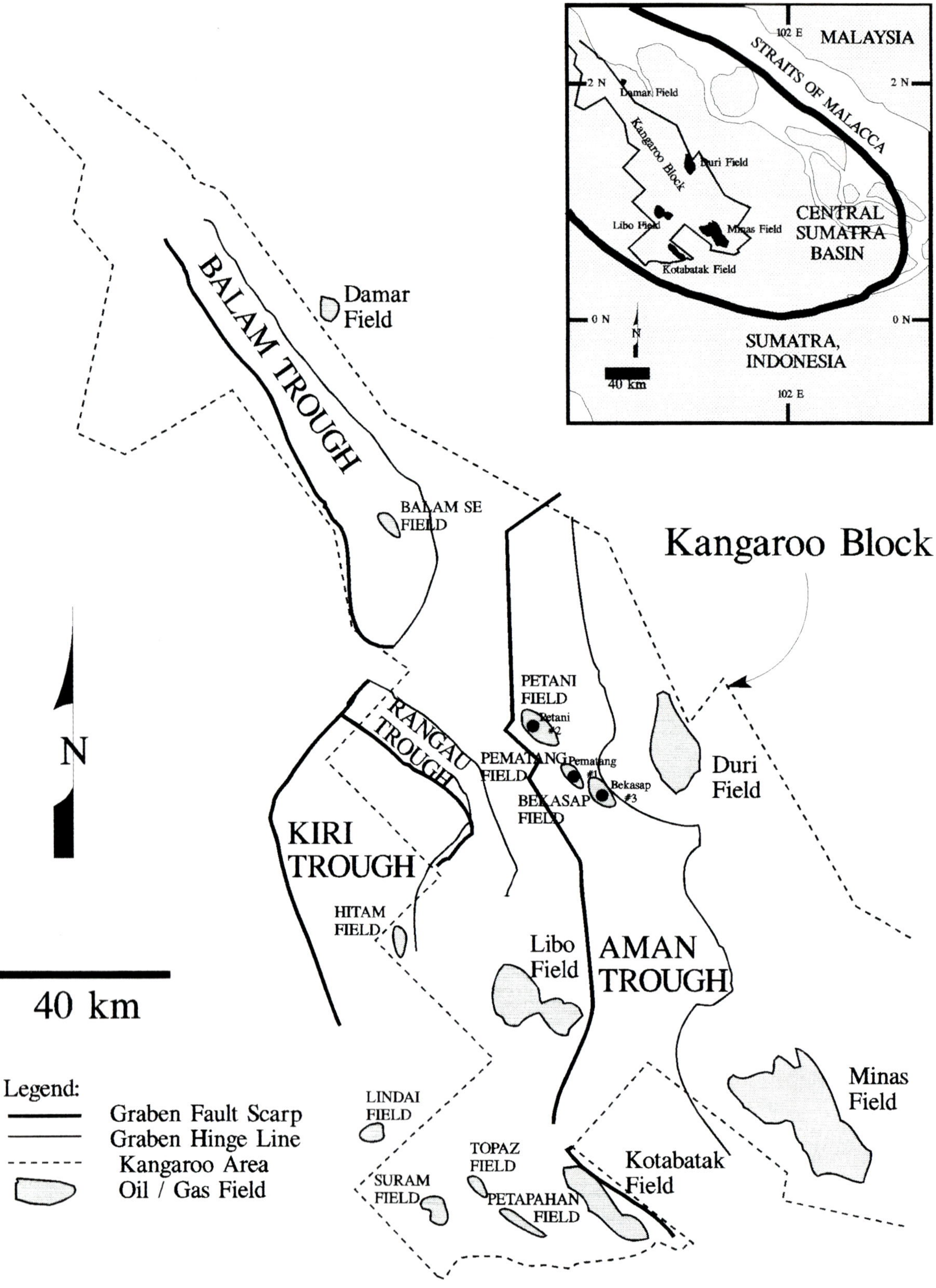

Fig. 1. Four Paleogene rift basins in Central Sumatra basin. (After Williams et al. 1985; Robinson and Kamal 1989; Katz and Mertani 1988)

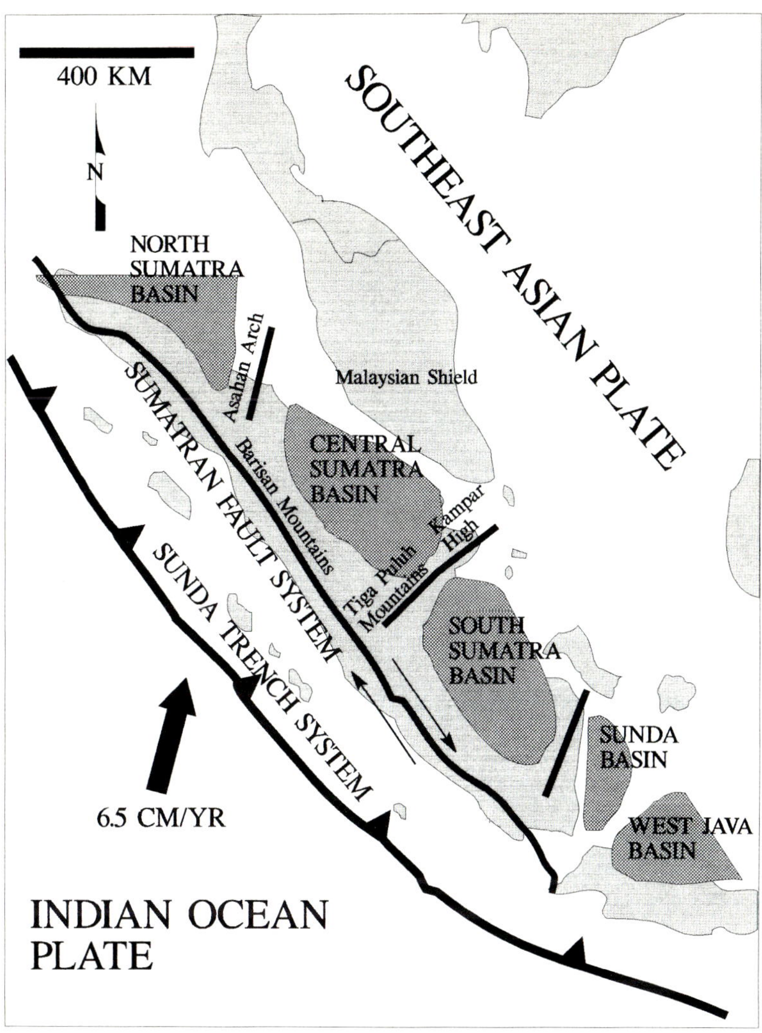

Fig. 2. Generalized geologic setting of Central Sumatra basin. (From Williams et al. 1985)

Stratigraphy

Three major stratigraphic sequences, separated by unconformities, are recognized:

1. Paleogene Pematang Group,
2. Lower Miocene Sihapas Group, and
3. Middle Miocene/Pliocene Petani Group.

(Roezin 1974; Mertosono 1975; Koning and Darmono 1984; Williams et al. 1985). The Paleogene Pematang Group consists of five formations (Figs. 3 and 4): Lower Red Beds, Brown Shale, Lake Fill, Coal Zone, and Fanglomerate. Age relationships of the Pematang are based on indirect evidence as few age diagnostic fossils have been identified in these lacustrine/continental sediments. Formation names are informal and first described by Williams et al. (1985). A brief summary of their description is given below.

The Lower Red Beds Formation consists of mudstones, siltstones, sandstones, and minor conglomerates. Shallow lacustrine or marsh/bog environments existed in the deepest basinal areas. The marginal basinal areas probably had deltaic and minor alluvial settings. Total thickness in the deepest trough exceeds 600 m.

The Brown Shale Formation conformably overlies or in some areas is laterally equivalent to the Lower Red Beds Formation. The formation consists of organic-rich well-laminated shale with siltstone stringers deposited in deep lake. Thin interbedded sandstone occurring in the deeper basinal areas are possible turbidites. Transverse delta sandstones interfinger with lacustrine shales along the hinge margin. Shallow water

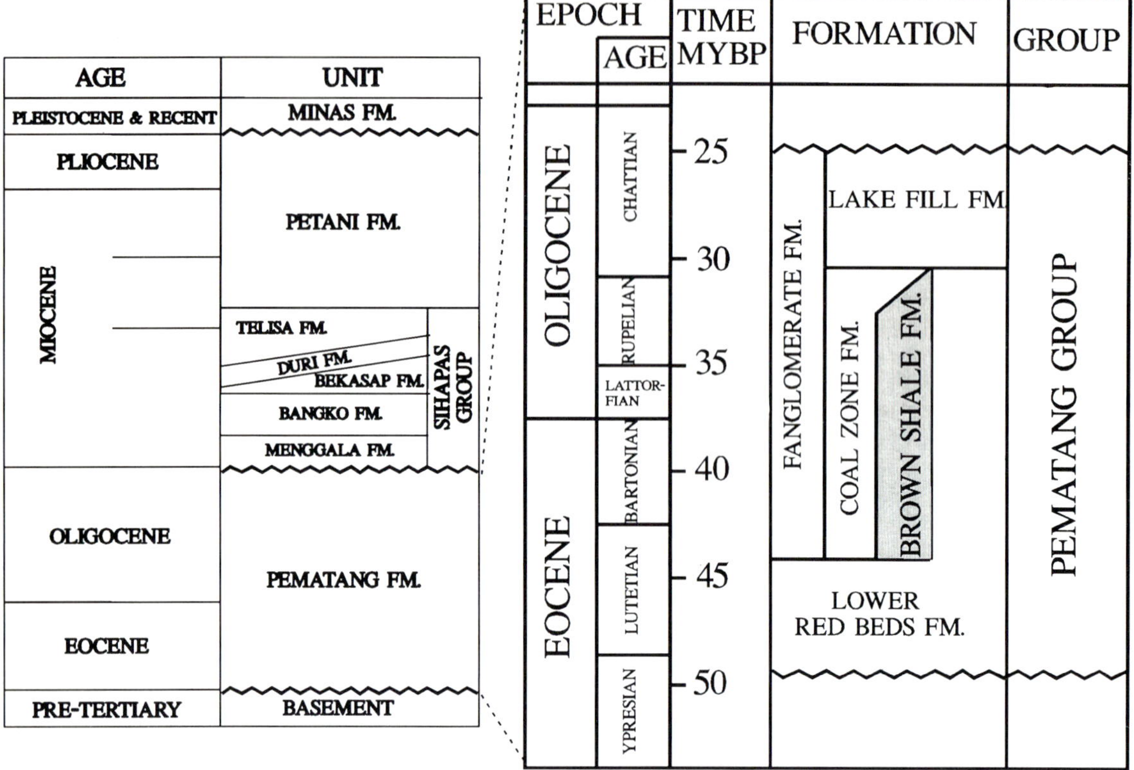

Fig. 3. Paleogene Pematang Group. (From Katz and Mertani 1989; Williams et al. 1985)

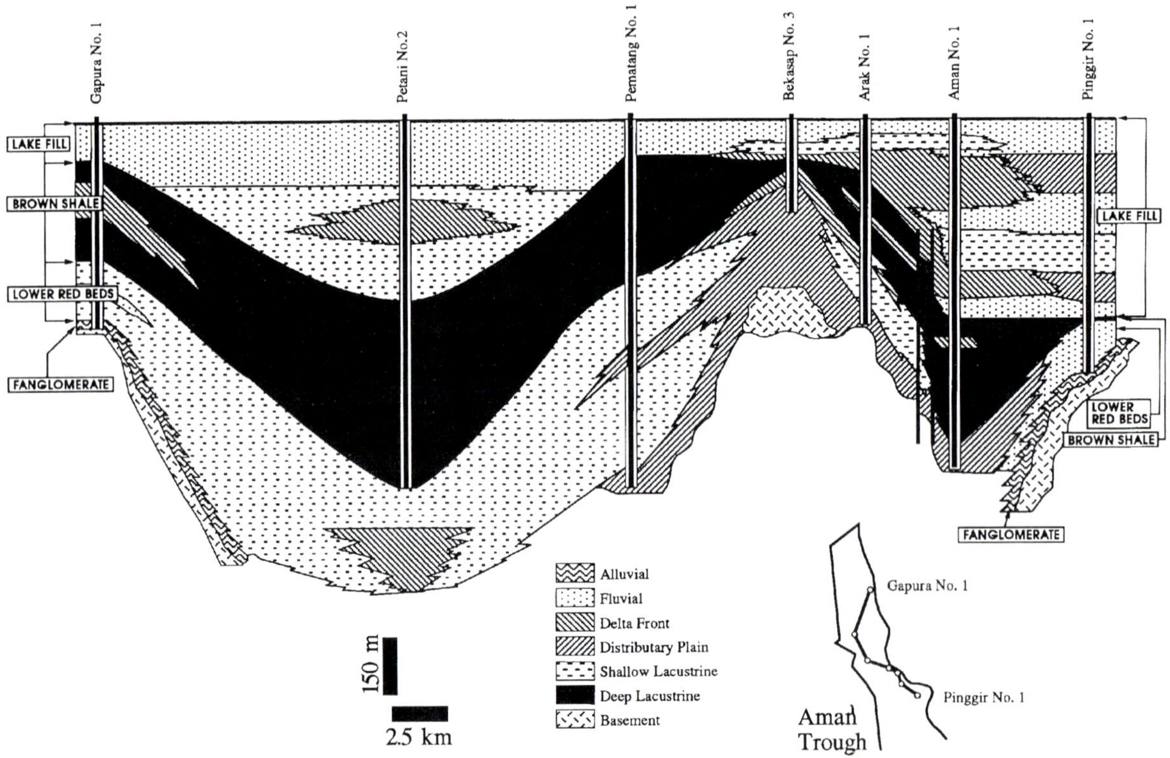

Fig. 4. Schematic cross section, Aman Trough

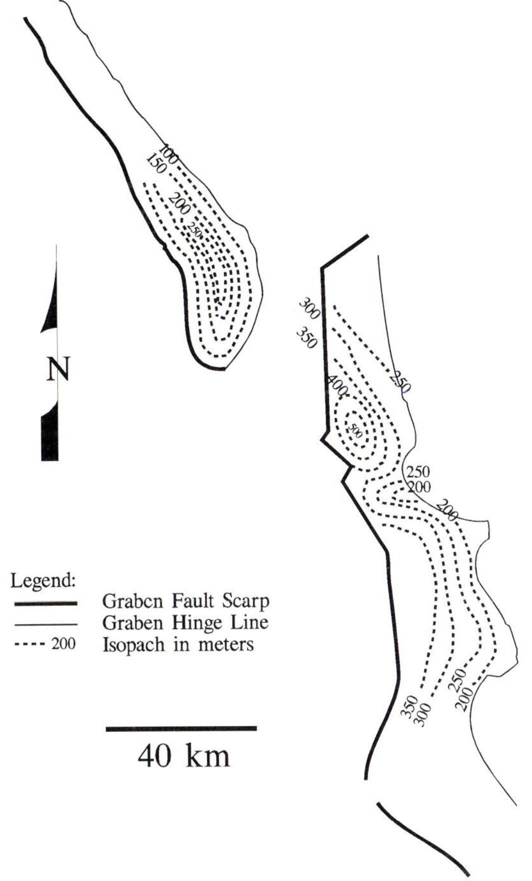

Fig. 5. Isopach map of Brown Shale

mixed carbonate-terrigenous mud flats developed within the Balam trough. Three sedimentary cycles occur at least in the deeper parts of the basins within the Brown Shale Formation which can be used as local time stratigraphic horizons within individual troughs. Total thickness ranges from 100 m to greater than 500 m (Fig. 5).

The Coal Zone Formation is, in part, laterally equivalent to the Brown Shale Formation and in part slightly younger. The formation does not occur in all the basins but is predominate in the Kiri trough. The lithology consists of interbedded shallow lacustrine shales, coals, and minor sandstones. The formation attains thickness of more than 600 m in the Kiri trough.

The Lake Fill Formation consists of fluvial and deltaic sandstones, conglomerates, and shallow lacustrine shales. Sediment input was predominately axially prograding fluvio-lacustrine delta systems with thicknesses exceeding 600 m locally. In deeper basinal areas, the Lake Fill Formation is conformable with the underlying Brown Shale or Coal Zone Formations. In some areas there is a strong downlap of the Lake Fill Formation onto the Brown Shale Formation.

The Fanglomerate Formation consists of sandstones, conglomerates and minor mudstones. The formation is restricted to the immediate area along the basin bounding fault scarps as a series of coalescing alluvial fans. The formation is laterally and vertically equivalent to the Lower Red Beds, Brown Shale, Coal Zone, and Lake Fill Formations. Maximum thickness attained exceeds 1800 m.

Structural Evolution

A four-stage structural evolution model by Williams et al. (1985) (Fig. 6) based on plate tectonic concepts is summarized here to present the structural evolution of the troughs and to illustrate its effect on sediment deposition.

Stage I – Pregraben (Early Eocene)

During the Early Eocene approximate north-south to northwest-southeast lines of weakness with complementary northeast-southwest shears developed as the Indian Ocean Plate moved N10 °E with little to no active subduction. These lines of weakness formed the hinge lines and fault scarps of the half graben structures. During the Early to Middle Eocene gentle crustal doming resulted from subduction as the angle of the plate convergence increased to N50 °E. Minor compressional component also began at this time (Karig et al. 1979). A shallow rift trough developed by incipient block rotation was filled with the Lower Red Beds Formation.

Stage II – Graben (Middle Eocene)

Accelerated thinning of continental crust and changing plate convergence to the present N20 °E resulted in rapid rift development. The deposition of the Brown Shale and Coal Zone Formations began during this stage.

Stage III – Pematang Structuring (Oligocene)

During Late Oligocene, episodic right-lateral wrench movements resulted in left-lateral transform

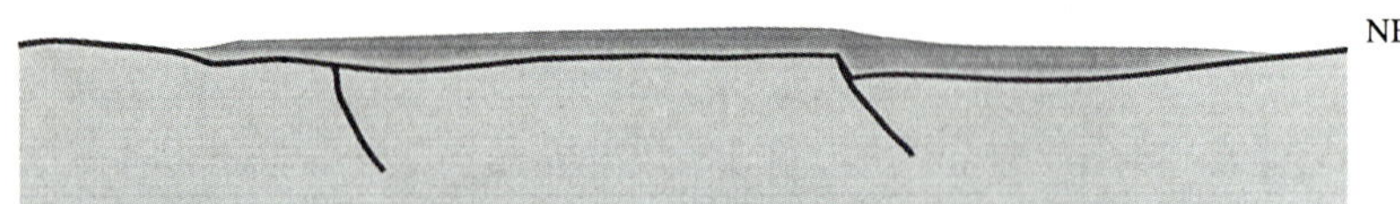

PREGRABEN (EARLY EOCENE): Formation of Zones of Weakness, Lower Red Beds Deposition

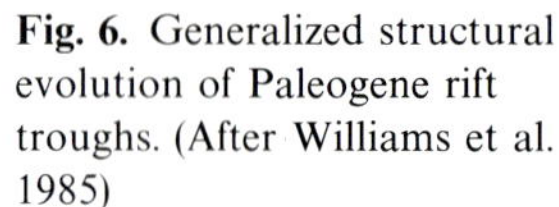

Fig. 6. Generalized structural evolution of Paleogene rift troughs. (After Williams et al. 1985)

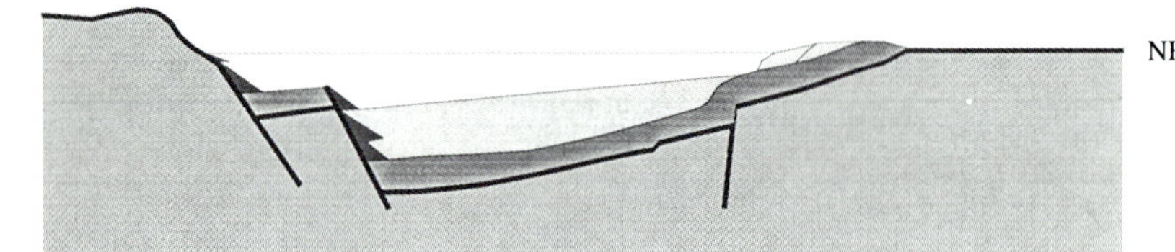

GRABEN (MID EOCENE): Block Rotation, Develop Deep Anoxic Lake Brown Shale Deposition

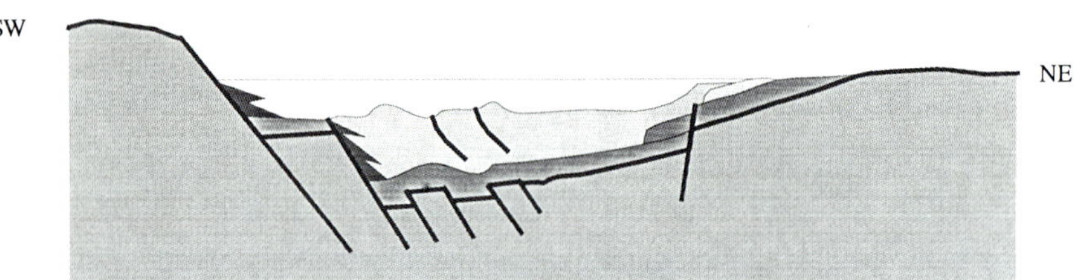

GRABEN STRUCTURING (OLIGOCENE): Compressional Phase, Uplift, & Erosion near Basin Margins

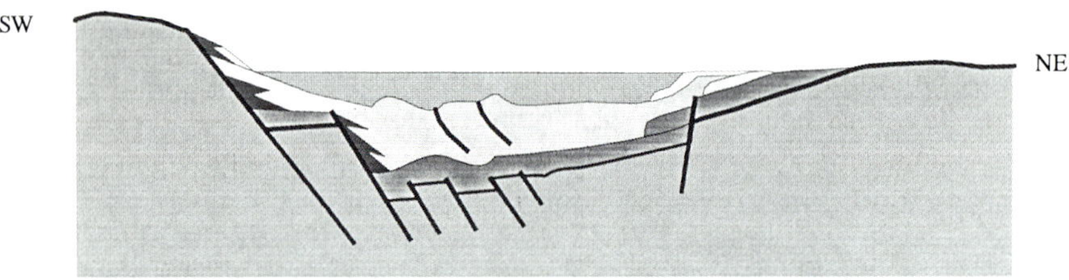

LAKE FILL (LATE OLIGOCENE - EARLY MIOCENE): Highland Areas Eroded, Prominantly Longitudinal With Some Transverse Sediment Input

movement along the northeast-southwest trend fault sets. Zones of compression associated with left-lateral transform movement resulted in the Pematang structuring.

Stage IV – Lake Fill (Late Oligocene-Early Miocene)

During Late Oligocene-Early Miocene rapid deposition of the Lake Fill Formation occurred and is characterized by uplift and rapid erosion of highland areas ending with the major unconformity at the end of Pematang deposition.

Regional subsidence in Early Miocene ended the rift trough development giving rise to the Miocene "sag basin". Later phases of tectonism have affected the rift trough sediments and complicated sequential interpretations.

Paleoclimate

Habicht (1979) and Frakes (1979) show Sumatra occupying an equatorial position, dominated by a humid tropical climate throughout the Paleogene. Other evidence supporting a humid tropical paleoclimate during the Paleogene are:

1. Numerical model output of climatic factors suggest humid tropical environment (Katz 1991).
2. Palynologic data suggest freshwater algae.
3. General lack of red beds (iron oxides) through the section.
4. Evaporites are virtually absent.

Age and Depositional Environment

The determination of age and depositional environment of the Pematang Group has been summarized by Williams et al. (1985). Nonmarine environment of deposition is suggested by (1) the lack of marine fossils, (2) the presence of freshwater gastropods of the genera *Viviparus* and *Thiaria*, (3) the occurrence of freshwater algae *Botroyococcus braunii* and *Pediastrum*, and (4) high carbon to sulfur ratios (Fig. 7). The age of the Pematang sediments is largely inferred. The presence of *Magnastriatities howardii* near the base of the Brown Shale Formation in the Balam trough suggests an age of Oligocene or younger (Williams et al. 1985). *Magnastriatities howardii* and *Florschuetzia trilobata* occur throughout the Lake Fill Formation and most of the Coal Zone Formation, suggesting an Oligocene age.

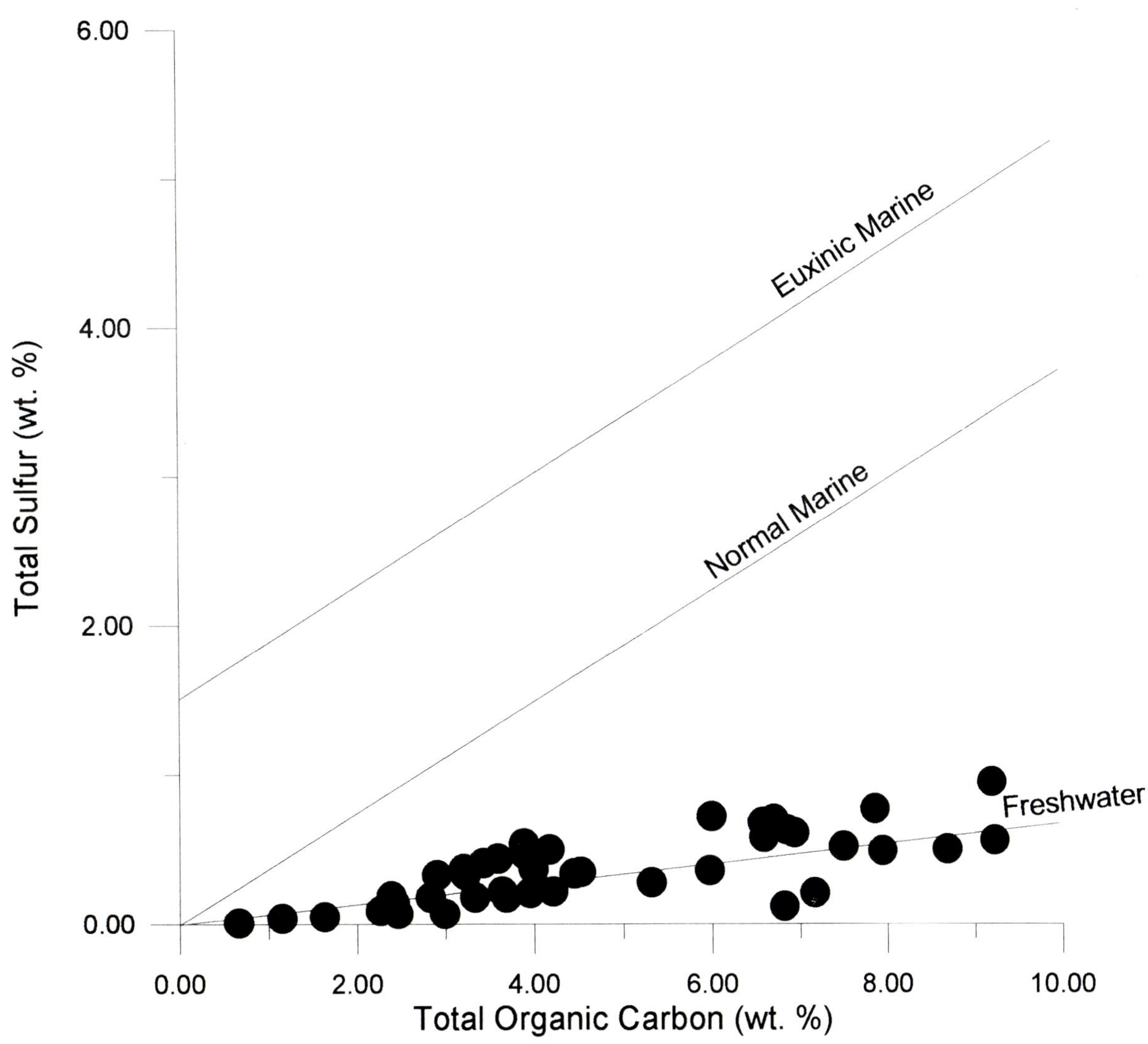

Fig. 7. Plot of total organic carbon versus total sulfur content of Brown Shale. (Leventhal 1983; Berner and Raiswell 1984)

Depositional Models

The sedimentology of the Pematang Group has been summarized by Williams et al. (1985). Sedimentation within the Pematang troughs is a function of the basin's tectonic development and the climatic conditions. Although the geometries of the basins are similar, two general basin types are recognized on the basis of their stratigraphy. The first type is a deep basin (Balam, Aman, and Rangau) and is characterized by a rapid rifting phase where sedimentation rates did not kept pace with subsidence. An anoxic lacustrine facies developed in the deep basin. The second type is a shallow basin (Kiri) and is characterized by a more continuous subsidence with sedimentation keeping pace with subsidence which restricted development of anoxic conditions.

The deposition of sediments within the basins is related to three structural events: pregraben (Lower Red Beds Formation), graben (Brown Shale and Coal Zone Formations), and lake fill (Lake Fill Formation). During the pregraben stage, the basins were undergoing initial rifting and the interior of the basins were characterized by swamps, distributary plains, and shallow lakes poorly interconnected by small fluvial channels and numerous deltas. Both the deep and shallow basin types appear to have similar pregraben sediments.

The graben stage is marked by intense rifting and characterized by the extensive development of lacustrine sediments with minor platform and delta front sediments. Variations in the tectonic development (rapid versus more continuous rifting) during the graben stage resulted in the two distinct basin sedimentary types: deep versus shallow troughs. The sediments of the shallow trough type, created by more continuous rifting, consist of shallow lacustrine shales with occasional coal stringers and minor sandstones. The organic material is primarily humic. In contrast, the deep basin type (Fig. 8), created by rapid initial rifting, consist primarily of organic-rich shales in the Aman and Rangau troughs and shales and marls in the Balam trough (Table 1). The organic material is algal-amorphous (oil-prone) deposited in deep anoxic lakes with little to no annual turnover and slow sedimentation rates of 4 to 5 cm/1000 years. Modern analogues to these sediments are the East

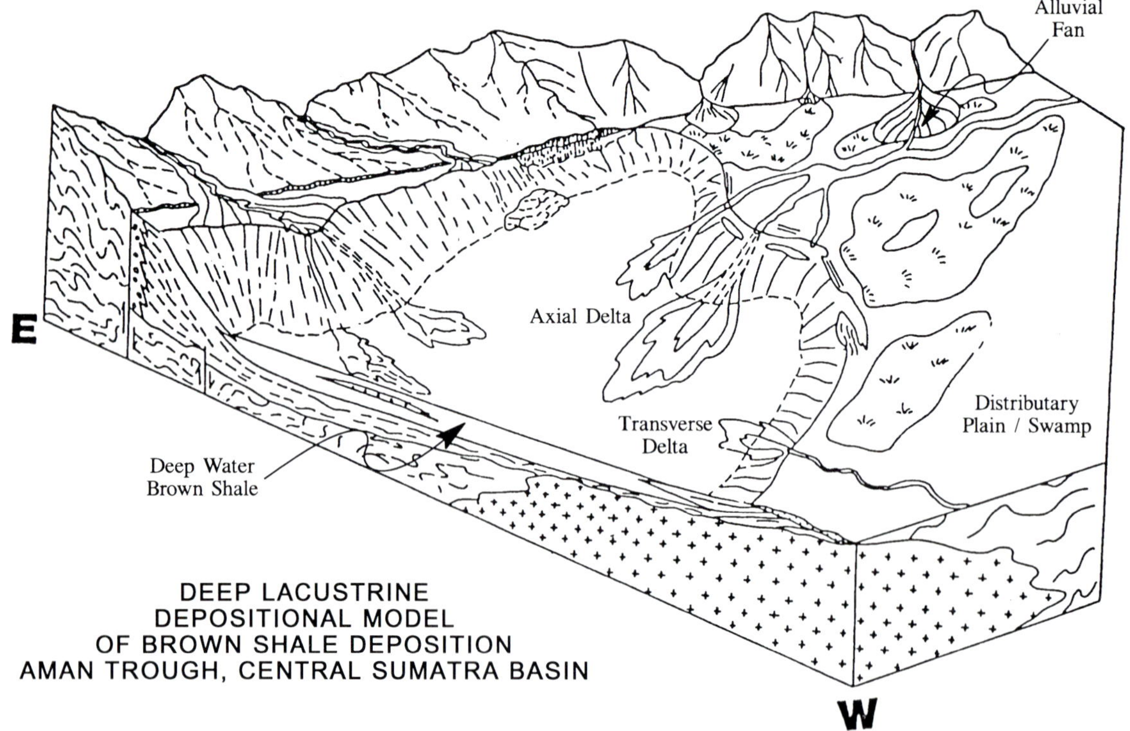

Fig. 8. Deep lacustrine depositional model of Brown Shale deposition, Aman trough, Central Sumatra. (After Williams et al. 1985)

Table 1. Summary of mineralogy data Pematang #1 and Bekasap #3 wells, Aman trough, and Pasada #1 well, Balam trough, Central Sumatra basin, Indonesia

Well		Depth (m)	Chlorite	Kaolinite	Illite	Montmorillonite	Mixed layer	% Expandability
Pematang #1	Top	1249	1.5	9.5	26.5	0.0	12.0	10
		1250	1.5	11.0	23.5	0.0	14.0	10
	Mid	1381	0.0	26.0	24.0	0.0	0.0	0
	Base	1478	0.0	29.5	10.5	0.0	10.5	10
		1483	1.5	8.0	33.0	0.0	7.5	10
Bekasap #3	Top	1004	1.5	18.5	14.5	0.0	15.5	10
		1005	0.0	19.5	16.0	0.0	14.5	10
	Base	1071	1.2	10.5	19.5	0.0	17.0	10
Pasada #1	Top	879	3.5	19.0	22.0	0.0	6.0	5
		883	2.5	16.0	31.0	0.0	0.5	5
		886	3.0	20.0	25.5	0.0	1.0	5
		888	2.5	20.0	22.0	0.0	5.5	5
	Base	1037	3.0	18.0	21.0	0.0	7.5	5
		1039	6.0	25.0	19.5	0.0	0.0	0
		1049	3.0	22.0	22.5	0.0	2.5	5

Well		Depth (m)	Total clay	Quartz	Kspar	Plagioclase	Pyrite
Pematang #1	Top	1249	20.5	0.0	0.5	0.0	2.0
		1250	23.0	0.5	0.5	0.0	0.5
	Mid	1381	47.0	0.0	0.0	0.0	0.0
	Base	1478	44.0	0.0	0.0	0.0	1.0
		1483	12.0	0.0	0.5	0.0	0.5
Bekasap #3	Top	1004	22.5	0.5	0.5	0.0	0.5
		1005	21.0	0.5	0.5	0.0	1.0
	Base	1071	17.0	0.0	0.5	0.0	0.5
Pasada #1	Top	879	5.5	3.5	0.0	0.0	0.0
		883	6.5	3.5	0.0	0.0	0.0
		886	7.5	5.5	0.0	0.0	0.0
		888	28.0	18.0	0.0	0.5	0.0
	Base	1037	28.0	17.5	0.5	0.5	0.0
		1039	37.0	5.5	0.0	0.5	2.5
		1049	33.5	13.0	0.0	0.5	0.5

Well		Depth (m)	Calcite	Dolomite	Siderite	Anhydrite	Gypsum	Halite
Pematang #1	Top	1249	0.0	0.0	0.0	0.0	27.0	0.5
		1250	0.0	0.0	0.0	0.0	25.5	0.0
	Mid	1381	0.0	0.0	0.0	0.0	3.0	0.0
	Base	1478	0.0	0.0	0.0	0.0	4.5	0.0
		1483	0.0	0.0	0.0	0.0	37.0	0.0
Bekasap #3	Top	1004	0.0	0.0	0.0	0.5	25.0	0.5
		1005	0.0	0.0	0.0	1.0	24.5	1.5
	Base	1071	0.0	0.0	0.0	0.0	32.0	0.5
Pasada #1	Top	879	40.5	0.5	0.0	0.0	0.0	0.0
		883	38.5	0.5	0.5	0.0	0.0	0.0
		886	36.0	0.5	0.0	0.0	0.0	1.0
		888	0.0	0.0	1.5	0.0	0.0	2.0
	Base	1037	0.0	0.0	1.0	0.0	0.0	1.5
		1039	0.0	0.0	2.0	0.0	0.0	2.5
		1049	0.0	0.0	2.0	0.0	0.0	0.0

Values given in percent

African rift lakes sediments (Katz and Kelley 1987).

The final stage of sedimentation, the lake fill stage, is characterized by shallow lacustrine, fluvial, distributary plain, delta front, and alluvial sediments. This stage was partially preceded by the Pematang structuring, which gave rise to local unconformities. Tectonism continued throughout the lake fill stage, finalizing the Pematang sedimentation with a regional Pematang unconformity.

Geothermal Gradient

The average geothermal gradient in the Central Sumatra basin is 60 °C/km (Eubank and Makki 1981). A BHT-depth cross plot shows a strong relationship indicating that most of the strata are in thermal equilibrium with the deep basement (Fig. 9). The best fit line is defined as the line from a surface temperature of 26.7 °C through the deep basement cluster of points. Shallow depth points falling significantly above the best fit line are interpreted as areas that are still cooling and have not yet reached equilibrium (after Plio-Pleistocene uplift). Deep depth points falling below the best fit line are interpreted as areas which are still warming up due to rapid deposition in the deep areas of the Pematang troughs.

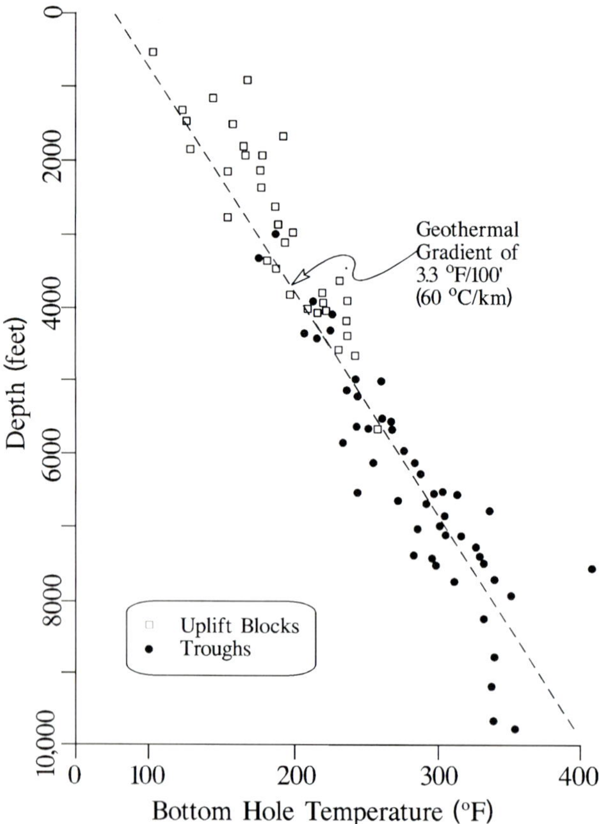

Fig. 9. Plot of bottom hole temperature versus depth for sediments in Central Sumatra basin. (After Eabank and Makki 1981)

Oil Character and Composition

There are more than ten billion barrels of oil in the Central Sumatra region (Root et al. 1987). The character and composition of Central Sumatra basin oils have been summarized by Williams et al. (1985), Robinson (1987), and Katz and Mertani (1989). The unaltered oils are largely waxy paraffin-based with an average composition of about 85% saturates, 10% aromatics, and 5% NSO + asphaltenes. API gravities range from 27° to 47° and pour points range from 26.7 °C to 48.9 °C. Biodegradation and water washing have altered some of the oils resulting in the lower API (< 27°) and pour points (4.4° to 18.3 °C). Sulfur contents for both the unaltered and altered oils are generally less than 0.2 wt. %.

The oils of the Rangau troughs are lighter (API gravities > 37°) than oils of the Balam and Aman troughs (API gravities 27° to 37°) (Fig. 10). This reflects the higher thermal maturity of the Brown Shale in the Rangau trough. The lower gravity oils (API gravities < 27°) northeast of the Aman basin are biodegraded and water-washed (see Fig. 11, Duri oil is biodegraded and water-washed while Minas oil is unaltered).

Variations in the carbon isotopes (Fig. 12) and the distribution of specific chemical compounds (*n*-paraffins, isoprenoids, terpanes, and steranes) of the oils are common among the oil fields (Seifert and Moldowan 1980; Moldowan et al. 1985; Williams et al. 1985; Katz and Mertani 1989). The pristane to phytane ratio is generally greater than 2 and as high as 7. Whole oil carbon isotopes varies from − 29 to − 22‰ relative to PDB standard (see Table 2 for selected oils). The oils are characterized by low sterane to terpane ratios (< 0.3), low gamacerane to hopane ratios (< 0.6), contain oleanane, high C_{30} 4-methyl sterane to C_{29} sterane ratios (> 0.5), and C_{27} steranes about half the C_{29} steranes (see Table 3 for Minas and Duri oils).

Fig. 10. Distribution of crude oil API gravities in Central Sumatra basin

Botryococcane has been found in Minas and Duri oils (Seifert and Moldowan 1980).

This has led to classification into several families and subfamilies (see Seifert and Moldowan 1980). Four concepts have been proposed to explain the family variations. The different families of oils were generated from:

1. different facies (lateral or vertical) within the Brown Shale of the same trough.
2. different facies of the Brown Shale from trough to trough.
3. the Brown Shale and overprinting to different degrees by bitumen in the carrier/reservoir system during migration, or
4. combination of (1), (2), and/or (3).

Although debate still exists as to the specific cause(s) of the uniqueness of each oil accumulation, all four concepts attribute the bulk of the oil to have been sourced from the Brown Shale.

Source Rock Characteristics

Three geochemical criteria are required for a prospective source rock to be classified as an "effective" source for oil (effective being defined as capable of generating and expelling commercial quantities of oil): organic enrichment, algal-amorphous (hydrogen-enriched) kerogen, and thermal maturity. Additionally, vertical and aerial extent are geologic requirements for effective source rocks. Specific cutoffs for the geochemical criteria can be found in Hunt (1979); Bissada (1982); Tissot and Welte (1984). Several prospective source rocks, the shales

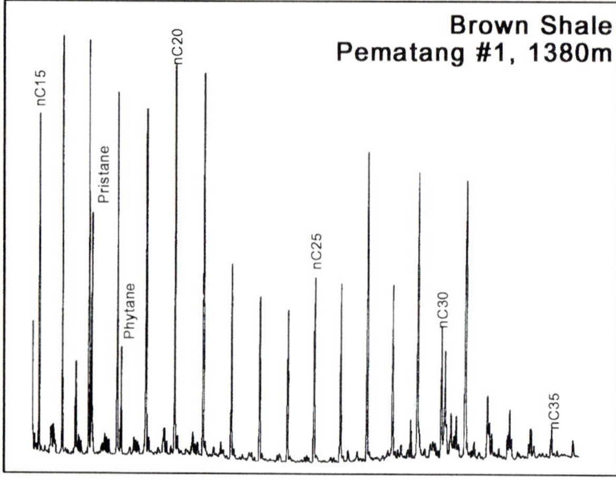

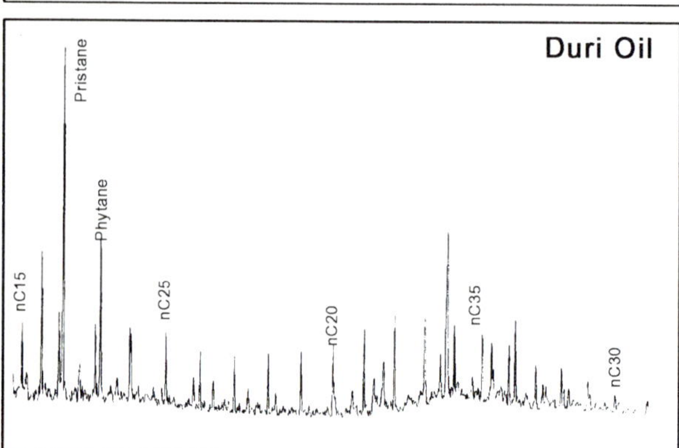

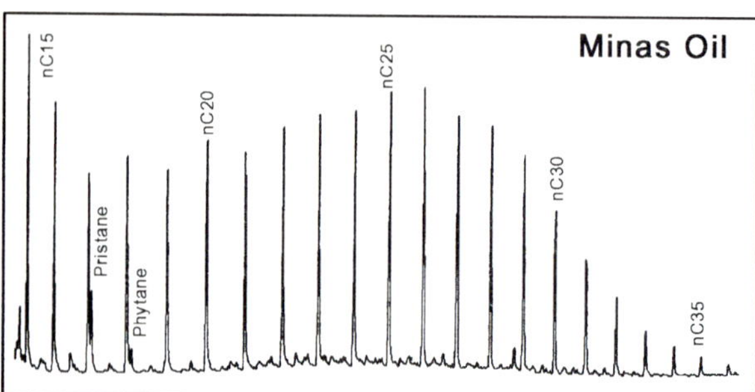

Fig. 11. C_{15+}saturate gas chromatograms of Brown Shale, Duri oil and Minas oil

of the Petani, Telisa, and Pematang (Sundararaman et al. 1988) and the coals of the Sihapas Group (MacGregor and McKenzie 1986) are present in the Central Sumatra basin. However, only the Brown Shale is an effective source for oils in the basin (Williams et al. 1985; Katz and Kelley 1987; and Katz and Kahle 1988).

Two general organic facies have been identified in the Brown Shale Formation in the Pematang troughs of Central Sumatra: an algal-amorphous facies (type I, I-II in Fig. 13) and a carbonaceous facies (type III, II-III in Fig. 13) (Williams et al. 1985; Katz and Mertani 1989). The algal-amorphous facies is oil-prone and present in the upper and middle portions of the Brown Shale in the Aman, Balam, and Rangau troughs. The carbonaceous facies is gas- and minor condensate/light oil-prone and present in the Kiri trough and basal portions of the Brown Shale in the Aman, Balam and Rangau troughs. A mixed facies is often present as the transition from algal-amorphous to carbonaceous facies in the Brown Shale in the Aman, Balam, and Rangau troughs. Tables 4 and 5 summarize the geochemical data of selected core intervals of the Brown Shale Formation for two wells in the Aman trough and one well in the Balam trough.

Algal-Amorphous Facies

The algal-amorphous facies is composed of alginite and fluorescent amorphous matter (type I and I-II kerogen) of algal and bacterial origin. The average organic richness for the samples of upper and middle portions of the Brown Shale in the Pematang #1 and Bekasap #3 wells is 5.8% TOC (total organic carbon) and for the samples of upper Brown Shale in the Pasada #1 well is 2.4% TOC; but is quite variable ranging from $< 1\%$ to $> 9\%$ TOC (see Tables 4 and 5). Average organic carbon content for Central Sumatra basin is between 2 to 4% (Williams et al. 1985) and for the Aman and Balam troughs 4.40% and 2.99% (Katz and Kelley 1987). The organic quality of this facies is oil-prone as characterized by high atomic hydrogen to carbon ratios (> 1.17 H/C, see upper and middle portions of Brown Shale in Fig. 13), predominance of algal-amorphous kerogen ($> 96\%$) and high hydrogen indices (greater than about 500). This oil-prone character is consistent with observations by Williams et al. (1985), Robinson (1987), Katz & Kelley (1987).

Considerable lateral and vertical variability exists within and between troughs and is attributed to overmaturity, variations in water chemistry and lake level, variations in algal species, and/or terrestrial input (Williams et al. 1985; Katz and Mertani 1989; Longley et al. 1990; Katz 1991). This is exemplified in the lower organic richness and quality in the Brown Shale in the Rangau trough and in the deepest areas of the Aman and Balam troughs.

Lower organic richness and poorer quality of the lower and middle portions of the Brown Shale in the Petani #2 well (deepest area of the Aman

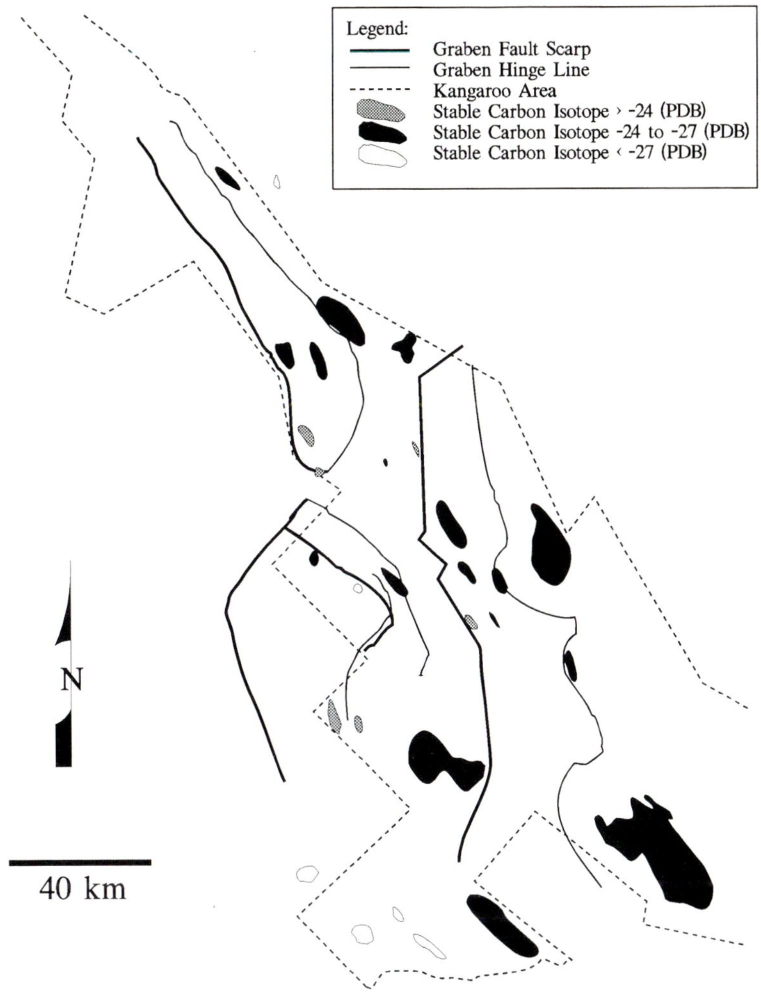

Fig. 12. Distribution of whole-oil stable carbon isotope, Central Sumatra basin

trough) is inferred to be caused by overmaturity, but suboxic depositional conditions also could have contributed. The maturity of the Brown Shale in the Petani #2 well (Fig. 15) is late stage (> 1.1% Ro, > 90% conversion of kerogen to oil). The kerogen in the lower potion of the Brown Shale in the Petani #2 well is nonfluorescent amorphous with minor exinites and solid bitumens. Vitrinite and inertinite (common in terrestrial kerogens) was not observed. Therefore, overmaturity is certainly a factor and terrestrial input is probably not a factor. Distinguishing lower quality amorphous kerogen (nonfluorescent) due to deposition in suboxic environment (change in lake level and/or water chemistry) from overmature high quality (fluorescent) algal-amorphous is very difficult to determine. However, volumetric considerations (Katz and Kahle 1988) suggest that a substantial portion of the lower Brown Shale in the Petani #2 well must have originally been high to have sourced the known volume of oil. Similar observations are seen for the entire Brown Shale section in the Rangau trough.

The lower organic richness and poorer quality of the lower portion of the Brown Shale in the Pematang #1 well (Table 4 and Figs. 13, 14) is inferred to be caused by suboxic depositional conditions. The organic kerogen is visually 98% weakly fluorescent algal-amorphous, but has hydrogen indices less than 365 and atomic H/C ratios less than 0.89. Further, the pristane/phytane ratio is less than 1.5 consistent with the upper algal-amorphous Brown Shale. These characteristics do not suggest terrestrial input. Maturity modeling suggests the lower portion of the Brown Shale to be thermally mature (25% < conversion of kerogen < 65%). By elimination of terrestrial input and overmaturity, suboxic depositional conditions appear the probable cause.

Table 2. Stable carbon isotopes, selected oils, and conventional cores, Central Sumatra basin, Indonesia

Central Sumatra oil field	Whole oil stable carbon isotopes (ppt)	Well Aman Trough	Depth (m)	Kerogen stable carbon isotopes (ppt)	Bitumen whole extract stable carbon isotopes (ppt)	Bitumen saturate fraction stable carbon isotopes (ppt)
Damar[a]	– 28.4	Petani #2	1880		– 25.7	– 27.7
Balam SE[a]	– 22.0		1883		– 25.6	– 27.3
Hitam[a]	– 22.3		1884	– 24.9		
Minas[a]	– 25.3	Pematang #1	1249	– 26.3	– 28.2	– 28.6
Duri[a]	– 24.7		1377	– 23.5		
Bakasap S.[a]	– 24.1		1380		– 24.5	– 25.6
Pematang[a]	– 22.1		1479	– 24.8	– 23.4	– 24.3
Suram[a]	– 28.8		1483		– 23.6	– 23.8
Petapahan[a]	– 28.5	Bekasap #3	1004	– 21.9		
			1062	– 22	– 25.7	– 26.2
			1071	– 23.7	– 26.9	– 29.7

[a] Seifert and Moldowan (1980)
Blank = not measured.

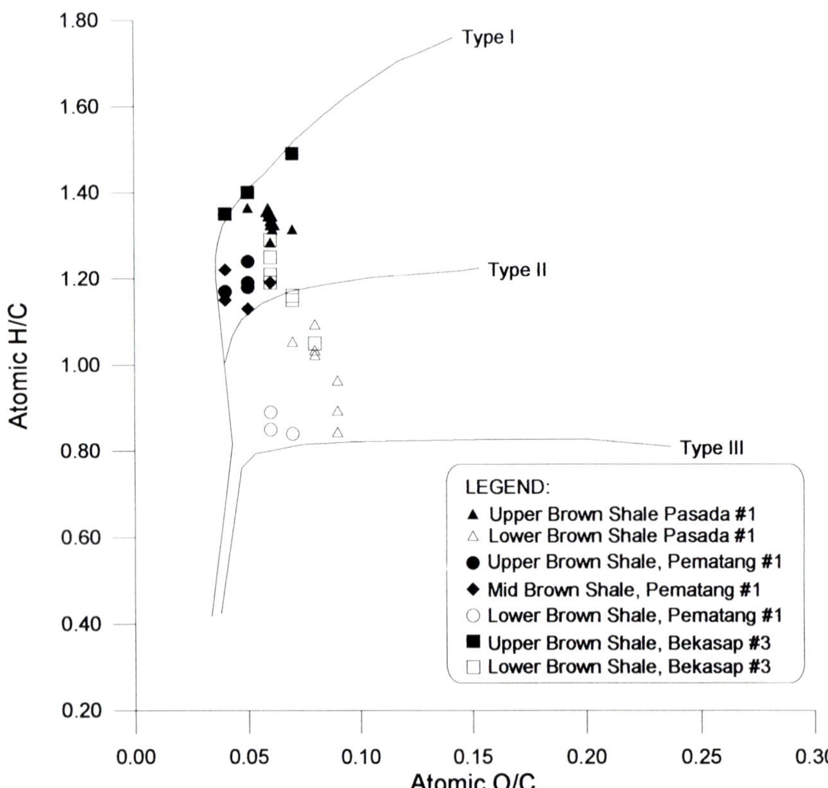

Fig. 13. Plot of atomic O/C versus atomic H/C of Brown Shale for selected cores from the Pasada #1, Pematang #1, and Bekasap #3 wells

Katz and Mertani (1989) attribute the cause for the lower organic quality of the lower portions of the Brown shale in the Balam trough (see lower Brown Shale of Pasada #1 well in Table 5) to a mix of water chemistry differences (suboxic depositional conditions) and terrestrial input. They show the upper portion of the Brown Shale in an unidentified well to be oil-prone while the lower

Table 3. Selected biomarker parameters from brown shale bitumen, Minas oil, and Duri oil Aman Trough, Central Sumatra Basin, Indonesia

	% TRI	26TRI/ 25TRI	TETRA/ 26TRI	Tm/Ts	NOR/ HOP	31HOMO/ HOP
Pematang #1 Brown Shale 1380 m	15.89%	1.82	0.26	0.63	0.85	0.45
Duri oil	11.11%	3.67	0.62	0.72	0.63	0.45
Minas oil	21.44%	2.04	0.71	0.27	0.60	0.50

	35HOMO/ 34HOMO	OL/HOP	BIS/HOP	GAM/HOP	HOM S/S + R	MOR/HOP
Pematang #1 Brown Shale 1380 m	0.56	0.04	0.03	0.06	0.55	0.12
Duri oil	0.46	0.07	0.04	0.04	0.59	0.08
Minas oil	0.62	0.13	0.03	0.06	0.57	0.17

	STE/TERP	27STE/ 29STE	28STE/ 29STE	30STE/ 29STE	DIA/STE	29STE AAS/ AAS + AAR
Pematang #1 Brown Shale 1380 m	0.13	0.47	0.53	1.54	0.32	0.36
Duri oil	0.12	0.61	0.55	0.81	0.51	0.62
Minas oil	0.26	0.55	0.60	0.61	0.92	0.68

	29STE BB/ BB + AA	27STE%	28STE%	29STE%		Pr/Ph
Pematang #1 Brown Shale 1380 m	0.46	23.47%	26.51%	50.01%		2.2
Duri oil	0.30	28.33%	25.30%	46.37%		2.2
Minas oil	0.34	25.48%	28.06%	46.46%		3.2

Key: Tm – 17a(H) -22,29,30 Trisnorhopane
Ts – 18a(H) -22,29,30 Trisnorhopane
BIS – 17a(H), 18a(H), 21B(H)-28,30 Bisnorhopane
NOR – 17a(H), 21b(H)-30 Norhopane
HOP – 17b(H), 21b(H) Hopane
GAM – Gammacerane
OL – Oleanane
MOR – 17b(H), 21a(H) Moretane
TRI – C19 Through C39 Tricyclics
C31 Trough C35-Homohopanes
C26T & C25T-Tricyclics
HOMO – C31 Through C35 17a(H),21b(H) Homohopnaes
TETRA – C24 Tetracyclic
%TRI – %Tricyclics in total identified peaks on M/Z 191 fragmentogram

TERP – Hopanes (total peaks on M/Z 191 fragmentogram)
DIA – Diasteranes (sum of C27, C28, C29, C30 BAS, BAR, ABS, ABR)
STE – Regular steranes (sum of C27, C28, C29, C30 AAS, AAR, BBS, BBR)
C##STE – sum of C## 5A,14A,17A 20S&R and 5A,14B,17B 20S&R cholestane
C##AAS – C# # 5A,14A,17A 20S Cholestane
C##AAR – C## 5A,14A,17A 20R Cholestane
C##BBS – C## 5A,14B,17B 20S Cholestane
C##BBR – C## 5A,14B,17B 20R Cholestane
C##BAS – C## 5A,14B,17A 20S Cholestane
C##BAR – C## 5A,14B,17A 20R Cholestane
C##ABS – C## 5A,14A,17B 20S Cholestane
C##ABR – C## 5A,14A,17B 20R Cholestane
Pr/Ph-Pristane / Phytane.

portion of the Brown Shale is gas-prone. They suggest that these differences are greater than can be explained by maturity differences and that the kerogen in the lower portion is composed mostly of nonfluorescent amorphous with varying amounts of fluorescent amorphous, vitrinite, and exinites.

Variations in the algal speciations is suggested by the different C_{15} + saturate gas-chromatograms

Table 4. Summary of geochemical data Pematang #1 and Bekasap #3 wells, Aman trough, Central Sumatra basin, Indonesia

Well		Depth (m)	TOC (wt. %)	Sulfur (wt. %)	S_1 (mg HC/g rock)	S_2 (mg HC/g rock)	S_3 (mg HC/g rock)	T_{max} (°C)
Pematang #1	Top	1247	3.89	0.47	4.94	22.71	1.34	440
		1248	3.88	0.54	4.10	19.27	1.49	441
		1249	4.52	0.35	4.93	33.05	1.34	444
		1250	3.64	0.22	9.34	18.99	0.81	433
		1250	3.42	0.41	5.59	19.36	1.19	438
		1251	3.59	0.44	5.76	21.49	1.37	439
		1251	3.68	0.18	10.38	20.66	0.71	432
		1252	5.97	0.36	12.25	47.19	1.26	437
	Mid	1377	9.22	0.56	14.78	71.92	1.61	440
		1378	8.68	0.50	14.30	67.87	1.67	439
		1379	3.22	0.33	4.68	18.19	1.23	439
		1379	6.56	0.68	14.39	58.71	1.68	437
		1380	6.69	0.70	9.33	52.05	1.90	438
		1381	6.93	0.61	14.56	57.73	2.22	437
		1383	6.84	0.63	15.26	53.45	1.72	439
		1382	3.99	0.37	6.64	19.59	1.40	444
	Base	1477	0.66	0.01	5.86	2.41	0.97	409
		1477	1.15	0.04	5.12	3.51	1.26	413
		1478	1.63	0.05	2.55	1.80	1.37	451
		1479	2.90	0.33	0.97	4.37	1.53	453
		1481	2.26	0.09	0.96	2.58	0.23	450
		1483	2.42	0.14	1.96	3.16	2.29	454
		1483	3.20	0.37	1.05	4.81	1.82	455
		1483	2.46	0.07	1.95	3.58	0.32	449
Bekasap #3	Top	1003	9.19	0.95	8.78	86.95	0.61	434
		1003	6.58	0.58	2.59	59.36	1.40	435
		1003	7.94	0.49	5.60	76.10	1.00	438
		1004	7.49	0.52	3.01	65.30	2.04	436
		1004	7.85	0.77	5.25	73.82	0.80	435
		1004	4.17	0.50	0.37	20.21	5.27	441
		1005	5.31	0.28	0.82	43.79	1.77	435
	Base	1061	4.21	0.22	0.65	19.42	4.30	437
		1061	2.83	0.18	0.47	15.34	2.09	438
		1062	3.33	0.18	0.66	17.40	2.59	435
		1065	2.38	0.19	0.25	4.34	4.24	437
		1067	5.99	0.72	0.65	22.18	1.88	426
		1068	4.45	0.34	0.92	39.44	1.76	420
		1070	3.95	0.21	0.85	23.28	1.00	437
		1070	2.99	0.07	0.60	7.69	0.78	431
		1070	7.16	0.21	2.68	38.55	0.62	428
		1071	6.81	0.12	3.94	40.88	0.78	423

KTR – kerogen transformation ratio. Pr – pristane. Re-MPI3 – vitrinite reflectance equivalent of methyl phenanthrene index #3. Amorphous – algal amorphous. Ph – phytane. %Ro – mean vitrinite reflectance.

and biomarker compositions in the extracts from the Aman and Balam troughs by Williams et al. (1985) and by the biomarker compositions and palynofacies from the troughs in Malacca Strait by Longley et al. (1990).

High amounts of high molecular weight paraffins ("wax", normal paraffins $> nC_{22}$) is characteristic of the oils and the bitumens in the Brown Shale (Moldowan et al. 1985; Williams et al. 1985; Robinson 1987; Katz and Mertani 1989). Although high molecular weight paraffins are present in higher plants as wax components (i.e., cuticles, waxy coating on leaves, etc.), high molecular components can be generated from fresh water algae as shown in

HI (S_2/TOC)	OI (S_3/TOC)	KTR ($S_1/S_1 + S_2$)	H/C (atomic)	O/C (atomic)	Visual assessment	Pr/Ph	Re-MPI3	%Ro	Kerogen carbon isotopes (ppt)
584	34	0.18	1.19	0.05					
497	38	0.18	1.18	0.05		1.55	0.85		
731	30	0.13	1.24	0.05	98% Algal	1.85	0.86		− 28.2
522	22	0.33	1.19	0.05					
566	35	0.22							
599	38	0.21	1.19	0.05					
561	19	0.33							
790	21	0.21	1.17	0.04					
780	17	0.17	1.22	0.04	98% Amorphous			0.38	− 23.5
782	19	0.17							
565	38	0.20	1.19	0.06					
895	26	0.20							
778	28	0.15	1.15	0.04		2.21	0.87		
833	32	0.20							
781	25	0.22	1.13	0.05					
491	35	0.25							
365	147	0.71				1.09			
305	110	0.59							
110	84	0.59				1.19	1.04		
151	53	0.18	0.84	0.07	98% Amorphous		1.03	0.83	− 24.8
114	10	0.27	0.89	0.06					
131	95	0.38	0.85	0.06		1.24			
150	57	0.18							
146	13	0.35							
946	7	0.09	1.35	0.04		3.93	0.87		
902	21	0.04							
958	13	0.07							
872	27	0.04	1.49	0.07	97% Algal				− 21.9
940	10	0.07							
485	126	0.02							
825	33	0.02	1.40	0.05					
461	102	0.03	1.19	0.06					
542	74	0.03	1.25	0.06					
523	78	0.04	1.21	0.06	96% Amorphous	3.51	0.95	0.41	− 22.0
182	178	0.05	1.16	0.07		3.59	0.94		
370	31	0.03				3.96	0.91		
886	40	0.02	1.29	0.06					
589	25	0.04	1.15	0.07					
257	26	0.07							
538	9	0.07							
600	11	0.09	1.05	0.08	42% Algal	5.04	0.92	0.51	− 23.7

the C_{15}+ saturate gas chromatogram of the bitumen (Fig. 11) and pyrolysis gas chromatogram of the kerogen (Fig. 15) from the Brown Shale at 1380 m in the Pematang #1 well. Further, Gelpi et al. (1970) has shown that high molecular weight paraffins have been generated from freshwater algae. Therefore, the high wax content in the oils of the Central Sumatra basin is believe to have been generated primarily from specific freshwater algae comprising the Brown Shale.

Carbonaceous Facies

The carbonaceous facies is composed predominantly of vitrinite (humic material and plant debris) with minor exinite (leaf cuticle and higher plant spores, pollen, and resin), inertinite (charcoal), and nonfluorescent amorphous matter (degraded plant debris). Organic richness is quite variable, ranging from 1 to 43% TOC. Organic quality is gas-prone with minor condensate/light oil as indicated by low

Table 5. Summary of Geochemical Data Pasada #1 well, Balam tough Central Sumatra basin, Indonesia

Well		Depth (m)	TOC (wt. %)	Sulfur (wt. %)	S_1 (mg HC/g rock)	S_2 (mg HC/g rock)	S_3 (mg HC/g rock)	T_{max} (°C)
Pasada #1	Top	879	3.55	0.22	0.92	33.96	0.36	441
		879	0.76	0.05	0.11	2.76	0.60	440
		880	0.97	0.09	0.26	3.92	0.64	439
		880	1.53	0.09	0.37	5.60	1.12	436
		880	2.75	0.16	0.71	22.67	0.48	440
		881	1.46	0.12	0.22	6.50	0.69	441
		882	2.45	0.13	0.93	10.62	1.66	438
		882	3.29	0.19	1.64	21.72	1.63	438
		882	1.79	0.08	0.44	7.62	1.43	435
		883	1.96	0.15	0.38	12.28	0.70	442
		883	1.00	0.05	0.47	1.66	2.87	439
		883	4.94	0.27	1.20	37.50	1.20	437
		884	4.72	0.21	1.07	43.38	0.64	440
		884	2.24	0.11	0.49	18.21	0.70	441
		885	1.39	0.08	0.20	4.69	0.98	440
		885	1.89	0.08	0.51	10.73	1.07	438
		885	2.36	0.14	0.59	21.19	0.77	445
		885	2.68	0.15	0.56	21.99	0.76	441
		885	3.36	0.16	0.94	33.23	0.48	438
		886	1.92	0.10	0.43	13.03	0.68	439
		887	1.88	0.07	0.38	15.50	0.57	440
		887	3.72	0.15	0.86	35.72	0.62	442
		888	1.10	0.00	0.15	4.09	0.73	438
		888	3.00	0.08	0.81	21.53	1.02	436
		888	3.86	0.13	1.64	21.54	2.28	434
	Base	1037	2.03	0.09	0.30	3.54	0.40	435
		1037	0.72	0.06	0.07	0.62	1.63	434
		1037	4.02	0.14	0.72	5.87	0.54	436
		1038	9.26	0.78	3.50	49.84	1.50	437
		1038	19.61	0.38	7.52	51.74	2.35	436
		1038	23.25	1.94	12.34	111.60	2.27	431
		1038	1.06	0.03	0.25	2.16	0.37	433
		1038	2.91	0.09	0.58	18.08	0.26	442
		1039	13.77	0.87	3.35	83.11	1.00	438
		1045	0.66	0.07	0.63	1.58	0.29	439
		1049	2.61	0.48	1.09	5.60	0.85	439

KTR – kerogen transformation ratio. Pr – pristane. Re-MPI3 – vitrinite reflectance equivalent of methyl phenanthrene index #3. Amorphous – algal amorphous. Ph – phytane. %Ro – mean vitrinite reflectance

hydrogen to carbon ratios (< 0.85) and hydrogen indices (< 350). This facies occurs in the Kiri trough as the Coal Zone Formation (equivalent in time to the Brown Shale Formation in the Aman and Balam troughs) and locally grades into true humic coals in the upper part of the Coal Zone and along this basin's hinge margin. Only local, light oils to condensates (> 44° API gravity) within the Coal Zone Formation of the Kiri trough are implied to have been source from this facies (Williams et al. 1985).

Organic Maturity

Organic maturity parameters and modeling indicate the Brown Shale Formation is at the peak maturity for hydrocarbon generation and expulsion in the Balam, Aman, and Kiri troughs (Williams et al. 1985; Katz and Kahle 1988) as exemplified by the BASINMOD model for the Petani #2 well (Fig. 16). The algal-amorphous facies in the Balam and Aman troughs is therefore

HI (S_2/TOC)	OI (S_3/TOC)	KTR ($S_1/S_1 + S_2$)	H/C (atomic)	O/C (atomic)	Visual assessment	Pr/Ph	Re-MPI3	%Ro	Kerogen carbon isotopes (ppt)
957	10	0.03	1.36	0.06	97% Algal	2.32	0.90	0.41	– 23.5
363	79	0.04	1.34	0.06					
404	66	0.06	1.31	0.06					
366	73	0.06				2.21	0.90		
824	17	0.03	1.32	0.06		2.27			
445	47	0.03	1.31	0.07					
433	68	0.08							
660	50	0.07							
426	80	0.05							
627	36	0.03	1.34	0.06		2.42	0.93		
166	287	0.22							
759	24	0.03							
919	14	0.02	1.28	0.06		2.28	0.98		
813	31	0.03							
337	71	0.04	1.34	0.06					
568	57	0.05				2.12			
898	33	0.03							
821	28	0.02							
989	14	0.03	1.35	0.06					
679	35	0.03	1.32	0.06					
824	30	0.02							
960	17	0.02	1.36	0.05		2.67			
372	66	0.04	1.33	0.06					
718	34	0.04							
558	59	0.07	1.34	0.06	96% Algal	2.75	0.95	0.39	– 22.1
174	20	0.08	0.89	0.09	63% Amorphous			0.58	– 26.8
86	226	0.10	1.03	0.08					
146	13	0.11							
538	16	0.07	1.05	0.07					
264	12	0.13	0.84	0.09		6.46	0.95		
480	10	0.10							
204	35	0.10	1.03	0.08					
621	9	0.03							
604	7	0.04	1.02	0.08		9.59	0.95		
239	44	0.29	1.09	0.08					
215	33	0.16	0.96	0.09					

generating and expelling oil, whereas the carbonaceous facies in the Kiri trough is generating and expelling gas and minor amounts of condensate/light oil. In the deeper parts of the Aman trough and in the entire Rangau trough, organic maturity is in the wet gas/condensate to dry gas generation and expulsion stages.

Maturity modeling suggests that hydrocarbon generation and expulsion from the Brown Shale and Coal Zone Formations in the Balam, Aman, and Kiri troughs began about 10 Ma B.P. followed by wet gas and condensate approximately 5 Ma B.P. The Brown Shale in the Rangau trough and in the deeper parts of the Aman trough began generating and expelling oil about 20 Ma B.P. followed by wet gas and condensate approximately 10 Ma B.P. and dry gas approximately 5 Ma B.P.

Crude Oil-Source Facies Correlation

The algal-amorphous facies showed good to excellent correlation with the crude oils of the Central Sumatra basin (Williams et al. 1985; Katz and Kahle 1988; Katz and Mertani 1989). The correlations

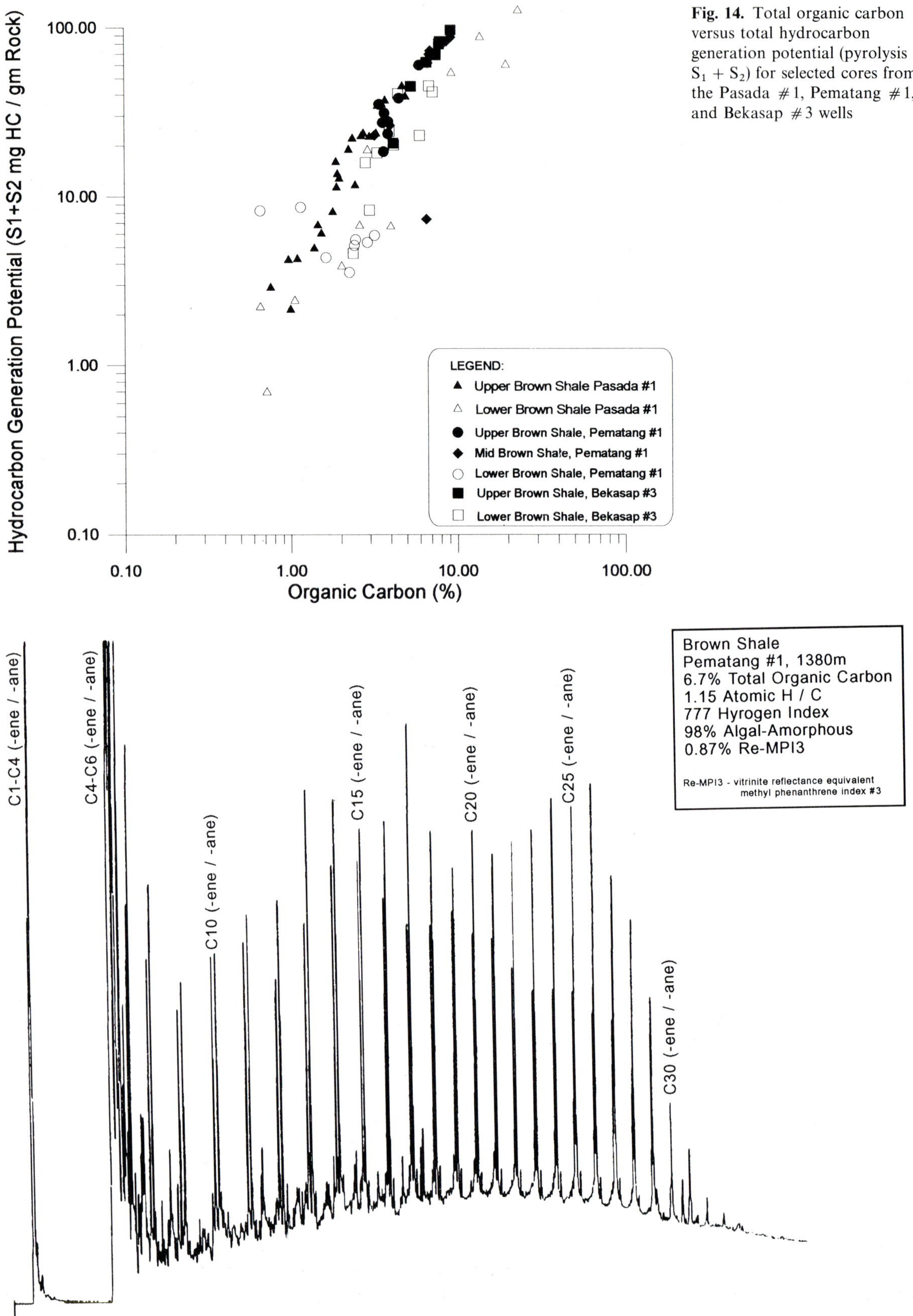

Fig. 14. Total organic carbon versus total hydrocarbon generation potential (pyrolysis $S_1 + S_2$) for selected cores from the Pasada #1, Pematang #1, and Bekasap #3 wells

Fig. 15. Pyrolysis gas chromatogram of Brown Shale kerogen in Pematang #1 well

MATURITY (VITRINITE) MODEL

KEROGEN CONVERSION MODEL
(BROWN SHALE)

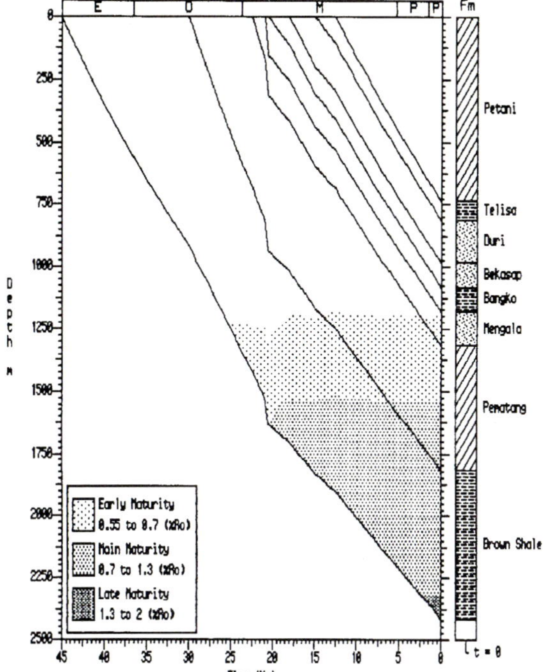

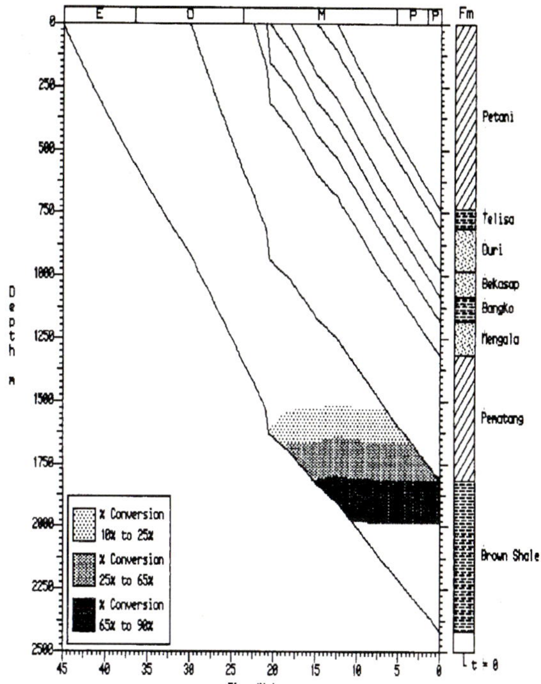

NOTES:
1) Modelled using BASINMOD (Platte River Assoc.)
2) Vitrinite kinetic parameters used in Maturity Model were defaults in BASINMOD (after LLNL)
3) Brown Shale kinetic parameters used in Kerogen Conversion Model obtained from Pyromat II and KINWIN (Lab Instruments)
4) Stratigraphy & Age from CPI files - analyses of cuttings and cores
5) Geothermal Gradient = 60 C / km, Suface Temperature = 25 C
6) Depths from depth corrected wireline logs

Fig. 16. Maturity modeling (BASINMOD) of Brown Shale in Petani #2 Well

of crude oils to source rocks were made on the basis of the isoprenoids pristane and phytane in the $C_{15}+$ saturate gas chromatograms, stable carbon isotopes, and biomarkers-terpanes (m/z 191) and steranes (m/z 217) (see Figs. 11, 17 and 18, and Tables 2 and 3 for representative data).

A given crude oil accumulation represents a mixture of oils generated from different oil-prone facies at different maturity affected by migration and in some cases altered in the reservoir. Further, the oils and the algal-amorphous Brown Shale facies show considered variance (previously discussed). Thus, correlations of a given crude oil accumulation to a specific source facies bitumens will show some similarities as well some dissimilarities.

All of the oils and algal-amorphous Brown Shale bitumens show similarities of low sulfur (<0.2), pristane to phytane ratios <4, Tm to Ts ratios <1.5, low sterane to terpane ratios (<0.3), low C_{27} sterane to C_{29} sterane ratios (<0.7), moderate to high C_{30} 4-methyl sterane to C_{29} steranes ratios (>0.6), and low gammacerane to hopane ratios (<0.6). These characteristics are indicative of lacustrine source rocks. Further, these observations are consistent with observations of deep lacustrine source in hydrocarbon productive basins in Indonesia (Robinson 1987).

Dissimilarities among the oils and algal-amorphous Brown Shale bitumens are evident in the wide range of stable carbon isotopes, tricyclic distributions, diasterane to sterane ratios, abundance

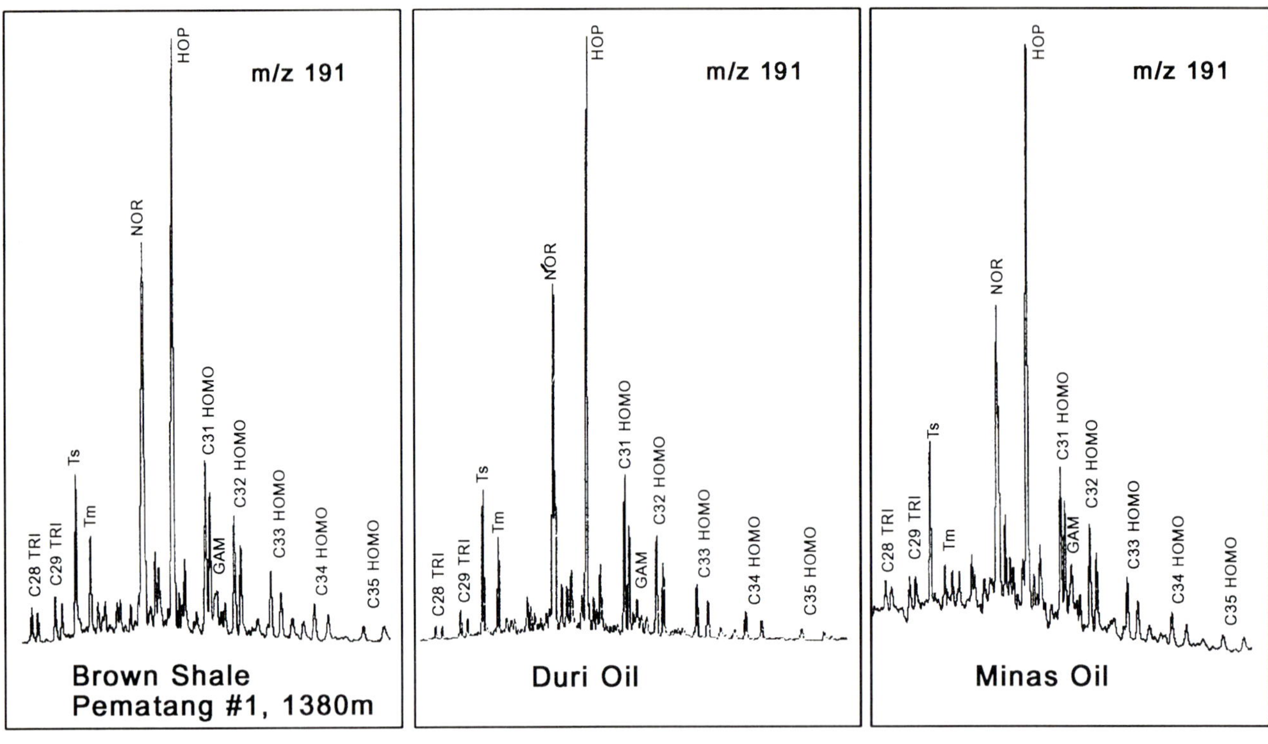

LEGEND:
Tm - 17a(H) -22,29,30 TRISNORHOPANE
Ts - 18a(H) -22,29,30 TRISNORHOPANE
NOR - 17a(H), 21b(H) -30 NORHOPANE
HOP - 17b(H), 21b(H) HOPANE
GAM - GAMMACERANE
MOR - 17b(H), 21a(H) MORETANE
TRI - TRICYCLICS
HOMO - HOMOHOPANES

Fig. 17. Terpane (m/z 191) fragmentograms of Brown Shale, Duri oil, and Minas oil

of tricyclics, and abundance of oleanane (Williams et al. 1985; Katz and Mertani 1989). These differences were attributed to sourcing from the Brown Shale either from variable facies within the same trough or from separate troughs.

Migration Routes

The asymmetry of the Pematang troughs governs the overall lateral direction of oil migration primarily towards the hinge margin of the basin. This is revealed in the overall oil discovery patterns which shows Pematang reservoirs in basin centers and Sihapas (Mengala/Telisa and Bekasap) reservoirs along the steep fault scarp margin and along the basin hinge margin (Fig. 19). Migration distances appear to be relatively short, being governed by structuring and average generally less than 10 km. The shales of the Bangko and Telisa Formations act as the ultimate seals for the entrapment of oil.

Focused migration into structural highs along the hinge margin is an important factor in the lateral movement of the bulk of the oil in the Aman basin as exemplified by the Minas and Duri Fields. Local faults and fractures appear to divert oil from the general lateral migration to vertical trends characterized by multiple pay fields. In general, timing of Pematang compressional structuring as well as tensional structuring, and with Plio-Pleistocene faulting, may have resulted in some vertical migration within the Pematang troughs. This suggests that the most laterally removed oil fields may have been sourced earlier than some fault associated oil fields.

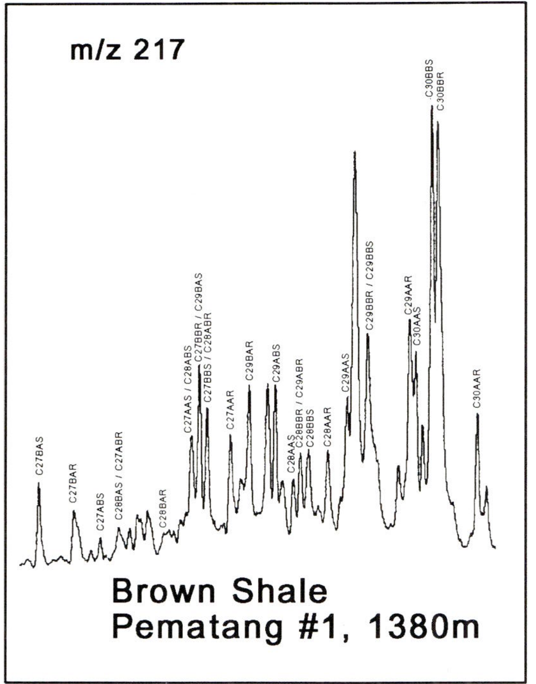

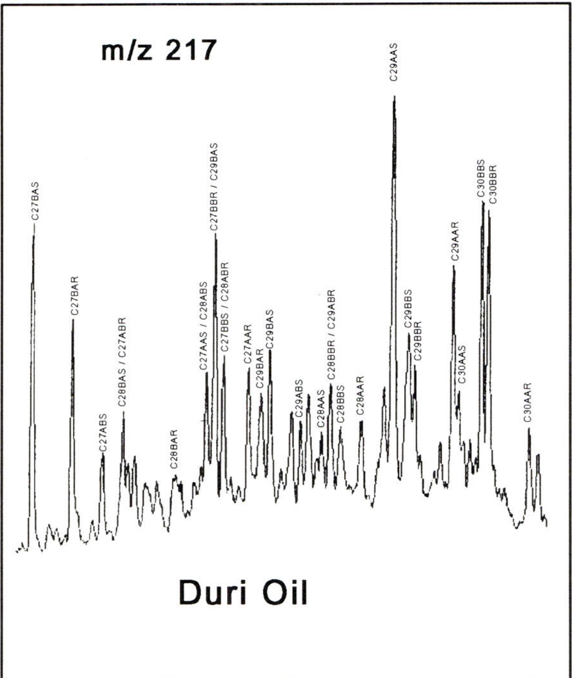

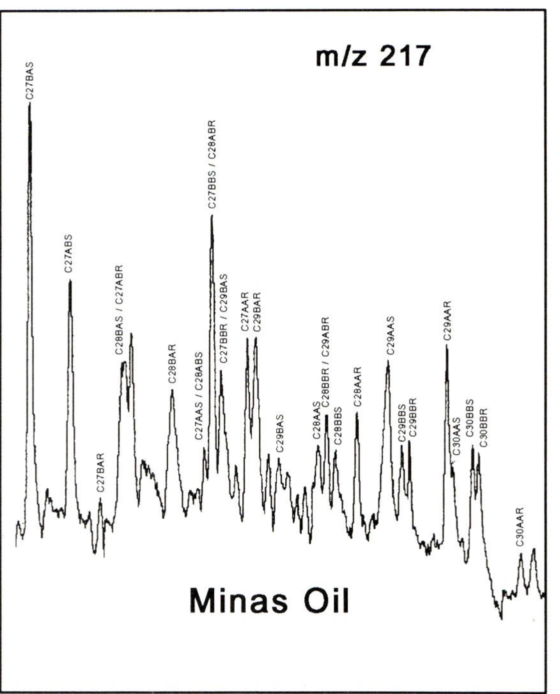

LEGEND:

C##BAS - C## 5A,14B,17A 20S CHOLESTANE	C##AAS - C## 5A,14A,17A 20S CHOLESTANE
C##BAR - C## 5A,14B,17A 20R CHOLESTANE	C##AAR - C## 5A,14A,17A 20R CHOLESTANE
C##ABS - C## 5A,14A,17B 20S CHOLESTANE	C##BBS - C## 5A,14B,17B 20S CHOLESTANE
C##ABR - C## 5A,14A,17B 20R CHOLESTANE	C##BBR - C## 5A,14B,17B 20R CHOLESTANE

Fig. 18. Sterane (m/z 217) fragmentograms of Brown Shale, Duri oil, and Minas oil

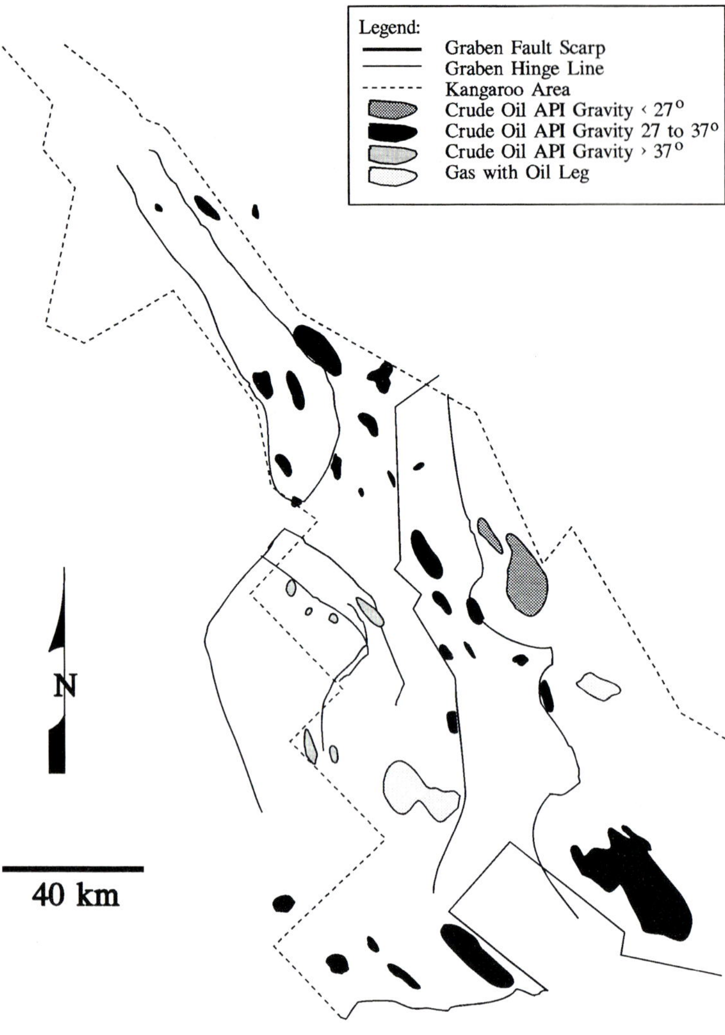

Fig. 19. Producing horizon of crude oils in Central Sumatra basin

Timing of trap/seal development in relation to oil generation and expulsion is important in controlling the amount and character of the accumulated oil around the Rangau trough. Significant amounts of oil probably were generated and expelled prior to development of effective traps and/or seals along the margins of the basin. Maturity modeling suggests that Brown Shale began generating and migrating in the Rangau trough approximately 20 MaB.P. followed by condensate approximately 10 MaB.P. (see section Organic Maturity this chapter). The Upper Red Beds Formation is the producing horizon and the shales of the Bangko Formation act as the ultimate seal but they were not deposited until approximately 25 and 20 Ma B.P., see Fig. 3. Recognizing that their effectiveness as a reservoir and seal developed following initial generation, the early normal oil generated from the Brown Shale was probably lost and only the later, lighter oils/condensates were trapped.

Summary

This chapter summarizes several earlier studies on the origin of oil in Central Sumatra Basin, Indonesia, by Williams et al. (1985); Katz and Kelley (1987); Robinson (1987); Sundararaman et al. (1988); Katz and Mertani (1989); Longley et al. (1990); and Katz (1991) supplemented with unpublished data. Specific summaries are:

1. The Pematang basins of Central Sumatra are part of a series of back-arc basins developed linearly along the leading edge of Sundaland, as a result of subduction of the Indian Ocean Plate

beneath the Southeast Asian Plate. The Pematang basins were formed during Eocene-Oligocene as a series of asymmetric half graben structures separated by horst blocks. The geometric dimensions of these half grabens are typically about 15 km wide by 40 to 100 km long. These half grabens were filled with non-marine clastics and lacustrine sediments of the Pematang Group which attains maximum thicknesses in excess of 1500 m.

2. The deposition of sediments within the basins is related to three structural events: Early Eocene pregraben (Lower Red Beds Formation), Mid Eocene to Early Oligocene graben (Brown Shale and Coal Zone Formations), and Late Oligocene to Early Miocene lake fill (Lake Fill Formation).

3. Two general basin types are recognized on the basis of their stratigraphy: a deep basin (Balam, Aman, and Rangau) characterized by a rapid rifting phase where sedimentation rates did not keep pace with subsidence and a shallow basin (Kiri) characterized by a more continuous subsidence with sedimentation keeping pace with subsidence.

4. The crude oils of Central Sumatra are largely waxy paraffin based. The API gravities of unaltered oils range from 27° to 47° and pour points range from 26.7° to 98.9°C. Biodegradation and water washing have altered some of the oils resulting in the lower API (< 27°) and pour points (4.4° to 18.3°C). Sulfur contents for both the unaltered and altered oils are generally less than 0.2 wt. %.

5. Two general organic facies have been identified in the Pematang basins of Central Sumatra: an algal-amorphous facies and a carbonaceous facies. The algal-amorphous facies is oil-prone and present as the Brown Shale Formation. The carbonaceous facies is gas and minor condensate/light oil-prone and present as the Coal Zone Formation.

6. The algal-amorphous facies showed good to excellent correlation with the crude oils and contains sufficient quantities of mature hydrogen rich organic matter to source the region's approximate ten billion barrels of oil. This nonmarine algae in the algal-amorphous facies is the source of the "waxy" character of the oils of Central Sumatra.

7. The asymmetry of the Pematang basins governs the overall lateral direction of oil migration primarily towards the hinge margin of the basin.

Acknowledgments. The authors wish to thank Pertamina, P.T.Caltex Indonesia, Texaco Inc, and Chevron Corp. for their permission to publish this work. The authors gratefully acknowledge the quality geochemical data provided by Texaco Exploration & Producing Technology Department, Chevron Petroleum Technology Company, Robertson Research (Singapore), and Core Laboratories (Singapore) which was used in this chapter. Thanks are extended to the drafting and secretarial staff of Texaco for their cooperation and effort.

References

Berner RA, Raiswell R (1984) C/S method for distinguishing freshwater from marine sedimentary rocks. Geol 12: 365–368

Bissada KK (1982) Geochemical constraints on petroleum and migration – a review. Proc. 2nd ASCOPE Conf. Manila, Oct, 1981, pp 69–87

Cameron NR (1983) The stratigraphy of the Sihapas Formation in the Northwest of the Central Sumatra basin. Proc 12th Indon Pet Assoc Ann Con, Jakarta, Oct, 1983, pp 43–65

de Coster GL (1974) The geology of the Central and South Sumatra basins. Proc 3rd Indon Pet Assoc Ann Con, Jakarta, June, pp 77–110

Eubank RT, Makki A (1981) Structural geology of the Central Sumatra back-arc Basin. Proc 10th Indon Pet Assoc Ann Con, Jakarta, Oct, 1: 153–196

Frakes LA (1979) Climates throughout geologic time. Elsevier New York, 310pp

Gelpi E, Schneider H, Mann J, and Oro J (1970) Hydrocarbon significance in microscopic algae. Phytochemistry 9: 603–612

Habicht JKA (1979) Paleoclimate, paleomagnetism, and continental drift. Am Assoc Pet Geol, Tulsa, Stud Geol 9, 31 pp

Hunt JM (1979) Petroleum geochemistry and geology. Freeman San Francisco, 617pp

Karig DE, Supraka S, Moore GF, Hehanussa PE (1979) Structure and Cenozoic evolution of the Sunda Arc in the Central Sumatra region. In: Watkins JS, Montadert L and Dickerson PW (eds) Geological and geophysical investigation of continental margins. Am Assoc Pet Geol, Tulsa, Memoir 29: 223–237

Katz BJ (1991) Controls on lacustrine source rock development: a model for Indonesia. Proc 20th Indon Pet Assoc Ann Con, Jakarta, Oct, 1: 587–619

Katz BJ, Kahle GM (1988) Basin evaluation: a supply-side approach to resource assessment. Proc 17th Indon Pet Assoc Ann Con, Jakarta, Oct, 1: 135–168.

Katz BJ, Kelley PA (1987) Central Sumatra and the East African rift lakes sediments: an organic geochemical comparison. Proc 16th Indon Pet Assoc Ann Con, Jakarta, Oct, 1: 259–289

Katz BJ, Mertani B (1989) Central Sumatra – a geochemical paradox. Proc 18th Indon Pet Assoc Ann Con, Jakarta, Oct, 1: 403–425

Koning T, Darmono FX (1984) The geology of the Beruk Northeast Field, Central Sumatra, oil production from Pre-Tertiary basement rocks. Proc 13th Indon Pet Assoc Ann Con, Jakarta, May, 1: 385–406

Leventhal JS (1983) An interpretation of carbon and sulfur relationships in Black Sea sediments as indicators of environments of deposition. Geochim Cosmochim Acta, 47: 133–137

Longley IM, Barraclough R, Bridden MA, Brown S (1990) Pematang lacustrine petroleum source rocks from the Malacca Strait PSC, Central Sumatra, Indonesia. Proc 19th IPA Conv, 1: 279–297

MacGregor DS, McKenzie AG (1986), Quantification of oil generation and migration in the Malacca Strait region Central Sumatra. Proc 15th Indon Pet Assoc Ann Con, Jakarta, Oct, 1: 305–320

Mertosono S (1975) Geology of Pungut and Tandung Oil Fields. Proc 4th Indon Pet Assoc Ann Con, Jakarta, June, 1: 165–179

Mertosono S, Nayoan GAS (1975) The Tertiary basinal area of Central Sumatra. Proc 4th Indon Pet Assoc Ann Con, Jakarta, June, pp 63–76

Moldowan JM, Seifert WK (1980) First discovery of botryococcane in petroleum. J Chem Soc Chem Commun, pp 912–914

Moldowan JM, Seifert WK, Gallegos EJ (1985) Relationship between petroleum composition and the environment of deposition of petroleum source rocks. Am Assoc Pet Geol Bull 69: 1255–1268

Robinson KM, Kamal A (1988) Hydrocarbon generation, migration and entrapment in the Kampar block, Central Sumatra. Proc 17th Indon Pet Assoc Ann Con, Jakarta, Oct, 1: 211–256

Robinson KM (1987) An overview of source rocks and oils in Indonesia. Proc 16th Indon Pet Assoc Ann Con, Jakarta, Oct, 1: 97–122

Roezin S (1974) The discovery and development of the Petapahan oil field, Central Sumatra. Proc 3rd Indon Pet Assoc Ann Con, Jakarta, June, pp 111–121

Root DH, Attanasi ED, Turner RM (1987) Statistics of petroleum exploration in the noncommunist world outside the United States and Canada. USGS Cir 931

Seifert WK, Moldowan JM (1980) Paleoreconstruction by biological markers. Proc 9th Indon Pet Assoc Ann Con, Jakarta, May, pp 189–212

Sundararaman P, Teerman SC, Mann RG, Mertani B (1988) Activation energy distribution: a key parameter in basin modeling and a geochemical technique for studying maturation and organic facies. Proc 17th Indon Pet Assoc Ann Con, Jakarta, Oct, 1: 169–185

Tissot BP, Welte DH (1984) Petroleum formation and occurrence 2nd edn. Springer, Berlin Heidelberg New York, 699pp

Williams HH, Kelley PA, Janks JS, Christensen RM (1985) The Paleogene rift basin source rocks of Central Sumatra. Proc 14th Indon Pet Assoc Ann Con, Jakarta, Oct, 2: 57–90

The Green River Shale: an Eocene Carbonate Lacustrine Source Rock

B.J. Katz[1]

Abstract

The Green River Formation was deposited in a series of saline, alkaline lake basins in the western United States which formed as a result of Laramide tectonics. Deposition occurred largely during the Eocene. Although there is considerable lithologic variability, the unit may be described, in general, as a dolomitic marl. The unit exhibits a wide range in organic enrichment (0.2 to 33.7 wt.% organic carbon) and hydrogen generation potential (up to 370.6 mg HC/g rock). Although the kerogen does display some variability, it generally displays a close affinity to the type I reference curve. The kerogen is isotopically depleted with most kerogen $\delta^{13}C$ values being less than $-30‰$ relative to PDB. Pyrolysis-gas chromatography indicates that the generated products would be high wax crudes.

Variability in the geochemical attributes are noted both stratigraphically and areally. The most organically enriched horizons, with the highest hydrocarbon generation potentials, are restricted to the Mahogany Zone within the Parachute Creek Member. Deposition of this material is thought to be associated with the transgressive lake phases (i.e., fresher water episodes). Areally the most organic enriched units, with the highest generation potential and the most hydrogen-enriched organic material are found within the Piceance Creek basin. Leaner and poorer quality organic material is found toward the orogenic belt.

Much of the unit is thermally immature and has not yet entered into the main phase of petroleum generation and release. Observational data and thermal maturity modeling suggest that the main phase of hydrocarbon generation and release is attained at a depth of approximately 2600 m in the vicinity of Bluebell-Altamont Field.

Although much of the unit is thermally immature, available data indicate that oil-in-place in excess of 1.9×10^{10} barrels can be attributed to this unit. An additional 1.5×10^{13} barrels of oil are believed to be bound within the immature oil shales.

The general exploration strategy for Green River-derived oils is directed toward the identification of stratigraphic traps in marginal lacustrine sandstones of the Green River Fm. or red bed facies of the Wasatch Fm. which are in close proximity to the thermally mature Green River.

Introduction

The Green River Formation was deposited in Lakes Gosiute and Uinta during the Eocene. The unit is preserved in several basins including the Green River, Washakie, Uinta and Piceance Creek (Fig. 1). The Green River was also deposited in an unnamed lake in western Wyoming and is preserved in a series of small 'fossil' basins. The Green River Fm. covers an area of approximately 62 150 km^2 (24 000 mi^2; Bradley 1929).

The Green River Formation displays considerable variations in thickness. This inter- and intrabasin variability in thickness is summarized in Table 1. The unit itself has been stratigraphically subdivided. The stratigraphic nomenclature in the Piceance Creek, Uinta, Green River, and Washakie basins is presented in Fig. 2 (Cole 1975; COSUNA 1988).

The unit has both conventional and nonconventional economic importance. An estimated 1.9×10^{10} barrels of oil have been attributed to the Green River source facies (Anders and Gerrild 1984). In addition to the conventional resources, the Green River Fm. is an oil shale with reserves estimated at 1.5×10^{13} barrels of oil equivalent (National Petroleum Council 1973).

This chapter attempts to review and summarize the organic geochemical characteristics of both the

[1]Texaco Inc., 3901 Briarpark, Houston, Texas 77041, USA

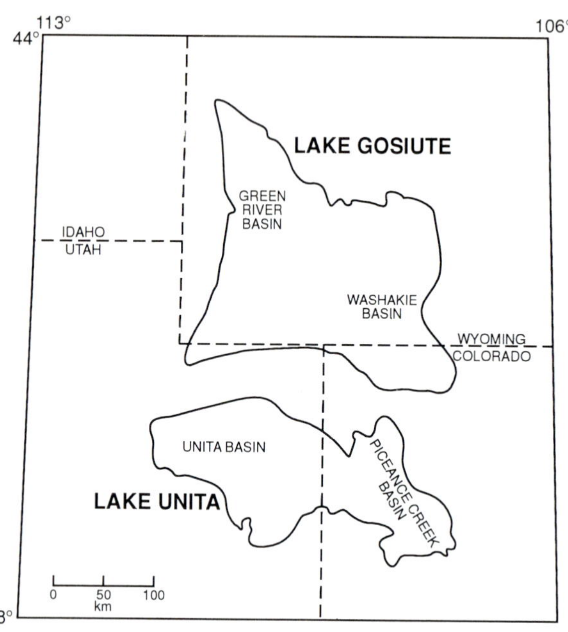

Fig. 1. Current basins containing the Green River Formation and outlines of paleolakes Uinta and Gosiute. (Eugster 1985)

Table 1. Summary of stratigraphic thicknesses. (COSUNA 1988)

Basin name	Minimum thickness m (ft)	Maximum thickness m (ft)
Piceance Creek	0 (0)	1500 (4920)
Eastern Uinta	750 (2460)	1515 (4970)
Western Uinta	1060 (3475)	1700 (5575)
Green River	540 (1770)	2892 (9480)
Washakie		1300 (4265)

Green River Formation source rock and its generated oils. It also presents a brief summary of the geologic setting in which these rocks were deposited.

Regional Setting

The lake basins in which the Green River Formation was deposited were formed as a direct response to Laramide tectonics. The lakes occupied foreland basins which developed during the latest Cretaceous and persisted for about 45 Ma into the early Tertiary. Green River deposition itself occurred over a 10–23 Ma period ranging locally from Paleocene to Eocene in age (Franczyk et al. 1990).

These foreland basins extended approximately 300 km (186 miles) beyond the Sevier orogenic belt to the west. The foreland basins are believed to have had an original dip of less than 0.4 m/km (2 ft/mile). Higher dips were present in the narrow belt immediately adjacent to the mountain front (Bradley 1964). The lake basins were at an elevation of approximately 300 m (1000 ft) above sea level (MacGinitie 1969).

Floral evidence suggests moderate temperatures within the region and that frost was probably absent during the time of Green River deposition (MacGinitie 1969). This is consistent with the results of paleoclimatic simulations using a general circulation model (GCM) which indicate that average winter air temperatures in the vicinity of the Green River lake basins were above freezing (Fig. 3).

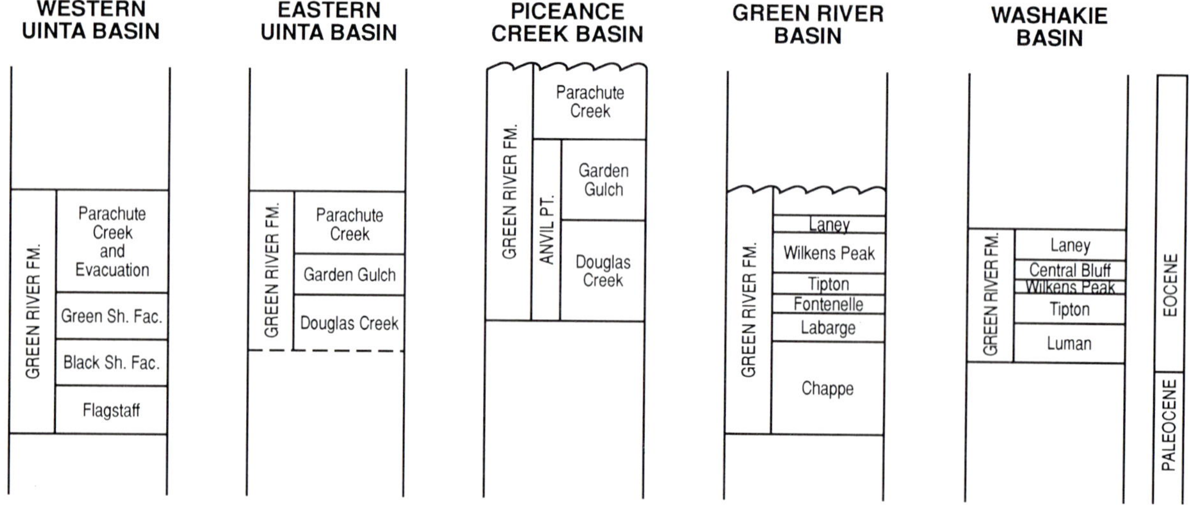

Fig. 2. Stratigraphic columns of the Green River Formation in the Piceance Creek, Green River, Uinta, and Washakie basins. (Cole 1975; COSUNA 1988)

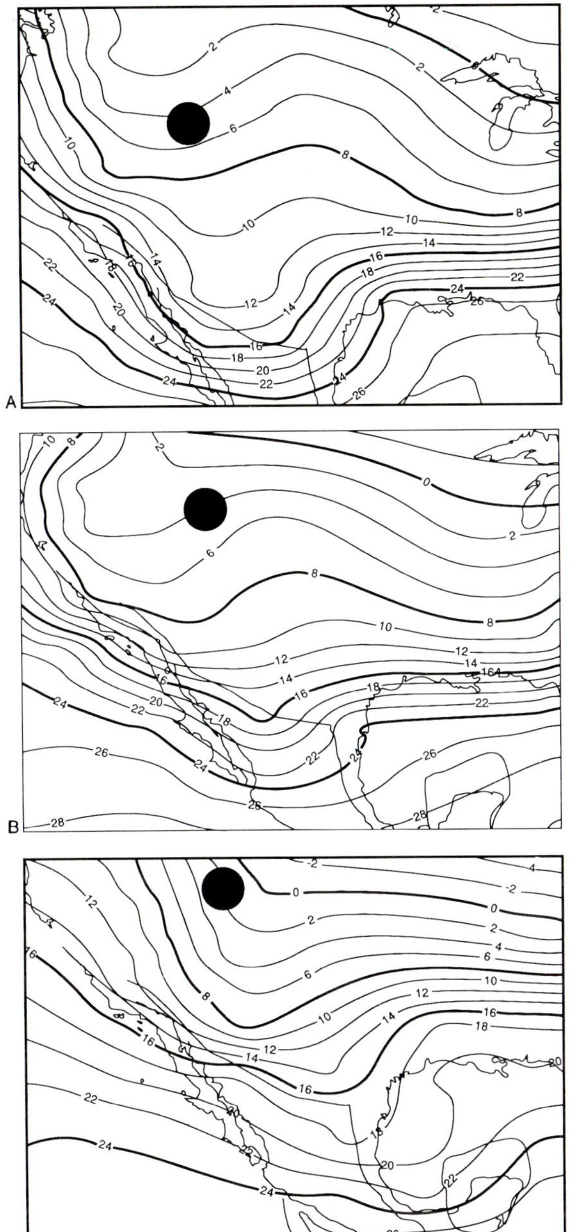

Fig. 3A-C. Modeled average winter surface-air temperature of the western United States for 55 Ma B.P. (**A**), 45 Ma B.P. (**B**) and 35 Ma B.P. (**C**). The *black dot* represents the approximate position of the Green River basins. These paleotemperatures are part of the output of a paleoclimatic simulation utilizing the National Center for Atmospheric Research's GCM, paleogeographic reconstructions of Ziegler et al. (1983), and atmospheric CO_2 partial pressures of Budyko et al. (1985)

Areally limited shallow water lacustrine sediments and coals were being deposited within the Laramide basins prior to the deposition of the Green River Formation. By the onset of Green River deposition these small lacustrine and paludal systems expanded to form Lakes Uinta and Gosiute. During the early Eocene, the lake levels rose above the Douglas Creek Arch, forming one large lake. By the close of the early Eocene, lake level decreased once again, forming two separate lake bodies. At the beginning of the middle Eocene lake expansion occurred once again. By the end of the middle Eocene (approximately 44 MaB.P.), Lake Gosiute was filled by volcaniclastic sedimentation. Approximately 3 Ma later much of Lake Uinta was filled, with the exception of the extreme western portion of the basin, which continued to receive sediment into the late Eocene (Tuttle 1991).

The sedimentary character within the lake basins varied in response to the changes in lake level. The organic-rich, largely dolomitic oil shales formed during the transgressive phases (Tuttle 1991). In contrast, during the more regressive phases there is evidence that as the lakes became saline evaporitic minerals were deposited. The evaporitic minerals deposited during these drier phases included halite, nahcolite, shortite, and trona.

In addition to these large-scale fluctuations in lake volume which impact the gross sedimentologic character, there are also small-scale variations in depositional conditions. The Green River Formation is varved. These varves represent seasonal changes in algal productivity. The summer algal blooms result in changes in water chemistry (a decrease in CO_2 as a result of increased photosynthetic activity; Dean 1981) which in turn results in the precipitation of carbonate and the dilution of organic matter.

The varves are bundled or grouped into sets differing in thickness or spacing of the organic-rich layers. The bundling is thought to represent sedimentary cycles associated with El Niño events (5.8 years), sunspot cycles (11 years), an unknown 30-year cycle, a 20000-year precessional cycle, and a 100000-year eccentricity cycle (Fischer and Roberts 1991).

A difference in the amount of clastic input into the basins is also observed. Higher percentages of clastic input are present closer to the orogenic belt (i.e., there is a greater percentage of clastic material in the Uinta basin compared to the Piceance Creek basin; Johnson 1985).

The depositional model for the Green River Formation remains problematic. Two primary depositional models exist. These include a playa lake model (Eugster and Surdam 1973) and a stratified lake model (Bradley 1929). Both models share some characteristics. These shared attributes are largely restricted to the water chemistry. The lake waters were temporarily alkaline, saline, and sodium-carbonate-dominated. Fresher water is considered to have been present during periods associated with higher river influx (Bradley 1929, 1931).

The playa lake model suggests that deposition occurred within a shallow closed lake basin alternating between oil shales and evaporites as a result of changes in relative aridity. The oil shales possibly resulted from blooms of blue-green algae (cyanobacteria) during humid phases, with trona deposition taking place during the dry phases.

The stratified lake model assumes a chemical stratification, with a highly saline lower water mass and a less saline upper water mass. The dissolved chemical load would support the growth of planktonic algae which would be preserved in the highly reducing, anoxic hypolimnion.

The readers are referred to Smith (1974), Eugster and Hardie (1975), Surdam and Wolfbauer (1975), Cole and Picard (1975), and Johnson (1981) for a review of the pros and cons for both of these depositional models.

Source Rock Characteristics

Organic Richness and Generation Potential

Two hundred and seventy four samples from the Green River, Uinta, and Piceance Creek basins were examined. The majority of these samples were obtained from the Parachute Creek and Wilkins Peak Members. (Note that the Mahogany Zone is located within the Parachute Creek Member.) Organic carbon (TOC) contents ranged from 0.2 to 33.7 wt.%, with a modal value of ~ 3 wt.% (Fig. 4). All but 12 of the samples exhibit above-average levels of organic enrichment (> 1.0 wt.%; Bissada 1982).

The total pyrolytic yield ($S_1 + S_2$ :: free + generable hydrocarbons) of the samples containing greater than 0.5 wt.% TOC ranged from 3.0 to 370.6 mg HC/g rock (Fig. 5). All of the pyrolyzed samples, therefore, display above-average generation potentials (> 2.5 mg HC/g rock; Bissada 1982), and all but three samples have yields consistent with good or excellent source rocks (> 6 mg HC/g rock; Tissot and Welte 1984). For comparison purposes, oil shales yields have a pyrolytic yield greater than 10 mg HC/g rock, which is equivalent to a Fischer assay oil yield of 2 gallons per ton or 8.6 l per metric ton.

These levels of organic enrichment and generation potential appear greater than those observed in most other lacustrine oil source rock systems. For example, data available from the Waltman Shale of the Wind River basin (Katz and Liro 1993) and the Towaco and Feltville Formations of the Newark basin (Katz et al. 1988) appear more typical of lacustrine source rock systems (Fig. 6).

Kerogen Character and Composition

Kerogen in the Green River Fm. typically accounts for ~ 80% of the organic matter (Tissot et al. 1978). Samples from the Green River Formation were originally utilized to establish the reference curve for type I organic matter on the van Krevelen and modified van Krevelen diagrams (Tissot et al. 1974; Espitalié et al. 1977). Thermally immature kerogens typically display H/C ratios > 1.5 and O/C ratios ≈ 0.05. There is, however, some variability when both the whole-rock hydrogen and oxygen indices (Fig. 7) and the atomic H/C and O/C ratios obtained on isolated kerogens (Fig. 8) are examined.

In general, the pyrolysis data appear to show the greatest deviation from the type I reference curve. Much of this appears as elevated oxygen index values and may not be reflecting so much the nature of the kerogen as the carbonate matrix that the kerogen is contained within (Katz 1983). It has been suggested that this is a particular problem when nahcolite is present as in the more saline facies of the Green River Formation (i.e., the Wilkins Peak Member in the Piceance Creek basin; Dean and Anders 1991).

The influence of the matrix interference is supported when the elemental data are compared with the Rock-Eval results. The atomic O/C ratios do not display the same magnitude of variability as the whole-rock oxygen indices. However, the atomic O/C ratios do display some variability, which indicates that there may be variations in the type of organic matter deposited, the degree of organic preservation, or both.

Variations are also present in both the hydrogen indices and the atomic H/C ratios. Katz (1988) noted that the percentage of residual carbon as determined through pyrolysis increased with increasing carbon content. This apparent increase in

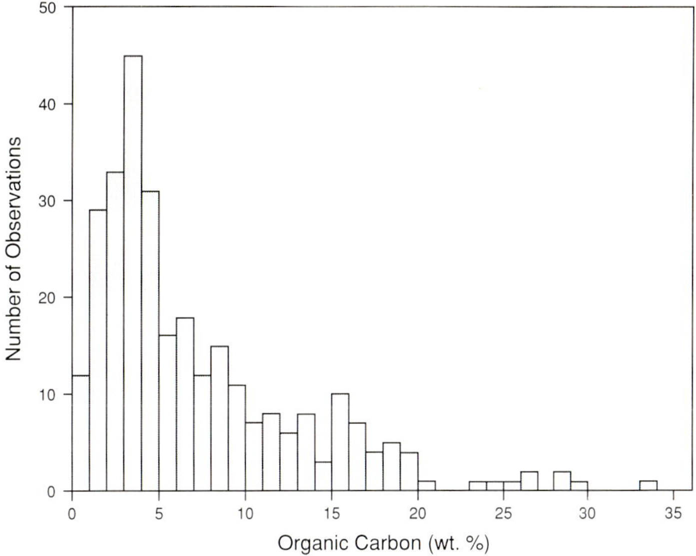

Fig. 4. Organic enrichment (wt.% organic carbon) of Green River Formation samples. (Includes data from Dean and Anders 1991)

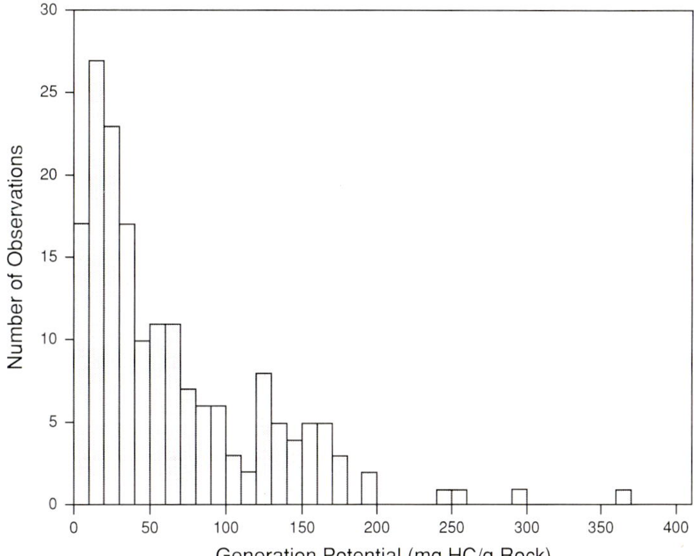

Fig. 5. Total pyrolytic yield ($S_1 + S_2$ = generation potential as determined by Rock-Eval) of Green River Formation samples. (Includes data from Dean and Anders 1991)

residual carbon was interpreted as being an artifact of the FID (flame ionization detector) saturation and not a reflection of an increase in the amount of inert material in the Green River kerogens. This interpretation is consistent with the observations of Dean and Anders (1991), who have suggested that the lower hydrogen indices are associated with organically leaner samples. A similar relationship between organic carbon content and atomic H/C ratios is observed (Fig. 9). Dean and Anders (1991) suggest that this may be a reflection in the degree of organic preservation and/or the introduction of hydrogen-depleted type III material associated with clastic material which would also result in the reduction in organic carbon content through dilution.

Pyrolysis-gas chromatography (PGC) also provides information on the nature of the Green River kerogen. A typical pyrolysis-gas chromatogram is presented in Fig. 10. The pyrolysate is dominated by a series of doublets of normal alkanes and alkenes. Such a pattern is typical of alginite (Larter and Douglas 1980). The long chain hydrocarbons ($> nC_{22}$) are important contributors to the total pyrolysate. Such a pyrolysis-gas chromatographic pattern would suggest that the generated oils would be classified as high wax, paraffinic crudes.

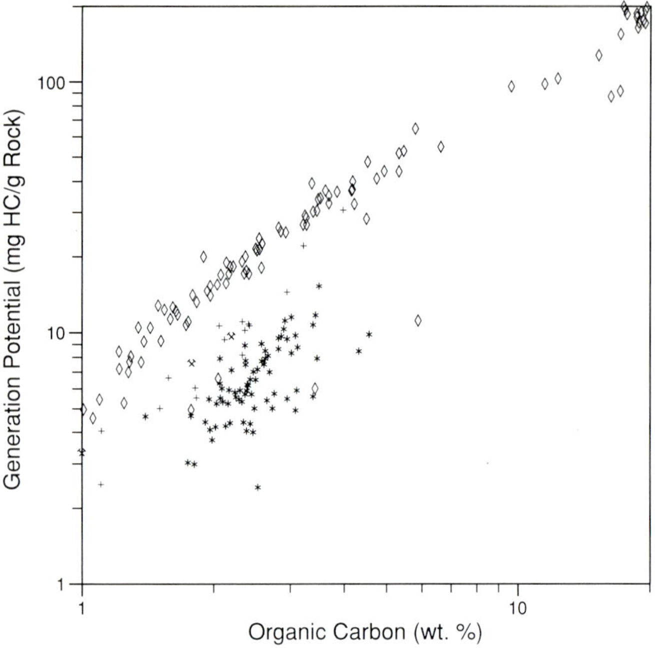

Fig. 6. Comparison of the relationship between total organic carbon content and total generation potential of several lacustrine source rock systems including the Green River Formation (◇), the Waltman Shale (*), and the Towaco (+) and Feltville Formations (×). (Includes data from Katz and Liro 1993 and Katz et al. 1988)

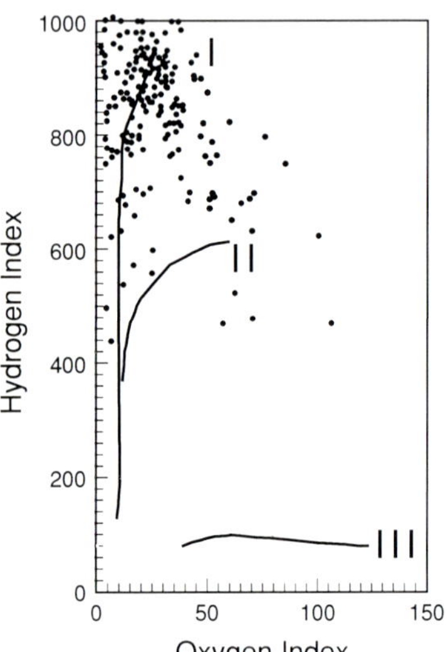

Fig. 7. Modified 'Rock-Eval' van Krevelen diagram using whole-rock samples. (Includes data from Dean and Anders 1991)

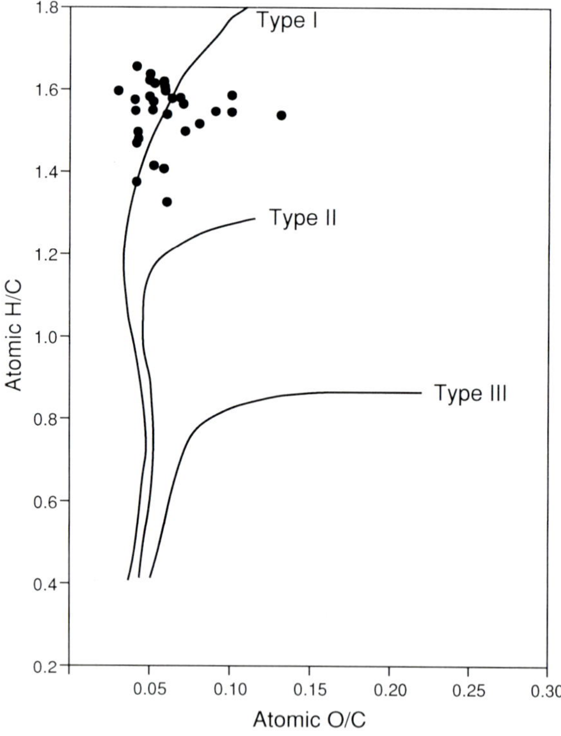

Fig. 8. Van Krevelen diagram based on the elemental composition of isolated kerogens

In addition, the pyrolysates typically display an even predominance in the $nC_{20}+$ range. Such an even preference and the observed dominance of nC_{22} is consistent with a saline, carbonate depositional setting for the kerogen (ten Haven et al. 1988).

The isotopic composition of the isolated kerogens is distinctly negative. The $\delta^{13}C$ values range from -41.5 to -28.3‰ relative to PDB (Collister and Hayes 1991; Dean and Anders 1991), with most values being isotopically lighter than -30‰ (Fig. 11). These values contrast with typical Tertiary-age

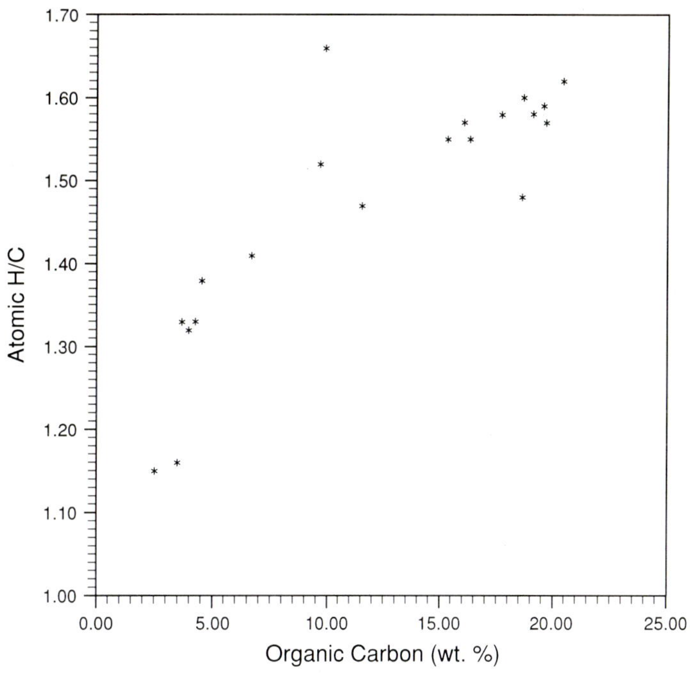

Fig. 9. Relationship between organic carbon content and atomic H/C ratios

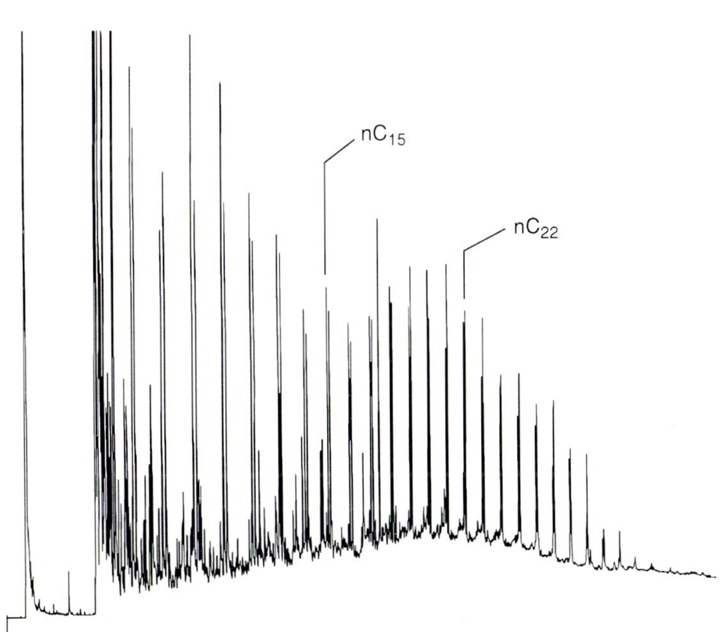

Fig. 10. Representative Green River Formation pyrolysis-gas chromatogram from the Piceance Creek basin

marine kerogen values of between −27 and −23‰ (Degens 1969). Collister and Hayes (1991) suggest that the carbon isotope depletion may, in part, be due to photosynthetic sulfur bacteria present at the chemocline and methane cycling by methylotrophs.

An analysis of the organic residues reveals the presence of bacteria, cyanobacteria, chlorophyte algae, euglenophytes, fungi, moss, fern spores, and pollens (Bradley 1931). This organic matter may represent in large part either algal or bacterial mats (Bucheim and Surdam 1977).

As was the case for the level of organic enrichment and the hydrocarbon generation potential, the apparent level of hydrogen enrichment of the Green River Fm. kerogens is rather atypical of

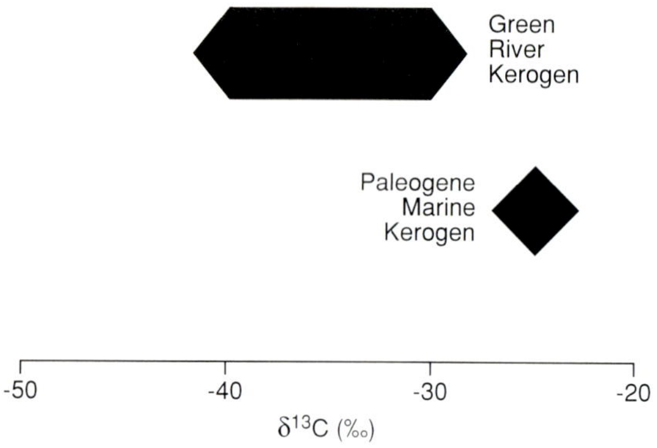

Fig. 11. Kerogen carbon isotopic range of the Green River Formation compared to Tertiary marine kerogen. Green River kerogen data sources include Collister and Hayes (1991) and Dean and Anders (1991). Marine kerogen range from Degens (1969)

many lacustrine systems. Many lacustrine source rocks display a greater affinity for type II kerogens than they do for type I kerogens. Only a few lacustrine sources such as the Lucaogou Fm. of the Junggar basin display atomic H/C and O/C ratios comparable to those of the Green River Fm. (Graham et al. 1990).

Bitumen Composition

The total organic extract yields of thermally immature Green River sediments were highly variable, ranging from ~ 350 to ~ 174 000 ppm. Most values were between 3000 and 15 000 ppm (Fig. 12).

As expected for thermally immature material, an examination of the relative abundance of saturated hydrocarbons, aromatic hydrocarbons and nonhydrocarbon components of these extracts reveals that the nonhydrocarbon fraction dominates (Fig. 13). However, saturated hydrocarbons dominate over aromatic hydrocarbons (i.e., saturate to aromatic ratios > 1.0). There is a progressive increase in abundance of saturated hydrocarbons with depth.

C_{15} +-saturated fraction chromatography of immature Green River extracts commonly reveals a bimodal pattern (Fig. 14). This is, in part, a reflection of the relative abundance of biomarker compounds. This bimodality decreases with increasing maturity (Fig. 15) (Tissot et al. 1978).

The samples display a broad range of pristane/phytane ratios, ranging from ~ 0.3 to ~ 2.5, with values commonly being less than 1.0. The dominance of phytane in most of the samples suggests deposition within a highly reducing environment. As a consequence of the low level of thermal maturity of most of the samples studied, the pristane/nC_{17} and phytane/nC_{18} ratios are all greater than 1.0. These ratios decrease with increasing thermal maturity.

An examination of the normalized *n*-alkane distributions for samples of comparable levels of thermal maturity reveals a wide range of variability

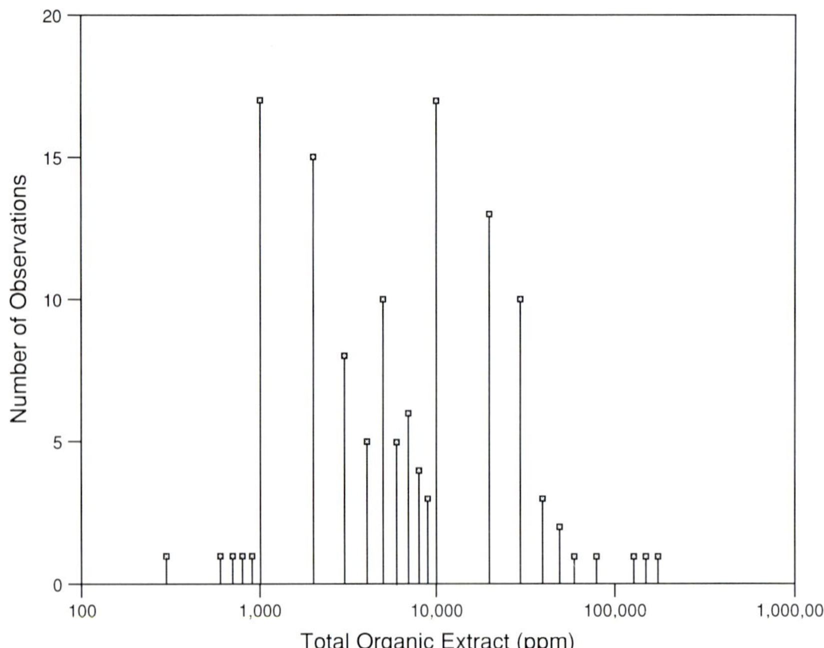

Fig. 12. Total bitumen yields of Green River Formation samples

(Fig. 16). This is manifested in several ways, including the abundance of nC_{22} + components and the relative abundance of nC_{22} compared to nC_{21} and nC_{23}.

Many of the samples contain significant quantities of gammacerane and β-carotane. The presence of these two compounds in such elevated quantities is indicative of hypersaline settings (Hall and Douglas 1983; ten Haven et al. 1988).

Gas chromatography-mass spectrometry reveals variations in the relative abundance of triterpanes, steranes, and diasteranes (Fig. 17). Such variability probably reflects differences in the relative importance of bacterial input, the nature of the mineral matrix, or the depositional environment's redox potential. Higher relative concentrations of triterpanes may be a reflection of abundant microbial input, with higher relative concentrations of

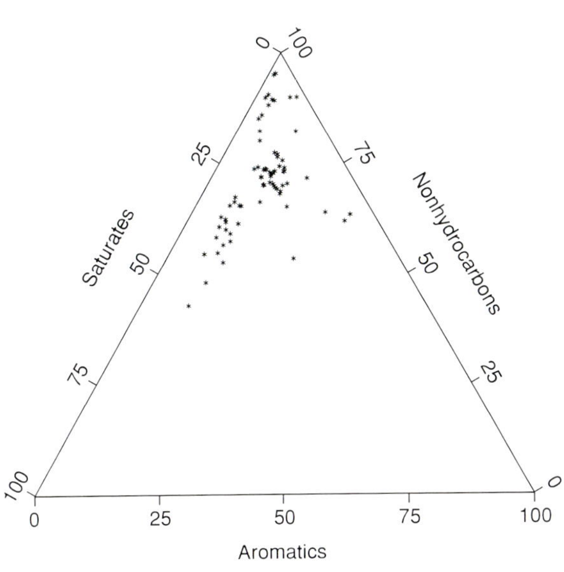

Fig. 13. Ternary diagram depicting bulk composition of organic extracts

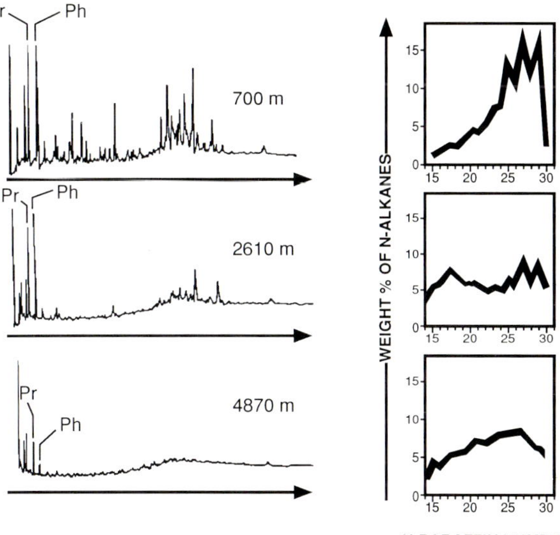

Fig. 15. Change in C_{15} + -saturated hydrocarbon fraction composition as a function of increasing thermal maturity. Chromatograms are for the combined isoprenoid and naphthene fractions. (Tissot et al. 1978). Note the decrease in relative abundance of the 'biomarker' compounds and the longer chain n-alkanes with increasing depth of burial

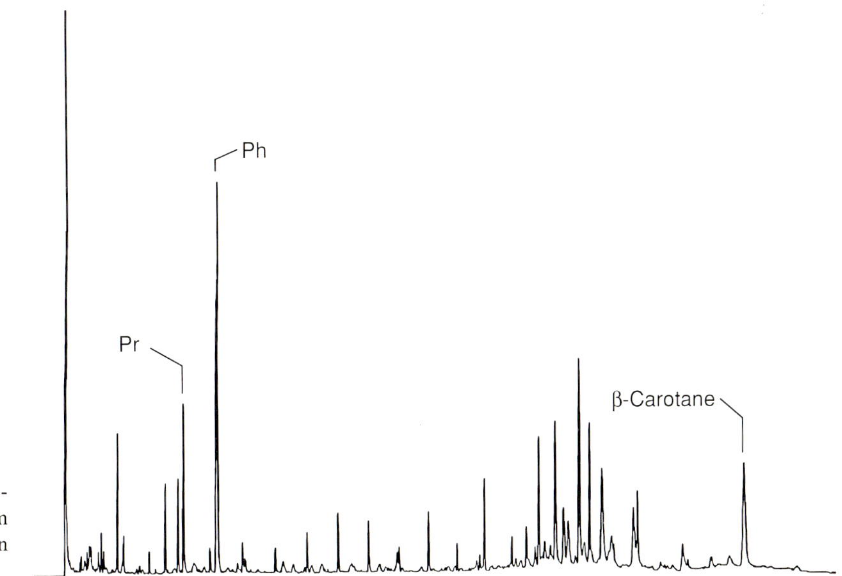

Fig. 14. Representative C_{15} + -saturated fraction gas chromatogram of immature Green River Formation extracts

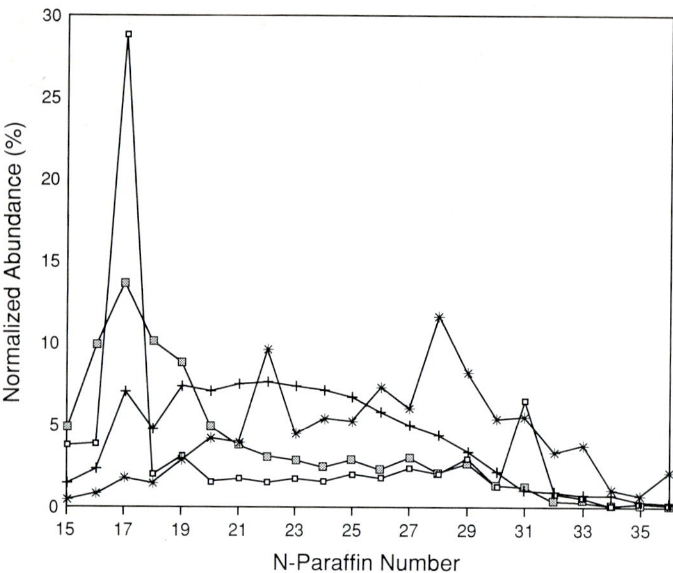

Fig. 16. Variability in *n*-alkane distribution in thermally immature Green River Formation extracts suggesting organic facies variations within the unit

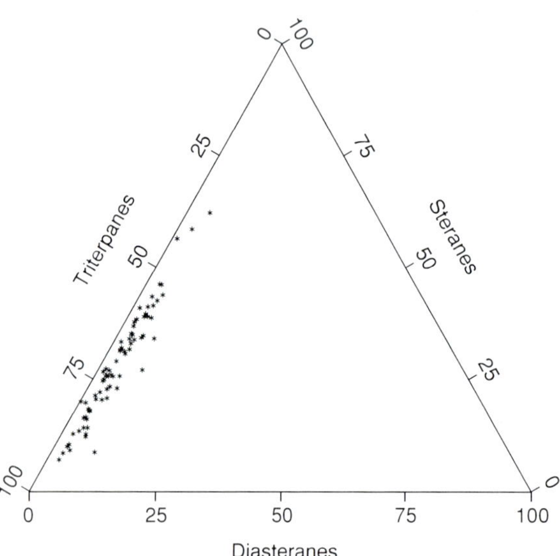

Fig. 17. Ternary diagram depicting the relative abundance of triterpanes, steranes, and diasteranes in Green River Formation extracts

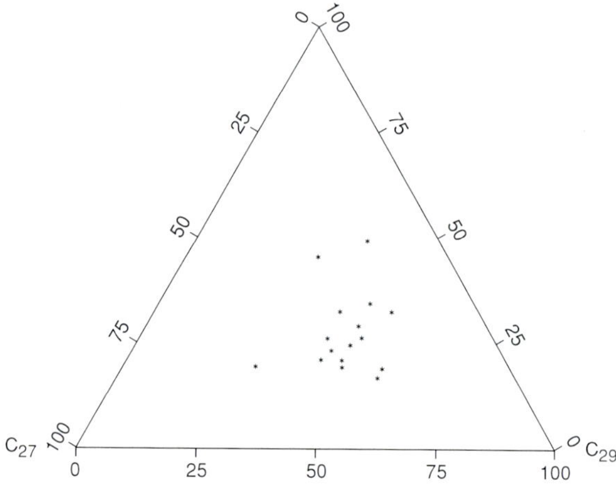

Fig. 18. Ternary diagram depicting the relative abundance of C_{27}, C_{28}, and C_{29} normal steranes in Green River Formation extracts

steranes indicating significant algal input (Mackenzie et al. 1984; Connan et al. 1986). Higher diasterane contents are typically associated with the presence of clay minerals which act as a catalyst for their formation (Hughes 1984). There are, however, a number of studies in which significant concentrations of diasteranes were reported within clay-free carbonate (Connan et al. 1986; Palacas et al. 1984). It has been suggested that the presence of diasteranes may actually be controlled by redox potential (Moldowan et al. 1986) and/or salinity (Clark and Philp 1989). An examination of the C_{27}, C_{28}, and C_{29} normal steranes reveals a general depletion of C_{27} steranes (Fig. 18).

Compound specific carbon isotope compositions of the bitumens further suggest that they were derived from a mixed biomass including primary photoautotrophs, methanotrophic bacteria, and other bacterial forms (Collister et al. 1992).

Mineralogy

As with the organic geochemistry, the mineralogy of the Green River Formation is highly variable. It is dominated by carbonate minerals and can be classified as a dolomitic marlstone. Quartz, albite, analcime, K-feldspar, and illite are the principal silica-bearing minerals. Albite, analcime, and K-feldspar are considered to be authigenic (Desborough 1978).

Desborough (1978), in his review of prior work on the carbonate mineralogy of the Green River, states that primary carbonate mineralogy of the unit includes calcite (including low Mg-calcite), aragonite, ordered dolomite, ankerite, and siderite. Of these, calcite and dolomite are generally the most common mineral species. An exception was noted by Smith and Robb (1966), who observed that ankerite was the dominate carbonate mineral in the Mahogany Zone of Colorado.

The Wilkins Peak Member contains abundant evaporite minerals including shortite, trona, dawsonite, and halite. This mineralogical assemblage was part of the evidence used by Eugster and Hardie (1975) to argue for Wilkins Peak deposition within a playa lake.

Thermal Maturity

Much of the Green River Formation appears to be thermally immature. This is manifested in most of the geochemical indices available. For example, the pyrolysis T_{max} temperatures are typically less than 440 °C, with production indices $[S_1/(S_1 + S_2)]$ less than 0.2 (Fig. 19). A T_{max} of 440 °C and a production index of 0.2 are commonly associated with the onset of the main stage of petroleum generation and expulsion (Bissada 1982). In addition, the atomic H/C ratios, which are typically greater than 1.5, and a dominance by nonhydrocarbons in organic extracts are consistent with a low level of thermal maturity.

Maturity profiles from within the Uinta basin suggest that the main phase oil generation and release is obtained at depths of approximately 2600 m (8500 ft; Anders and Gerrild 1984). These results are consistent with the thermal modeling results presented by Sweeney et al. (1986) who assumed an activation energy of 52.4 kcal/mol for hydrocarbon generation from Green River kerogen.

Areal and Stratigraphic Variability

The highest levels of organic enrichment and the greatest pyrolytic yields were determined on samples from the Parachute Creek Member (Mahogany Zone) of the Piceance Creek basin. In general, the Mahogany Zone samples were more organic-rich and capable of yielding higher quantities of hydrocarbons. In addition, the organic matter appears to be less concentrated toward the orogenic

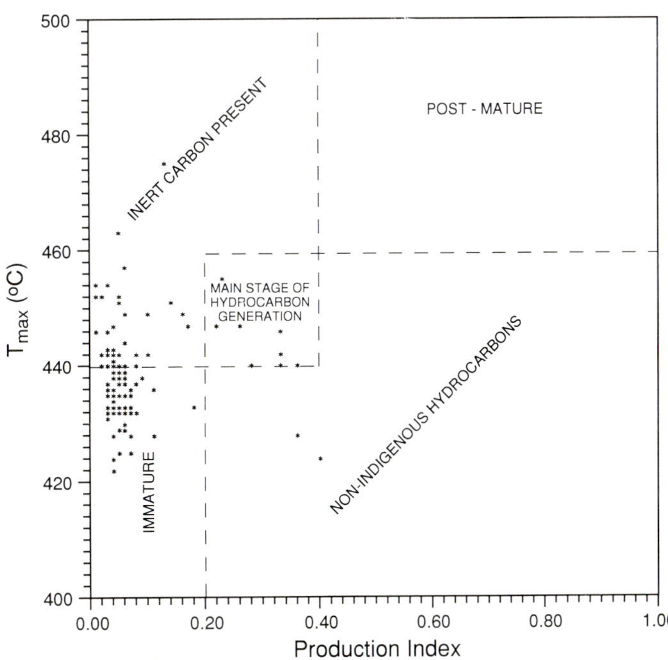

Fig. 19. Relationship between T_{max} and the production index $[S_1/(S_1 + S_2)]$

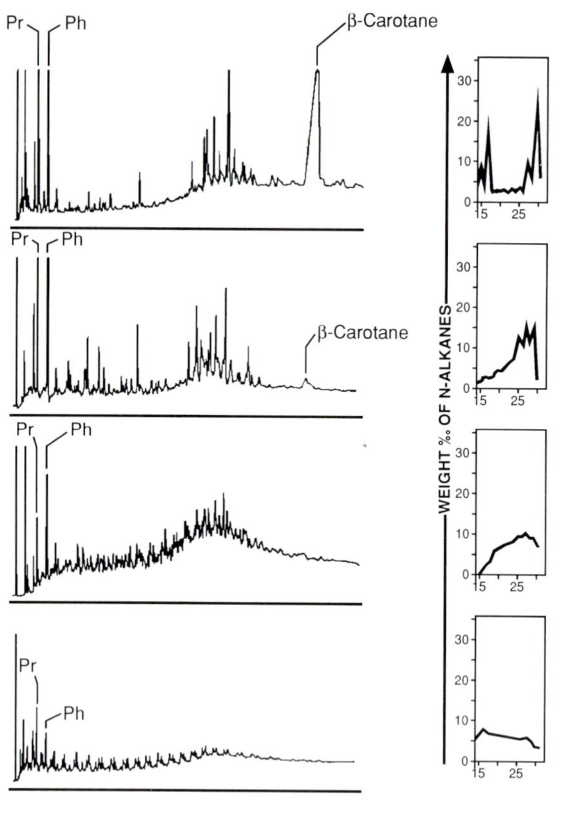

Fig. 20. C_{15} + -saturated hydrocarbon fraction compositions representative of the different organic facies present in the Green River Formation. Chromatograms are for the combined isoprenoid and naphthene fractions. (Tissot et al. 1978)

belt, possibly as a result of clastic dilution (Dean and Anders 1991).

Tissot et al. (1978) suggested that the variability in hydrocarbon composition of extracted bitumens displays a distinct stratigraphic pattern (Fig. 20). The lower units display the greatest percentage of normal alkanes, a rather flat *n*-alkane distribution (i.e., no dominance by nC_{17}, no distinct odd-carbon predominance in the longer-chain *n*-alkanes), an absence of significant quantities of β-carotane, and low relative concentrations of steranes and triterpanes. This compares with the samples from younger horizons which display a greater relative abundance of isoalkanes, a dominance of nC_{17}, an odd-carbon predominance in the longer-chain *n*-alkanes, the presence of significant quantities of β-carotane, and high relative abundances of steranes and triterpanes.

Tissot et al. (1978) concluded that these differences are a reflection of differences in the intensity of bacterial reworking. Samples from the oldest horizons display the greatest degree of reworking. Most of the geochemical fossils derived from higher land plant debris, algae, and other planktonic material have been destroyed. Samples from the youngest intervals display the greatest molecular diversity indicating higher degrees of preservation and less bacterial reworking.

Hydrocarbon Characteristics

The oils attributed to the Green River Formation usually display API gravity values from about 18 to >40°, with most values being greater than 35°. The oils display pour points commonly between 35 and 46 °C (95 to 115 °F). The oils are typically dominated by saturated hydrocarbons, with paraffins commonly accounting for >75% of the C_9 + fraction (Fig. 21). Those oils with the lower concentrations of saturated hydrocarbons appear to be biodegraded and display the lowest API gravity values. A further examination of the saturated hydrocarbon fraction reveals that both the relative abundance of *n*-alkanes, isoalkanes, and cycloalkanes, as well as the distribution of *n*-alkanes vary in the same manner as the extracted bitumens. That is, oils derived from stratigraphically deeper zones display a flat *n*-alkane distribution, contain more

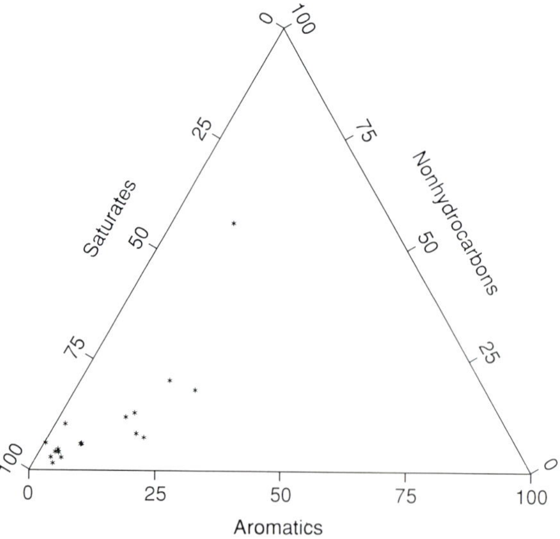

Fig. 21. Ternary diagram depicting the C_9 + bulk oil compositions of Green River Formation-derived oils

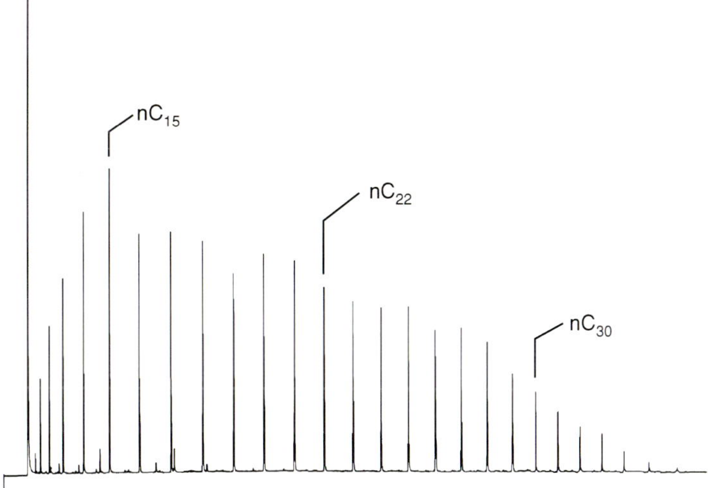

Fig. 22. Representative $C_{15}+$-saturated fraction gas chromatogram of Green River Formation-derived oils

n-alkanes, and lack *β*-carotane, while oils derived from shallow zones display a dominance by nC_{17}, an odd predominance in the longer chain *n*-alkanes, as well as a lower proportion of *n*-alkanes (Tissot et al. 1978).

An examination of the saturated fraction gas chromatograms reveal, that unlike the immature extracts, the oils normally display elevated pristane/phytane ratios (> 1.4, and commonly > 2.0). The oils also display pristane/nC_{17} and phytane/nC_{18} ratios less than 1, and commonly less than 0.2. These chromatograms are dominated by *n*-alkanes and do not generally display the bimodality observed in the extracts as a consequence of the abundance of the biomarker compounds (Fig. 22).

The isotopic composition of the whole-oils ranges from approximately − 33 to − 26‰ relative to PDB. As would be expected because of the dominance of saturated hydrocarbons in the oils, these values are closely aligned with the isotopic composition of the saturated fraction. These data suggest that some isotopically heavier kerogens than observed are present within the Green River Fm. and that not all of the kerogens examined have actually contributed to the conventional resource base.

Just as in the extracts, an examination of the biomarker compositions of these oils reveals differences in the relative abundances of triterpanes, steranes, and diasteranes. However, the observed variability is much greater and the relative abundance of diasteranes is greater in the oils than in the extracts (Fig. 23). As with the extracts there is also considerable variation in the normal sterane composition. However, the relative depletion of C_{27} steranes is not as pronounced in the oils as in the extracts (Fig. 24).

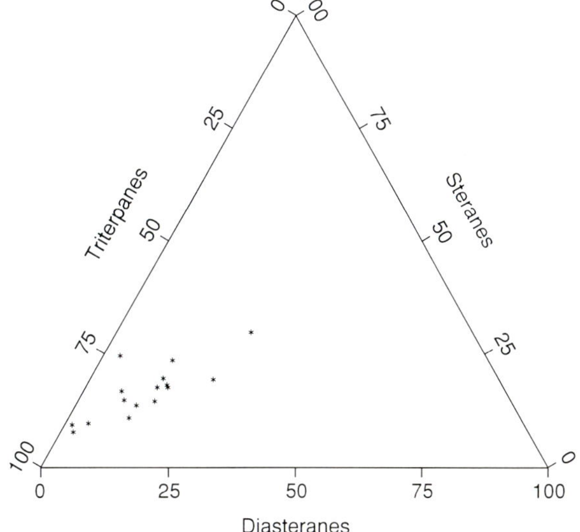

Fig. 23. Ternary diagram depicting the relative abundance of triterpanes, steranes, and diasteranes in Green River Formation-derived oils

Exploration Strategy

Conventional hydrocarbon exploration strategies for Green River-derived oils appear to be associated with the identification of thermally mature

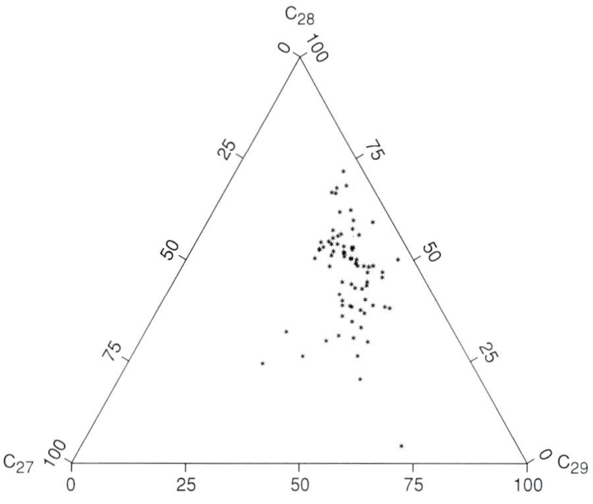

Fig. 24. Ternary diagram depicting the relative abundance of the C_{27}, C_{28}, and C_{29} normal steranes in oils derived from the Green River Formation

generative sub-basins and the identification of potential reservoirs in close proximity to these mature regions. Migration is largely up-dip in a bed-parallel fashion. Tissot et al. (1978) note that cross-formation migration was negligible.

Many of the reservoir targets are associated with the lacustrine barrier-beach and fluvial-deltaic facies within the Green River Fm. and the red bed facies of the Wasatch Fm. (e.g., Red Wash, Walker Hollow, and Wonsists Valley fields; Castle 1990). It is within this environment that largely fine- to medium-grained and moderately to well-sorted sandstones developed. These sands are mineralogically mature with detrital quartz accounting for 70 to 90% of the rock fabric. The trapping mechanism for these sandstone reservoirs appears to be largely stratigraphic, resulting from porosity pinchouts associated with facies changes (Baker and Lucas 1972) caused by lacustrine transgressions.

Other targets for Green River-derived oils include low-permeability, fractured sandstone reservoirs (e.g., Bluebell-Altamont field; Lucas and Drexler 1976).

Summary

The Green River Formation is an example of a carbonate lacustrine source rock. It was deposited within long-lived tectonic lacustrine basins which formed as a result of Laramide tectonics. The unit is currently preserved in the Green River, Piceance Creek, Uinta, and Washakie basins covering an area of 62 150 km^2 and obtains a thickness in excess of 2000 m.

Although the character of the lake is not fully understood, it is clear that the lake was saline, alkaline, and stratified.

The levels of measured organic carbon and hydrogen enrichment, as well as generation potential, are atypical of most lacustrine source rock systems. The unit displays elevated values of organic enrichment and hydrocarbon generation potential relative to other known lacustrine source rocks. The products generated are paraffinic and are commonly dominated by *n*-alkanes. The oils and bitumens are isotopically light compared to marine-derived products.

Areal and stratigraphic variability exists within the organic facies of the Green River. These differences are due to changes in lake level, the amount of detrital input, as well as variations in the intensity of bacterial reworking.

Exploration for conventional Green River-derived oils focuses on the identification of thermally mature sub-basins and the identification of effective reservoir facies up-dip, in close proximity to thermally mature regions. Many of the exploration targets are stratigraphic traps associated with porosity pinchouts.

Acknowledgments. Support of the National Center for Atmospheric Research's GCM was provided by the Earth Systems Science Center of Pennsylvania State University under the direction of Dr. Eric Barron. Assistance with the preparation of this manuscript was provided by M. Hill, J. Mulvaney, and G. Novez. An earlier version of this chapter was reviewed by Drs. V. D. Robison and W. Dean. Permission for publication of this work was provided by Texaco Inc.

References

Anders DE, Gerrild PM (1984) Hydrocarbon generation in lacustrine rocks of Tertiary age, Uinta basin, Utah – organic carbon, pyrolysis yield, and light hydrocarbons. In: Woodward J, Meisnner FF, Clayton JL (eds) Hydrocarbon source rocks of the greater Rocky Mountain region. Rocky Mt Assoc Geol, Denver, pp 513–529

Baker DA, Lucas PT (1972) Strat trap production may cover 280 + square miles. World Oil 174: 65–68

Bissada KK (1982) Geochemical constraints on petroleum generation and migration – a review. Proc 2nd ASCOPE Conf, Manila, Oct, 1981, pp 69–87

Bradley WH (1929) The varves and climate of the Green River epoch. US Geol Surv Prof Pap 158E: 87–110

Bradley WH (1931) The origin of the oil shale and its microfossils of the Green River Formation of Colorado and Utah. US Geol Surv Prof Pap 168, 58

Bradley WH (1964) Geology of the Green River Formation and associated Eocene rocks in southwestern Wyoming and adjacent parts of Colorado and Utah. US Geol Surv Prof Pap 496-A: 86

Bucheim HP, Surdam RC (1977) Fossil catfish and the depositional environment of the Green River Formation, Wyoming. Geology 5: 196–198

Budyko MI, Ronov AB, Yanshin AL (1985) History of the Earth's atmosphere. Springer, Berlin Heidelberg New York, 139 pp

Castle CW (1990) Sedimentation in Eocene Lake Uinta (lower Green River Formation), northeastern Uinta basin, Utah. In: Katz, BJ (ed) Lacustrine basin exploration – case studies and modern analogs. Am Assoc Petrol Geol, Tulsa, Mem 50: 243–263

Clark JP, Philp RP (1989) Geochemical characterization of evaporite and carbonate depositional environments and correlation of associated crude oils in the Black Creek basin, Alberta. Bull Can Petrol Geol 37: 401–416

Cole RD (1975) Sedimentology and sulfur isotope geochemistry of Green River Formation (Eocene), Uinta basin, Utah, Piceance Creek basin. PhD Diss, Univ Utah, Salt Lake City, 274 pp

Cole RD, Picard MD (1975) Primary and secondary sedimentary structures in oil shale and other fine-grained rocks, Green River Formation (Eocene), Utah and Colorado. Utah Geology 2: 49–67

Collister JW, Hayes JM (1991) A preliminary study of the carbon and nitrogen isotopic biogeochemistry of lacustrine sedimentary rocks from the Green River Formation, Wyoming, Utah, and Colorado. In: Tuttle, ML (ed) Geochemical, biogeochemical, and sedimentological studies of the Green River Formation, Wyoming, Utah, and Colorado. US Geol Surv Bull 1973-A-G: C1–C16

Collister JW, Summons RE, Lichtfouse E, Hayes JM (1992) An isotopic biogeochemical study of the Green River oil shale. Org Geochem 19: 265–276

Connan J, Bouroullec J, Dessort D, Albrecht P (1986) The microbial input in carbonate-anhydrite facies of a sabkha paleoenvironment from Guatemala: a molecular approach. In: Leythaeuser D, Rullkötter J (eds) Advances in organic geochemistry, 1985. Pergamon Press, Oxford, pp 29–50

COSUNA (Correlation of stratigraphic Units of North America) (1988) Central and Southern Rockies Region. Am Assoc Petrol Geol, Tulsa, 1 sheet

Dean WE (1981) Carbonate minerals and organic matter in sediments of modern north temperate hard-water lakes. In: Ethridge FG, Flores RM (eds) Recent and ancient nonmarine depositional environments: models for exploration. Soc Econ Paleontol Mineral, Tulsa, Spec Publ 31: 213–231

Dean WE, Anders DE (1991) Effect of source, depositional environment, and diagenesis on characteristics of organic matter in oil shale form the Green River Formation, Wyoming, Utah, and Colorado. In: Tuttle ML (ed) Geochemical, biogeochemical, and sedimentological studies of the Green River Formation, Wyoming, Utah, and Colorado. US Geol Surv Bull 1973-A-G: F1-F16

Degens ET (1969) Biogeochemistry of stable carbon isotopes. In: Eglinton G, Murphy MTJ (eds) Organic geochemistry. Springer, Berlin Heidelberg New York, pp 304–329

Desborough GA (1978) A biogenic-chemical stratified lake model for the origin of the oil shale of the Green River Formation – an alternative to the playa-lake model. Geol Soc Am Bull 89: 961–971

Espitalié J, Laporte LJ, Madec M, Marquis F, Leplat PJ, Paulet J, Boutedeu A (1977) Méthode rapide de caractérization des roches mères de leur potential pétrolier et de leur degré d'évolution. Rev Inst Français du Pétrole 32: 32–42

Eugster HP (1985) Oil shales, evaporites and ore deposits. Geochim Cosmochim Acta 49: 619–635

Eugster HP, Hardie LA (1975) Sedimentation in an ancient playa-lake complex: The Wilkins Peak Member of the Green River Formation of Wyoming. Geol Soc Am Bull 86: 319–334

Eugster HP, Surdam RC (1973) Depositional environment of the Green River Formation of Wyoming–a preliminary report. Geol Soc Am Bull 86: 319–334

Fischer AG, Roberts LT (1991) Cyclicity in the Green River Formation (lacustrine Eocene) of Wyoming. J Sediment Petrol 61: 1146–1154

Franczyk KJ, Pitman JK, Nichols DJ (1990) Sedimentology, mineralogy, palynology, and depositional history of some Uppermost Cretaceous and Lowermost Tertiary rocks along the Utah Book and Roan Cliffs east of the Green River. US Geol Surv Bull 1787: 27

Graham SA, Brassell S, Carroll AR, Xiao X, Demaison G, Mcknight CL, Liang Y, Chu J, Hendrix MS (1990) Characteristics of selected petroleum source rocks, Xianjiang Uygur Autonomous Region, northwest China. Am Assoc Petrol Geol Bull 74: 493–512

Hall P B, Douglas A G (1983) The distribution of cyclic alkanes in two lacustrine deposits. In: Bjorøy M, Bjørlykke K, Eggen S, Elvsborg A, Finstaed KG, Grønneberg T, Haegh T, Skaar FE (eds) Advances in organic geochemistry, 1981. Wiley, Chichester, pp 576–587

Hughes WB (1984) Use of thiophene organosulfur compounds in characterizing crude oils derived from carbonate versus siliciclastic sources. In: Palacas JG (ed) Geochemistry and source rock potential of carbonate rocks. Am Assoc Petrol Geol, Tulsa, Stud Geol 18: 181–196

Johnson RC (1981) Stratigraphic evidence for a deep Eocene Lake Uinta, Piceance Creek basin, Colorado. Geology 9: 55–62

Johnson RC (1985) Early Cenozoic history of the Uinta and Piceance Creek basins, Utah and Colorado, with special reference to the development of Eocene Lake Uinta. In: Flores RM, Kaplan SS (eds) Cenozoic paleogeography of Western United States. Rocky Mt Sec Soc Econ Paleontol Mineral, Denver, pp 247–276

Katz BJ (1983) Limitations of 'Rock-Eval' pyrolysis for typing organic matter. Org Geochem 4: 195–199

Katz BJ (1988) Clastic and carbonate lacustrine systems: an organic geochemical comparison (Green River Formation and East African lake sediments). In: Fleet AJ, Kelts K, Talbot MR (eds) Lacustrine petroleum source rocks. Geol Soc, London, Spec Publ 40: 81–90

Katz BJ, Liro LM (1993) The Waltman Shale Member, Fort Union Formation, Wind River basin: a Paleocene clastic lacustrine source system. In: Keefer WR, Metzger WJ,

Godwin LH (eds) Oil and gas and other resources of the Wind River basin, Wyoming. Wyo Geol Assoc, Cheyenne, pp 163–174

Katz BJ, Robison CR, Jorjorian T, Foley FD (1988) The level of organic maturity within the Newark basin and its associated implications. In: Manspeizer W (ed) Triassic-Jurassic rifting: continental breakup and the origin of the Atlantic Ocean and passive margins. Elsevier, Amsterdam, Part B: 683–696

Larter SR, Douglas AG (1980) A pyrolysis-gas chromatographic method for kerogen typing. In: Douglas AG, Maxwell JR (eds) Advances in organic geochemistry, 1979. Pergamon Press, New York, pp 579–583

Lucas PT, Drexler JM (1976) Altamont-Bluebell–a major, naturally fractured stratigraphic trap. In: Braunstein J (ed) North American oil and gas fields. Am Assoc Petrol Geol, Tulsa, Mem 24: 121–135

MacGinitie HD (1969) Eocene Green River flora of northwestern Colorado and northeastern Utah. Univ Calif Press, Berkeley, Univ Calif Publ Geol Sci 83, 203 pp

Mackenzie AS, Maxwell JR, Coleman ML, Deegan CE (1984) Biological marker and isotope studies of North Sea crude oils and sediments. Proc 11th World Petrol Congr, Wiley, Chichester, 2: 1–12

Moldowan JM, Sundararaman P, Schoell M (1986) Sensitivity of biomarker properties to depositional environment and/or source input in the Lower Toarcian of SW Germany. Org Geochem 10: 915–926

National Petroleum Council (1973) U. S. energy outlook–oil shale availability. Oil Shale Task Group of the Other Energy Resources Subcommittee of the National Petroleum Council's Committee on US Energy Outlook, AE Kelley, chairman, 87 pp

Palacas JG, Anders DE, King JD (1984) South Florida basin–a prime example of carbonate source rocks of petroleum. In: Palacas JG (ed) Geochemistry and source rock potential of carbonate rocks. Am Assoc Petrol Geol, Tulsa, Stud Geol 18: 71–96

Smith JW (1974) Geochemistry of oil-shale genesis in Colorado's Piceance Creek basin. In: Murray DK (ed) Rocky Mountain association of geologists guidebook to the energy resources of the Piceance Creek basin, Colorado, 25th Field Conference. Rocky Mt Assoc Geol, Denver, pp 71–79

Smith JW, Robb WA 1966 Ankerite in the Green River Formation's Mahogany Zone. J Sediment Petrol 36: 486–490

Surdam RC, Wolfbauer CA (1975) Green River Formation, Wyoming: a playa lake complex. Geol Soc Am Bull 86: 335–345

Sweeney JJ, Burnham AK, Braun RL (1986) A model for hydrocarbon maturation in the Uinta basin, Utah, U.S.A. In: Burrus J (ed) Thermal modeling in sedimentary basins. Éditions Technip, Paris, pp 547–561

ten Haven HL, de Leeuw JW, Sinninghe-Damsté JS, Schenck PA, Palmer SE, Zumberge JE (1988) Applications of biological markers in recognition of paleohypersaline environments. In: Fleet AJ, Kelts K, Talbot MR (eds) Lacustrine petroleum source rocks. Geol Soc, London, Spec Publ 40: 123–130

Tissot BP, Welte DH (1984) Petroleum formation and occurrence, 2nd edn. Springer, Berlin Heidelberg New York, 699 pp

Tissot BP, Durand B, Espitalié J, Combaz A (1974) Influence of nature and diagenesis of organic matter in formation of petroleum. Am Assoc Petrol Geol Bull 58: 499–506

Tissot B, Derro G, Hood A (1978) Geochemical study of the Uinta basin: formation of petroleum from the Green River Formation. Geochim Cosmochim Acta 42: 1569–1485

Tuttle ML (1991) Introduction. In: Tuttle ML (ed) Geochemical, biogeochemical, and sedimentological studies of the Green River Formation, Wyoming, Utah, and Colorado. US Geol Surv Bull 1973-A-G: A1–A11

Ziegler AM, Scotese CR, Barrett SF, (1983). Mesozoic and Cenozoic paleogeographic maps. In: Brosche P, Sündermann J (eds) Tidal friction and the earth's rotation II. Springer, Berlin Heidelberg New York, pp 240–252

Subject Index

Springer-Verlag and the Environment

We at Springer-Verlag firmly believe that an international science publisher has a special obligation to the environment, and our corporate policies consistently reflect this conviction.

We also expect our business partners – paper mills, printers, packaging manufacturers, etc. – to commit themselves to using environmentally friendly materials and production processes.

The paper in this book is made from low- or no-chlorine pulp and is acid free, in conformance with international standards for paper permanency.